Biology

of

Female
Cancers

Biology
of
Female Cancers

Edited by

Simon P. Langdon, Ph.D.
Imperial Cancer Research Fund
Medical Oncology Unit
Western General Hospital
Edinburgh, Scotland

William R. Miller, Ph.D., D.Sc.
Imperial Cancer Research Fund
Medical Oncology Unit
Western General Hospital
Edinburgh, Scotland

Andrew Berchuck, Ph.D.
Department of Obstetrics and Gynecology
Division of Gynecologic Oncology
Duke University Medical Center
Durham, North Carolina

CRC Press
Boca Raton   New York

Publisher: Robert B. Stern
Project Editor: Renee Taub
Marketing Manager: Susie Carlisle
Direct Marketing Manager: Becky McEldowney
Cover Designer: Denise Craig
PrePress: Carlos Esser
Manufacturing: Sheri Schwartz

Library of Congress Cataloging-in-Publication Data

Biology of female cancers / edited by Simon P. Langdon, William R.
   Miller, Andrew Berchuck.
                         p.   cm.
      Includes bibliographical references and index.
      ISBN 0-8493-9443-0 (alk. paper)
      1. Generative organs, Female—Cancer—Molecular aspects.
   I. Langdon, Simon P.   II. Miller, William R.   III. Berchuck,
   Andrew.
      [DNLM: 1. Genital Neoplasms, Female.   WP 145 B615 1997]
   RC280.G5B56   1997
   616.99′465—dc20
   DNLM/DLC
   for Library of Congress   96-34494
                                                                    CIP

TABLE OF CONTENTS

PREFACE

In the past decade, there have been major advances in our understanding of the molecular pathogenesis and biology of human cancers. In this book we summarize the current knowledge of the mechanisms of growth regulation and genetic features associated with the common forms of human female cancers. These cancers include malignancies of the breast, ovary, uterus, cervix, vulva, and gestational trophoblastic disease.

For each of the major tumor types, an introductory chapter describes the general aspects of the disease including etiology, incidence, pathology, staging, and treatment. The risk of developing these tumor types and their growth may be strongly influenced by hormones, growth factors, and cytokines and these effects are reviewed. The discovery of oncogenes, tumor suppressor genes, and more recently DNA repair genes has expanded our knowledge of cell growth regulation and proliferation, and mechanisms describing how these genes become aberrant in the process of malignant transformation are rapidly being defined. The roles of these genes in the specific female cancer types are discussed. Finally the applications for therapy which have arisen from an understanding of the biology and possible new translational approaches are considered. The final chapter reviews the hereditary links associated with gynecologic malignancies. Major developments have recently occurred in this field as the molecular details of several hereditary cancer susceptibility genes have been characterized and these advances are discussed.

THE EDITORS

Simon Langdon, M.A., Ph.D., is an ICRF staff scientist working in the ICRF Medical Oncology Unit at the Western General Hospital in Edinburgh. He obtained a B.A. and subsequently M.A. in the Department of Chemistry at Oxford University and his Ph.D. in the CRC Chemotherapy Research Group at Aston University. He continued as a CRC research fellow at Aston University and came to the ICRF Medical Oncology Unit in 1985 as an ICRF research fellow. In 1989, he was appointed to the position of staff scientist and is an honorary fellow of the University of Edinburgh. His research interests include the cell biology of ovarian cancer and the development of therapeutic strategies exploiting inhibition of cell signaling pathways. He has published over 70 scientific papers and is a member of the Editorial Board of *Oncology Reports*.

William Miller, Ph.D., D.Sc. holds a Personal Chair in Experimental Oncology in the Department of Oncology of the University of Edinburgh. He obtained his B.Sc. in the Department of Biochemistry and his Ph.D. in the Department of Medicine, University of Leeds. He came to Edinburgh in 1969 to take up an appointment as research fellow in the Department of Clinical Surgery, Royal Infirmary, Edinburgh. He was awarded a D.Sc. by the University of Edinburgh for his work on biochemical aspects of the breast, its secretions, and tumors. In 1980, he was appointed lecturer in the Department of Clinical Surgery and promoted to reader in the same department in 1987. In 1988, his readership became shared with the Department of Oncology and he was appointed Honorary Deputy Director of the ICRF Unit of Medical Oncology at the Western General Hospital, being responsible for the laboratory research program within the unit. Current research interests include the involvement of hormones, growth factors, and second messenger systems in the development, progression, and prognosis of breast and ovarian cancer. Professor Miller has published over 150 scientific papers and is on the Editorial Board of the *British Journal of Cancer*, *The Breast*, and *Expert Opinion on Therapeutic Patents*. He is a former Hon. Secretary of the British Association for Cancer Research.

Andrew Berchuck, M.D., is an Associate Professor of Gynecologic Oncology in the Department of Obstetrics and Gynecology, Duke University and Director of Gynecologic Cancer Research at the Comprehensive Cancer Center. He was trained in Obstetrics and Gynecology at Case Western Reserve University in Cleveland, Ohio. After a laboratory research fellowship at the Unversity of Texas Southwestern Medical School, he completed his clinical training in Gynecologic Oncology at Memorial Sloan-Kettering Cancer Center in New York City. For the past 9 years, he has been on the full time faculty in the Department of Obstetrics and Gynecology and the Division of Gynecologic Oncology at Duke University Medical Center in Durham, North Carolina. During this time he has been actively involved in the treatment of women with gynecologic cancers. In addition, his laboratory has sought to elucidate the molecular alterations involved in the development and growth regulation of endometrial and ovarian cancers. His research has been supported by the National Cancer Institute, the American Cancer Society, and American Gynecological and Obstetrical Society. In 1993 he was awarded the R. Wayne Rundles Award For Excellence in Cancer Research.

CONTRIBUTORS

Awatif Al-Nafussi
Department of Pathology
University of Edinburgh
Edinburgh, Scotland

Mark J. Arends
Department of Pathology
University of Edinburgh
Edinburgh, Scotland

John M. S. Bartlett
Department of Surgery
Royal Infirmary
Glasgow, Scotland

Jacqueline S. Beesley
Department of Medical Oncology
Fox Chase Cancer Center
Philadelphia, Pennsylvania

Andrew Berchuck
Department of Obstetrics and Gynecology
Division of Gynecologic Oncology
Duke University Medical Center
Durham, North Carolina

Jeff Boyd
Department of Obstetrics and Gynecology
University of Pennsylvania Medical Center
Philadelphia, Pennsylvania

Camille Busby-Earle
Department of Obstetrics and Gynaecology
University of Edinburgh
Edinburgh, Scotland

David A. Cameron
ICRF Medical Oncology Unit
Western General Hospital
Edinburgh, Scotland

Anthony C. Evans
Department of Obstetrics and Gynecology
Duke University Medical Center
Durham, North Carolina

Katherine V. Ferry
Department of Medical Oncology
Fox Chase Cancer Center
Philadelphia, Pennsylvania

Sir A. Patrick Forrest
Professor Emeritus
University of Edinburgh
Edinburgh, Scotland

Hani Gabra
ICRF Medical Oncology Unit
Western General Hospital
Edinburgh, Scotland

Andrew K. Godwin
Department of Medical Oncology
Fox Chase Cancer Center
Philadelphia, Pennsylvania

Vivian E. von Gruenigen
Department of Obstetrics and Gynecology
Division of Gynecologic Oncology
University of Texas Southwestern Medical Center
Dallas, Texas

Thomas C. Hamilton
Department of Medical Oncology
Fox Chase Cancer Center
Philadelphia, Pennsylvania

Stephen W. Johnson
Department of Medical Oncology
Fox Chase Cancer Center
Philadelphia, Pennsylvania

Beth Y. Karlan
Department of Obstetrics and Gynecology
Cedars-Sinai Medical Center and UCLA School of
  Medicine
Los Angeles, California

Roger J. B. King
School of Biological Sciences
University of Surrey
Guildford, England

Simon P. Langdon
ICRF Medical Oncology Unit
Western General Hospital
Edinburgh, Scotland

Robin Leake
Department of Biochemistry
Glasgow University
Glasgow, Scotland

Robert C. F. Leonard
Department of Clinical Oncology
Western General Hospital
Edinburgh, Scotland

J. Michael Mathis
Department of Obstetrics and Gynecology
Division of Gynecologic Oncology
University of Texas Southwestern Medical Center
Dallas, Texas

David Scott Miller
Department of Obstetrics and Gynecology
Division of Gynecologic Oncology
University of Texas Southwestern Medical Center
Dallas, Texas

William R. Miller
ICRF Medical Oncology Unit
Western General Hospital
Edinburgh, Scotland

Carolyn Y. Muller
Department of Obstetrics and Gynecology
Division of Gynecologic Oncology
University of Texas Southwestern Medical Center
Dallas, Texas

George Mutter
Department of Pathology
Brigham and Women's Hospital
and
Harvard Medical School
Boston, Massachusetts

Pondichery G. Satyaswaroop
Department of Obstetrics and Gynecology
The Milton S. Hershey Medical Center
Hershey, Pennsylvania

Barbara J. B. Simpson
ICRF Medical Oncology Unit
Western General Hospital
Edinburgh, Scotland

Anne P. Shapter
Department of Obstetrics and Gynecology
Cedars-Sinai Medical Center
and
UCLA School of Medicine
Los Angeles, California

George E. Smart
Department of Obstetrics and Gynaecology
Royal Infirmary
Edinburgh, Scotland

John F. Smyth
ICRF Medical Oncology Unit
Western General Hospital
Edinburgh, Scotland

C. Michael Steel
School of Biological and Medical Sciences
University of St. Andrews
St. Andrews, Scotland

Siamak Tabibzadeh
Department of Pathology
The Moffitt Cancer Center
Tampa, Florida

Rosemary A. Walker
Breast Cancer Research Unit
School of Medicine
University of Leicester
Leicester, England

Tjoung Won Park
Universitats-Frauenklinik
Freiburg, Germany

Thomas C. Wright
Department of Pathology
College of Physicians and Surgeons of Columbia
  University
New York, New York

Part I.
Introduction

Chapter 1

Introduction to the Cell and Molecular Biology of Cancer

Simon P. Langdon and William R. Miller

CONTENTS

1. INTRODUCTION

Cancer is characterized by the properties of unregulated growth, invasion and metastasis and distinct cell and molecular processes are associated with each of these events. It is now clear that cancer arises as a result of progressive alterations in the genome, and understanding of the early molecular events in cancer has improved rapidly over the last two decades, particularly with the identification of two families of genes that are critical to cancer development — namely, the oncogenes and tumor suppressor genes. Growth of cancer cells is dependent both on the influences of the products of oncogenes and tumor suppressor genes, and on other regulators, such as hormones, growth factors, and cytokines. The aim of this chapter is to provide a brief introduction to the genetics and growth regulation of cancer as a prelude to the detail that is presented for specific female cancers in subsequent chapters.

2. THE MULTISTEP NATURE OF CANCER

The evidence that the genesis of cancer is a multistage process comes from a variety of sources, including (1) investigations of the actions of carcinogens in mice (1940s) and (2) studies of the age-related incidence of common cancers (1950s). The characterization of oncogenes (1970s) and tumor-suppressor genes (1980s), subsequently led to the identification of the specific molecules involved.

2.1 Experimental Carcinogenesis Studies

In the early 1940s, animal models of skin cancer provided the first experimental evidence that carcinogenesis could be divided into several distinct phases.[1,2] Thus, cancer-inducing chemicals could be categorised into two types of agents — incomplete and complete carcinogens. Incomplete carcinogens only produced tumors when treated cells were exposed to a second type of agent termed a promoter while complete carcinogens produced tumors without assistance. These observations led to the hypothesis that two steps were involved — "initiation" and "promotion". Complete carcinogens, such as the polycyclic aromatic hydrocarbons, were capable of acting both as initiators and promoters while an agent such as urethane initiated in the skin model, but required the application of a promoter (for example, a phorbol ester) to produce a tumor. A single exposure to the initiator was often sufficient to complete the initiation event, while promoters required prolonged and repeated exposure and did not produce tumors when given alone. The initiation event was shown to produce stable mutations of genomic DNA. In the absence of initiation, promoters did not have permanent effects on the cell. The tumors produced in these experiments were predominantly benign papillomas that only occasionally progressed to malignancy. However, cells re-exposed to an initiator after the initiation-promotion step produced many more carcinomas. This indicated a second mutational step which has been referred to as "conversion".[3] The model is summarized in Figure 1.

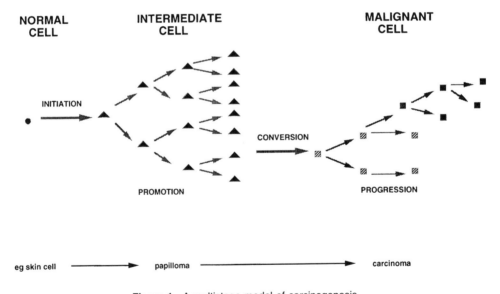

Figure 1 A multistage model of carcinogenesis.

2.2 Epidemiological Support for Multistage Carcinogenesis

Analysis of the changes of cancer incidence with age led Nordling[4] in 1953 and Armitage and Doll[5] in 1954 to suggest that carcinogenesis involved five to seven stages. Therefore, incidence rates for many common cancers increase dramatically with age, which would be consistent with a multihit hypothesis in which it may take many decades to accumulate the number of mutations necessary to cause malignancy. Subsequently, a two-stage model was proposed in which the first stage comprised exponential cell growth, and the second stage involved escape from normal control and malignant transformation.[6] While these models explained certain features of cancer incidence, they did not account for either the patterns displayed by pediatric tumors that by definition peak in childhood, or the shape of incidence curves for breast and ovarian cancer which plateau at the age of menopause. To account for these characteristics and the data on experimental carcinogenesis, Knudson put forward the following model of tumor development.[7]

2.3 The Knudson Model of Human Carcinogenesis

In 1981, Knudson together with his co-worker, S.H. Moolgavkar, presented a model of carcinogenesis proposing that a mutational event occurred in a normal cell giving rise to an intermediate cell, which underwent population expansion and perpetuated the mutation.[7] A second mutational event then occurred that converted the intermediate cell into a malignant cell (see Figure 1). This model was allied to the "classical" initiation-promotion-conversion system of experimental skin carcinogenesis. With appropriate adjustments, the mathematical formulation of the model could be used to describe the log-log plots of incidence of the common solid tumors, the "hooked" incidence plots found for breast and ovarian cancer, and the early incidence rates in pediatric tumors such as retinoblastoma. Knudson based this model largely on previous studies of retinoblastoma[8,9] — a prototype for inherited cancers. Retinoblastoma occurs in two forms — a hereditary form that represents 40% of the disease and frequently develops in both eyes, and a sporadic form, representing 60% of the disease, characteristically occurring in only one eye. Individuals who possess the germ-line mutation develop an average of 3 to 4 tumors while the incidence of the sporadic disease is extremely rare, occurring in 1 in 30,000 children. Because of this, Knudson proposed that two mutational events were required for the disease to develop.[8,9] In hereditary retinoblastoma, the first mutation is inherited in the germ line and is clonally expanded during the development of the retinoblasts. A single further mutational event ("single-hit") is then sufficient to produce a tumor. In the sporadic form of the disease, both mutations are somatic. The nature of the genetic events involved in this disease have now been identified and are described later (Section 3.2.2).

2.4 The Single-Cell Origin and Clonal Evolution of Tumors

Analysis of solid tumors commonly reveals heterogeneity in a range of characteristics at the level of both DNA and protein. To account for this genotypic and phenotypic diversity, Nowell proposed a model of clonal evolution in which tumor cell populations evolved from a single cell, and a series of stepwise genetic alterations occurred during the progression of the tumor.[10] Support for single cell, or monoclonal evolution of cancer came initially from analyses of specific markers in leukemias and lymphomas. Three sets of observations were particularly informative: (1) cytogenetic studies indicated an abnormal karyotype could be displayed by all cells in a primary cancer suggesting a common origin (e.g., the Philadelphia chromosome of chronic myelocytic leukemia,[11] (2) studies in women whose cells were heterozygous for the X-chromosome-linked isoenzymes of glucose-6-phosphate dehydrogenase indicated that tumors expressed only one of the two possible isoenzyme types while normal tissues demonstrated a mix of the two forms,[12] (3) the immunoglobulin production within plasma cell tumors is homogeneous in all cells.[13] Evidence for the monoclonal origin of solid tumors has now been obtained. For example, analyses of loss of heterozygosity, p53 mutations, and X-chromosome inactivation in ovarian tumors have clearly demonstrated that these characteristics are constant between primary tumor and metastatic deposits consistent with a unifocal origin.[14,15]

It would seem that the change produced by the carcinogen in the susceptible normal cell provided it with a selective growth advantage over its unaffected neighbors. Other cells may also be "initiated" by the carcinogen, but either they do not proliferate or are destroyed before they progress to a detectable tumor. Transformed cells appear to be genetically less stable than their normal counterparts (see Section 3.3). Consequently, as neoplastic cells proliferate and progress, genetically variant sublines may emerge. Consistent with this concept, many tumors become increasingly aneuploid as they progress. Favorable variants preferentially survive and their descendants become the predominant subpopulation until a yet more favorable variant is selected. Biological characteristics of tumor progression, (e.g., metastasis and drug-resistance, parallel these genetic changes.

3. CANCER GENETICS

The view that cancer arises as the result of the accumulation of multiple genetic changes is not new, and originated as long ago as 1914 with the studies of Boveri.[16] However, until the first oncogene was identified in 1970,[17] the evidence that cancer was a genetic disease relied on the following: (1) the monoclonal origin of most cancers, (2) the inherited nature of some cancers, and (3) the mutagenic nature of most carcinogens. Direct proof that genetic mutations lead to malignancy has now been obtained. Two families of genes are particularly important — the oncogenes and tumor suppressor genes. It has been suggested that the proliferation of normal cells is regulated by the balance between the growth-promoting proto-oncogenes and the growth-regulating tumor suppressor genes.[18] Activation of proto-oncogenes and/or loss or inactivation of tumor suppressor genes appear to be the underlying causes of a cell becoming malignant. These concepts are developed below.

3.1 Oncogenes and Proto-Oncogenes

To date, approximately 70 genes have been identified which upon transfer into appropriate recipient cells, cause malignant transformation.[18,19] These transforming genes have been named "oncogenes". In 1976, one of these genes, *src* (for sarcoma), identified within the genome of a virus, was reported to be present in the genome of normal chicken cells.[20] This observation indicated that certain normal cellular genes had oncogenic potential and these genes were named "proto-oncogenes".

Two main approaches have been used to identify oncogenes — the use of retroviruses and transfection assays (see Figure 2). The majority of oncogenes have been discovered from studies of animal retroviruses that produce cancer in animals. The structures of these tumor viruses, many of which contain only three or four genes, were sufficiently simple to allow identification of oncogenes which had become incorporated into the viral genome. Oncogenes identified in this manner have generally been named after the virus (v-onc) in which they were found (see Table 1). Transfection experiments, involving transfer of genetic material from malignant cells into normal cells, resulting in the development of malignant characteristics in the recipient cells, have also helped identify a number of oncogenes.

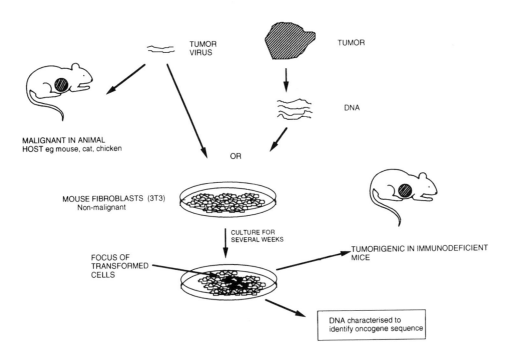

Figure 2   Assays used to identify oncogenes.

3.1.1 Proto-Oncogene Function

Proto-oncogenes are believed to be involved in normal growth and development. In general, they encode proteins (oncoproteins) which are components of signaling pathways involved in growth-factor regulation of cell growth.[21] These signaling pathways are depicted in Figure 3, and following growth-factor binding to a specific membrane receptor, a cascade of intracellular signals are triggered, eventually resulting in transcription of specific genes linked to proliferation (as described in Section 4.1). Examples of these oncogenes and their cellular locations are illustrated in Figure 4. Proto-oncogenes can be classified into four groups; growth factors, growth factor receptors, signal transducers, and nuclear transcription factors.

Table 1   Oncogenes and Oncoproteins

Oncogene	Chromosome	Source/Means of Initial Identification	Cell Location	Oncogene Protein
sis	22q	Simian sarcoma	Secreted	Truncated PDGF
hst	11q	NIH/3T3 transfection	Secreted	Growth factor
int-2	11q	Retrovirus insertion	Secreted	Growth factor
fgf-5	4q	NIH/3T3 transfection	Secreted	Growth factor
erbB	7p	Chicken erythroblastosis	Plasma membrane	Truncated receptor
fms	5q	Cat sarcoma	Plasma membrane	M-CSF receptor
ros	6q	Chicken sarcoma	Plasma membrane	Receptor tyrosine kinase
erbB-2	17q	NIH/3T3 transfection	Plasma membrane	Receptor tyrosine kinase
kit	4p	Cat sarcoma	Plasma membrane	Receptor tyrosine kinase
met	7q	Mouse osteosarcoma	Plasma membrane	Protein tyrosine kinase
ret	10q	NIH/3T3 transfection	Plasma membrane	Protein tyrosine kinase
rab	11p	X-hybridization	Cytoplasm	GTP binding protein
Ha-ras	11p	Rat sarcoma/erythroleukemia	Cytoplasm	GTP binding protein
Ki-ras	12p	Rat sarcoma/erythroleukemia	Cytoplasm	GTP binding protein
N-ras	1p	NIH transfection	Cytoplasm	GTP binding protein
ral	7p	X-hybridization	Cytoplasm	GTP binding protein
abl	9q	Mouse leukemia	Cytoplasm	Protein tyrosine kinase
fes	15q	Cat sarcoma	Cytoplasm	Protein tyrosine kinase
fgr	1p	Cat sarcoma	Cytoplasm	Protein tyrosine kinase
sea	11q	Chicken sarcoma	Cytoplasm	Protein tyrosine kinase
src	20q	Chicken sarcoma	Cytoplasm	Protein tyrosine kinase
yes	18q	Chicken sarcoma	Cytoplasm	Protein tyrosine kinase
raf	3p	Mouse sarcoma	Cytoplasm	Protein ser/thr kinase
mos	8q	Mouse sarcoma	Cytoplasm	Protein ser/thr kinase
pim-1	6p	RV insertion	Cytoplasm	Protein serine kinase
erbA	17p	Chicken erythroblastosis	Nuclear	Thyroid hormone receptor
rel	2p	Turkey reticuloendotheliosis	Cytoplasm/nucleus	Transcription factor
fos	14q	Mouse osteosarcoma	Nucleus	Transcription factor
jun	1p	Chicken sarcoma	Nucleus	Transcription factor
ski	1q	Chicken squamous ca	Nucleus	Transcription factor
c-myc	8q	Chicken sarcoma/carcinoma	Nucleus	Transcription factor
N-myc	2p	X-hybridization	Nucleus	Transcription factor
L-myc	1p	X-hybridization	Nucleus	Transcription factor
myb	6q	Chicken myeloblastosis	Nucleus	Transcription factor
ets-1	11q	Chicken	Nucleus	Transcription factor
ets-2	21q	Chicken	Nucleus	Transcription factor
bcl-1	11q	Translocation breakpoint	Nucleus	Cyclin
bcl-2	18q	Translocation breakpoint	Cytoplasm	Apoptotic inhibitor

The first group of proto-oncogenes encode proteins homologous to known growth factors. For example, the sis product is analogous to a subunit of platelet-derived growth factor, while int-2, hst, and fgf-5 are members of the fibroblast growth-factor family.

The second group of proto-oncogene products are strongly associated with growth-factor receptors. Several are truncated forms of receptors. For example, the *erbB* gene product is a reduced form of the epidermal growth-factor (EGF) receptor, while the kit product is a truncated stem-cell receptor. Other examples include the erbB-2 protein, which shares homology with the EGF receptor while the *fms* product is the receptor for the colony-stimulating-1 factor. All four of these oncoproteins have protein tyrosine kinase activity.

The third group of oncoproteins are components of the intracellular signaling pathways. Signals are transmitted from the cell-membrane receptors to the nucleus by second messengers. Cascades of protein kinases are involved in mediating these signals and many of these kinases are proto-oncogene products. One of the best understood is the *ras* gene which encodes the ras-p21 protein, a G protein, whose function is to bind guanosine triphosphate (GTP).

The fourth group are the transcription factors. These proto-oncogene products include myc, fos, and jun, which bind to DNA in a sequence-specific manner and activate gene expression.

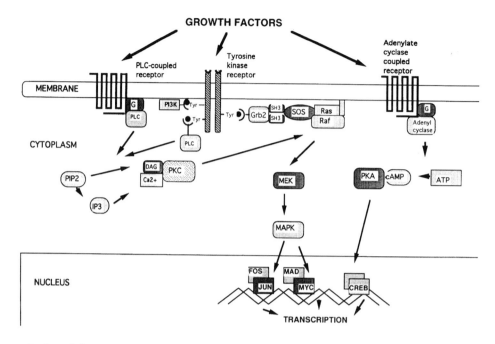

Figure 3 Growth-factor stimulated cell signaling. Abbreviations used: G, G protein; PLC, phospholipase C; PI3K, phosphatidylinositol-3′-kinase; Tyr, tyrosine residue; Grb2, growth-factor binding 2; SH3, src-homology region 3; sos, son-of-sevenless; PIP2, phosphatidylinositol 4,5-diphosphate; IP3, inositol (1,4,5)-triphosphate; DAG, diacylglycerol; PKC, protein kinase C; MEK, MAP kinase kinase; MAPK, mitogen-activated protein kinase; PKA, protein kinase A; ●, indicates phosphorylated form of kinase; ●)-, SH2 (src homology 2) domain.

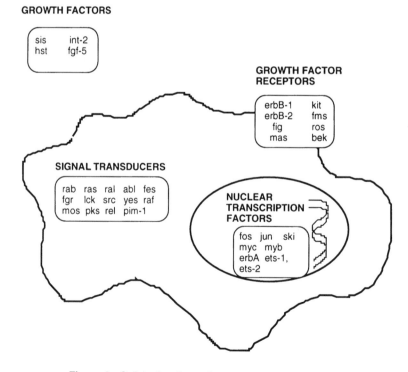

Figure 4 Cellular locations of proto-oncogene products.

3.1.2 Oncogene Activation

Three major mechanisms by which proto-oncogenes can be converted into oncogenes have been identified — amplification, mutation, and translocation. Amplification of proto-oncogenes most simply involves the presence of multiple copies of the gene which results in overproduction of the gene product; however overexpression of the gene product may also result from abnormalities in the regulatory regions of the gene. High levels of gene amplification (10- to 100-fold) are found in homogeneously staining regions of chromosomes, in which the amplification occurs as a contiguous element, and in double-minute chromosomes (essentially minichromosomes).[22] Low-level amplification can arise as the result of extra copies of a complete chromosome (trisomy). Trisomy of chromosomes 8, 9, 12 and 21 is commonly found in a number of hematological malignancies, although trisomy of chromosome 7 is the only commonly observed form in solid tumors.[22] Examples of oncogene amplification include that of the *N-myc* oncogene (up to 1000-fold) in advanced-stage neuroblastoma[23] and the *erbB-2* gene in approximately 25% of breast and ovarian cancers.[24]

Mutations in proto-oncogenes can produce abnormal protein products which may have modified activity. The best described example is that of the ras p21 protein wherein amino acid substitutions at positions 12, 13, and 61 result in changed function.[25,26] These oncogenic changes invariably produce a signaling molecule with decreased deactivating responsiveness, i.e., the mutated ras p21 protein produces a prolonged signal relative to the normal form of p21. Such point mutations in ras have been identified in many solid tumors including lung, breast, and colon cancer.[26] Translocations can result in oncogene activation by one of two major mechanisms.[22] The first involves deregulating an oncogene by placing it under the influence of an active enhancer or promoter sequence. This is typified by c-*myc* activation in Burkitt's lymphoma.[27] In 90% of cases of this disease, the *myc* gene on chromosome 8 is moved next to the enhancer sequences of immunoglobulin loci on chromosome 14 and comes under their influence (see Figure 5). The normal regulation of the *myc* gene then no longer operates, and the gene product is overproduced. Alternatively, translocation may result in the fusion of two genes. The classic example of this is the 9;22 translocation found in chronic myelogenous leukemia (CML). The Philadelphia chromosome of CML was the first consistent chromosome abberation identified in human neoplasia[11] and in 1973 was demonstrated to be a translocation.[28] This rearrangement moves the *Abl* oncogene on chromosome 9 next to the *BCR* sequence on chromosome 22 resulting in a fusion gene (see Figure 5) which produces a protein with abnormal tyrosine kinase activity.[29,30] Many examples of activating translocations and products of fusion proteins have now been identified (as reviewed in Reference 31).

3.1.3 Oncogene Collaboration

In the early 1980s, the view developed that while a single oncogene acting alone was unable to convert a truly normal cell into a malignant cell, pairs of oncogenes were able to achieve this.[32,33] For example, *ras* required the help of a nuclear oncogene such as *myc* to produce complete transformation.[32] Extension of these studies led to the conclusion that collaboration could be achieved by combining cytoplasmic oncogenes (e.g., *ras*, *raf*, *src*, *erbB*) with nuclear oncogenes (e.g., *myc*, *myb*, *erbA*).[18] A number of implications follow from these observations.[18] First, cellular proto-oncogenes may have evolved in such a way as to preclude any one of them from acquiring the ability to transform cells after mutation; this would help safeguard the cell. Second, it is possible that the multistep nature of carcinogenesis might be explained by the need of a tumor cell to sustain multiple mutations in distinct proto-oncogenes, each of which represents a step towards the fully malignant state. Although the above results argue for the requirement of a minimum of two genetic changes for cancer development, single genes may be oncogenic in specific situations. For example, normal cells into which individual oncogenes have been transfected, together with a drug-resistance gene, only expanded rapidly once the selecting drug was applied, killing all the untransfected cells, implying that normal cells apply strong inhibitory effects. These observations suggest that an early step in tumorigenesis is mutation of an oncogene, but the single mutated cell may be unable to expand because of the inhibitory effects of its neighbors. Only when inhibitory effects are overcome can the altered cell proliferate and produce a tumor.[18] Such inhibitory effects point to the existence of another group of genes important in tumorigenesis — the tumor-suppressor genes (discussed below).

3.1.4 Oncogenes and Apoptosis

Maintenance of the steady-state in tissues is a balance of cell growth, differentiation, and death (see Figure 6). While the importance of cell proliferation has long been recognized in the process of tumorigenesis, the suppression of programmed cell death has only recently been the focus of attention.

A. SCHEMATIC VIEW OF THE 8;14 TRANSLOCATION IN BURKITT'S LYMPHOMA

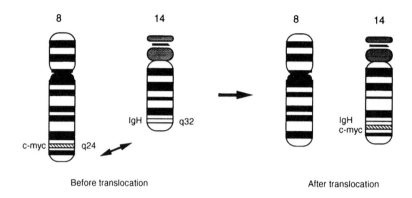

B. SCHEMATIC VIEW OF THE 9;22 TRANSLOCATION IN CML

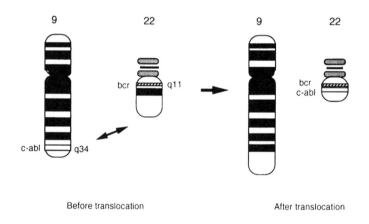

Figure 5 Examples of translocations resulting in oncogene activation.

Certain oncogenes appear to be linked with programmed cell death or apoptosis. Indeed an oncogene may drive either cell growth or cell "suicide", depending upon circumstances.[34] For example, the c-*myc* oncogene will induce apoptosis as a normal function unless prevented by "survival" factors. In their presence, c-*myc* programs for proliferation.[34] One such survival factor is the *Bcl-2* proto-oncogene.[35,36] The Bcl-2 protein can heterodimerize with its partner Bax to block apoptosis while Bax-Bax homodimers promote cell death.[35, 36] It is unclear at present whether Bcl-2 blocks a Bax-dependent suicide pathway or whether Bax blocks a Bcl-2 dependent survival function.[34] Coexpression of c-myc and Bcl-2 gives rise to a markedly increased tumor incidence.[37, 38] The c-*fos* gene may also be involved in apoptosis. Thus, sustained expression of c-*fos* has been linked to apoptosis in mouse embryos while deregulated expression of either c-*fos* or c-*myc* produces apoptosis in serum-deprived fibroblasts.[34,39]

3.2 Tumor-Suppressor Genes

The progression of many tumors to full malignancy requires more than oncogene activation and involves a second family of genes, the tumor-suppressor genes.[40-42] These genes, which have also been called anti-oncogenes or recessive oncogenes, act as growth regulators in normal cells, and when inactivated

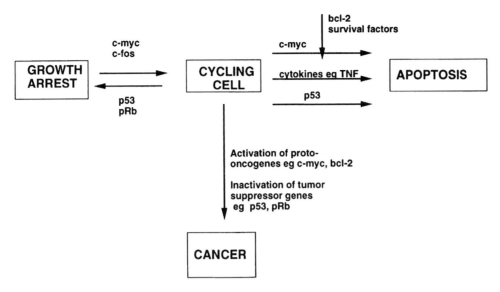

Figure 6 Factors regulating the processes of cell growth, apoptosis, and transformation.

Table 2 Tumor Suppressor Genes

Tumor Suppressor Gene	Chromosome Location	Tumor Suppressor Protein	Associated Tumors	Ref.
Retinoblastoma (RB-1)	13q	105-kDa Nuclear protein	Retinoblastoma, osteosarcomas, small-cell lung cancer, sarcomas	40, 42-45
p53	17p	53-kDa Nuclear protein	Most cancers including breast colorectal, lung, ovarian, cervical, brain, bladder	40, 42, 46-49
Wilms tumor (WT-1)	11p	Zinc finger nuclear protein	Wilms tumor	50, 51
Adenomatous polyposis coli (APC)	5q		Colorectal cancer	52-55
Deleted in colon carcinoma (DCC)	18q	190-kDa Transmembrane protein	Colorectal cancer	56, 57
Neurofibromatosis 1 (NF1)	17q	p21^{ras} Interacting protein	Neurofibrosarcomas	58, 59
Neurofibromatosis 2 (NF2)	22q	Membrane-organizing protein	Schwannomas	60, 61
Von Hippel-Lindau Syndrome (VHL)	3p	Unknown	CNS and retinal hemangioblastomas, renal cell cancer	62
PTPα	3p	Receptor-type protein, tyrosine phosphatase	Renal cell, lung cancer	63
MCC	5q	G-protein coupled m3 muscarinic acetylcholine receptor	Colorectal cancer	64
nm 23	17q	Nucleoside diphosphate kinase	Breast, colorectal cancer	65

or lost, allow the unconstrained growth found in cancer cells. To date, approximately a dozen tumor-suppressor genes have been identified in human tumors (see Table 2).[40-65]

Evidence for the existence of tumor-suppressor genes came initially from studies of somatic cell hybridization in which fusion of tumor cells with normal cells produced nontumorigenic hybrids.[66] These experiments demonstrated that normal cells contained genes capable of suppressing the neoplastic phenotype. These hybrids were often unstable, and as chromosomal material was lost, the hybrid became malignant, suggesting that the lost material contained tumor-suppressor functions. Furthermore, fusion of two malignant cells could produce nonmalignant hybrids. These studies provided the first indication that cancer cells might lose growth-regulatory controls as they progressed to malignancy.[40]

The next indication for the existence of these genes was provided by studies of the retinoblastoma gene. As described above, Knudson had deduced from studies of the incidence of this disease that two successive alterations were required to produce retinoblastoma.[8,9] The locus of the retinoblastoma gene (*RB1*) was first mapped to band q14 of chromosome 13 by karyotypic analyses,[67] and this was confirmed by studies investigating loss of heterozygosity which demonstrated loss of allelic fragments of chromosome 13 in retinoblastoma tumors.[68] Eventually, the gene was cloned[69-71] and Knudson's hypothesis was confirmed by the demonstration of a single mutation in the fibroblasts of a retinoblastoma patient with bilateral disease and a second mutation in the tumor itself.[71] Retinoblastoma, therefore, only occurs when both copies of the gene are either lost or altered. Loss of heterozygozity studies continue to provide a useful means of identifying new tumor-suppressor genes.

The third line of evidence for these genes derives from studies of the fruit fly, *drosophila*, in which eight tumor-suppressor genes have been identified.[41] Mutations in most of these genes led to hyperplastic or neoplastic overgrowth of specific tissues, resulting in death. Several of these genes are homologous to human genes, and their investigation may help reveal new human tumor-suppressor genes.

3.2.1 Function of Tumor-Suppressor Genes

Just as the proto-oncogenes encode components of the cell-signaling pathways associated with growth control, the tumor-suppressor gene products function as negative control elements in these signaling cascades. These products have roles in the cell membrane, the cytoplasm, and the nucleus.

The *DCC* (deleted in colorectal cancer) product, a 190-kDa transmembrane phosphoprotein has the characteristics of a cell-surface signal-transducing receptor.[40,56] It possesses sequence homology with cell adhesion molecules and may function as such.[72] At the next level within the cell, the PTPα protein encodes a phosphatase that counteracts the actions of receptor tyrosine kinases.[63] Within the signal transduction pathways, the *NF-1* gene is a GTP-ase activating protein and has been proposed to be a possible down-regulator of p21^{ras}.[40] Several tumor-suppressor genes, including *WT-1*, *RB1*, and *p53*, function as transcription factors.[40] The *WT-1* gene encodes a 345-amino-acid protein containing four zinc finger domains indicative of a sequence-specific DNA binding domain.[50,51] The *RB* and *p53* genes are described in detail below.

3.2.2 Retinoblastoma Gene

The retinoblastoma (*RB*) gene encodes a 105 kDa nuclear phosphoprotein named pRb.[40,44] The major function of pRb is to associate with a number of proteins, in particular the E2F family of transcription factors.[73,74] E2F sites are found in the promoters of genes encoding thymidine kinase, myc, myb, dihydrofolate reductase, and DNA polymerase-α, the products of which are involved in DNA synthesis during S phase.[73] Binding of pRb to the E2F protein prevents the transcriptional activation of these genes. Several other pRb-related proteins have been identified, including p107 and p130, and these have also been shown to associate with E2F and related proteins.[75,76]

In the G$_1$ phase of the cell cycle, pRb is hypophosphorylated and, in this form, associates with E2F and suppresses growth (see Figure 7). In late G$_1$, pRb becomes hyperphosphorylated and this modification is maintained through the S, G$_2$ and M phases of the cell cycle until it is reversed by a phosphatase in late M phase.[43] The phosphorylation of pRB is controlled by the cyclin-dependent kinases (Cdks) complexes and the probable species involved are cyclin D/Cdk4 and cyclin E/Cdk2.[45]

In cells transformed by DNA tumor viruses, e.g., SV40 (which encodes large T antigen), adenovirus (encoding E1A) and the human papillomavirus HPV (encoding E7), the virus-encoded oncoproteins complex with pRb.[40,45] The mechanism of action of these oncoproteins is believed to be that of nullifying pRb by formation of these complexes and quiescent cells exposed to these viruses will enter S phase as a result of pRb inactivation.

Germline heterozygosity at the *RB* locus has been demonstrated not only in retinoblastomas but also in osteosarcomas. *RB* gene inactivation has been found in a variety of somatic tumors, including sporadic retinoblastomas, sarcomas, small-cell lung cancer (SCLC), non-SCLC, bladder, and breast cancer.[40]

The introduction of mutated forms of the *RB* gene into the mouse genome has helped define the role of the gene and its mutated forms *in vivo*.[77-79] Mice that are heterozygous for a defective RB allele develop pituitary tumors, confirming that RB encodes a tumor-suppressor gene.[77-79] Mice that are homozygous for the null allele of RB die at about the 14th day of gestation due to defects in erythropoiesis and

PHASE OF CELL CYCLE

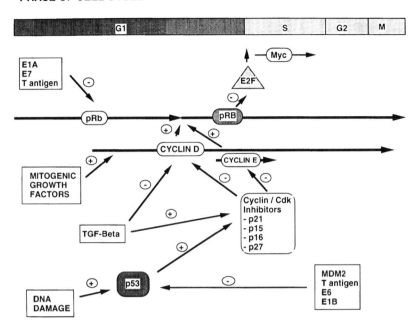

Figure 7 Interactions of growth factors, p53, pRB and, cyclins with the cell cycle.

neural development. Interestingly, there is a marked increase in apoptosis in these affected tissues and the significance of this is discussed below.

3.2.3 p53 Gene

p53 is the most frequently mutated gene identified to date in human tumors.[40] The protein has been termed "the guardian of the genome" and "a molecular policeman" in recognition of its role as a mediator of the cell's response to DNA damage.[46] DNA damage induces a rapid increase of p53, which then leads to either growth arrest in the G_1 phase of the cell cycle to allow repair to take place, or causes the cell to enter apoptosis.[47,48] The protein is a sequence-specific transcription factor which binds to p53 recognition sites and activates its target genes.[80-82] Of the target genes in which transcription is induced by p53, one in particular has attracted a great deal of attention recently. This is the *WAF1/CIP1* gene which encodes a 21 kDa protein, which may itself be a tumor-suppressor gene.[83] This protein binds to and inhibits complexes of G_1 cyclins and their partner CdK subunits which are regulators of the cell cycle.[84,85] Key targets of the Cdk/cyclin complexes include the retinoblastoma family proteins which, as described above, regulate transit through G_1 and entry into S phase is permitted. Therefore, as shown in Figure 7, it is conceivable that p53 produces G_1 arrest by inducing p21 which in turn inactivates Cdk/cyclin complexes leading to underphosphorylation of the retinoblastoma proteins. In addition to acting as a transcription factor, p53 can act as a transcription inhibitor by virtue of its ability to associate with the TATA-binding protein found within transcription complexes.[86-88]

As well as producing a G_1 arrest, p53 can induce apoptosis.[47,48] The decision as to whether p53 produces a G_1 arrest or apoptosis appears to depend on the status of the pRb family proteins.[48,49] If pRb proteins are inactivated by the human papillomavirus (HPV), E7 or adenovirus EIA proteins, then apoptosis results.[89,90] As described above, many of the growth-inhibitory effects of the pRb family are mediated by binding transcription factors of the E2F group. In support of the pRb family mediating the effect of p53, it has been demonstrated that constitutive expression of E2F (which substitutes for pRb inactivation) together with excess p53 induces apoptosis.[91] pRb proteins may therefore antagonize p53-dependent apoptosis[47] but it is unclear as to which are the positive mediators of the apoptotic process. One candidate is the *bax* protein, induced by p53, and known to promote cell death.[92] Together, these observations suggest a model wherein p53 will induce a G_1 arrest if the pRb proteins are functional; if these proteins are not functional, then the cell will fail to growth arrest and apoptosis will ensue.[48,49]

As mentioned above, p53 mutation is the most common cancer-related change at the gene level.[40] In addition to point mutations, allelic loss, rearrangements, and deletions of the p53 gene have been detected in human tumors.[93] Particularly common are mutations in the evolutionarily conserved codons of the gene which encompass exons 5 through exon 8.[93] Analysis of these mutations has provided clues to the function of specific regions of p53. The wild-type form of p53 binds to DNA as a tetramer.[80,94-96] In tumors, many of the mutant forms of p53, including point-mutated proteins, have been shown to have lost sequence-specific DNA-binding function.[94,96,97] These can act as dominant negatives to inhibit the activity of the wild-type p53.[96,98] This is achieved by the formation of inactive hetero-oligomers between the mutant and wild-type proteins resulting in reduced or lost DNA-binding function. As a result, when DNA is damaged in tumors with defective p53, replication is not switched off to allow repair to take place. Consequently, mutations and other chromosomal changes accumulate, leading to selection of increasingly malignant clones. A number of tumor virus oncoproteins complex with p53 to inactivate it. These include SV40 large T antigen, adenovirus E1B, and papilloma virus E6;[46] the oncoprotein MDM2 acts similarly.[46]

3.3 DNA Repair Genes

A third family of genes, the DNA mismatch repair genes, is attracting increasing interest for its role in tumorigenesis. Tumor evolution has been proposed to be associated with increased genetic instability.[10] The suggestion has been put forward that cancer cells exhibit a "mutator phenotype" based on the greater-than-expected numbers of mutations observed in tumor cells compared to normal cells.[99,100] As cancer genomes become increasingly unstable, more mutations arise, thereby increasing the probability that critical mutations may occur in oncogenes or tumor-suppressor genes, allowing rapid tumor progression. Recent molecular evidence supports this concept. This is based on microsatellite instability in both inherited and sporadic tumors.[101-102] The genome is punctuated with repetitive sequences between genes consisting of mono-, di-, tri-, and tetranucleotide sequences called microsatellites. These are relatively constant in normal cells, but may vary in length in tumors; mutations within these sequences are also markers of increased replication errors. Variations in the numbers of repetitive sequences in each microsatellite (as a result of mutation) can be demonstrated by changes in the DNA fingerprint. The importance of microsatellites, therefore, lies in their ability to provide a very sensitive indicator of genetic instability. Interestingly, a mutator phenotype was demonstrated in a familial form of colorectal cancer — hereditary nonpolyposis colon cancer (HNPCC).[103-105] A family of genes responsible for mismatch repair have been found in *E. coli* — namely *mut*S, *mut*L and *mut*H. and homologous human genes *hMSH*, *hMLH1*, *hPMS1*, and *hPMS2* have been identified.[106-111] These genes are mutated in HNPCC tumors and are believed to be responsible for the mismatch repair and microsatellite instability observed. Microsatellite instability has also been demonstrated in a substantial percentage of sporadic colon, stomach, and endometrial cancers.[101] It seems likely, therefore, that mutations in genes responsible for genetic stability could be an important early event in tumorigenesis.

3.4 Telomerase and Cancer

A recently identified mechanism that may have an important role in carcinogenesis involves the reactivation of the enzyme telomerase.[112] Telomeres are repeat sequences $(TTAGGG)_n$ placed at the ends of chromosomes which provide stability to the chromosome. DNA polymerase is unable to complete the replication of chromosome ends and this is achieved by telomerase which synthesizes telomere repeat sequences *de novo* onto chromosome ends. Telomerase is present in both male and female germline tissues, but not in normal somatic cells (with the possible exception of certain hematopoietic cells).[113] Furthermore, experimental studies demonstrated that, in the absence of telomerase, telomeres shortened with each cell division, while in immortalized human cell lines and in human tumors, telomerase is activated and telomere length is maintained. This has led to the hypothesis that telomere length is associated with the control of cell lifespan; repression of telomerase early in development is involved in the replicative senescence of normal cells while reactivation of telomerase is a necessary event in cell immortalization and cancer.[113] In support of a role for telomerase in cancer, 90 of 101 primary tumors investigated in a recent study, including those of breast, brain, colon, prostate, lung, and uterus, were positive for telomerase activity, while normal tissues were negative.[112,114]

3.5 Colorectal Cancer: A Paradigm of Multistage Carcinogenesis Involving Oncogenes, Tumor-Suppressor Genes, and DNA Repair Genes

The best-defined model of carcinogenesis at the molecular level is that provided by colorectal cancer which evolves through a series of well-defined pathological stages and in which a number of specific mutations have been linked with each of these steps.[115-117] Conversion of normal to hyperproliferative colorectal epithelium is associated with mutation at the *APC* tumor-suppressor gene found on chromosome 5.[53-55] This mutation may occur somatically and eventually result in production of a single colorectal tumor,[53] or be present in the germ line,[54,55] when it results in a predisposition to the disease (familial adenomatous polyposis). Patients who possess this particular germ-line mutation, can develop thousands of tumors throughout the colon. While mutation leads to the formation of benign adenomas, further development to the intermediate adenoma stage is associated with mutations of the ras oncogene[115] and development to late adenomas with mutations in the *DCC* tumor-suppressor gene.[56] The final progression to the malignant carcinoma stage is linked with mutations of the p53 tumor-suppressor gene.[118,119] The changes identified in this model elegantly link the concept of multistage carcinogenesis with the proposed roles of oncogenes and tumor-suppressor genes.

Tumors arising in HNPCC patients also show somatic mutations of the same genes, suggesting that the above pathway is followed but progression is assisted by the associated genetic instability.[111] High levels of telomerase activity have been reported in over 90% of colon carcinoma samples, while no activity has been detected to date in normal colon or in polyps, supporting the view described above that reactivation of telomerase is an important event in carcinogenesis.[120]

4. GROWTH CONTROL IN CANCER CELLS

Like that of their normal counterparts, the growth of cancer cells is under the regulation of growth factors, cytokines, and hormones. Interaction of these factors with their respective receptors initiates cascades of intracellular signals leading eventually to the activation or repression of specific genes.[121] In malignant cells, these signaling pathways are commonly dysregulated by oncogene activation or tumor-suppressor gene inactivation. The major pathways of growth factor, cytokine, and hormonal control will first be described, and then the mechanisms for their dysregulation by oncogenes will be considered.

4.1 Growth-Factor Signaling

Polypeptide growth factors are an important class of pleiotropic signaling molecules which bind to cell-surface receptors in target cells, initiating events leading predominantly to cell division, although some also lead to growth inhibition (see Figure 2). Characteristics of some of these growth factors are summarized in Table 3.

Table 3 Growth Factors

Growth Factor	Size (Amino Acids)	Chromosomal Location	Receptor Type
Epidermal growth factor (EGF)	53	4q	Protein tyrosine kinase
Transforming growth factor alpha (TFGα)	50	2p	Protein tyrosine kinase
Transforming growth factor-beta TGF-β1	112	19q	Protein serine/threonine kinase
TGF-β2	112	1q	Protein serine/threonine kinase
TGF-β3	112	14q	Protein serine/threonine kinase
Insulin-like growth factor I (IGF-I)	70	12q	Protein tyrosine kinase
Insulin-like growth factor II (IGF-II)	67	11p	Protein tyrosine kinase
Platelet-derived growth factor (PDGF) A chain	110	7p	Protein tyrosine kinase
B chain	109	11q	Protein tyrosine kinase
Fibroblast growth factor — acidic	140	5	Protein tyrosine kinase
— basic	154	4	
Gastrin-releasing peptide (GRP)	27	18q	G-protein-linked to PLC
Erythropoietin (EPO)	166	7q	JAK-2 activator
Interleukin 3 (IL-3, multi-CSF)	133	5q	JAK-2 activator
Granulocyte-macrophage (GM-CSF) colony-stimulating factor	127	5q	JAK-2 activator
Granulocyte colony-stimulating factor (G-CSF)	174	17q	JAK-2 activator

The receptors to which growth factors bind are predominantly of two types, namely, those with intrinsic cytosolic tyrosine kinase domains and those with seven transmembrane domains. The former bind growth factors, such as epidermal growth factor (EGF) and platelet-derived growth factor (PDGF).[122,123] Interaction of the growth factor with the extracellular binding domain produces a conformational change in the receptor which encourages receptor dimerization.[124] The close proximity of the cytosolic domains within receptor dimers allows transautophosphorylation of the tyrosine residues on these domains. When phosphorylated, tyrosine phosphoprotein sites attract the SH2 (src homology 2) domains of signaling molecules, such as phospholipase C, phosphatidylinositol-3′-kinase (PI3 kinase), and p21 rasGTPase-activating protein (GAP).[125] One of these SH2-containing molecules is Grb2 (for growth-factor binding), an adaptor protein, which contains another type of domain, SH3, which binds proline-rich motifs.[125] The SH3 sites of Grb2 interact with a molecule named sos ("son of sevenless") which in turn activates the ras product, p21. A signaling cascade through a series of kinases is then initiated: ras activates raf-1; raf-1 activates MEK; MEK activates MAP kinase which translocates to the nucleus and activates the transcription factors myc, fos, and jun.[126-128] Interaction of fos and jun produces the transcription complex AP-1 (activator protein 1) which switches on a number of genes associated with proliferation,[129] while myc heterodimerizes with its partner max to activate transcription.[130]

The other major family of receptors to which mitogenic peptides bind are receptors containing a 7 transmembrane domain motif. These receptors are linked via G-proteins to either phospholipase C or adenylate cyclase. Mitogenic peptides such as bombesin or vasopressin interact with receptors resulting in activation of phospholipase C.[131] This catalyzes the hydrolysis of phosphatidylinositol 4,5-biphosphate (PIP$_2$) to diacylglycerol and inositol 1,4,5-triphosphate (IP$_3$).[132] IP$_3$ causes the release of intracellular Ca^{2+} which together with 1,2-diacylglycerol activates protein kinase C (PKC). Activation induces translocation of PKC from the cytosol to the membrane placing the enzyme close to its substrate. Among the substrates phosphorylated by PKC is raf-1 which feeds into the signal cascade mentioned above.[133]

Stimulation of receptors linked to adenylate cyclase generates cyclic AMP (cAMP). Increased cAMP in turn activates cAMP-dependent kinases (also known as protein kinase A; PKA), serine-threonine kinases, which exist as two major isozymes, type I and type II PKAs.[134] These isoforms, although possessing identical catalytic subunits, are distinguished by containing different regulatory subunits, RI and RII. The two isozymes have been claimed to have different roles — enhanced expression of type I PKA is associated with cell growth and transformation, while type II PKA is linked with growth inhibition and differentiation.[134] Binding of cAMP to the regulatory subunits releases the catalytic subunit which either phosphorylates cytoplasmic targets or translocates to the nucleus and phosphorylates the serine residue of a number of nuclear proteins.[135] These nuclear factors, which may be either activators or repressors, bind to the cAMP-responsive element present in the promoter region of cAMP-inducible genes; they may also heterodimerize with fos and jun, giving rise to the possibility of transcriptional cross-talk.[136,137]

Transforming growth-factor-β (TGF-β), which is classified as both a growth factor and cytokine, acts primarily as a growth inhibitor in most cell types. Its signaling pathway within the cell has recently been described.[138,139] Members of the TGF-β family bind to the type II TGF-β receptor. This ligand-bound receptor forms a complex with the type I TGF-β receptor which is phosphorylated at serine and threonine residues by the ligand-activated type II receptor.[138,139] TGF-βs suppress expression of cyclins and cyclin-dependent kinases; furthermore these factors can induce expression of p15 and p27, inhibitors of the cyclin D/Cdk4 complex.[140-143] This will, in turn, prevent phosphorylation of pRb and block activation of the E2F transcription factors (described above), resulting in the cell-cycle block in late G$_1$ observed after TGF-β treatment.

4.2 Cytokine Regulation

Cytokines are a diverse group of regulatory proteins which act as the local controllers of intercellular communication, playing major roles in cell immunity, growth, development, and apoptosis. They can modulate all these systems both positively and negatively. For example, certain cytokines can stimulate cell growth in a manner similar to that of growth factors, or act as survival factors to block apoptosis, while others can inhibit cell growth or initiate apoptosis. While some can mediate destruction of the tumor vasculature, others can stimulate angiogenesis. Each cytokine can have multiple activities and the interactions between cytokines are therefore complex in both normal and malignant cells. Examples of cytokines are listed in Table 4.

Recently, rapid progress has been made in our understanding of cytokine signaling within the cell (see Figure 8).[145,146] Cytokines interact with receptors of the cytokine receptor superfamily. Most cytokine

Table 4 Cytokine Families

Interleukins	Colony-Stimulating Factors	Interferons	Growth Factors
IL-1α	GM-CSF	IFN-α	EGF
IL-1β	G-CSF	IFN-β	TGF-α
IL-2	M-CSF	IFN-γ	a-FGF
IL-3	EPO		b-FGF
IL-4	TNF-α		KGF
IL-5	TNF-β	**TGF-β Family**	PDGF-A
IL-6	LIF		PDGF-B
IL-7		TGF-β1	NGF-β
IL-9	**Chemotactic Factors**	TGF-β2	IGF-I
IL-10		TGF-β3	IGF-II
IL-11	IL-8	Inhibin	
	MCP-1	Activin	

Source: Modified from Miyajima, A. et al., *Annu. Rev. Immunol.*, 10, 295, 1992.

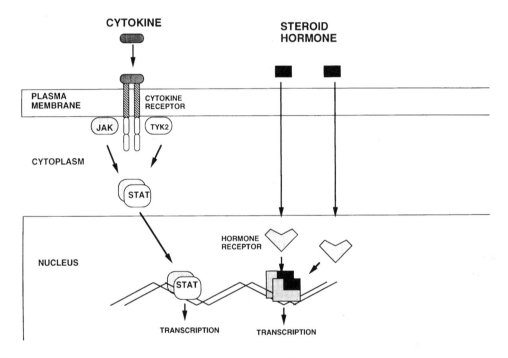

Figure 8 Cytokine and steroid hormone signaling pathways. Cytokines interact with receptors of the cytokine-receptor superfamily, whereupon they dimerize and associate with members of the JAK family of kinases (JAK 1, 2, or 3 and TYK2). Activated JAKs then phosphorylate members of the STAT family which dimerize and activate transcription. Steroid hormones bind to nuclear receptors, which also dimerize and activate transcription.

receptors consist of a multichain complex — a ligand-specific receptor and a class-specific signal transducer.[147] Upon cytokine binding, these receptors dimerize and associate with members of the JAK (Janus kinase) family of cytoplasmic kinases.[145,146] This family consists of four known members (JAK1, JAK2, JAK3, and TYK2) which have protein tyrosine kinase activity, and upon binding to the cytokine receptor, cross-phosphorylation of the auto-phosphorylation site on JAKs occurs to activate the kinase. The activated JAK then phosphorylates the receptor and also cellular substrates. Key targets include the STAT (*s*ignal *t*ransducing and *a*ctivators of *t*ranscription) proteins which are involved in the regulation of gene transcription.[145,146] The STAT family consists of six known members (STAT 1–6) and the tyrosine phosphorylation (by JAKs) which is required for nuclear translocation and DNA binding. Dimerization of STAT molecules has also been proposed to be essential for nuclear transport and/or DNA binding.

While the JAK-STAT pathway is the predominant pathway for cytokine signaling, the ras/MAP kinase pathway can also be activated by a number of cytokines.[145]

4.3 Hormonal Regulation

Hormones are key regulators of the growth and development of many normal and malignant cell types. Estrogen, in particular, is a sufficiently potent mitogen of certain malignant tumors that its deprivation can produce effective growth control. In contrast to the growth factors and cytokines, which interact with plasma membrane receptors and require intermediate molecules to transmit the message to the nucleus, the steroid hormones are sufficiently small and lipophilic to reach the nucleus where they interact with receptors that are capable of functioning directly as transcription factors (see Figure 8).[148] Steroid interaction activates the receptors by causing disassociation of complexed heat-shock proteins and induces key phosphorylations thus allowing dimerization of the receptor molecules to occur. (see Figure 9). These dimers then bind to specific response elements located in the vicinity of hormone-responsive genes. Both the estrogen and progesterone receptors have two types of transcriptional activation functions (TAFs).[148] TAF-1 is constitutively active and can operate independently of the ligand-binding domain, while TAF-2 is only active in ligand-receptor complexes.[148]

Estrogen may regulate growth-factor genes and increase expression of growth-stimulatory growth factors, including TGF-α, IGF-II, and PDGF, while reducing expression of the inhibitory growth factor TGF-β.[149] In addition to modulating growth-factor genes, steroids may increase the production of proteases such as pro-cathepsin D which can enhance the invasiveness of tumor cells (described below).[150]

4.4 Oncogene Dysregulation of Growth-Factor Pathways

That oncogenes encode proteins which are components of the growth factor signaling pathways suggests a number of possible mechanisms by which activation could produce abnormal growth control. For example, normal cells require growth-factors in order to undergo mitogenesis and to proceed through the cell cycle.[121] "Competence factors," such as EGF, PDGF, and FGF, are required to traverse through the G_1 phase of the cell cycle while "progression factors," such as insulin-like growth factor 1 and insulin, are needed to commit cells to DNA synthesis.[121] Growth factors may be provided via the systemic circulation (endocrine control) or locally (paracrine control). Without growth factors, normal cells remain quiescent. By contrast, many malignant cells grow autonomously in the absence or presence of reduced growth factors. This observation has led to the concept of "autocrine" driven growth whereby tumor cells produce their own growth-factors resulting in self-stimulated growth and independence from their surroundings.[151] The same result, i.e., independence of the need for an exogenous supply of growth factors, can be achieved by oncogene activation in a number of ways,[21] including the oncogene product being a growth factor (e.g., the *sis* product, PDGF, could produce cell-signaling without the need for an external supply), or overexpression of the growth factor receptors (as exemplified by the *erbB-2* oncogene). In the case of erbB, a truncated version of the EGF receptor in which the ligand binding site is absent, the receptor may be constitutively active in the absence of a growth factor. At the signal transduction level, mutated *ras* p21 protein, may prolong a growth factor-generated signal. Finally, constitutive expression of transcription factors such as myc, fos, or jun would clearly obviate the requirement for prior signaling.

4.5 Cell-Cycle Control and Cancer

Transition of a cell through the phases of the cell cycle is regulated at checkpoints by the cyclin-dependent kinases (Cdks) and their partners, the cyclins. In cancer, it has been suggested that deregulation occurs at these checkpoints and this may arise as a result of aberrant expression of the cyclins or the loss of their inhibitor proteins, the cyclin-dependent kinase inhibitors (CDIs).[152]

Of the cyclins, the cyclin-D (1, 2, and 3) family is most closely linked to the initiation of the cell cycle at START (the restriction point) in late G_1 and these regulators have been suggested to act as growth factor sensors.[152] Cyclin D is associated with both oncogene and tumor-suppressor gene pathways; myc induces cyclin D1, while D-type cyclins phosphorylate pRb. Cyclin-D1 is encoded by the *CCND1* gene which has been identified as the *PRAD1* proto-oncogene and is a candidate for the *BCL1* proto-oncogene.[153,154] Overexpression of this cyclin is found in a percentage of breast, gastric, and esophageal carcinomas.[152]

The recently identified CDIs can potentially act as tumor suppressors. Four inhibitors (p15, p16, p21, and p27) have been identified; p15 and p16 bind and inhibit Cdk4 and Cdk6, while both p21 and p27 inhibit a variety of cyclin/Cdk complexes, including cyclin D/Cdk4, cyclin E/Cdk2, and cyclin A/Cdk2.

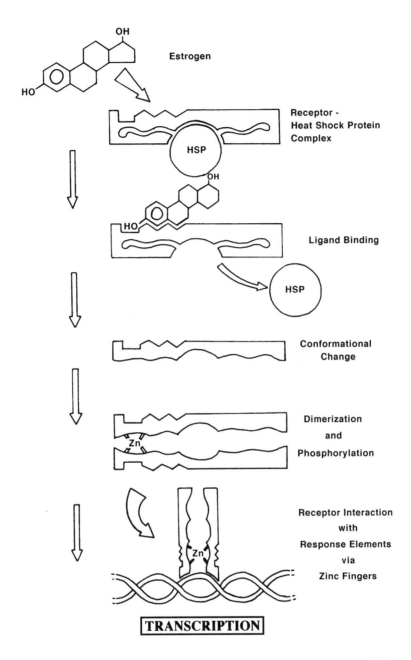

Figure 9 Estrogen regulation of transcription. Estrogen binds to a nuclear receptor which contains a heat-shock protein (HSP) molecule. Upon steroid binding, the HSP disassociates from the complex allowing dimerization of receptors to occur. These dimers then bind to specific response elements in DNA and activate transcription.

The p16 locus is rearranged, deleted, or mutated in many tumor cell lines and may correspond to the the putative tumor-suppressor gene (MTS1) that maps to the same region of chromosome 9.[155,156] The inhibitors, p15 and p27, are induced by TGF-β and are probably the mediators of its growth inhibitory effects.[142,143] Many transformed cells are insensitive to the effects of TGF-ß and it has been suggested that this might be due to altered p15 or p27 function during oncogenesis.[152] Inhibitor p21 is a target for p53 and is essential for p53-mediated arrest of growth in G_1.[83-85] The inactivation of p53 results in reduced expression of p21 and this is consistent with the observed low expression of p21 in transformed cells.

5. CANCER INVASION AND METASTASIS

The fundamental characteristics that distinguish a malignant tumor from its benign counterpart and result in life-threatening consequences are the properties of invasion and metastasis. Metastasis has been defined as the ability of tumor cells to cross tissue compartment boundaries and intermix with cells on the other side of the boundary, a characteristic that separates invasive tumors from benign tumors and carcinomas *in situ*.[157] The process of invasion is a cascade of several steps (see Figure 10).[158-160] First, tumor cells attach to the basement membrane; this is followed by secretion and activation of cellular proteinases to degrade the basement membrane. Once the barrier has been breached, tumor cells will migrate through the matrix. These processes of attachment, proteolysis, and migration are found not only in malignancy, but also have roles in trophoblast implantation, embryonic morphogenesis, and tissue remodeling. Within cancer there is a either a quantitative escalation or perhaps loss of regulation of the invasive process.[158,161]

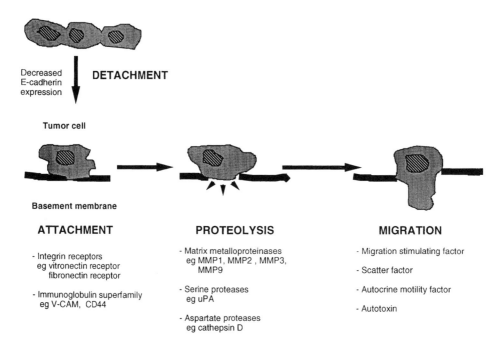

Figure 10 Stages of invasion and metastasis. Invasion and metastasis involve a number of steps including detachment, attachment, proteolysis, and migration. The major families of molecules involved in each of these processes are shown.

At this stage, the tumor has not developed its own vasculature, and nutrients and oxygenation are provided via diffusion. This limits the tumor growth rate. For tumors to expand beyond about 2 mm, new capillary blood vessels are required.[160,162] Angiogenesis is required for this expansion to occur and the new blood vessels formed help provide access of individual cells into the circulation. Once vascularization occurs, the growth rate of the tumor accelerates and metastasis can take place more rapidly. Cells moving into the vascular or lymphatic channels may circulate until they arrest in the capillary beds of other organs.[160] The cell or cell aggregates will then invade into the organ parenchyma and, as for the primary tumor, angiogenesis will be needed for this metastatic deposit to expand beyond a certain size. This process of tumor spread is considered to be relatively inefficient since experiments using radiolabeled murine melanoma cells indicate that <0.1% of cells injected directly into the venous circulation will successfully form secondary deposits.[160]

The molecular events underlying these cellular processes are currently being defined and a brief summary of the present view is now given.

5.1 Molecular Events in Invasion
5.1.1 Tumor-Cell Detachment

Before a tumor cell can invade, it has to detach from its neighbors. For this to happen, changes in cell-cell contacts must occur.[160] An important group of cell-cell connective molecules are the cadherins which help provide the intercellular "cement" that binds cells of the same type together.[163] Epithelial cadherin (E-cadherin) acts in this manner in epithelial cells and the down-regulation or loss of E-cadherin expression has been shown to relate to invasiveness in carcinomas.[163] Experimental modulation of E-cadherin in model systems has confirmed this link with invasiveness and has led to the suggestion that this molecule functions as an invasion suppressor-gene product.[164,165]

5.1.2 Tumor-Cell Adhesion

Several families of molecules have been associated with tumor-cell adhesion to the basement membrane prior to invasion. Cellular attachment to the extracellular matrix (ECM) is mediated via receptors of which the integrins are perhaps the best described. Integrins are heterodimeric receptors consisting of α and β subunits which combine to give at least 20 different receptors.[166] They link cytoskeletal microfilaments (including actin, vinculin, and talin) to proteins such as laminin, fibronectin, collagen, and vitronectin in the ECM. Integrin binding utilizes a specific tripeptide motif, Arg-Gly-Asp (RGD), and in support of the integrins having a role in invasion, synthetic RGD-containing peptides which interfere with this binding inhibit invasion both *in vitro* and *in vivo*.[167,168] Further evidence of integrin involvement is provided by observations of overexpression of certain integrin receptors (e.g., the vitronectin receptor and the fibronectin receptor) being related to increasing tumor progression.[158]

Other cell-adhesion molecules thought to be involved in invasion are V-CAM and CD44, members of the immunoglobulin superfamily.[158,160] V-CAM was identified on endothelial cells and interacts with the integrin VLA-4 ($\alpha 4 \beta 1$) which has been shown to be expressed on malignant melanoma cells.[169] This has been speculated to facilitate interaction of circulating melanoma cells with the endothelium in advance of tumor-cell extravasation. Enhanced attachment of circulating malignant cells may also be helped by the presence of CD44 (a lymphocyte homing molecule). As with V-CAM, overexpression of CD44 in malignant cells correlated with increased metastasis.[170]

5.1.3 Proteolysis

Proteolysis of tissue barriers is achieved by the production of proteolytic enzymes. Four classes of enzyme have been identified — cysteinyl-, serine-, aspartyl-, and metalloproteinases.[158,171] Inhibitors of cysteinyl, serine, and metalloproteinases have been shown to block tumor-cell invasion, suggesting roles for these enzymes in the metastatic process. Since all of these enzyme types have functions in normal cells, it seems likely that it is the inappropriate expression of these enzymes that is critical in the invasive process.

A major group of proteolytic enzymes involved in this process are the matrix metalloproteinases (MMPs).[172,173] Each member of this family of degradative enzymes has a distinct substrate specificity; therefore, MMP-1 degrades interstitial collagens (types I, II, and III collagen), MMP-2 and MMP-9 degrade type IV collagen (a component of the basement membrane), gelatin and fibronectin, while MMP-3 (stromelysin) degrades proteoglycans. These enzymes are secreted as inactive proenzymes and require cleavage for activation. The activity of latent and active MMPs is controlled by natural tissue inhibitors (TIMPs) and it is thought that proteolysis is a result of the balance between MMP and TIMP activities. Further, both MMP and TIMP activities are regulated by cytokines.[173]

Of the serine proteases, urokinase-type plasminogen activator (uPA) has attracted much interest as overexpression has been related to enhanced invasion and metastasis.[174] This enzyme converts inactive plasminogen to plasmin, an enzyme capable of degrading collagen, fibronectin, fibrin, and laminin. Just as the activities of the MMPs are influenced by the TIMPs, inhibitors for uPA have been identified (PAIs).[175]

The cathepsins are aspartate proteases normally found inside lysosomes. Both cathepsin B and D have been suggested to have roles in the invasion and metastasis of breast cancer. [176,177]

5.1.4 Migration

A number of factors have been identified that influence cell motility.[171] These include migration-stimulating factor, scatter factor, autocrine-motility factor and autotoxin. Migration-stimulating factor is secreted by fetal or malignant fibroblasts and stimulates migration of normal fibroblasts;[178] scatter factor is also produced by fibroblasts but acts only on epithelial cells.[179] Autocrine motility factor is produced by tumor-cells and stimulates their motility via receptors that are localized at the leading and trailing edges of the tumor-cells.[180] Autotoxin is another autocrine-motility factor demonstrated to have activity in melanoma cells.[181]

5.1.5 Angiogenesis

For a tumor to grow beyond a certain size, it must develop a vasculature; this is indicated by observations that tumors without neovascularization rarely metastasize.[182] Many factors are capable of inducing apoptosis, including acidic and basic FGFs, platelet derived (PDECGF) and vascular endothelial growth factor (VEGF), TGF-α, TGF-β, and PDGF.[183] Certain of these agents stimulate endothelial cell proliferation, while others stimulate endothelial cell migration.

5.2 Metastatic Suppressor Genes

Just as tumor progression is believed to arise as a result of changes in oncogenes and tumor-suppressor genes, certain genetic malfunctions have been proposed to be linked to metastasis. The nm (nonmetastatic) 23 gene is one such putative, metastatic suppressor gene. Initially identified in cell-line model systems, wherein reduced expression is found in highly metastatic lines,[184] evidence has been obtained that expression of nm23 was reduced in breast cancer patients with evidence of metastasis to the lymph nodes.[185] Furthermore, loss of mRNA is strongly associated with poor survival in breast cancer patients.[186] Consistent with this function, transfection of the nm23 cDNA into metastatic murine melanoma cells reduced the metastatic potential of these cells.[187] The nm23 gene product is a nucleoside diphosphate kinase which participates in at least two key cellular processes — microtubule assembly/disassembly and signal transduction through G proteins.[172] The suggestion has been made that the increased aneuploidy (and genetic instability) observed in metastatic cells may result from reduced expression/mutation of nm23, thereby allowing aberrant mitosis on a compromised spindle.[172] Other genes, such as the mta1 and Tiam-1 genes, show increased expression in invasive and metastatic disease.[188,189] As for the oncogenes and tumor-suppressor genes, these genes encode signal-transducing molecules.[188,189]

REFERENCES

1. **Berenblum, I.,** The mechanism of carcinogenesis: a study of the significance of carcinogenic action and related phenomena, *Cancer Res.*, 1, 807, 1941.
2. **Friedewald, W.F. and Rous, P.,** The initiating and promoting elements in tumor production: an analysis of the effects of tar, benzpyrene and methylcholanthrene on rabbit skin, *J. Exp. Med.*, 80, 101, 1944.
3. **Hennings, H., Shores, R., Wenk, M.L., Spangler, E.F., Tarone, R., and Yuspa, S.H.,** Malignant conversion of mouse skin tumours is increased by tumour initiators and unaffected by tumour promoters, *Nature*, 304, 67, 1983.
4. **Nordling, C.O.,** A new theory on cancer inducing mechanisms, *Br. J. Cancer*, 7, 68, 1953.
5. **Armitage, P. and Doll, R.,** The age distribution of cancer and a multistage theory of carcinogenesis, *Br. J. Cancer*, 8, 1, 1954.
6. **Armitage, P. and Doll, R.,** A two-stage theory of carcinogenesis in relation to the age distribution of human cancer, *Br. J. Cancer*, 11, 161, 1957.
7. **Moolgavkar, S.H. and Knudson, A.G.,** Mutation and cancer: a model for human carcinogenesis, *J. Natl. Cancer Inst.*, 66, 1037, 1981.
8. **Knudson, A.G.,** Mutation and cancer: statistical study of retinoblastoma, *Proc. Natl. Acad. Sci. U.S.A.*, 68, 820, 1971.
9. **Knudson, A.G.,** Retinoblastoma: a prototypic hereditary neoplasm, *Semin. Oncol.*, 5, 57, 1978.
10. **Nowell, P.C.,** The clonal evolution of tumor-cell populations, *Science*, 194, 23, 1976.
11. **Nowell, P.C. and Hungerford, D.A.,** Chromosome studies on normal and leukemic human leukocytes, *J. Natl. Cancer Inst.*, 25, 85, 1960.
12. **Fialkow, P.J.,** The origin and development of human tumors studied with cell markers, *N. Engl. J. Med.*, 291, 26, 1974.
13. **Linder, D. and Garter, S.M.,** Glucose-6-phosphate dehydrogenase mosaicism: utilization as a cell marker in the study of leiomyomas, *Science*, 150, 67, 1965.
14. **Mok, C.-H., Tsao, S.-W., Knapp, R.C., Fishbaugh, P.M., and Lau, C.C.,** Unifocal origin of advanced human epithelial ovarian cancers, *Cancer Res.*, 52, 5119, 1992.
15. **Jacobs, I.J., Kohler, M.F., Wiseman, R.W., Marks, J.R., Whitaker, R., Kerns, B.A.J., Humphrey, P., Berchuck, A., Ponder, B.A.J., and Bast, R.C.,** Clonal origin of epithelial ovarian carcinoma: analysis by loss of heterozygosity, p53 mutation, and X-chromosome inactivation, *J. Natl. Cancer Inst.*, 84, 1793, 1992.
16. **Boveri, T.,** Zur Frage der Estehung Maligner Tumoren, Fischer, Jena, Germany, 1914.
17. **Martin, G.S.,** Rous sarcoma virus: a function required for the maintenance of the transformed state, *Nature*, 227, 1021, 1970.

18. **Weinberg, R.A.,** Oncogenes, tumor-suppressor genes and cell transformation: trying to put it all together, in *Origins of Human Cancer: A Comprehensive Review,* Brugge, J., Curran, T., Harlow, E., and McCormick, F., Eds., Cold Spring Harbor Laboratory, 1991, 1.

19. **Marx, J.,** Oncogenes reach a milestone, *Science,* 266, 1942, 1994.

20. **Stehelin, D., Varmus, H.E., Bishop, J.M., and Vogt, P.K.,** DNA related to the transforming gene(s) of avian sarcoma viruses is present in normal avian DNA, *Nature,* 260, 170, 1976.

21. **Cantley, L.C., Auger, K.R., Carpenter, C., Duckworth, B., Graziani, A., Kapeller, R., and Soltoff, S.,** Oncogenes and signal transduction, *Cell,* 64, 281, 1991.

22. **Solomon, E., Borrow, J., and Goddard, A.D.,** Chromosome aberrations and cancer, *Science,* 254, 1153, 1991.

23. **Brodeur, G.M., Seeger, R.C., Swhabo, M., Varmus, H.E., and Bishop, J.M.,** Amplification of N-myc in untreated human neuroblastomas correlates with advanced disease stage, *Science,* 224, 1121, 1984.

24. **Slamon, D.J., Godolphin, W., Jones, L.A., Holt, J.A., Wong, S.G., Keith, D.E., Levin, W.J., Stuart, S.G., Udove, J., Ullrich, A., and Press, M.F.,** Studies of HER-2/neu proto-oncogene in human breast and ovarian cancer, *Science,* 244, 707, 1987.

25. **Tabin, C.J., Bradley, S.M., Bargmann, C.I., Weinberg, R.A., Papageorge, A.G., Scolnick, E.M., Dhar, R., Lowy, D.R., and Chang, E.H.,** Mechanism of activation of an oncogene, *Nature,* 300, 143, 1982.

26. **Barbacid, M.,** Ras genes, *Annu. Rev. Biochem.,* 56, 779, 1987.

27. **Leder, P., Battey, J., Lenoir, G., Mouldong, C., Murphy, W., Potter, H., Stewart, T., and Taub, R.,** Translocations among antibody genes in human cancer, *Science,* 222, 765, 1983.

28. **Rowley, J.D.,** A new consistent chromosomal abnormality in chronic myelogenous leukemia identified by quinacrine fluorescence and giemsa staining, *Nature,* 343, 290, 1973.

29. **Kurzrock, R., Gutterman, J.U., and Talpaz, M.,** The molecular genetics of Philadelphia chromosome positive leukemias, *N. Engl. J. Med.,* 319, 990, 1988.

30. **Konopka, J.B., Watanabe, S.M., and Witte, O.N.,** An alteration of the human c-abl protein in K562 leukemia cells unmasks associated tyrosine kinase activity, *Cell,* 37, 1035, 1984.

31. **Rabbits, T.H.,** Chromosomal translocations in human cancer, *Nature,* 372, 143, 1994.

32. **Land, H., Parada, L.F., and Weinberg, R.A.,** Tumorigenic conversion of primary embryo fibroblasts requires at least two cooperating oncogenes, *Nature,* 304, 596, 1983.

33. **Ruley, H.E.,** Adenovirus early region 1A enables viral and cellular transforming genes to transform primary cells in culture, *Nature,* 304, 602, 1983.

34. **Harrington, E.A., Fanidi, A., and Evan, G.I.,** Oncogenes and cell death, *Curr. Opin. Genet. Dev.,* 4, 120, 1994.

35. **Oltvai, Z.N. and Korsmeyer, S.J.,** Checkpoints of duelling dimers foil death wishes, *Cell,* 79, 189, 1994.

36. **Oltvai, Z.N., Milliman, C.L., and Korsmeyer, S.J.,** Bcl-2 heterodimerizes in vivo with a conserved homolog, bax, that accelerates programmed cell death, *Cell,* 74, 609, 1993.

37. **Bissonnette, R., Echeverri, F., Mahboubi, A., and Green, D.,** Apoptotic cell death induced by c-myc is inhibited by bcl-2, *Nature,* 359, 552, 1992.

38. **Fanidi, A., Harrington, E., and Evan, G.,** Co-operative interaction between c-myc and bcl-2 proto-oncogenes, *Nature,* 359, 554, 1992.

39. **Smeyne, R., Vendrell, M., Hayward, M., Baker, S., Miao, G., Schilling, K., Robertson, L., Curran, T., and Morgan, J.,** Continuous c-fos expression precedes programmed cell death in vitro, *Nature,* 363, 166, 1993.

40. **Weinberg, R.A.,** Tumor suppressor genes, *Science,* 254, 1138, 1991.

41. **Bryant, P.J.,** Towards the cellular functions of tumour suppressor genes, *Trends Cell. Biol.,* 3, 31, 1993.

42. **Hinds, P.W. and Weinberg, R.A.,** Tumor suppressor genes, *Curr. Opin. Genet. Dev.,* 4, 135, 1994.

43. **Hinds, P.W.,** The retinoblastoma tumor-suppressor gene, *Curr. Opin. Genet. Dev.,* 5, 79, 1995.

44. **Cobrinik, D., Dowdy, S.F., Hinds, P.W., Mittnacht, S., and Weinberg, R.A.,** The retinoblastoma protein and the regulation of cell cycling, *Trends Biochem. Sci.,* 17, 312, 1992.

45. **Scherr, C.J.,** The ins and outs of Rb: coupling gene expression to the cell cycle clock, *Trends Cell. Biol.,* 4, 15, 1994.

46. **Lane, D.,** p53, guardian of the genome, *Nature,* 358, 15, 1992.

47. **Donehower, L.A. and Bradley, A.,** The tumor-suppressor p53, *Biochim. Biophys. Acta.,* 1155, 181, 1993.

48. **Haffner, R. and Oren, M.,** Biochemical properties and biological effects of p53, *Curr. Opin. Genet. Dev.,* 5, 84, 1995.

49. **White, E.,** p53, guardian of Rb, *Nature,* 371, 21, 1994.

50. **Wang, Z.Y., Qiu, Q.Q., and Deuel, T.F.,** The Wilm's tumor gene product WT activates or suppresses transcription through separate functional domains, *J. Biol. Chem.,* 268, 9172, 1992.

51. **Van Heynigen, V. and Hastie, N.D.,** Wilm's tumor: reconciling genetics and biology, *Trends Genet.,* 8, 16, 1992.

52. **Bodmer, W.F., Bailey, C.J., Bodmer, J., Bussey, H.J.R., Ellis, A., Gorman, P., Lucibello, F.C., Murday, V.A., Rider, S.H., Scambler, P., Scher, D., Solomon, E., and Spurr, N.,** Localization of the gene for familial adenomatous polyposis on chromosome 5, *Nature,* 328, 614, 1987.

53. **Powell, S.M., Zilz, N., Beazer-Barclay, Y., Bryan, T.M., Hamilton, S.R., Thibodeau, S.N., Vogelstein, B., and Kinzler, K.W.,** APC mutations occur early during colorectal tumorigenesis, *Nature,* 359, 235, 1992.

54. **Nishisho, I., Nakamura, Y., Miyoshi, Y., Miki, Y., Ando, H., Horii, A., Koyama, K., Utsunomiya, J., Baba, S., Hedge, P., Markmam, A., Krush, A.J., Petersen, G., Hamilton, S.R., Nilbert, M.C., Levy, D.B., Bryan, T.M., Preisinger, A.C., Smith, K.J., Su, L.-K., Kinzler, K.W., and Vogelstein, B.,** Mutations of chromosome 5q21 genes in FAP and colorectal cancer patients, *Science,* 253, 665, 1991.

55. **Groden, J., Thilivers, A., Samowitz, W., Carlson, M., Gelbert, L., Albertsen, H., Joslyn, G., Stevens, J., Spirio, L., Robertson, M., Sargeant, L., Krapcho, K., Wolff, E., Burt, R., Hughes, J.P., Warrington, J., McPherson, J., Wasmutt, J., Le Paslier, D., Abderrahim, H., Cohen, D., Leppert, M., and White, R.,** Identification and characterization of the familial adenomatous polyposis coli gene, *Cell*, 66, 589, 1991.

56. **Fearon, E.R., Cho, K.R., Nigro, J.M., Kern, S.E., Simons, J.W., Ruppert, J.M., Hamilton, S.R., Preisinger, A.C., Thomas, G., Kinzler, K.W., and Vogelstein, B.,** Identification of a chromosome 18q gene that is altered in colorectal cancers, *Science*, 247, 49, 1990.

57. **Klingelhultz, A.J., Smith, P.P., Garrett, L.R., and McDougall, J.K.,** Alteration of the DCC tumor-suppressor gene in tumorigenic HPV-18 immortalized human keratinocytes transformed by nitrosomethylurea, *Oncogene*, 8, 95, 1993.

58. **Seizenger, B.R.,** NF1: A prevalent cause of tumorigenesis in human cancers, *Nature Genet.*, 3, 97, 1993.

59. **Legius, E., Marchuk, D.A., Collins, F.S., and Glover, T.W.,** Somatic deletion of the neurofibromatosis type-1 gene in a neurofibrosarcoma supports a tumour suppressor gene hypothesis, *Nature Genet.*, 3, 122, 1993.

60. **Trofatter, J.A., MacCollin, M.M., Rutter, J.L., Murrell, J.R., Duyao, M.P., Parry, D.M., Eldridge, R., Kley, N., Memon, A.G., Pulaski, K., and Haase, V.H.,** A novel moesin-, ezrin-, radixin-like gene is a candidate for the neurofibromatosis 2 tumor suppressor, *Cell*, 72, 791, 1993.

61. **Rouleau, G.A., Merel, P., Lutchman, M., Sanson, M., Zucman, J., Marineau, C., Hoang-Xuan, K., Demczuk, S., Desmaze, C., and Plougastel, B.,** Alteration in a new gene encoding a putative membrane-organizing protein causes neuro-fibromatosis type 2, *Nature*, 363, 515, 1993.

62. **Latif, F., Tory, K., Gnarra, J., Yao, M., Duh, F.M., Orcutt, M.L., Stackhouse, T., Kuzmin, I., Modi, W., and Geil, L.,** Identification of the von Hippel-Lindau disease tumor-suppressor gene, *Science*, 260, 1317, 1993.

63. **Laforgia, S., Morse, B., Levy, J., Barnen, G., Cannizzaro, L.A., Li, F., Nowell, P.C., Boghosian-Sell, L., Glick, J., Weston, A., Harris, C.C., Drabkin, H., Patterson, D., Croce, C.M., Schlessinger, J., and Huebner, K.,** Receptor protein tyrosine phosphatase c is a candidate tumor-suppressor gene at human chromosome region 3p21, *Proc. Natl. Acad. Sci. U.S.A.*, 88, 5036, 1991.

64. **Kinzler, K.W., Nilbert, M.E.F., Vogelstein, B., Bryan, T.M., Levy, D.B., Smith, K.J., Preisinger, A.C., Hamilton, S.R., Hedge, P., Markham, A., Carlson, M., Joslyn, G., Groden, J., White, R., Miki, Y., Miyoshi, Y., Nishisho, I., and Nakamura, Y.,** Identification of a gene located at chromosome 5q21 that is mutated in colorectal cancers, *Science*, 251, 1366, 1991.

65. **Steeg, P.S., Cohn, K.H., and Leone, A.,** Tumor metastasis and NM23 — current concepts, *Cancer Cells*, 3, 257, 1991.

66. **Harris, H.,** The analysis of malignancy by cell fusion: the position in 1988, *Cancer Res.*, 48, 3302, 1988.

67. **Yunis, J.J. and Ramsay, N.,** Retinoblastoma and subband deletion of Chromosome 13, *Am. J. Dis. Child.*, 132, 161, 1978.

68. **Cavanee, W.K., Dryja, T.P., Phillips, R.A., Benedict, W.F., Godbout, R., Gallie, B.L., Murphree, A.L., Strong, L.C., and White, R.L.,** Expression of recessive alleles by chromosomal mechanisms in retinoblastoma, *Nature*, 305, 779, 1983.

69. **Friend, S.H., Bernards, R., Rogelj, S., Weinberg, R.A., Rapaport, J.M., Albert, D.M., and Dryja, T.P.,** A human DNA segment with properties of the gene that predisposes to retinoblastoma and osteosarcoma, *Nature*, 323, 643, 1986.

70. **Lee, W.H., Bookstein, R., Hong, F., Yoing, L.J., Shew, J.H., and Lee, W.Y.H.P.,** Human retinoblastoma susceptibility gene: cloning, identification and sequence, *Science*, 235, 1394, 1987.

71. **Fung, Y.K.T., Murphree, A.L., T'ang, A., Qian, J., Hinrichs, H.S., and Benedict, W.F.,** Structural evidence for the authenticity of the human retinoblastoma gene, *Science*, 236, 1657, 1987.

72. **Hedrick, L., Cho, K.R., and Vogelstein, B.,** Cell adhesion molecules as tumour suppressors, *Trends Cell. Biol.*, 3, 36, 1993.

73. **Nevins, J.R.,** E2F — A link between the Rb tumour suppressor protein and viral oncoproteins, *Science*, 258, 424, 1992.

74. **Weintraub, S.J., Prater, C.A., and Dean, D.C.,** Retinoblastoma protein switches the E2F site from a positive to a negative element, *Nature*, 358, 259, 1992.

75. **Zhu, L., van den Heuvel, S., Helin, K., Fattney, A., Ewan, M., Livingston, D., Dyson, N., and Harlow, E.,** Inhibition of cell proliferation by p107, a relative of the retinoblastoma protein, *Genes Dev.*, 7, 1111, 1993.

76. **Cobrinik, D., Whyte, P., Peeper, D.S., Jacks, T., and Weinberg, R.A.,** Cell cycle-specific association of E2F with the p130 E1A-binding protein, *Genes Dev.*, 7, 2392, 1993.

77. **Lee, E.H.Y.P., Chang, C.Y., Hu, N.P., Wang, Y.C.I., Lai, C.C., Herrup, K., Lee, W.H., and Bradley, A.,** Mice deficient for RB are nonviable and show defects in neurogenesis and haematopoiesis, *Nature*, 359, 288, 1992.

78. **Jacks, T., Fazeli, A., Schmitt, E.M., Bronson, R.T., Goodell, M.A., and Weinberg, R.A.,** Effects of an Rb mutation in the mouse, *Nature*, 359, 295, 1992.

79. **Clarke, A.R., Maandag, E.R., Vanroon, M., Vanderlugt, N.M.T., Hooper, M.L., Verns, A., and Riele, H.T.,** Requirement for a functional Rb-1 gene in murine development, *Nature*, 359, 328, 1992.

80. **El-Deiry, W.S., Kern, S.E., Pietenpol, J.A., Kinzler, K.W., and Vogelstein, B.,** Definition of a consensus binding site for p53, *Nature Genet.*, 1, 45, 1992.

81. **Zambetti, G.P., Bargonetti, J., Walker, K., Prives, C., and Levine, A.J.,** Wild-type p53 mediates positive regulation of gene expression through a sequence specific DNA sequence element, *Genes Dev.*, 6, 1143, 1992.

82. **Funk, W.D., Pak, D.T., Karas, R.H., Wright, W.E., and Shay, J.W.,** A transcriptionally active DNA-binding site for human p53, *Mol. Cell. Biol.*, 12, 2866, 1992.

83. **El-Deiry, W.S., Tokino, T., Velculescu, V., Levy, D.B., Parsons, R., Trent, J.M., Lin, D., Mercer, W.E., Kinzler, K.W., and Vogelstein, B.,** WAF1, a potential mediator of p53 tumor suppression, *Cell*, 75, 817, 1993.

84. **Harper, J.W., Adami, G.R., Wei, N., Keyomarsi, K., and Elledge, S.J.,** The p21 Cdk-interacting protein cip 1 is a potent inhibitor of G1 cyclin-dependent kinases, *Cell*, 75, 805, 1993.

85. **Xiong, Y., Hannon, G.J., Zhang, H., Casso, D., Kobayashi, R., and Beach, D.,** p21 is a universal inhibitor of cyclin kinases, *Science*, 366, 701, 1993.

86. **Truant, R., Xiao, H., Ingles, C.J., and Grenblatt, J.,** Direct interaction between the transcriptional activation domain of human p53 and the TATA box-binding complex, *J. Biol. Chem.*, 268, 2284, 1993.

87. **Martin, D.W., Munoz, R.M., Subler, M.A., and Deb, S.,** p53 binds to the TATA-binding protein-TATA complex, *J. Biol. Chem.*, 268, 13062, 1993.

88. **Seto, E., Usheva, A., Zambetti, G.P., Momand, J., Horikoshi, N., Weinmann, R., Levine, A.J., and Shenk, T.,** Wild-type p53 binds to the TATA-binding protein and represses transcription, *Proc. Natl. Acad. Sci. U.S.A.*, 89, 12028, 1992.

89. **Debbas, M. and White, E.,** Wild-type p53 mediates apoptosis by EIA, which is inhibited by EIB, *Genes Dev.*, 7, 546, 1993.

90. **White, A.E., Livanos, E.M., and Tlsty, T.D.,** Differential disruption of genomic integrity and cell cycle regulation in normal human fibroblasts by the HPV oncoproteins, *Genes Dev.*, 8, 666, 1994.

91. **Wu, X.W. and Levine, A.J.,** p53 and E2F-1 cooperate to mediate apoptosis, *Proc. Natl. Acad. Sci. U.S.A.*, 91, 3602, 1994.

92. **Selvakumaran, M., Lin, H.K., Miyashita, T., Wang, H.G., Krajewski, S., Reed, J.C., Hoffman, B., and Liebermann, D.,** Immediate early up-regulation of bax expression by p53 but not TGFß 1; a paradigm for distinct apoptotic pathways, *Oncogene*, 9, 1791, 1994.

93. **Hollstein, M., Sidransky, D., Vogelstein, B., and Harris, C.C.,** p53 mutations in human cancers, *Science*, 253, 49, 1991.

94. **Cho, Y.J., Gorina, S., Jeffrey, P.D., and Pavletich, N.P.,** Crystal structure of a p53 tumor suppressor-DNA complex: understanding tumorigenic mutations, *Science*, 265, 346, 1994.

95. **Clore, G.M., Omichinski, J.G., Sakaguchi, K., Zambrano, N., Sakamoto, H., Appella, E., and Grnenborn, A.M.,** High resolution structure of the oligomerization domain of p53 by multidimensional NMR, *Science*, 265, 386, 1994.

96. **Kern, S.E., Pientepol, J.A., Thiagalingam, S., Seymour, A., Kinzler, K.W., and Vogelstein, B.,** Oncogenic forms of p53 inhibit p53 regulated gene expression, *Science*, 256, 827, 1992.

97. **Pavletich, N.P., Chambers, K.A., and Pabo, C.O.,** The DNA-binding domain of p53 contains the 4 conserved regions and the major mutation hot spots, *Genes Dev.*, 7, 2556, 1993.

98. **Shaulian, E., Zauberman, A., Ginsberg, D., and Oren, M.,** Identification of a minimal transforming domain of p53: negative dominance through abrogation of sequence-specific DNA binding, *Mol. Cell. Biol.*, 12, 5581, 1992.

99. **Loeb, L.A.,** Mutator phenotype may be required for multistage carcinogenesis, *Cancer Res.*, 51, 3075, 1991.

100. **Loeb, L.A.,** Microsatellite instability: marker of a mutator phenotype in cancer, *Cancer Res.*, 54, 5059, 1994.

101. **Eshleman, J.R. and Markowitz, S.D.,** Microsatellite instability in inherited and sporadic neoplasms, *Curr. Opin. Oncol.*, 7, 83, 1995.

102. **Thibodeau, S.N., Bren, G., and Schaid, D.,** Microsatellite instability in cancer of the proximal colon, *Science*, 260, 816, 1993.

103. **Peltomaki, P., Lothe, R.A., Aaltonen, L.A., Pylkkanen, L., Nystrom-Lahti, M., Seruca, R., David, L., Holm, R., Ryberg, D., Haugen, A., Brogger, A., Borresen, A.-L., and de la Chapelle, A.,** Microsatellite instability is associated with tumors that characterize the hereditary nonpolyposis colorectal carcinoma syndrome, *Cancer Res.*, 53, 5853, 1993.

104. **Ionov, Y., Peinado, M.A., Malkhosyan, S., Shibita, D., and Perucho, M.,** Ubiquitous somatic mutations in simple repeated sequences reveal a new mechanism for colonic carcinogenesis, *Nature*, 363, 558, 1993.

105. **Thibidou, S.N., Bren, G., and Schaid, D.,** Microsatellite instability in cancer of the proximal colon, *Science*, 260, 816, 1993.

106. **Fishel, R., Lescoe, M.K., Rao, M.R.S., Copeland, N.G., Jenkins, N.A., Garber, J., Kane, M., and Kolodner, R.,** The human mutator gene homolog MSH2 and its association with hereditary nonpolyposis colon cancer, *Cell*, 75, 1027, 1993.

107. **Leach, F.S., Nicolaides, N.C., Papadopolous, N., Liu, B., Jen, J., Parsons, R., Peltouchi, P., Sistonen, P., Aaltonen, L.A., Nystom-Lahti, M., Guan, X.-Y., Zhang, J., Meltzer, P.S., Yu, S.-W., Kao, F.-T., Chen, D.J., Cerosaletti, K.M., Fournier, R.E., Todd, S., Lewis, T., Leach, R.J., Naylor, S.L., Weissenbach, J., Mechlin, J.-P., Jarvinen, H., Peternen, G.M., Hamilton, S.R., Green, J., Jurr, J., Watson, P., Lynch, H.T., Trent, J.M., de la Chapelle, A., Kinzler, K., and Vogelstein, B.,** Mutations of a mutS homolog in hereditary nonpolyposis colorectal cancer, *Cell*, 75, 1215, 1993.

108. **Bronner, C.E., Baker, S.M., Morrison, P.T., Warren, G., Smith, L.G., Lescoe, M.K., Kane, M., Earabino, C., Lipford, J., Lindblom, A., Tannergard, P., Bollag, R.J., Godwin, A.R., Ward, D.G., Nordenskjold, M., Fishel, R., Kolodner, R., and Liskay, R.M.,** Mutation in the DNA mismatch repair gene homologue hMLH1 is associated with hereditary nonpolyposis colon cancer, *Nature*, 368, 258, 1994.

109. **Papadopolous, N., Nicolaides, N.C., Wei, Y.-F., Ruben, S.M., Carter, K.C., Rosen, C.A., Haseltine, W.A., Fleischmann, R.D., Fraser, C.M., Adams, M.D., Venter, J.C., Hamilton, S.R., Petersen, G.M., Watson, P., Lynch, H.T., Peltomaki, P., Mecklin, J.-P., de la Chapelle, A., Kinzler, K.W., and Vogelstein, B.,** Mutation of a mutL homolog in hereditary colon cancer, *Science*, 263, 1625, 1994.

110. **Nicolaides, N.C., Papadopolous, N., Liu, B., Wei, Y.-F., Carter, K.C., Ruben, S.M., Rosen, C.A., Haseltine, W.A., Fleischmann, R.D., Fraser, C.M., Adams, M.D., Venter, J.C., Dunlop, M.G., Hamilton, S.R., Petersen, G.M., de la Chapelle, A., Kinzler, K.W., and Vogelstein, B.,** Mutation of two fPMSFf homologues in hereditary nonpolyposis colon cancer, *Nature*, 371, 75, 1994.

111. **Kinzler, K.W.,** Genetics of hereditary colorectal cancer, *Proc. Am. Assoc. Cancer Res.*, 36, 668, 1995.

112. **Shay, J.W., Piatyszek, M.A., Word, R.A., Gazdar, A.F., Wright, W.E., Kim, N.W., Weinrich, S.L., Prowse, K.R., Harley, C.B., Hiyama, E., Mehle, C., Roos, G., Sommerfeld, H.J., Meeker, A.K., and Coffey, D.S.,** You haven't heard the end of it: telomeres, telomerase and tumors, *Proc. Am. Assoc. Cancer Res.*, 36, 673, 1995.

113. **Harley, C.B., Andrews, W., Chiu, C.-P., Feng, J., Funk, W., Gaeta, F., Hirsch, K., Kim, N.W., Kozlowski, M., Wang, S.-S., Weinrich, S.L., West, M.D., Avilion, A., Le, S., Greider, C.W., and Villeponteau, B.,** Human telomerase inhibition and cancer, *Proc. Am. Assoc. Cancer Res.*, 36, 671, 1995.

114. **Kim, N.W., Piatyszek, M.A., Prowse, K.R., Harley, C.B., West, M.D., Ho, P.L.C., Coviello, G.M., Wright, W.E., Weinrich, S.L., and Shay, J.W.,** Specific association of human telomerase activity with immortal cells and cancer, *Science*, 266, 2011, 1994.

115. **Vogelstein, B. and Kinzler, K.W.,** The multistep nature of cancer, *Trends Genet.*, 9, 138, 1993.

116. **Fearon, E.R. and Vogelstein, B.,** A genetic model for colorectal tumorigenesis, *Cell*, 61, 759, 1990.

117. **Vogelstein, B., Fearon, E.R., Hamilton, S.R., Kern, S.E., Preisenger, A.C., Leppert, M., Nakamura, Y., White, R., Smits, A.M.M., and Bos, J.L.,** Genetic alterations during colorectal tumor development, *N. Engl. J. Med.*, 319, 525, 1988.

118. **Baker, S.J., Markowitz, S., Fearon, E.R., Willson, J.K.V., and Vogelstein, B.,** Suppression of human colorectal carcinoma cell growth by wild-type p53, *Science*, 249, 912, 1990.

119. **Baker, S.J., Fearon, E.R., Nigro, J.M., Hamilton, S.R., Preisinger, A.C., Jessup, J.M., van Tuinen, P., Ledbetter, D.H., Barker, D.R., Nakamura, Y., White, R., and Vogelstein, B.,** Chromosome 17 deletions and p53 gene mutations in colorectal carcinomas, *Science*, 244, 217, 1989.

120. **Bacchetti, S., Counter, C.M., Chadeneau, C., Gupta, J., Harley, C.B., Leber, B., Gallinger, S., Hirte, H.W., Siegel, P., and Muller, W.J.,** Telomerase activity in human and murine somatic tissues and tumours, *Proc. Am. Assoc. Cancer Res.*, 36, 674, 1995.

121. **Aaronson, S.A.,** Growth factors and cancer, *Science*, 254, 1146, 1991.

122. **Yarden, Y. and Ullrich, A.,** Molecular analysis of signal transduction by growth factors, *Biochemistry*, 27, 3114, 1988.

123. **Ullrich, A. and Schlessinger, J.,** Signal transduction by receptors with tyrosine kinase activity, *Cell*, 61, 203, 1990.

124. **Yarden, Y. and Schlessinger, J.,** Self-phosphorylation of epidermal growth factor receptor: evidence for a model of intermolecular allosteric activation, *Biochemistry*, 26, 1434, 1987.

125. **Pawson, T. and Schlessinger, J.,** SH2 and SH3 domains, *Curr. Biol.*, 3, 434, 1993.

126. **Pelech, S.L.,** Networking with protein kinases, *Curr. Biol.*, 3, 513, 1993.

127. **Marx, J.,** Forging a path to the nucleus, *Science*, 260, 1588, 1993.

128. **McCormick, F.,** How receptors turn ras on, *Nature*, 363, 15, 1993.

129. **Curran, T. and Franza, B.R., Jr.,** Fos and jun: the AP-1 connection, *Cell*, 55, 395, 1988.

130. **Amati, B. and Land, H.,** Myc-max-mad: a transcription factor network controlling cell cycle progression, differentiation and death, *Curr. Opin. Genet. Dev.*, 4, 102, 1994.

131. **Taylor, S.J., Chae, H.Z., Rhee, S.G., and Exton, J.A.,** Activation of the B1 isozyme of phospholipase C by a subunits of the Gq class of G protein, *Nature*, 380, 516, 1991.

132. **Meldrum, E., Parker, P.J., and Carozzi, A.,** The PtdIns-PLC superfamily and signal transduction, *Biochim. Biophys. Acta*, 1092, 49, 1991.

133. **Kolch, W., Heidecker, G., Kochs, G., Hummel, R., Vahidid, H., Mischak, H., Finkenzeller, G., Marme, D., and Rapp, U.R.,** PKCa activates raf-1 by direct phosphorylation, *Nature*, 364: 249, 1993.

134. **Cho-Chung, Y.S.,** Role of cyclic AMP receptor proteins in growth, differentiation and suppression of malignancy: new approaches to therapy, *Cancer Res.*, 50, 7093, 1990.

135. **Lalli, E. and Sassone-Corsi, P.,** Signal transduction and gene regulation: the nuclear response to cAMP, *J. Biol. Chem.*, 269, 17359, 1994.

136. **Hai, T. and Curran, T.,** Cross-family dimerisation of transcription factors fos/jun and ATF/CREB alters DNA binding specificity, *Proc. Natl. Acad. Sci. U.S.A.*, 88, 3720, 1991.

137. **Masquilier, D. and Sassone-Corsi, P.,** Transcriptional cross-talk: Nuclear factors CREM and CREB bind to AP-1 sites and inhibit activation by jun, *J. Biol. Chem.*, 267, 22460, 1992.

138. **Massague, J. and Polyak, K.,** Mammalian antiproliferative signals and their targets, *Curr. Opin. Genet. Dev.*, 5, 91, 1995.

139. **Wrana, J.L., Attisano, L., Wieser, R., Ventura, F., and Massague, J.,** Mechanism of activation of the TGF-β receptor, *Nature*, 370, 341, 1994.

140. **Ewen, M.E., Sluss, H.K., Whitehouse, L.L., and Livingston, D.M.,** TGFß inhibition of Cdk4 synthesis is linked to cell cycle arrest, *Cell*, 74, 1009, 1993.

141. **Geng, Y. and Weinberg, R.A.,** Transforming growth factor ß effects on expression of G1 cyclins and cyclin-dependent protein kinases, *Proc. Natl. Acad. Sci. U.S.A.*, 90, 10315, 1993.

142. **Polyak, K., Kato, J.-Y., Solomon, M.J., Scherr, C.J., Massague, J., Roberts, J.M., and Koff, A.,** p27Kip1, a cyclin-Cdk inhibitor, links transforming growth factor-ß and contact inhibition to cell arrest, *Genes Dev.*, 8, 9, 1994.

143. **Hannon, G.J. and Beach, D.,** p15^{INK4B} is a potential effector of TGF-ß induced cell cycle arrest, *Nature*, 371, 257, 1994.

144. **Miyajima, A., Kitamura, T., Harada, N., Yokota, T., and Arai, K.,** Cytokine receptors and signal transduction, *Annu. Rev. Immunol.*, 10, 295, 1992.

145. **Ihle, J.N., Witthuhn, B.A., Quelle, F.W., Yamamoto, K., Thierfelder, W.E., Kreider, B., and Silvennoinen, O.,** Signaling by the cytokine receptor superfamily: JAKS and STATs., *Trends Biochem. Soc.*, 19, 222, 1994.

146. **Darnell, J.E., Kerr, I.M., and Stark, G.R.,** Jak-STAT pathways and transcriptional activation in response to IFNs and other extracellular signaling proteins, *Science*, 264, 1415, 1994.

147. **Kishimoto, T., Taga, T., and Akira, S.,** Cytokine signal transduction, *Cell*, 76, 253, 1994.

148. **Green, S. and Chambon, P.,** Nuclear receptors enhance our understanding of transcription regulation, *Trends Genet.*, 4, 309, 1988.

149. **Cullen, K.J. and Lippman, M.E.,** Estrogen regulation of protein synthesis and cell growth, *Vitam. Horm.*, 45, 127, 1989.

150. **Morisset, M., Caponey, F., and Rochefort, H.,** Processing and estrogen regulation of the 52-kDa protein inside MCF7 breast cancer cells, *Endocrinology*, 119, 2773, 1986.

151. **Sporn, M.B. and Roberts, A.B.,** Autocrine growth factors and cancer, *Nature*, 313, 745, 1985.

152. **Hunter, T. and Pines, J.,** Cyclins and Cancer II: cyclin D and CDK inhibitors come of age, *Cell*, 79, 573, 1994.

153. **Motokura, T., Bloom, T., Kim, H.G., Juppner, H., Ruderman, J.V., Kronenberg, H.M., and Arnold, A.,** A novel cyclin encoded by a bcl-linked candidate oncogene, *Nature*, 350, 512, 1991.

154. **Withers, D.A., Harvey, R.C., Faust, J.B., Melnyk, O., Carey, K., and Meeker, T.C.,** Characterization of a candidate bcl-1 gene, *Mol. Cell. Biol.*, 11, 4846, 1991.

155. **Kamb, A., Gruis, N.A., Weaver-Feldhaus, J., Liu, Q., Harshman, K., Tavtigian, S.V., Stockert, E., Day, R.S.I., Johnson, B.E., and Skolnick, M.H.,** A cell cycle regulator potentially involved in genesis of many tumor types, *Science*, 264, 436, 1994.

156. **Nobori, T., Miura, K., Wu, D.J., Lois, A., Takabayashi, K., and Carson, D.A.,** Deletions of the cyclin-dependent kinase-4 inhibitor gene in multiple human cancers, *Nature*, 368, 753, 1994.

157. **Liotta, L.A. and Stetler-Stevenson, W.G.,** Tumor invasion and metastasis: an imbalance of positive and negative regulation, *Cancer Res.*, 51, 5054s, 1991.

158. **Aznavoorian, S., Murphy, A.N., Stetler-Stevenson, W.G., and Liotta, L.A.,** Molecular aspects of tumor-cell invasion and metastasis, *Cancer*, 71, 1368, 1993.

159. **Fidler, I.J. and Hart, I.R.,** Biological diversity in metastatic neoplasms, *Science*, 217, 998, 1982.

160. **Hart, I.R. and Saini, A.,** Biology of tumour metastasis. *Lancet*, 339, 1453, 1992.

161. **Stetler-Stevenson, W.G., Aznavoorian, S., and Liotta, L.A.,** Tumor cell interactions with the extracellular matrix during invasion and metastasis, *Annu. Rev. Cell. Biol.*, 9, 541, 1993.

162. **Folkman, J., Watson, K., Ingber, D., and Hanahan, D.,** Induction of angiogenesis during the transition from hyperplasia to neoplasia, *Nature*, 339, 58, 1989.

163. **Takeichi, M.,** Cadherin cell adhesion receptors as a morphogenetic regulator, *Science*, 251, 1451, 1991.

164. **Vlemincki, K., Vakaert, L., Mareel, M., and Fiers, W., Van Roy, F.,** Genetic manipulation of E-cadherin expression by epithelial tumor cells reveals an invasion suppressor role, *Cell*, 66, 107, 1991.

165. **Hedrick, L., Cho, K.R., and Vogelstein, B.,** Cell adhesion molecules as tumour suppressors, *Trends Cell. Biol.*, 3, 36, 1993.

166. **Hynes, R.O.,** Integrins: a family of cell-surface receptors, *Cell*, 48, 549, 1987.

167. **Humphries, M.J., Olden, K., and Yamada, K.M.,** A synthetic peptide from fibronectin inhibits experimental metastasis of murine melanoma cells, *Science*, 233, 467, 1986.

168. **Humphries, M.J., Yamada, K.M., and Olden, K.,** Investigation of the biological effects of anti-cell adhesion synthetic peptides that inhibit experimental metastasis of B16-F10 murine melanoma cells, *J. Clin. Invest.*, 81, 782, 1988.

169. **Albelda, S.M., Mette, S.A., Elder, D.E., Stewart, R., Damjanovich, L., Herlyn, M., and Buck, C.A.,** Integrin distribution in malignant melanoma: association of the ß3 subunit with tumor progression, *Cancer Res.*, 50, 6757, 1990.

170. **Gunthert, U., Hofmann, M., and Rudy, W.,** A new variant of glycoprotein CD44 confers metastatic potential to rat carcinoma cells, *Cell*, 65, 13, 1991.

171. **MacDonald, N.J. and Steeg, P.S.,** Molecular basis of tumour metastasis, *Cancer Surv.*, 16, 175, 1993.

172. **Liotta, L.A., Steeg, P.S., and Stetler-Stevenson, W.G.,** Cancer metastasis and angiogenesis: an imbalance of positive and negative regulation, *Cell*, 64, 327, 1991.

173. **Matrisian, L.M.,** The matrix-degrading metalloproteinases, *Bioessays*, 14, 455, 1992.

174. **Dano, K., Andreasen, P., Grondahl-Hansen, J., Kristensen, P., Nielsen, L., and Skriver, L.,** Plasminogen activators, tissue degradation and cancer, *Adv. Cancer Res.*, 44, 139, 1985.

175. **Baker, M., Bleakley, P., Woodrow, G., and Doe, W.,** Inhibition of cancer cell urokinase by its specific inhibitors PAI-2 and subsequent effects on extracellular matrix degradation, *Cancer Res.*, 50, 4676, 1990.

176. **Sloane, B., Moin, K., Krepela, E., and Rozhin, J.,** Cathepsin B and endogenous inhibitors: the role in tumor malignancy, *Cancer Metast. Rev.*, 9, 333, 1990.

177. **Tandon, A., Clark, G., Chambiss, G., Chirgwin, J., and McGuire, W.,** Cathepsin D and prognosis in breast cancer, *N. Engl. J. Med.*, 332, 3904, 1990.

178. **Schor, S., Schor, A., Grey, A., and Rushton, G.,** Foetal and cancer patients fibroblasts produce an autocrine migration-stimulating factor not made by normal adult cells, *J. Cell Science*, 90, 391, 1988.

179. **Stoker, M., Gherardi, E., Perryman, M., and Gary, J.,** Scatter factor is fibroblast-derived modulator of epithelial cell motility, *Nature*, 327, 239, 1987.

180. **Nabi, I.R., Watanabe, H., and Raz, A.,** Autocrine motility factor and its receptor: role in cell locomotion and metastasis, *Cancer Metast. Rev.*, 11, 5, 1992.

181. **Stracke, M., Krutzch, H., and Unsworth, E.,** Identification, purification and partial sequencing analysis of autotoxin, a novel motility-stimulating protein, *J. Biol. Chem.*, 267, 2524, 1992.

182. **Srivastava, A., Laidler, P., Davies, R., and Horgan, K.,** The prognostic significance of tumour vascularity in intermediate thickness (0.76-4.0 mm thick) skin melanoma: a quantitative histologic study, *Am. J. Pathol.*, 133, 419, 1988.

183. **Folkman, J. and Klagsbrun, M.,** Angiogenic factors, *Science*, 235, 442, 1987.

184. **Steeg, P.S., Bevilacqua, G., Kopper, L., Thorgeirsson, U.P., Talmadge, J.E., Liotta, L.A., and Sobel, M.E.,** Evidence for a novel gene associated with low metastatic potential, *J. Natl. Cancer Inst.*, 80, 200, 1988.

185. **Bevilacqua, G., Sobel, M.E., Liotta, L.A., and Steeg, P.S.,** Association of low nm23 RNA levels in human primary infiltrating ductal breast carcinomas with lymph node involvement and other histopathological indicators of high metastatic potential, *Cancer Res.*, 49, 5185, 1989.

186. **Hennessy, C., Henry, J.A., May, F.E.B., Westely, B., Angus, B., and Lennard, T.W.J.,** Expression of the antimetastatic gene nm 23 in human breast cancer: an association with good prognosis. *J. Natl. Cancer Inst.*, 83, 281, 1991.

187. **Leone, A., Flatow, U., King, C.R. Sandeen, M.A., Margulies, I.M.K., Liotta, L.A., and Steeg, P.S.,** Reduced tumor incidence metastatic potential and cytokine responsiveness of nm23-transfected cells, *Cell*, 65, 25, 1991.

188. **Nicolson, G.L., Toh, Y., Pencil, S.D., and Moustafa, A.S.,** Differentially expressed metastasis-associated genes: the novel gene mta1 is overexpressed in highly metastatic mammary adenocarcinoma cell lines and similarly mta 1 is overexpressed in metastatic human breast cancer cells, *Proc. Am. Assoc. Cancer Res.*, 36, 694, 1995.

189. **Collard, J.G.,** Identification and functional analysis of the invasion-inducing tiam-1 gene, *Proc. Am. Assoc. Cancer Res.*, 36, 693, 1995.

Part II.
Breast Cancer

Chapter 2

Introduction to Breast Cancer

A. Patrick Forrest

CONTENTS

1. INTRODUCTION

Cancer of the breast is the commonest cancer affecting females living in the westernized world. It is also their commonest cancer death. Its cause is unknown, but its great rarity among males indicates an etiological role for the female sex hormones, while its varying geographic distribution points to the importance of environmental factors. Although generally slow growing, the tumor develops invasive properties early in its life-time, so that by the time it has become clinically apparent, it is likely to have already metastasized to distant sites. It is this pattern which accounts for the failure of purely local treatment to control the disease. Detection of the disease at an early preclinical, nonmetastatic stage by mammographic screening, and the treatment, by systemic hormonal or chemotherapy of clinically established disease, are currently the most effective methods of improving outcome. Its ultimate control, however, must await identification of its cause. In this chapter these various issues will be addressed.

2. ORIGINS OF BREAST CANCER

2.1. The Normal Breast

The breast is a secretory organ, composed of glandular tissue and ducts contained in a stroma of supporting fat and fibrous tissue. The primary secreting units consist of groups of terminal ductules with sac-like ends, which are embedded in a fine specialized connective tissue to form the breast lobules, or terminal duct lobular units (TDLU). These are clearly distinguishable microscopically from the surrounding coarse fibrous stroma and fat. During lactation, these become dilated to form alveoli which secrete milk. A coalescing system of ducts of gradually increasing size drain these terminal duct lobular units, ending in some 6 to 10 main lactiferous ducts which emerge on to the surface at the nipple.[1-3]

The characteristic shape of the breast is due to a system of fibrous septa which attach it to the overlying skin, and enclose large deposits of fat. As the breast is only loosely attached to the underlying chest wall, this arrangement allows mobility with preservation of shape, thereby facilitating suckling. The arrangement of the lobules and ducts within the breast is not, as frequently depicted, segmental, but haphazard. Lobules are distributed throughout the whole of the breast.

0-8493-9443-0/97/$0.00+$.50

With the exception of the lactiferous ducts close to the nipple, the whole of the duct system of the breast is lined by a single layer of epithelial cells. A second layer of cells, the myoepithelial cells, surrounds the epithelium with contractile processes, which when stimulated by the neurohypophyseal hormone oxytocin, released during suckling, drive milk towards the nipple. Unlike some mammals, the human breast does not contain reservoirs for milk storage. Outside the lobules the ducts are clothed with elastic tissue which, during lactation, allows them to stretch. The myoepithelial cells may have other functions, such as maintaining the integrity of the basement membrane.

Although the breast only reaches functional maturity with the secretion of milk during lactation, there is a constant turn-over of fluid which can be aspirated by suction from the nipple.[4-7] This is rich in electrolytes, steroid hormones, and immunoglobulins. It may also have mutagenic activity, and contain potentially carcinogenic metabolites which may affect the environment of the epithelial cells. So, also, may the known paraendocrine activity of the fibrous and fatty stroma contribute. These facts may have relevance to the genesis of cancer, for it is now firmly believed that breast cancer starts in the epithelium which lines the terminal ductules within the lobule.

2.2 Development of Cancer

Breast lobules normally involute following the menopause, but in a breast which contains a cancer, they persist. Further, in the contralateral breast of a patient with breast cancer, the number of lobular units is increased, compared with that normally found. Many now consider that the presence of actively functioning lobules is an essential prerequisite to the development of cancer.[8]

In cancer of the colon, which shares some etiological associations with breast cancer, a well defined premalignant phase has been recognized in the form of the localized adenomatous polyp. Some suspect that a similar premalignant phase may exist for breast cancer, but unlike lesions in the bowel, there is no simple method for its detection. Proliferative change is known to occur in the breast lobules, the epithelial cells enlarging and increasing in number to become several layers thick (mild to moderate hyperplasia) and later to develop abnormal features and a disturbed pattern of growth (atypical hyperplasia). Such atypical hyperplasia is known to be associated with an increased incidence of cancer, particularly in women with a family history of the disease.[9] However, it is not known whether such a change regularly precedes the disease.

Once an epithelial cell has been transformed into a malignant phenotype, it is no longer subject to normal growth-controlling mechanisms or immunological surveillance. To become truly malignant to its host, it must also develop the capacity to invade and metastasize. These changes need not occur simultaneously. A breast cancer cell may be noninvasive and unable to penetrate the basement membrane so that it remains confined within the terminal ductules in the lobule. This is the stage of "*in situ* cancer".

Two main types of *in situ* breast cancer are recognized microscopically; lobular carcinoma *in situ* (LCIS) and ductal carcinoma *in situ* (DCIS). In LCIS, one or more breast lobules are packed with uniform but frankly malignant cells, a microscopic appearance which is not associated with physical or mammographic features. LCIS is purely a histopathological condition. DCIS is of two main histopathological types — solid and cribriform. In the solid form, the lumen is packed with malignant cells which necrose in the center to form "worms" of white tissue — the so-called "comedo" DCIS. The cribriform or micropapillary form is characterized by small cells growing in a lace-like pattern or as delicate fronds into the duct lumen. Unlike LCIS, DCIS is a clinical entity. Before the availability of mammography, it presented as a breast mass or bloody discharge from the nipple, but now it is more commonly detected radiologically when its presence is indicated by distortion of breast architecture, asymmetrical increase in density or, most commonly, linear or punctate microcalcifications.

As the diagnosis of LCIS and DCIS can only be made following removal of the lesion, the natural history of untreated disease can only be assessed through the behavior of the long-term follow-up of those patients in whom it had been treated by local excision alone. Such studies suggest that whereas LCIS is but a marker of instability of the epithelium of both breasts, DCIS is a preinvasive lesion, particularly when of comedo type. Thirty to 50 percent of women with DCIS treated by local excision alone develop invasive cancer of the affected breast within 10 to 18 years.[10,11]

The histopathological appearances of an invasive breast cancer are distinctive. A small proportion of tumors show a regular structure, and are classified as being of special type — cribriform, papillary, medullary, or lobular, names stemming from their histological appearance. The majority (70 to 80%) are of "no special type," the scene being one of disorder. Groups of large irregular and obviously malignant cells infiltrate a fibrous stroma with primitive attempts to form glandular structures. There is increased collagen, streaks of elastica and an infiltrate of lymphocytes, and in many, obvious infiltration of lymph or blood vascular channels. It is this invasion early in the natural history of the disease which leads to the deposition of tumor-cells in lymph nodes and systemic sites, particularly bones, liver, lungs, the abdominal cavity, and brain. Although in some patients breast cancer is fatal on account of its local manifestations, in the majority it is organ failure from systemic metastases which causes death. As was indicated by Keynes, "the invaders are bound to gain the upper hand in the end."[12]

3. MOLECULAR EVENTS

It is now known that a large number of genetic defects can be identified in an established breast cancer. These include overexpression of the oncogenes c-myc, int-2, c-erbB2, and loss of heterozygosity on chromosomes 1p, 1q, 3p, 11p, 13q, 17p, 17q, and 18q, of which loss or mutation of p53 on the short arm of chromosome 17 may be the most important.[13] In the Li-Fraumeni syndrome, of which breast cancer is a common component, loss of p53 is consistently found. The identification, on chromosome 17, of a gene associated with risk in families carrying a dominant inheritance pattern for breast cancer (BRCA1) represents a large step forward. A second gene (BRCA2) is known to exist on chromosome 13.[14,15] Not only may these prove to be associated with risk of sporadic disease but, as alteration of one "hit" in the transformation process may prove sufficient to reverse tumor formation, it opens up the potential for gene therapy. For the development of full malignant potential, i.e., the capacity to invade, further genetic "hits" are required, of which increased expression of metalloproteinases and loss of the metastasis-suppressor gene nm 23 are but two examples.[16, 17] The role of oncogenes and tumor-suppressor genes in this disease will be discussed in detail in Chapter 4.

4. INCIDENCE AND ETIOLOGY

The age-standardized incidence rates of breast cancer in the U.K., North America, Australasia, and Scandinavia approximate 60 to 70 per 100,000 women per annum.[18] In the U.K. alone, 25,000 new cases are diagnosed each year. Mortality rates also are high and parallel those of incidence (see Figure 1). In the U.K. 15,000 women die annually from breast cancer.[19] The gap between the incidence and mortality rates has been taken to represent a 'cured' group who survive their disease, but this conclusion may be fallacious, as strict definition of diagnosis (for example between invasive and noninvasive cancer) and cause of death is not always available. Long-term follow-up studies of women treated for clinically evident breast cancer by local treatment alone have indicated that excess mortality from metastatic disease persists for over 30 years (see Figure 2). From a number of such studies, it appears that less than 30% of women with clinically apparent breast cancer treated only by local surgery and/or radio-therapy can expect freedom from relapse during their lifetime.[20-25]

Although the cause of breast cancer is unknown, a number of factors which affect the risk of developing the disease have been identified. Of greatest importance are to be female, to live in a Western environment, and to live past middle-age.

Less than one in each hundred cases of breast cancer occurs in males. It is predominantly a disease of women who have had a normal span of reproductive life. Functioning ovaries are necessary for the initiation of breast cancer, both in experimental animals and women. Surgical removal of the ovaries, if performed before the age of 35 years, reduces the incidence of breast cancer by half, an effect which lasts for the remainder of the woman's life.[26,27] Prolongation of the duration of a woman's reproductive life, either by an early menarche or late menopause, increases risk.[28] Yet the administration of exogenous ovarian hormones, other than when taken in early life, has, at most, only a modest effect.[29-31]

Compared to single women, those who are married enjoy relative protection against breast cancer. So also do those who bear children, particularly in early life.[32,33] A woman with a full-term pregnancy before the age of 20 years has one third to one half of the risk of a woman whose first pregnancy does not occur until age 35 years or more. Experimental evidence suggests that stem cell differentiation during late pregnancy may reduce susceptibility to carcinogens.[34]

Women living in what were "developing" countries, such as Japan, Malaysia, and Indonesia have long been known to have a lower incidence of the disease than do their counterparts in Europe, America, or Australia, but on migration to the West they assume the Western incidence.[35] Although hormone-related and racial effects may account for some of these differences, it is apparent that environment, particularly urban, is of overriding importance. A positive correlation between the intake of dietary fat and breast cancer incidence and mortality has been described, but this may be but a marker of increasing "development" rather than a causative factor.[36] Some suggest that diets rich in vegetables, such as are

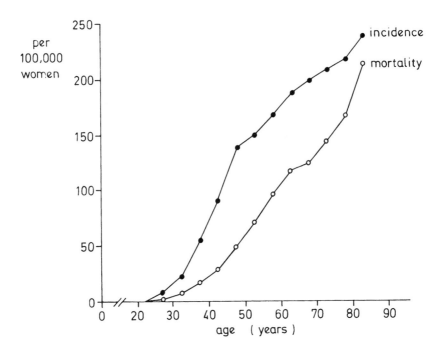

Figure 1 Age-specific incidence rate (1982) and mortality rate (1985) for breast cancer, U.K. (Redrawn from Breast Cancer Screening. Report to Health Ministers of England, Wales, Scotland and Northern Ireland, HMSO, 1986. Crown copyright is reproduced with the permission of the controller of HMSO.)

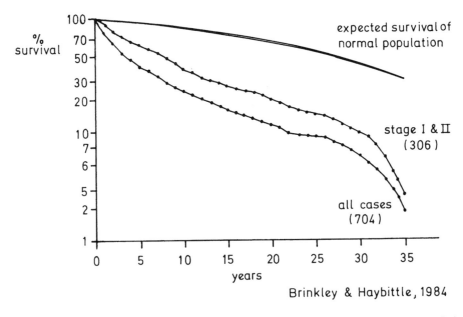

Figure 2 Long-term survival of 704 patients with breast cancer treated by local surgery or radiotherapy. (Redrawn from Brinkley, D. and Haybittle, J.L., *Lancet*, 1, 11-18, 1984. With permission.)

consumed in the East, may be protective as a result of limiting the availability of biologically active estrogen.[37] A case-control study from Singapore indicated a protective effect for soya products.[38] Breast cancer incidence in the East is now reported to be rising, although it is still relatively low, compared to the West.[18]

The effect of age on breast cancer incidence is presumably environment dependent. Rare before the age of 25, its incidence increases steadily during the premenopausal years. Following a downward "hook", it continues to climb with advancing years, albeit at a slower rate.[39] One third of breast cancers are diagnosed at the age of 70 years or more, but because of variations in age distribution, its greatest prevalence in the West occurs at 55 years.

Genetic factors may influence breast cancer risk. An affected first-degree relative increases risk, and there are well-established kindreds in which the disease appears as a dominant hereditary trait.[40] It is from molecular genetic studies of these families that the BRCA1 and 2 genes have been identified. A history of a previous biopsy for benign disease of the breast and radiation during childhood are also risk factors for the disease.[41-43]

5. MANAGEMENT OF BREAST CANCER

There are three steps in the management of a patient with any cancer; first to diagnose the disease, second to determine its extent, and third to treat local and systemic manifestations.

5.1 Diagnosis

Most breast cancers still present to the surgeon as a palpable breast mass when a firm diagnosis can be made in over 95% of cases by combining physical examination with mammographic (or ultrasonic) imaging, and fine-needle aspiration cytology or core-biopsy — the so-called "triple approach".[44] Reaching a firm preoperative diagnosis allows further investigations to be conducted and treatment planned for a patient who is fully informed and can receive appropriate counseling. Similar diagnostic procedures are used also in screening programs (see Section 6).

5.2 Staging

Traditionally, the determination of the stage of a breast cancer was based on the TNM system, which defined tumor size (T), clinical stage of regional lymph nodes (N) and evidence of metastases (M).[45] The importance of defining the extent of the local disease for the operating surgeon is obvious, for it is well recognized that there is a stage when surgical treatment is harmful. For breast cancer this is when it involves the overlying skin, or underlying muscles, or has signs of substantial lymphatic involvement, causing satellite nodules of tumor, lymphoedema of the skin, or gross involvement of the axillary lymph nodes. However, traditional clinical tests for metastatic involvement, such as radiographs of chest and pelvis, scintiscans of the skeleton, and biochemical tests of liver function can only detect substantial disease. Newer methods of imaging, such as computerized tomography (CT), magnetic resonance imaging (MRI), and positron-emission tomography (PET), have greater sensitivity, but still cannot detect microscopic deposits of tumor. For this reason, examination of the axillary lymph nodes for metastatic deposits of tumor is regarded as an essential step in the staging procedure. Assays of tumor secretory products offer the best hope of detecting minimal metastatic disease in other sites, but to be of value they must be specific for breast cancer and detectable in the circulating blood in nanomole concentrations. Markers with this degree of specificity or sensitivity are not available.

The lack of good physical, biochemical, or immunological markers of tumor burden has led to the exploration of biological factors which mark the aggressiveness of an established tumor, and therefore the likelihood that it has metastasized. Longest established is the histological grade, which reflects the degree of differentiation of the tumor and its cells.[46] Flow cytometric estimates of ploidy and S-phase fraction provide additional information.[47] With recognition that the concentration of estrogen and progesterone receptor protein in a tumor is a determinant of prognosis, other biological factors concerned with tumor growth are being studied. These include matrix-degrading proteins, and the products of oncogene, tumor-suppressor gene, and metastases suppressor gene expression (see Table 1).[48-50] While these biological products reflect the metastases-producing potential of the tumor, their relevance to the management of an individual patient has still to be determined. Prognostic indices formed by relatively simple determinants such as tumor size, grade, and extent of lymph node involvement can in the meantime, offer greater guidance to the practicing clinician.[51,52]

5.3 Treatment

Ideally, treatment regimes for breast cancer should relate to the aggressiveness and likely extent of the tumor. This is not yet practicable. However, much guidance to appropriate treatment has come from controlled, randomized trials of local and systemic therapy.

Table 1 Some Factors which Are Known to Influence the Prognosis of Breast Cancer

Time Dependent	**Growth Factors, Oncogenes, and Receptors**
Tumor size	Estrogen receptor
Axillary node status	Progesterone receptor
Systemic micro- and macrometastases	Epidermal growth-factor receptor
	erb-B2 protein
Histopathological	H-ras protein
Histological type	c-myc protein
Cytological and nuclear grade	p21
Lympho-vascular invasion	pS2
	Heat-shock proteins
Markers of Proliferation	**Growth-Suppressor Genes**
DNA content (ploidy)	p53
S-phase fraction	
Monoclonal antibodies	**Factors for Invasion and Metastasis**
Ki67, KiSi, MIB-1	Proteinases
	Cathepsin-D
Markers of Apoptosis	Stromelysin-3
bcl-2	Tenascin
	Laminin receptors
	Angiogenesis factor (bFGF)
	nm 23 gene product

For recent reviews and sources of references see References 48 to 52.

5.3.1 Local Therapy

Local surgery or radiotherapy can only cure if the tumor has not metastasized. In most cases, its purpose is to control the local disease. As local relapse is both unpleasant and life-threatening, it is logical to use methods of local treatment which, as far as possible, guarantee freedom from further local disease.

During the first half of this century, the standard treatment of breast cancer was by radical mastectomy — removing the breast, the underlying muscles of the chest wall, and clearing the axilla of lymph nodes.[53,54] This extensive and mutilating operation was developed with the objective of achieving control of what now would be regarded as advanced local disease, and although it was then also considered to give the best chance of cure, it has little place today. Less radical forms of mastectomy, with preservation of the pectoral muscles, or reliance on radiation to treat lymph-node areas are now practiced.[55,56]

In 1937, Keynes reported that results equal to those of the radical operation could be achieved by local excision of the tumor followed by radical radiotherapy to the breast and lymph node areas, allowing the breast to be preserved.[57] Others followed his example, an action now fully vindicated by the results of a number of controlled randomized trials from which it is apparent that in suitable cases, treatment conserving the breast can now be offered as an option to mastectomy. These are tumors less than 4 cm in size and without either multifocal spread or an extensive *in situ* component.[58-60]

During this period of change, the psycho-sexual morbidity associated with breast cancer, and particularly with mastectomy, was being defined.[61,62] Although in some women mastectomy adds an intolerable burden, which can be offset by reconstructive surgery,[63] it is the fear of the disease, lack of confidence in their treatment, guilt, and withdrawal of support which are the main causes of distress. In dedicated breast units counselling by trained professionals has become an indispensable part of therapy.

5.3.2 Systemic Therapy

It is almost 100 years since Beatson reported that removal of the ovaries could alter the course of recurrent breast cancer, an effect believed to be due to the reduction in circulating estrogens.[64] During the years which followed, other surgical and pharmacological methods of estrogen deprivation were shown to affect the progress of advanced disease in one third to one half of patients. Most notable has been the development of the antiestrogen tamoxifen.[65]

During this period, it was found that the concentration of estrogen-receptor protein in the tumor was a predictor of response to antiestrogen treatment. Most tumors which were receptor poor failed to benefit, whereas 50 to 60% of receptor-rich tumors did so.[66] The alternative approach to the successful palliation of advanced systemic disease was to use combinations of cytotoxic agents. Since the 1950s, a variety of cytotoxic drugs have been used for the palliation of locally advanced or metastatic breast cancer. Of these, doxorubicin was found to have greatest potency.[67] Concern with the modest responses achieved led to exploration of the use of a combinations of non-cross resistant drugs, including cyclophosphamide, methotrexate, 5-fluorouracil, doxorubicin, vincristine, mitoxantrone, mitomycin C, and prednisone, with improved results. These regimes are toxic, but the development of the antiemetic serotonin 5-HT$_3$ receptor antagonist drugs has reduced troublesome nausea and vomiting.[68] In some centers, the doses of the agents are escalated to maximum tolerance; bone marrow rescue is then required.[69]

For adjuvant therapy of primary operable breast cancer, the usual regime now given is six 28-day cycles of cyclophosphamide, 5-fluorouracil, and methotrexate, based on that used in the first controlled randomized trial of combination chemotherapy reported in early disease.

In 1953, Ralston Paterson initiated the first controlled randomized trial of systemic therapy in operable breast cancer, to determine the effect of ovarian radiation.[70] This, and subsequent trials of systemic adjuvant therapy have been included in a meta-analysis reported by the Early Breast Cancer Trialists Co-operative Group, which includes, in treatment and control groups, over 75,000 women.[71] This analysis unequivocally indicates that antiestrogen therapy by ovarian ablation in premenopausal women, tamoxifen in postmenopausal women, and multiagent cytotoxic chemotherapy in both pre- and post-menopausal women reduces mortality from primary operable disease. The recent Scottish trial in premenopausal women with proven lymph-node metastases has indicated that ovarian ablation and chemotherapy give comparable benefit. It was found that this was related to the estrogen receptor (ER) status of the tumor; women with ER-rich tumors derived preferential benefit from ovarian ablation, confirming an indication of this effect in an international trial.[72,73] Although in absolute terms the mortality differences in the overview analysis are small (see Figure 3 and Table 2), they have persisted for 10 years, and worldwide, represent an annual reduction of 100,000 deaths for each million women treated for breast cancer!

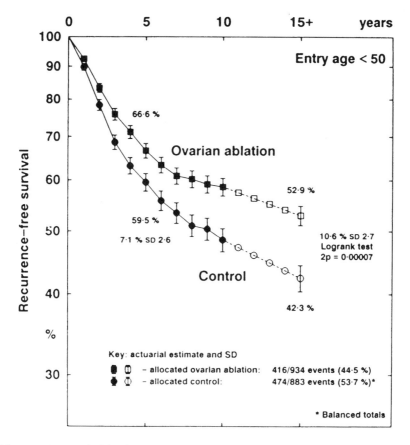

Figure 3   15-year outcome of trials of ovarian ablation in 2000 women with operable breast cancer under 50 years of age when randomized. (From Early Breast Cancer Triallists' Collaborative Group, *Lancet*, 339, 1-15, 71-85, 1992. With permission.)

Table 2 Reduction in Annual Mortality Rates and Absolute Reduction in Mortality at 10 Years in Women with Operable Breast Cancer in which Allocated Forms of Adjuvant Systemic Therapy Have Been Compared with No Similar Treatment

Allocated Treatment	Number of Patients	Proportional Reduction in Annual Mortality (% ± S.D.)	Absolute Reduction in 10-year Mortality (% ± S.D.)
Age < 50			
Ovarian ablation	1817	25 ± 7	10.6 ± 2.7
Age > 50			
Polychemotherapy	11450	12 ± 4	6.3 ± 1.4
Tamoxifen	30081	17 ± 2	6.2 ± 0.9

From Early Breast Cancer Triallists' Collaborative Group, *Lancet*, 339:1-15, 71-85, 1992. With permission.

The selection of patients for systemic therapy is not precise, but the identification of micrometastases in the axillary nodes is an obvious indication that the disease is systemic. In node-negative cases, such indices as the size of the tumor, its histological type and grade, and signs of vascular invasion are taken into account in determining the need for therapy.

6. MAMMOGRAPHIC SCREENING

In 1913 Salomon first imaged the breast radiologically, in order to study the pattern of spread of breast cancer in mastectomy specimens.[74] The first attempt to radiograph the breast of a living person was reported in 1927 and, in 1930, Warren reported 119 examinations.[75,76] Although during the next 10 years a number of isolated reports of breast radiology appeared, it was only following detailed technological studies by Egan, Gross, Gershon-Cohen, and others in the 1960s that modern film-screen mammography became available. Recognition that occult nonpalpable breast cancer could be visualized radiologically led Shapiro and Strax to initiate the first controlled randomized trial of breast cancer screening in the State of New York.[77-80]

Nine randomized trials and one large comparative controlled study, five case-control studies and a large demonstration project have now been reported which indicate unequivocally that the screening of well women by mammography can result in a significant reduction in breast cancer mortality. The 7 to 10-year results of the ten controlled randomized trials which include 330,000 women and a meta-analysis of Swedish trials, have recently been reported and were reviewed at a consensus meeting in Europe.[81] These indicate a reduction in deaths averaging 20% (see Figure 4). It should be noted that a significant reduction in mortality has been observed only in women over 50 years of age. Although there are suggestions that with longer follow-up a mortality reduction is beginning to emerge in younger women, this is insufficient to recommend a change of policy in those countries which have instituted programs of population screening. It should also be stressed that to be effective, a program of screening must define the target population, encourage them to attend by personal invitation, monitor quality, and evaluate results. The program which was initiated in the U.K. by the National Health Service in 1988, by which women of 50 to 64 years of age are invited for mammographic screening every 3 years, has paid attention to these needs.[19] Guidelines have been laid down for the conduct of screening and there is both local and central monitoring of quality. Of particular importance was the decision that the further diagnosis of mammographic abnormalities should not be left to individual clinicians, but should be arranged through the screening service by multidisciplinary assessment teams who apply the "triple diagnostic approach" to nonpalpable as well as palpable lesions. Opportunistic screening in a radiologist's office may benefit individual women, but only at the price of unnecessary biopsies, worry, and expense.

7. PREVENTATIVE STRATEGIES

It is clear that until the cause of breast cancer has been identified, true prevention is a forlorn hope. However, because of its known role in promoting breast carcinogenesis, attention has focused on methods to inhibit ovarian function or, in older women, the availability of estrogenic hormones. From the results

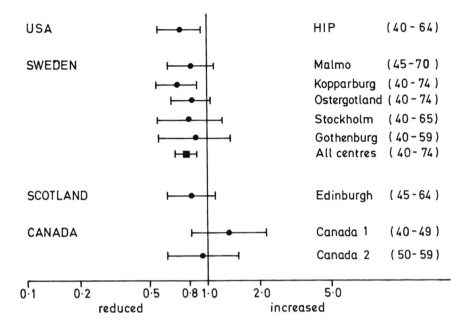

Figure 4 Seven to twelve years mortality in all randomized trials of population screening by mammography. Odds of death with 95% confidence limits are given for each trial and for the overview of Swedish trials. (Constructed from results given in References 33 to 36. From Dixon, M. and Sainsbury, R., *Handbook of Diseases of the Breast*, Churchill Livingstone, London, 1993. With permission.)

of trials of tamoxifen, given as adjuvant systemic therapy, it is apparent that the development of second tumors in the contralateral breast is delayed in those women given tamoxifen, and this has led to interest in the use of the drug in a prophylactic setting.[82,83] A number of large, controlled, randomized trials have now been instituted in which the drug is being given to women considered to be at high risk of the disease; for example by having a positive family history or a previous breast biopsy. It is known that continuous tamoxifen therapy does not affect bone density or blood lipid profiles adversely, because of estrogenic action on osteoblasts and hepatocytes. However, it is estrogenic effects on the endometrium which are causing concern, there being some evidence of an increased incidence of endometrial cancer in tamoxifen users.[84] Of great interest is the demonstration of a reduction of fatal myocardial infarcts and of hospital admissions for ischaemic heart disease and other cardiovascular conditions in those given tamoxifen in the Scottish and Swedish randomized trials.[85-87] The continuous administration of tamoxifen may have greater benefits than those related to breast cancer.

Reversible ablation of ovarian function by gonadotrophin-releasing hormone analogues has also been proposed as a method of inhibiting hormonal stimulation of the breast. Knowing that the greatest proliferative activity of breast epithelium occurs during the luteal phase of the ovarian cycle, this has the advantage that the availability of progesterone as well as estrogen is reduced.[88] On the assumption that unopposed estrogen is not a risk factor for breast cancer, Pike and his colleagues suggest that the undesirable effects of pharmacological inhibition of ovarian function can safely be prevented by estrogen-replacement therapy.[89] To counteract an increased risk to the endometrium, he proposes intermittent courses of small-dose progesterone. It is suggested that were this regime to be used in young women, not only would it produce effective contraception without the dangers of the pill, but it would reduce the incidence of breast cancer by half, and that of ovarian cancer by two-thirds.

Reduction of dietary fat intake is known to reduce available estrogens. In a group of 73 postmenopausal women, reduction of fat intake from 40% to 20% of total kilocalories over a period of 10 to 22 weeks reduced total and weakly bound estradiol by 17%, an effect which, based on data from an Edinburgh study of estrogen levels in women with breast cancer, was regarded as likely to reduce incidence significantly.[90,91] Dietary intervention studies are now underway in the U.S., but the design of these has been criticized.[92] Currently, there is considerable interest in the possibility that lignans and phytoestrogens in vegetable products may act as antiestrogens, or direct the hydroxylation of estrone to less-active metabolites.[93] Further evidence is required before these effects can be extended to a preventive strategy.

8. CONCLUSION

It is apparent that breast cancer is a complex disease. The advent of molecular genetics is unquestionably unraveling many of its mysteries, and it can only be hoped that in time these will lead to more direct methods of prevention than are currently being contemplated. Meantime one can only ensure that all women are aware of the advances which have been made and are encouraged to participate in screening programs to ensure that appropriate treatment is given. For there is no doubt that the early detection of the disease and the correct use of systemic, as well as local therapy has produced substantial dividends.

REFERENCES

1. **Forrest, A.P.M., Hawkins, R.A., and Miller, W.R.,** Breast, in *Jamieson and Kay's Textbook of Surgical Physiology*, Ledingham, I.B., Ed., Churchill Livingstone, Edinburgh and London, 1989, 85–94.
2. **Azzopardi, J.G.,** *Problems in Breast Pathology*, J.B. Saunders, London, 1979.
3. **Page, D.L. and Anderson, T.J.,** *Diagnostic Histopathology of the Breast*, Churchill-Livingston, Edinburgh, 1988.
4. **Petrakis, N.L., Emster, V.L., and Sacks, S.T.,** Epidemiology of breast fluid secretion associated with breast cancer risk factors and cerumen type, *J. Natl. Cancer Inst.*, 67, 277, 1981.
5. **Miller, W.R., Humenik, V., and Forrest, A.P.M.,** Factors affecting dehydroepiandrosterone sulphate levels in human breast secetion, *Breast Cancer Res. Treat.*, 1, 267, 1981.
6. **Yap, P.L., Miller, W.R., Humenik, J., Pryde, E.A.D., Mirtle, C., and McLelland, D.B.L.,** Milk protein concentrations in the mammary secretions of lactating women, *J. Reproduct. Immunol.*, 3, 49, 1981.
7. **Petrakis, N.L.,** Oestrogens and other cytological and biochemical components in nipple aspirates of breast fluid: relationship of 15 risk factors for disease, *Proc. R. Soc. Edinburgh*, 95B, 169, 1989.
8. **Wellings, S.R., Jensen, H.M., and Marcum, R.G.,** An atlas of subgross pathology of human breast with special reference to possible pre-cancerous lesion, *J. Natl. Cancer Inst.*, 55, 231, 1975.
9. **Dupont, W.D. and Page, D.L.,** Risk factors for breast cancer in women with proliferative disease, *N. Engl. J. Med.*, 312, 1055, 1985.
10. **Page, D.L., Steel, C.M., and Dixon, J.M.,** ABC of breast disease: carcinoma in situ and patients at high risk of breast cancer, *Br. Med. J.*, 310, 39, 1995.
11. **Sloan, J.P.,** Borderline lesions detected by breast screening, *Br. J. Surg.*, 81, 481, 1994.
12. **Keynes, G.,** Carcinoma of the breast: the unorthodox view, *Proc. Cardiff Med. Soc.*, 4, 40, 1953.
13. **Callahan, R., Cropp, C.S., Merlo, G.R., Liscia, D.S., Cappa, A.P.M., and Lidereau, R.,** Somatic mutations and human breast cancer, *Cancer*, 69, 1582, 1992.
14. **Miki, Y., Swensen, J., Shattuck-Eidens, D., Futreal, P.A., Harshman, K., Tavtigian, S., Liu, Q., Cochran, C., Bennett, L.M., and Ding, W.,** A strong candidate for the breast and ovarian cancer susceptibility gene BRCA1, *Science*, 266, 66, 1994.
15. **Wooster, R., Neuhausen, S.L., Mangion, J., Quirk, Y., Ford, D., Collins, N., Nguyen, K., Seal, S., Tran, T., Averill, D., Fields, P., Marshall, G., Narod, S., Lenoir, G., Lynch, H., Feunteun, J., Devilee, P., Cornelisse, C., Menko, F., Daly, P., Ormiston, W., McManus, R., Pye, C., Lewis, C., Cannon-Albright, L., Peto, J., Ponder, B., Skolnick, M., Easton, D., Goldgar, D., and Stratton, M.,** Localization of a breast cancer susceptibility gene, BRCA2, to chromosome 13q12-13, *Science*, 265, 2088, 1994.
16. **Matrisiam, L.M.,** The matrix-degrading metalloproteinases, *Bioessays*, 14, 455, 1992.
17. **Steeg, P.S., Bevilaqua, G., Kopper, L., Thorgeisson, W.P., Liotta, L.A., and Sobel, M.E.,** Evidence for a novel gene associated with low tumour metastatic potential, *J. Natl. Cancer Inst.*, 80, 200, 1988.
18. **Parkin, D.M., Muir, C.S., Whelan, S.L., Gao, Y.T., Ferlay, J., and Powell, J.,** *Cancer Incidence in Five Continents*, Vol. 6. International Agency for Research on Cancer, Lyon, 1992.
19. Breast Cancer Screening, Report to the Health Ministers of England, Wales, Scotland and Northern Ireland by a Working Group chaired by Sir Patrick Forrest, Her Majesty's Stationery Office, 1987.
20. **Brinkley, D. and Haybittle, J.L.,** Long-term survival of women with breast cancer, *Lancet*, 1, 1118, 1984.
21. **Adair, F., Berg, J., Joubert, L., and Robbins, G.F.,** Long-term follow-up of breast cancer patients: the thirty-year report, *Cancer*, 33, 1145, 1974.
22. **Mueller, C., Ames, F., and Anderson, G.D.,** Breast cancer in 3558 women: age as a significant determinant in the role of dying and causes of death, *Surgery*, 83, 123, 1978.
23. **Langlands, A.O., Pocock, S.J., Kerr, J.R., and Gore, S.M.,** Long-term survival of patients with breast cancer; a study of the curability of the disease, *Br. Med. J.*, 2, 1247, 1979.
24. **Le, M.G., Hill, C., Rezvani, A., Sarrazin, A., Contesso, G., and Lacour, J.,** Long-term survival of women with breast cancer; a study of the curability of the disease, *Lancet*, 2, 922, 1984.
25. **Rutqvist, L.E. and Wallgren, A.,** Long-term survival of 458 young breast cancer patients, *Cancer*, 55, 658, 1985.

26. **Feinleib, M.,** Breast cancer and artificial menopause, *J. Natl. Cancer Inst.*, 41, 315, 1968.
27. **Trichopoulos, D., MacMahon, B., and Cole, P.,** Menopause and breast cancer risk, *J. Natl. Cancer Inst.*, 48, 605, 1972.
28. **Kvale, G.,** Reproductive factors in breast cancer epidemiology, *Acta Oncol.*, 31, 187, 1992.
29. **Hunt, K. and Vessey, M.,** Long-term effects of postmenopausal hormone therapy, *Br. J. Hosp. Med.*, 38, 450, 1987.
30. **Pike, M.C., Henderson, B.E., Krailo, M.D., Duke, M.D., Duke, A., and Roy, S.,** Breast cancer in young women and use of oral contraceptives: possible modifying effects of formulation and age of use, *Lancet*, 2, 926, 1983.
31. UK national case-control study group, Oral contraceptive use and breast cancer risk in young women, *Lancet*, 1, 973, 1989.
32. **MacMahon, B., Cole, P., Lin, M., Lowe, C.R., Mirra, A.P., and Ravnihar, B.,** Age at first birth and breast cancer risk, *Bull WHO*, 43, 209, 1970.
33. **Brinton, L., Williams, R.R., and Hoover, R.N.,** Breast cancer risk factors among screening programme participants, *J. Natl. Cancer Inst.*, 62, 37, 1979.
34. **Russo, J., Tay, L.K., and Russo, I.H.,** Differentiation of the mammary gland and susceptibility to carcinogenesis, *Breast Cancer Res. Treat.*, 2, 5, 1982.
35. **Hoover, R.N., Pike, M.C., Hildesheim, A., Nomura, A.M.Y., and West, D.W.,** Migration patterns and breast cancer risk in Asian-American women, *J. Natl. Cancer Inst.*, 85, 1819, 1993.
36. **Willett, W.C. and MacMahon, B.,** Diet and cancer: an overview, *N. Engl. J. Med.*, 310, 697, 1984.
37. **Aldecreutz, H., Mousavi, Y., and Hockerstedt, K.,** Diet and human breast cancer, *Acta Oncol.*, 31, 175, 1992.
38. **Lee, H.P., Gourlet, L., Duffy, S.W., Esteve, J., Lee, J., and Day, N.E.,** Dietary effects in breast cancer risk in Singapore, *Lancet*, 337, 1197, 1991.
39. **Clemmenson, J.,** On the aetiology of some human cancers, *J. Natl. Cancer Inst.*, 12, 1, 1951.
40. **Kalache, A.,** Risk factors for breast cancer; a tabular summary of the epidemiological literature, *Br. J. Surg.*, 68, 797, 1981.
41. **McPherson, K., Steel, C.M., and Dixon, J.M.,** ABC of breast diseases: breast cancer — epidemiology, risk factors and genetics, *Br. Med. J.*, 309, 1003, 1994.
42. **Kamada, N., Shigeta, C., and Kuramoto, A.,** Acute and late effects of A-bomb irradiation: studies in a group of young girls with a defined condition at the time of bombing, *J. Radiat. Res.*, 30, 218, 1989.
43. **Modan, B., Chetrit, A., and Alfaridarg, E.,** Increased risk of breast cancer after low-dose irradiation, *Lancet*, 1, 629, 1989.
44. **Dixon, J.M. and Mansel, R.E.,** ABC of breast disease: symptoms, assessment and guidelines for referral, *Br. Med. J.*, 309, 722, 1994.
45. **Hermanek, P. and Sobin, L.H.,** *TNM Classification of Malignant Tumours*, 4th ed., International Union against Cancer, Berlin, 1987.
46. **Bloom, H.J.B. and Richardson, W.W.,** Histological grading prognosis in breast cancer: study of 409 cases of which 359 have been followed up for 15 years, *Br. J. Cancer*, 11, 35, 1957.
47. **Osborne, C.D.,** DNA flow cytometry in early breast cancer: a step in the right direction, *J. Natl. Cancer Inst.*, 94, 60, 1989.
48. **McGuire, W.L., Clark, G.M., Fisher, E.R., and Henderson, I.C.,** Predicting recurrence and survival in breast cancer, *Breast Cancer Res. Treat.*, 9, 27, 1987.
49. **Gasparini, G., Pozza, F., and Harris, A.L.,** Evaluating the potential usefulness of new prognostic and predictive factors in node negative breast cancer patients, *J. Natl. Cancer Inst.*, 85, 1206, 1993.
50. **Miller, W.R., Ellis, I.O., Sainsbury, J.R.C., and Dixon, J.M.,** ABC of breast disease: prognostic factors, *Br. Med. J.*, 309, 722, 1994.
51. **Haybittle, J.L., Blamey, R.W., Elston, C.W., Johnson, J., Doyle, P.J., and Campbell, F.C.,** A prognostic index in primary breast cancer, *Br. J. Cancer*, 45, 361, 1982.
52. **Brown, J.M., Bensen, E.A., and Jones, M.,** Confirmation of a long-term prognostic index in breast cancer, *The Breast*, 2, 144, 1993.
53. **Halsted, W.S.,** The results of radical operations for the cure of carcinoma of the breast, *Ann. Surg.*, 46, 1, 1907.
54. **Meyer, W.,** Carcinoma of the breast — 10 years experience with my method of radical operation, *J. Am. Med. Assoc.*, 65, 292, 1907.
55. **McWhirter, R.,** Simple mastectomy and radiotherapy in the treatment of breast cancer, *Br. J. Radiol.*, 28, 128, 1955.
56. **Patey, D.H. and Dyson, W.H.,** The prognosis of carcinoma of the breast in relation to the type of operation performed, *Br. J. Cancer*, 2, 7, 1948.
57. **Keynes, G.,** Conservative treatment of cancer of the breast, *Br. Med. J.*, 2, 643, 1937.
58. **Veronesi, U. Saccozzi, R., Del Vecchio, M., Banfi, A., Clemente, C., and DeLena, M.,** Comparing radical mastectomy with quadrantectomy, axillary dissection and radiotherapy in patients with small cancers of the breast, *N. Engl. J. Med.*, 305, 6, 1981.
59. **Veronesi, U., Greco, M., Luini, A., Merson, M., Sacchini, V., and Zurrida, S.,** Conservative treatment of breast cancer: the Milan experience, in *Breast Cancer: Controversies in Management*, Wise, L., Johnston, H., Jr., Eds., Futuras Publishing Company Inc, Armonk, NY 1994, 139–145.

60. **Fisher, B., Bauer, M., Magolese, R., Poisson, R., Pilch, Y., and Redmond, C.,** Five-year results of a randomised clinical trial comparing total mastectomy and segmental mastectomy with or without radiation in the treatment of breast cancer, *N. Engl. J. Med.*, 312, 665, 1985.

61. **Morris, T., Steven-Greer, H., and White, P.,** Psychological and social adjustment to mastectomy (a two-year follow-up study), *Cancer*, 40, 2381, 1977.

62. **Macguire, G.P., Lee, E.G., and Bevington, D.J.,** Psychiatric problems in the year after mastectomy, *Br. Med. J.*, 1, 963, 1978.

63. **Dean, C., Chetty, U., and Forrest, A.P.M.,** Effects of immediate breast reconstruction on psychological morbidity associated with a diagnosis of early breast cancer, *Lancet*, 1, 459, 1983.

64. **Beatson, G.T.,** On the treatment of inoperable cases of carcinoma of the mamma, *Lancet*, 2, 104, 162, 1896.

65. **Harper, M.J. and Walpole, A.L.,** Contrasting endocrine activities of cis- and trans-isomers in a series of substituted triphenyl ethylenes, *Nature*, 212, 87, 1966.

66. **Jensen, E.V. and Jacobsen, H.I.,** Basic guidelines to the mechanism of estrogen action, *Prog. Hormone Res.*, 8, 387, 1962.

67. **Williams, C.J. and Buchanan, R.B.,** *The Medical Management of Breast Cancer*, Castle House Publications Ltd, Tunbridge Wells, 1987.

68. **Seynaeve, C.,** *Ondansetran: Clinical Studies of Cytotoxic Chemotherapy*, IGG Printing, Dordrecht, 1991.

69. **Peters, W.P., Shapali, E.J., Jones, R.B., and Olsen, G.A.,** High dose combination alkylating agents with bone marrow support as initial treatment for metastatic breast cancer, *J. Clin. Oncol.*, 6, 1368, 1988.

70. **Paterson, R. and Russell, M.H.,** Clinical trials in malignant disease: value of irradiation of the ovaries, *J. Fac. Radiol.*, 10, 130, 1959.

71. Early Breast Cancer Triallists' Collaborative Group. Systemic treatment of early breast cancer by hormonal cytotoxic or immune therapy: randomised trials involving 31,000 recurrences and 24,000 deaths among 75,000 women, *Lancet*, 339, 1, 71, 1992.

72. Scottish Cancer Trials Breast Group and ICRF Breast Unit, Guy's Hospital, London, Adjuvant ovarian ablation versus CMF chemotherapy in premenopausal women with pathological stage II breast carcinoma, *Lancet*, 341, 1293, 1993.

73. International Breast Cancer Study Group, Late effects of adjuvant oophorectomy and chemotherapy upon premenopausal breast cancer patient, *Acta Oncol.*, 1, 30, 1990.

74. **Salomon, A.,** Beitrage zur pathologie und klinic der mammacarcinom, *Arch. Klin. Chir.*, 101, 573.

75. **Egan, R.L.,** *Breast Imaging: Diagnosis and Morphology of Breast Disease*, WB Saunders, Philadelphia, 1988.

76. **Warren, S.L.,** Roentgenologic study of the breast, *Am. J. Roentgenol.*, 24, 113, 1930.

77. **Egan, R.L.,** Experience with mammography in a tumour institution. Evaluation of 1000 studies, *Radiology*, 75, 894, 1960.

78. **Gros, C.M.,** Methodologie: symposium sur le sein, *J. Radiol. Electr.*, 48, 638, 1987.

79. **Gershon-Cohen, J. and Strickler, A.,** Roentgenological examination of the normal breast: its evaluation in demonstrating early neoplastic change, *Am. J. Roentgenol.*, 40, 189, 1938.

80. **Shapiro, S., Venet, W., Strax, P., and Venet, L.,** Periodic screening for breast cancer. The Health Insurance plan project and its sequelae 1963–1986. The Johns Hopkins University Press, Baltimore, 1988.

81. **Wald, N.J., Chamberlain, J., and Hackshaw, A.,** On behalf of the Evaluation Committee. Report of the European Society for Mastology Breast Cancer Screening Evaluation Committee, *The Breast*, 2, 209, 1993.

82. **Powles, T.J., Hardy, J.R., Ashley, S.E., Farrington, G.M., Cosgrove, D., and Davey, J.B.,** A pilot trial to evaluate the acute toxicity and feasibility of tamoxifen for the prevention of breast cancer, *Br. J. Cancer*, 14, 23, 1989.

83. **Love, R.R.,** Prospects for antiestrogen chemoprevention of breast cancer, *J. Natl. Cancer Inst.*, 82, 18, 1990.

84. **Fornander, T., Rutqvist, L.E., Cedermark, B., Glas, W., Mattson, A., and Silfsward, C.,** Adjuvant tamoxifen in early breast cancer; occurrence of new primary cancers, *Lancet*, 1, 117, 1989.

85. **McDonald, C.C. and Stewart, H.J.,** for the Scottish Breast Cancer Committee. Fatal myocardial infarction in the Scottish adjuvant tamoxifen trial, *Br. Med. J.*, 303, 435, 1991.

86. **Rutqvist, L.E. and Mattson, A.,** for the Scotckholm Breast Cancer Study Group. Cardiac and thromboembolic morbidity amongst postmenopausal women with early stage breast cancer in a randomised trial of adjuvant tamoxifen, *J. Natl. Cancer Inst.*, 85, 1398, 1993.

87. **McDonald, C.C., Alexander, F.E., Whyte, B.W., Forrest, A.P., and Stewart, H.J.,** Morbidity in women receiving adjuvant tamoxifen for breast cancer — cardiac and vascular effects, *Br. Med. J.*, 1995, in press.

88. **Anderson, T.J. and Battersby, S.,** The involvement of oestrogen in the development and function of the normal breast: histological evidence, *Proc. R. Soc. Edinburgh*, 95B, 23, 1989.

89. **Pike, M.C., Ross, R.K., Lobo, R.A., Key, T.J.A., Potts, M., and Henderson, B.E.,** LHRH agonists and the prevention of breast and ovarian cancer, *Br. J. Cancer*, 60, 142, 1989.

90. **Prentice, R., Thompson, D., Clifford, C., Gorbach, S., Goldin, B., and Byar, D.,** Dietary fat reduction and plasma oestradiol concentration in healthy postmenopausal women, *J. Natl. Cancer Inst.*, 82, 129, 1990.

91. **MacFadyen, I.J., Prescott, R., Groom, G.V., and Forrest, A.P.M.,** Plasma steroid levels in women with breast cancer, *Lancet*, 1, 100, 1976.

92. **Hunter, D. and Willet, W.,** Dietary fat and breast cancer, *J. Natl. Cancer Inst.*, 85, 1776, 1993.

93. **Bradlow, H.L. and Michnovicz, J.T.,** A new approach to the development of breast cancer, *Proc. R. Soc. Edinburgh*, 95B, 77, 1989.

Chapter 3

Hormonal, Growth Factor, and Cytokine Control of Breast Cancer

William R. Miller and Simon P. Langdon

CONTENTS

1. INTRODUCTION

The breast is a complex organ which is responsive to a variety of regulatory influences including steroid hormones, growth factors, and cytokines. The recognition that steroid hormones have a critical role in the development of both normal and malignant breasts was made in the last century without a knowledge of their nature. However, in recent years the molecular details of the processes involved in steroid regulation have been defined. It has also become evident that polypeptide growth factors may influence the proliferation of mammary epithelial cells and that certain oncogenes implicated in the natural history of breast cancer encode proteins which are either growth factors or elements of their signaling pathways. Cytokines which may be produced not only by mammary cells, but also by cells infiltrating the breast add to the complexity of trophic effects.

The aims of this chapter are therefore to review the evidence that steroid hormones, growth factors, and cytokines influence (1) the proliferation of the normal breast, (2) susceptibility to malignant transformation and the risk of developing breast cancer, and (3) the behavior and prognosis of established cancers. Where possible, the processes underlying these effects will be described.

2. STEROID HORMONES

The steroid hormone family comprises four major classes (estrogens, progestogens, androgens, and corticosteroids) which have varied biological activities; however, the influences of estrogen on the breast appear disproportionate to those of other steroid hormones. The following review therefore emphasizes

43

the central role of estrogen but the effects of other steroids can be important and will be mentioned where appropriate.

2.1 The Normal Breast

That the human female breast is under the primary control of hormones is self-evident. The organ develops at puberty but only achieves full functionality during the course of pregnancy. Conversely, involutionary changes occur *postpartum* and glandular structures show progressive atrophy after the menopause. These observations point unequivocally to the involvement of ovarian and placental hormones in breast development. The role of estrogen appears central; in girls with gonadal dysgenesis, full breast development may be induced by the simple expedient of administering estrogen,[1] although in individuals whose gonadal failure is secondary to pituitary deficiency, it is sometimes necessary to provide pituitary hormones in addition to estrogen. These observations are compatible with reports that estrogen-secreting tumors in prepubescent girls cause precocious breast development[2] and that administered estrogen may stimulate breast-cell proliferation and preserve glandular structure in postmenopausal women.[3] However, it seems unlikely that estrogen alone accounts for breast development; at puberty, breast development in girls usually precedes the secretion of major estrogen by the ovary;[4] indeed, levels of estrogen in pubertal girls with substantial breast development may be no more than in males.[1] Additionally, a good correlation has not been demonstrated between degree of breast development and circulating levels of estrogen.[5] These observations in general relate to the external appearance of the breast. To study the parenchymal structures, it has been necessary to investigate terminal duct lobular units (TDLUs) obtained from women undergoing biopsies for benign conditions. These studies have shown that during the menstrual cycle, (1) normal breast TDLU epithelium exhibits greater proliferation in the luteal than in the follicular phase, an effect opposite to that found in endometrium,[6-9] and (2) when estrogen levels are high, mitotic activity can be low.[10] These observations suggest that if estrogen stimulates breast epithelial proliferation, the route is not direct. One possible explanation of these findings is that in the presence of permissive amounts of estrogen, progestins play an important role in regulating proliferation in the breast,[11] a view supported by the observations that oral contraceptives (particularly progestin only) increase luteal-phase proliferation,[6-8] and, consistent with this, a marker of progestin induction (fatty acid synthetase) increases in parallel with proliferation in human breast epithelium.[12]

In summary, while estrogen has a central role in the development of the normal breast, this appears in part to be permissive, and other factors seem to modify and enhance its action.

2.2 Estrogens and the Risk of Breast Cancer

The link between estrogens and the risk of breast cancer has been derived from multidisciplinary research including epidemiology and endocrinology studies. Several of the more significant observations are described.

2.2.1 Epidemiological Evidence

Epidemiological evidence comes from two areas: (1) the associations between reproductive/menstrual history and breast cancer and (2) the effects of exogenous estrogens on the incidence of breast cancer.

The etiology of breast cancer has a strong hormonal component (Table 1) and menstrual status and reproductive history are important determinants of risk to breast cancer.[13] The disease does not occur before puberty and an extended reproductive life, whether by an early menarche or late menopause, increases risk.[14] There is also a downward inflection in the incidence rate for breast cancer at the time of the menopause. This suggests that the rise in premenopausal women is the consequence of ovarian sex hormone production and the menopause saves some who would have otherwise developed breast cancer. Other epidemiological observations support this concept; women with a menopause at 55 years of age or older have almost twice the risk of those whose menopause occurred naturally before 45 years of age while castration early in reproductive life markedly reduces risk (by two thirds if performed before 35 years of age).[15]

Age at menarche is also an influential factor associated with risk. Those who start to menstruate early have an increased risk and cumulative protection occurs with each year that menarche is delayed. The effects can be significant, and it is suggested that the later age at menarche in Japanese girls as compared with their American counterparts might account for at least part of the difference in breast cancer rates between the countries.[16]

The most likely explanation for the association between an increased risk of breast cancer and extended reproductive life is the exposure of the breast to persistent trophic ovarian hormones. As

Table 1 Risk Factors for Breast Cancer

Factor	High Risk	Low Risk	Relative Risk
Gender	Female	Male	100×
Age	Elderly	Young	>10×
Geography	USA/UK	Japan	5×
Age at menarche	<12 years	>13 years	2×
Age at menopause	>50 years	<35 years	3×
Parity	Parous	Nulliparous	1.5×

indicated, estrogens and progestogens may act in concert to promote the proliferation of cells, making them susceptible to the effects of carcinogens. If cumulative ovulatory menstrual cycles increase risk, the protective influence of pregnancy (especially if pregnancy was early in life and followed by prolonged lactational amenorrhea) might result from a disruption in the regular pattern of menstruation. In addition to breaking up cyclic exposure of the breast to trophic influences, placental hormones are thought to cause stem cells to differentiate.

Body weight is another risk factor which might be mediated by estrogen. While body weight has little influence on the risk of premenopausal breast cancer, heavier women seem to be at greater risk in the postmenopausal period.[17] It is thus relevant in postmenopausal women, that estrogen is largely derived from the extraglandular conversion of adrenal androgens,[18] and the rate and degree of this conversion increases with body weight.[19] It is also pertinent that the geographical differences in breast cancer risk are most marked in the postmenopausal period[20] and postmenopausal women in westernized societies tend to be more overweight than their counterparts in less industrialized countries.[16,21] Indeed, using a theoretical model, Pike suggested that this factor might account for 85% of the difference in breast cancer rates between Japanese and American postmenopausal women.[22]

If estrogens influence risk of breast cancer, it might be expected that women given exogenous estrogen would have a higher incidence of cancer. Estrogens have been administered to women on a relatively large scale: (a) during pregnancy, (b) at and after the menopause, and (c) during child-bearing years.

During the 1950s and 1960s, very high doses of diethylstilbestrol were given to pregnant women in the U.S. who either had a history of abortion, were threatening to abort, or had other complications. Subsequently, it was shown that in addition to an increased incidence of clear-cell adenocarcinoma of the vagina and cervix in the daughters, there was an increased risk of breast cancer (by about 35%) in the mothers themselves.[23]

Treatment to relieve osteoporosis and hot flushes at the menopause by estrogen replacement therapy (ERT) usually involves use of conjugated estrogens such as premarin. While there is unequivocal evidence that such therapy can cause endometrial cancer,[24] effects on breast cancer are much more controversial despite a considerable amount of study. Although some case-control studies have shown an increased risk with estrogen replacement therapy,[25,26] and one investigation demonstrated that ERT obliterated the protective effect of bilateral oophorectomy,[27] these reports are somewhat exceptional and several recent meta-analyses[27-30] have failed to confirm increased risk, with the possible exception of women with a longer duration of exposure to estrogen.

Estrogens administered in the form of oral contraceptive pills usually consist of synthetic estradiol 17β derivatives, (ethinyl estradiol or its 3-methyl ether, mestranol) and one of a number of progestogens. Oral contraceptive pills were first introduced in the U.S. in the 1960s and in the UK shortly thereafter. The large number of studies on these women have been reassuring, in that none have suggested any significant increase in risk, apart from the occasional report of a positive association between breast cancer risk and total duration of oral contraceptive use. However, anxieties have been expressed about possible harmful effects of oral contraceptive when used either before first full-term pregnancy or before age 25[31-33] and there is now evidence of positive trends between risk and duration of oral contraceptive use.[33,34]

The epidemiological evidence linking estrogens with risk of breast cancer is therefore provocative but not conclusive. Menarche and menopause are influential in the etiology of breast cancer and their effects are likely to be mediated by estrogen. Similarly, findings relating to exposure to exogenous estrogens during pregnancy and at the menopause are suggestive, but not unequivocal, of increased risk of breast cancer; the data on the oral contraceptives, while largely reassuring, are still too immature to be totally convincing, but long-term use in young women is a matter of concern. These influences are summarized in Figure 1, along with other factors likely to be involved in the development of this disease.

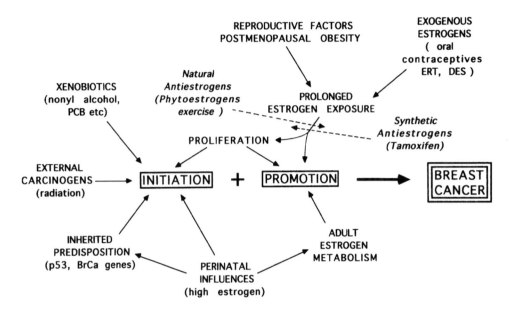

Figure 1 Interactions between estrogens and other factors in the development of breast cancer.

2.2.2 Endocrinological Evidence

Considerable efforts have been expended in attempts to show that women at high risk of breast cancer and/or those who subsequently develop the disease, have abnormal endocrine profiles. Initially, measurements of endogenous steroids were hampered by the lack of sensitive assays, and early studies were performed on urine which allowed the extraction of large sample volumes. However, population studies of high- and low-risk groups and familial studies did not yield consistent evidence that any individual urinary estrogen was associated with increased risk.[35] Interestingly, the evidence that subsequent breast cancer can be predicted by urinary steroids is stronger for androgens than for estrogens. Of particular importance was the prospective study initiated by Bulbrook in 1961 on the Island of Guernsey in which urine specimens were collected from 5,000 normal women.[36] When a case of breast cancer was subsequently diagnosed, the patients specimen was retrieved from storage together with up to ten matched controls. Assays for androsterone, etiocholanolone, and dehydroepiandrosterone were then carried out. Within 10 years, 27 women had developed breast cancer. Their excretion of androsterone and etiocholanolone was significantly lower than that of the controls, this being present up to 9 years before onset of the disease. The abnormality was found mainly in premenopausal women and a similar study by De Waard and Banders-van-Halewun[37] failed to show subnormal androgen excretion in postmenopausal women. Twenty-five years of follow-up are now available in Guernsey and it is apparent that while cases of breast cancer occurring in the early years of the study had a low excretion of androgen metabolites, women in whom tumors appeared late in life had a higher excretion of these steroids.[38] In other words, androgen excretion was related to age at diagnosis, and not to absolute risk.[39]

With the advent of radioimmunoassays, estrogens could be measured in blood. This provided a more direct estimate of the levels of hormones to which target tissues were exposed, and allowed a reevaluation of the hypothesis that the breasts of women who were at increased risk to cancer may be exposed to more estrogen than those who were not. Hormones such as estradiol and testosterone in the blood are largely bound with high affinity to sex hormone binding globulin (SHBG), with most of the remainder bound nonspecifically with low affinity to albumin, and only a small fraction is free. Attention has focused on this latter fraction since there is evidence that free steroids readily diffuse into cells, whereas those bound to SHBG do not. Perhaps the most illuminating observation has come from the investigation carried out on the Island of Guernsey (referred to earlier). Blood was obtained from the same 5,000 normal subjects. Preliminary results on the first thirteen women to develop breast cancer showed that they had a much higher proportion of their serum estradiol in the nonprotein-bound fraction than the controls, at a significance level of $p = 0.00005$.[40] Despite the absolute amounts of estrogen involved being small, the conclusion drawn from these findings was that women with high amounts of biologically available estradiol in their blood were at greater risk of breast cancer. However, as the study matured, the new cases of breast cancer were found to have SHBG levels within the normal range, and eventually the difference between cases and controls became insignificant. It was therefore proposed that the time between a negative screen and the diagnosis of breast cancer reflected tumor growth rates.[41] A high proportion of free estradiol was associated with rapidly growing tumors which appeared in the first few years of the study.

The levels of steroids in urine and plasma only provide an indirect, and not necessarily accurate indication of those within the breast, and more information may be obtained on the estrogenic environment by performing measurements in breast-derived fluids and tissue. However, with the exception of certain subsets of cyst fluids, there is little to suggest that measurements of estrogen within breast-derived material will help identify women at particular risk of subsequent breast cancer.

From the above considerations, it is difficult to come to any other conclusion than that there is no consistent evidence derived from measurements of endogenous estrogens to support the concept that abnormal levels are associated with increased risk to subsequent breast cancer (although it may be that a hyperestrogenic environment can stimulate occult tumors so that they become quickly apparent). Nevertheless, it is still necessary to consider that (1) any hormone abnormality may be transient, being present only at certain critical inductive times; (2) cumulative exposure is important and small differences in hormone levels, which are difficult to detect, become critical if maintained over a long period of time; and (3) pattern of presentation may be so crucial that cyclicity or periodicity assumes more importance than absolute levels.

2.3 Role of Hormones on the Behavior and Prognosis of Established Tumors

While the majority of data linking steroid hormones to the risk of breast cancer comes from epidemiology and endocrinology studies, the most convincing evidence to link hormones to the behavior of breast cancer stems from direct clinical and laboratory studies, although interesting perspectives can be extrapolated from considerations of epidemiological and endocrinological results.

2.3.1 Clinical Results

Clinical experience shows that breast cancer may regress following a variety of procedures which reduce the trophic effects of steroid hormones. These are discussed in detail in Chapter 5 and include ovariectomy, and use of LHRH analogs in premenopausal women and hypophysectomy, adrenalectomy, antiestrogens and aromatase inhibition in postmenopausal women. In general, these procedures have been associated with benefits in 30 to 40% of patients. Conversely, there is evidence that administration of steroid hormones may cause increased tumor growth. For example, Dao and co-workers showed an increase in thymidine labeling index in skin metastases from breast cancers following treatment with ethinyl estradiol and progesterone.[42] Indeed, these observations have given rise to the concept of using steroids clinically to recruit tumor-cells into cell division before the use of chemotherapy.[43] However, it should be noted that not all data support the concept that increased exposure of breast tumors to steroids produces stimulation of growth. For example, the epidemiological studies investigating the influence of pregnancy (during which estrogen levels are at least tenfold higher than in nonpregnant women) and the administration of hormone replacement therapy (HRT) in peri- and postmenopausal women failed to show that these conditions are consistently associated with accelerated growth of overt tumors or the early appearance of recurrent disease.[44] It may, however, be that the increased levels of steroid hormones during pregnancy and HRT are confounded by other variables.

Although many of the endocrine manipulations involved in the clinical management of breast cancer have effects on several hormones, there is reason to implicate estrogen in particular. Thus, all procedures have in common the ability to reduce tumor levels of estrogen or antagonize its mechanism of action (Table 2). Additionally, the major beneficial effects of such therapy are associated with cancers which possess high-affinity receptors for estrogen.

2.3.2 Laboratory Studies

The particular involvement of steroid hormones in the growth control of breast cancer has been better defined by the use of malignant cell lines. These have generally been derived from patient pleural effusions or ascitic fluids and are maintained either in culture or as xenografts in immunosuppressed animals. Many lines possess specific receptors for steroids and may respond to such hormones (Table 3).[45-51] For example, estrogen-responsive lines include ZR-75-1 and T-47D cells,[46,46] although the best characterized is probably the MCF-7 line.[47] This cell line is ER-positive and its growth in culture is stimulated by physiological concentrations of estrogens, an effect which may be blocked by antiestrogens

Table 2 Major Endocrine Therapies — Effects on Estrogen Biosynthesis/Action

Therapy	Potential Mechanism of Action
Ovariectomy	Ablation of major source of estrogen in premenopausal women
Adrenalectomy	Eliminates major source of androgen precursor in postmenopausal women
Hypophysectomy	Removes pituitary hormones trophic to ovarian and adrenal biosynthesis of estrogenic hormones
LHRH-agonists	Down-regulate LHRH drive for ovarian production of estrogens
Aromatase inhibitors	Prevent biosynthesis of estrogens from androgens
Antiestrogens	Block estrogen at its receptor
Pharmacological doses of steroids (e.g., diethylstilbestrol)	Down-regulate estrogen receptors

Table 3 Breast Cancer Model Systems

Cell Line Model	Histopathology	Receptor Status			Growth Responsive		Ref.
		ER	PR	EGF-R	Estrogen	EGF	
MCF-7	IDC	+	++	+	+	+	45
ZR-75-1	IDC	+	+	+	+	+	46
T47-D	IDC	+	+	+	+	+	47
MDA-MB-134	IDC	+	+	+	NK	NK	48
PCMC 42	AC	+	+	+	+	+	49
CAM-1	MC	+	+	−	+	−	50
MDA-MB-231	AC	−	−	+	−	−	51
Hs 578 T	IDC	−	−	+	−	−	51

Note: ER = Estrogen receptor, PR = progesterone receptor, EGF-R = EGF receptor, IDC = Infiltrating duct carcinoma, AC = Adenocarcinoma, MC = Medullary carcinoma, NK = Not known.

such as tamoxifen. Estrogen is also required for optimal tumor formation and growth in athymic mice.[52] In these systems it can be shown that estrogen may induce a variety of other responses including stimulation of enzymes associated with DNA synthesis, the secretion of peptide growth factors and proteases, and the expression of markers such as the progesterone receptor and heat-shock proteins.[53] In contrast, the growth and response of ER-negative or ER-poor cell lines such as MDA-MB-231 appear independent of steroid hormones and their antagonists.[51]

Although most attention has been directed towards the effects of estrogen on cellular proliferation, other steroid hormones may be influential in certain circumstances. Some cell lines, such as the T-47D, express high levels of progesterone receptors; physiological concentrations of progesterone stimulate these cells to proliferate, whereas pharmacological levels and antiprogestins, may be inhibitory.[54] Effects of progestins may also differ according to the estrogenic environment. Similar complexities relate to the effects of androgens. Thus, MCF-7 cells may be stimulated or inhibited by several types of androgen such as 5α-dihydrotestosterone and Δ5-androstenediol.[55,56] It is unclear, however, whether the effects are primarily mediated via the androgen receptor or are "estrogenic," resulting from interaction via the estrogen receptor.[56]

2.4 Progression

In so far as many of the proteins induced by steroids in hormone-sensitive cancers, such as cathepsin D and TGF-α, may be implicated in the process of tumor spread, it seems reasonable to assume that these are directly involved in the progression of hormone-dependent breast cancer. Additionally, it is tempting to speculate that with the acquisition of hormone independence, such factors are constitutively produced. In support of a steroid involvement in tumor progression, it is worth noting that the *in vitro* invasiveness of ER-positive breast cancer cell lines such as MCF-7 and T-47D can be stimulated by 17 β-estradiol, whereas no effects are seen on the ER-negative MDA-MB-231 line.[57] There is, however, little direct clinical evidence relating estrogen stimulation of hormone sensitive tumors to *in vivo* spread and progression. If estrogen were involved in the metastasis of ER-positive tumors, it might be expected that the proportion of ER-positive cells in axillary node metastases would be higher than in primary tumors because of a selection process. However, most clinical studies show no appreciable difference in receptor content between primary tumors and axillary metastases. It has also been suggested that the poorer prognosis of patients offered surgery in the follicular phase of the cycle is caused by unopposed estrogen promoting metastatic spread.[58,59] However, the basic observation between prognosis and stage of the cycle is still controversial and not underpinned by evidence of metastatic spread.

Paradoxically, the evidence for (1) hormone sensitivity of breast cancer changing with clinical advancement, and (2) the process being accelerated by the use of hormone deprivation therapy are more convincing. Thus, ER-positive cancers tend to be at an earlier stage and better differentiated; incidence of ER-positivity in breast cancer decreases both with increasing size of the primary tumor and advancing clinical stage.[60] In general, more advanced tumors are also less likely to respond to hormonally based treatment. However, the question remains as to whether hormonal autonomy leads to a general aggressive phenotype, or vice-versa, or whether there are simply common underlying mechanisms. The evolution of cell lines which model breast cancer at various stages of progression have been informative in this respect. Thus, studies have utilized (1) wild-type cells with differing inherent hormone sensitivities, (2) sublines derived from continuous growth/passage under selective pressures, and (3) genetically engineered cells. These models can mimic the spectrum of hormone sensitivities which occur clinically and include phenotypes of (1) dependency/sensitivity, (2) independence/sensitivity, and (3) total autonomy. Comparison of these models suggest that progression to hormone independence may occur by multiple routes; for example, estrogen receptor function may be normal, abnormal, or lost.[61] However, the data also strongly implicate altered regulation of specific subsets of estrogen-regulated genes.[62] Certain of these genes, e.g., EGF receptor, laminin receptor, and cathepsin D, appear central to the mechanisms of other processes such as angiogenesis, cell mobility, invasion, and metastasis. It is therefore of interest that subclones of MCF-7 human breast cancers which have developed hormone insensitivity and independence show increased invasiveness and *in vivo* metastatic potential.[62]

2.5 Mechanism of Action

As described in Chapter 1, Section 4.3, the major effects of steroid hormones appear to be mediated through specific intracellular receptors which are predominantly located in the cell nucleus and act as nuclear transcription factors. Consequently, new species of RNAs that did not exist prior to stimulation accumulate.[63] Many encode products involved in the development and progression of hormone-dependent breast cancers and include oncogenes/tumor-suppressor gene products, proteolytic enzymes, and polypeptide growth factors.[53,64] In cell lines and xenografts of breast cancer, exposure to estrogen has been shown to influence the expression of several genes thought to be involved both in the genesis and behavior of breast cancer; these include c-myc,[65] c-erbB2,[66] and p53.[67] Proteases such as plasminogen activator and precursors of lysozymal cathepsins (most notably cathepsin D) are also induced.[68]

Estrogen may influence the production and secretion of polypeptide growth factors and these have been suggested to be the mediators in the mitogenic regulation of breast cancer.[53,64] Such controls may be both distant and local. In terms of the former, estrogens are known to act on the pituitary to stimulate synthesis and secretion of growth factors such as prolactin and IGF-I.[69] These estromedin-like substances may act on the breast in an endocrine manner. Recently, however, the major interest has surrounded the possibility that growth factors are synthesized and secreted locally within the breast, thereby acting in an autocrine or paracrine manner (Figure 2).[70] The evidence is based on the observations that media from estrogen receptor-positive cells grown in the presence of estrogen, after the removal of estrogen, may stimulate (1) proliferation of other breast cancer cells in monolayer culture beyond that of media from cells cultured in the absence of estrogen, and (2) tumor formation *in vivo* in athymic nude mice.[71] Such results suggest that estrogens induce the secretion of growth factors which specifically influence the proliferation of breast cancer cells (Figure 3).[53,64] Transforming growth-factor alfa (TGF-α),[72,73] insulin-like growth factors (IGFs),[53,74,75] and fibroblast growth factors (FGFs) have all been implicated since these may be produced by breast cancer cells and secretion may be influenced by estrogen in hormone-sensitive cell lines. For example, a TGF-α like peptide is induced by estradiol in the ER-positive MCF-7, T-47D and ZR-75-1 cell lines.[73] Similarly, an IGF-I-like growth factor (whose precise identity is still controversial) is induced three- to sixfold following treatment of MCF-7 breast cancer cells with estrogen;[53] conversely secretion is inhibited by antiestrogens.[74] A role for IGF-II in estradiol-regulated breast cancer growth has been suggested as a result of the observation that, estradiol treatment increases IGF-II mRNA in estrogen-dependent T47D human breast cancer cells, whereas in the T61 human breast cancer xenograft (whose growth is inhibited by estrogen), estrogen down-regulates IGF-II.[75] Estradiol may also induce platelet-derived growth-factor (PDGF)[76] and while tumor-cells do not express PDGF receptors,[77] PDGF can stimulate IGF-I production in fibroblasts.[78] Thus, an additional paracrine stimulation for breast cancer cells could operate through the effects of estrogen-induced, tumor-

AUTOCRINE   PARACRINE

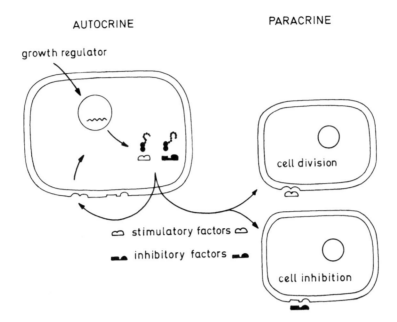

Figure 2 Autocrine and paracrine mechanisms of growth control.

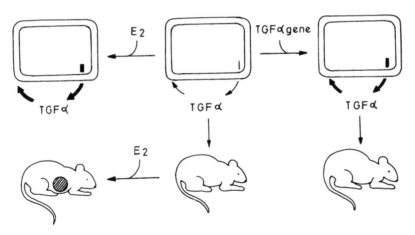

Figure 3 Estrogen regulation of TGF-α growth control in breast cancer. Estrogen induces the secretion of growth factors, including TGF-α, which may stimulate the proliferation of breast cancer cells. Similarly, overexpression of TGF-α may stimulate the growth of breast cancer cells. However, estrogen is essential for the *in vivo* growth of many ER-positive breast cancer models and overexpression of TGF-α does not result in tumor growth *in vivo*.

derived growth factors stimulating stromal cells to secrete IGFs, which in turn stimulate adjacent tumor-cells.

In contrast to the stimulatory effects of growth factors, the secretion of TGF-β in MCF-7 breast cancer cells is inhibited by treatment with estrogens, whereas growth-inhibitory anti-estrogens strongly stimulate secretion.[79] Furthermore, TGF-β and medium from antiestrogen treated MCF-7 cells inhibit the growth of ER-negative cells such as the MDA-MB-231 cancer cell line.[80] Since breast cancers can exist as mixtures of ER-positive and ER-negative tumor-cells, this opens up the possibility that antiestrogens may inhibit ER-negative clones as a consequence of TGF-β induction in ER-positive cells.

From these observations, it is tempting to speculate that the mitogenic effect of estrogen is indirect and mediated by changes in growth-factor secretion. However, the central role of growth factors in developmental processes demands that they are considered in more detail with regard to the development and progression of breast cancer.

3. GROWTH FACTORS

3.1 Growth Factors in the Normal Breast

The development of the normal breast is influenced by growth factors.[53] Thus, for example, IGF concentrations peak at puberty and correlate more closely with breast size than with chronological age.[81]

Specific IGF receptors can also be found in normal breast parenchyma.[81] However, the significance of these observations has been questioned as others have shown that normal breasts can develop in girls with congenital IGF-I deficiency.[82] Breast cells may also be growth stimulated by factors operating through the EGF receptor which is expressed by normal epithelium. TGF-α mRNA and protein and EGF receptors can be detected in proliferating human mammary epithelium, and an anti-EGF receptor antibody has been shown to block TGF-α-induced proliferation of normal human mammary epithelial cells in culture.[83] Both EGF and TGF-α have been reported to increase indices of proliferation in explants of normal breast maintained in culture.[84] Similarly, EGF may influence total phosphorylation and that of specific proteins in particulate fractions derived from normal breast.[85] Other factors, such as MDGF1, may inhibit growth of breast epithelium in culture.[86] Whether such factors are active *in vivo* is still to be defined.

Interestingly, breast cyst fluids may contain remarkably high levels of growth factors such as EGF and IGF-I.[87] Levels of EGF are significantly higher in type I cysts which have been reported to be associated with increased risk of breast cancer.[88] This same population of cysts also contains high concentrations of estrogen conjugates but lower levels of cytokines such as interleukin-6 (IL-6) or IL-8.[89] Although these concentrations of estrogen and EGF would be sufficient to induce proliferation *in vitro*, their relevance *in vivo* remains to be determined.

3.2 Growth Factors in the Behavior and Prognosis of Established Cancers

The concept that the genesis and progression of breast cancer is associated with increasing abnormalities in local growth-factor controls is currently very much in vogue, but is largely based on evidence obtained from model systems. It is therefore important to put these results in the context of observations from clinically derived materials.

Perhaps the most convincing evidence that growth factor control may be abnormal in breast cancer are the findings involving EGF receptor and the related protein p185 encoded by the c-erbB2 oncogene. Both receptors are overexpressed in a minority of breast cancers and are associated with poor prognosis.[90-93] Both at the level of mRNA and protein, TGF-α which binds to the EGF receptor, has been detected in the majority (about 70%) of breast cancers.[94] This finding compares with the detection rates in benign breast lesions of about 30%.[95] In terms of other growth-factor systems, IGF-I receptor expression appear to be higher in breast tumors than in adjacent normal tissue;[96] in breast cancers, binding capacity is exclusively located in the proliferative epithelial compartment.[97] Consequently, it has been suggested that overexpression of IGF-I receptors is associated with malignant transformation within the breast and also that breast cancers may be more sensitive to IGF-like factors. The level of IGF receptors also relates to those of estrogen and progesterone receptors which suggests that the receptors are associated with similar growth controls, all of which confer good prognosis.[96]

Data from model systems are much more expansive but strongly implicate deranged growth-factor mechanisms as primary influences on the behavior of breast cancer cells. Evidence has been put forward to suggest that breast cancer cells (1) synthesize and secrete increased quantities of specific growth factors, (2) overexpress the cognate growth-factor receptors, and (3) induce inappropriate expression of mitogenic factors in adjacent nonmalignant cells and stroma (Figure 4). Therefore, the pathological features of breast cancer may result from the consequent autocrine and paracrine growth regulatory loops as reviewed below.

The most extensively studied autocrine growth-factor loop is that of the TGF-α/EGF receptor system. The majority of breast cancer cell lines, including MCF-7 cells, produces TGF-α-like material, although the molecular weight of the immunologically active peptide is higher than that of classical TGF-α.[53] Antibodies directed against either TGF-α or the EGF receptor produce growth inhibition in MCF-7 cells grown as anchorage-independent colonies or high-density monolayer cultures.[98] Autocrine growth control has also been demonstrated in the MDA-MB-468 cell line which has high TGF-α expression and an amplified level of EGF receptors.[99]

Transfection of the TGF-α gene into the immortalized breast epithelial cell line MCF-10 (which possesses high levels of EGF receptor) produces transformation.[100] However, similar manipulation of MCF-7 breast cancer cells fails to produce substantial changes in growth characteristics both *in vitro*

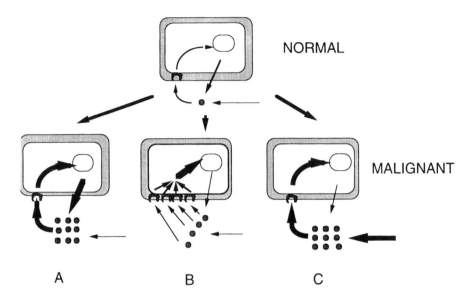

Figure 4 Possible mechanisms of abnormal growth factor regulation found in malignant cells. A. Synthesis and secretion of increased quantities of growth factors; B. Overexpression of growth factor receptors; C. Inappropriate expression of mitogenic growth factors in adjacent nonmalignant cells and stroma.

and *in vivo*.[101] (Figure 3). This may be because MCF-7 cells have markedly lower levels of EGF receptor but equally it may be that production of TGF-α in this model system is already enhanced. There is also evidence to suggest that overexpression of EGF receptors may be particularly critical for dependency on EGF-like factors such as TGF-α. For example, Madsen et al.[102] showed that in immortalized human breast epithelial cells, the acquisition of EGF-independent proliferation was associated with a tenfold increase in EGF receptor expression and only a small increase in TGF-α expression. Co-expression of TGF-α and EGF receptor was insufficient to free cells from a need for exogenous EGF and growth factor independence only occurred when receptor levels were increased. Similarly, certain breast cancer cell lines express both TGF-α and EGF receptor and yet are still dependent on exogenous EGF for growth.[103] There are further considerations in that (1) the EGF receptor is activated by other ligands such as amphiregulin or cripto and may dimerize with related receptors (e.g., c-erbB-2) and (2) *in vivo* there may be paracrine interactions with nonmalignant components.[103,104]

The IGF system may also regulate breast cancer proliferation.[53] Breast cancer cells respond to exogenous IGFs[105] and IGF-I like mRNA has been detected in some of the same cell lines.[106] Rigorous characterization has revealed that this mRNA is not that of IGF-I and it has proven difficult to demonstrate authentic IGF-I in conditioned media from breast cancer cells.[107] However, both normal and benign breast cells and tissues are positive for IGF-I mRNA, and IGF-I is synthesized by stromal fibroblasts.[107] IGF-II mRNA is also evident in stromal fibroblasts derived from breast malignancies.[108] It is thus possible that IGF-dependent breast cancer cells may be driven by a paracrine mechanism involving stromal factors.

In terms of growth stimulation, paracrine loops involving interactive growth factor production by both epithelial and stromal components of breast cancer may be important. Many breast cancers have a pronounced stromal "reaction" and stromal cells may be regulated by growth factors supplied by malignant epithelium. For example, fibroblasts may be stimulated by EGF-like factors, FGFs and PDGF, and endothelial cells by heparin-binding growth factors such as FGFs and endothelins;[104] Many of these factors are produced by malignant cells. In return, nonmalignant cells may supply factors such as IGFs or provide the vasculature for enhanced malignant growth.

The possibility of inhibitory growth controls has largely surrounded the role of TGF-β. TGF-βs, which occur as three isoforms, inhibit the growth of both normal and malignant breast epithelial cells.[109,110] The mRNAs for these TGF-βs are widely but selectively expressed.[110] The concept that the autoinhibitory function of TGF-β disappear with tumor progression will be discussed in more detail in the following section. Other growth inhibitory factors for breast epithelial cells include mammastatin[111] and MDGFI;[86] however, their relevance to growth regulation of breast cancer is still poorly defined.

3.3 Growth Factors and Tumor Progression

Given the known metastatic and angiogenic properties of many growth factors, it is an attractive hypothesis that tumor progression might be associated with enhanced expression of growth factors, and more specifically that hormone regulated production of factors may be supplanted by constitutive synthesis during the transition from hormone dependence to independence (Figure 5).

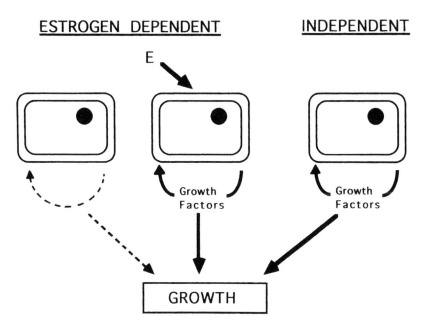

Figure 5 Changes in growth factor regulation in the transition from hormone dependent to independent states. In the estrogen-regulated cell, expression of growth factors is low in the absence of estrogen and upregulated by exposure to the hormone. In the hormone-independent state constitutive expression of growth factors is high and thereby unaffected by exposure to estrogen.

Among growth factors, the link with tumor progression is probably strongest for TGF-β. Paradoxically, for an inhibitor of epithelial cells, the association tends to be positive. Thus, TGF-β immunoreactivity is greater in invasive cancers when compared with *in situ* malignancies.[112] Within invasive cancers, immunohistochemical staining for TGF-β increases with higher tumor stage. Others have shown that TGF-β1 is positively associated with rate of disease progression and this may be independent of stage, node status, and ER status.[113] These observations are consistent with TGF-β's ability to (1) enhance proteolytic activity and metastatic potential in experimental systems, (2) inhibit both humoral and cellular immunity, and (3) modulate angiogenesis and promote neovascularization in both primary and metastatic breast cancers. TGF-β may also be involved in the progression to hormone independence. High constitutive levels of TGF-β activity are reported in media conditioned by hormone-independent breast cancer cells.[114] Loss of estrogen sensitivity in T-47D cells is associated with acquired sensitivity to stimulation by TGF-β1 and with a marked increase in TGF-β mRNA.[115] Upregulation may occur in primary breast cancer and it has been reported that whereas only 13% of ER/PR-positive tumors have high levels of TGF-β1 protein, 78% of ER-positive/PR-negative cancers do so.[115] Unexpectedly high levels of TGF-β1 mRNA have also been found in patients whose tumors were unresponsive to tamoxifen.[116]

Hormone independence may also be associated with changes in other growth factors and their receptors. Thus, high levels of TGF-α have been observed in 65% of breast cancers, with hormone-unresponsive tumors more frequently showing high levels of TGF-α immunoreactivity.[117] Although expression of TGF-α has not been convincingly linked with that of EGF receptors, tumors having EGF receptors are more likely to be resistant to tamoxifen therapy. Many groups have reported an inverse relationship between EGF receptor and ER expression.[118,119]

4. CYTOKINES

Several cytokine families are involved in the growth control of breast tumors and these include the interleukins, the interferons, and tumor necrosis factor. In addition to these cytokines having direct effects on breast cancer epithelial cells and associated infiltrating cells, they can also modulate estrogen and growth factor control. Since the effects of cytokines result primarily in growth inhibition, the therapeutic potential of several of these has been investigated in clinical trials. These topics are explored below.

4.1 Cytokine Expression in Normal Breast Tissues and Fluids

Cytokine profiles have been investigated in normal human mammary gland epithelia and in milk. Normal mammary gland epithelia is reported to express IL-6 and IL-8 but not IL-1 or any of the forms of interferon (α, β or γ).[120,121] In a study investigating the expression of TNF-α and its two receptors (TNF-R p55 and TNF-p75) in normal breast and nonmalignant breast tissue adjacent to tumors, TNF-α and TNF-R p75 were undetectable by immunohistochemistry while TNF-R p55 was expressed in occasional stromal cells.[122]

Human breast milk has important immunoprotective and immunosuppressive functions for an infant and cytokines are likely to contribute to these properties. Substantial amounts of IL-6, IL-8 and interferon-γ are present in milk, while levels of TNF-α, interferon α, and β and IL-1 are undetectable.[121] Concentrations of interferon-γ are higher in milk compared to blood levels, but other measured cytokines are lower.[123]

Analysis of human colostrum revealed significant levels of IL-1β while concentrations of IL-α and TNF were negligible.[124] Because IL-1 increases resistance to infection, the presence of this cytokine has been suggested to represent a beneficial aspect of breast feeding. The concentration of IL-6 is also elevated in colostrum relative to milk taken at a month after parturition.[125]

4.2 Cytokine Expression and Function in Malignant Breast Tissues

Expression patterns of certain cytokines are changed in malignant breast when compared to normal breast epithelia. Hence, TNF-α, which is reported absent in normal breast epithelium, is expressed focally in 50% of tumors (primarily in macrophage-like cells in the stroma), and both TNF-R p55 and TNF-R p75 are expressed in most tumors,[122] while expression of IL-6, which is at a high level in normal breast tissues and fluids, is abolished.[121]

The effects of cytokines on the growth of breast cancers are likely to be complex, since a specific cytokine may have one effect on epithelial cells but the opposite effect on infiltrating lymphocytes. Of the interleukins, IL-1,[126] IL-4,[127] and IL-6[128] have demonstrated direct inhibitory effects on cultured breast cancer cells, while IL-2 has been reported to be stimulatory.[129] TNF and the interferons are also growth inhibitory to breast cancer cells.[130,131] In addition to their effects on growth, both IL-1 and IL-6 increase motility of breast cancer cells.[132]

4.3 Cytokine Regulation of Estrogen and Growth Factors

In many breast tumors, the local synthesis of estrogen from androgens provides the major source of the hormone. Synthesis involves conversion of androstenedione to estrone (E_1), a reaction catalyzed by aromatase, and conversion of E_1, to produce the more potent estrogen 17β-estradiol (E_2,), catalyzed by 17β-hydroxysteroid dehydrogenase (E_2DH). Several cytokines, including TNF-α, IL-1, and IL-6 stimulate both breast fibroblast aromatase activity and breast carcinoma E_2DH activity.[133-135] A role for IL-6 in regulating aromatase activity in breast tissues is suggested by the correlation between IL-6 levels and aromatase activity in breast adipose tissue adjacent to breast tumors. IL-3 reduces E_2DH activity and as this cytokine is found in male fibroblasts but not female breast fibroblasts, it has been speculated to be a protective factor in males.[136] It should be noted though that while the cytokines IL-1 and IL-6 enhance production of estrogens via modulation of these enzymes, they inhibit breast cancer cell growth directly and down-regulate ER concentrations.[137,138] TNF-α also antagonizes estrogen-stimulation and down-regulates ER content.[139]

Cytokines may not only regulate estrogen synthesis but also the production and function of growth factors; for example, IL-1 increases TGF-β secretion and its inhibitory growth effects may be mediated by this factor.[138] TNF-α has been shown to block the growth-stimulatory effects of EGF, bFGF, and IGF-1 on a breast cancer cell line (T47D) while the same growth factors can partially antagonize the growth inhibition induced by TNF-α.[139,140] These data suggest that the growth of breast cancer cells may be at least partially regulated by the opposing effects of cytokines and growth factors.

4.4 Clinical Studies with Cytokines

The therapeutic potential of several cytokines has been investigated in clinical studies and a number of these have included patients with breast cancer. The rationale behind these therapies has been based both on direct growth inhibition of epithelial cells (e.g., studies involving TNF-α or IFN-α) and stimulation of lymphocyte cytotoxicity (e.g., the use of IL-2).

Although some cytokines have proven disappointing, for example, TNF, which produced no responses in a study of 19 breast cancer patients,[141] others have shown some evidence of growth inhibition in this disease. In a phase I study of IL-2, a partial response in a breast cancer patient was documented.[142] Subsequently, IL-2 was studied in combination with other cytokines, and combination biotherapy of IL-2 and interferon-α has produced responses in 4 of 23 breast cancer patients in one trial.[143] Interferon-α has also been investigated in combination with tamoxifen; of seven patients with advanced, heavily pretreated breast cancer, four responded to this combination. Furthermore, it was noted in this study that interferon treatment consistently increased ER expression.[144] These data suggest that further studies with cytokines are warranted.

REFERENCES

1. **Laron, Z., Kanli, R., and Pertzelan, A.,** Clinical evidence on the role of oestrogens in the development of the breasts, *Proc. R. Soc. Edinburgh,* 95B, 13, 1988.
2. **Howell, A.,** Clinical evidence for the involvement of oestrogens in the development and progression of breast cancer, *Proc. R. Soc. Edinburgh,* 95B, 49, 1988.
3. **Husseby, R.A. and Thomas, L.B.,** Histological and histochemical alterations in the normal breast tissues of patients with advanced breast cancer being treated with estrogenic hormones, *Cancer,* 7, 54, 1954.
4. **Stoll, B.A.,** Breast cancer risk in Japanese women with special reference to the growth hormone insulin-like growth factor axis, *Jpn. J. Clin. Oncol.,* 22, 1, 1992.
5. **Bidlingmaier, F. and Knorr, D.,** Oestrogens — Physiological and clinical aspects, *Pediatric and Adolescent Endocrinology,* Vol. 4, Karger, Basle, 6, 1978.
6. **Going, J.J., Anderson, T.J., Battersby, S., and MacIntyre, C.,** Proliferative and secretory activity in human breast during natural and artificial menstrual cycles, *Am. J. Pathol.,* 130, 193, 1988.
7. **Anderson, T.J., Battersby, S., King, R.J.B., McPherson, K., and Going, J.J.,** Oral contraceptive use influences resting breast proliferation, *Hum. Pathol.,* 20, 1139, 1989.
8. **Williams, G., Anderson, E., Howell, A., Watson, R., Coyne, J., Roberts, S.A., and Potten, C.S.,** Oral contraceptive (OCP) use increases proliferation and decreases oestrogen receptor content of epithelial cells in the normal human breast, *Int. J. Cancer,* 48, 206, 1991.
9. **Ferenczy, A., Bertrand, G., and Gelfand, M.M.,** Proliferation kinetics of human endometrium during the normal menstrual cycle, *Am. J. Obstet. Gynecol.,* 133, 859, 1988.
10. **Anderson, T.J. and Battersby, S.,** The involvement of oestrogen in the development and function of the normal breast: histological evidence, *Proc. R. Soc. Edinburgh,* 95B, 23, 1988.
11. **King, R.J.B.,** A discussion of the roles of oestrogen and progestin in human mammary carcinogenesis, *J. Steroid Biochem. Mol. Biol.,* 39 (5B), 811, 1991.
12. **Chalbos, D., Escot, C., Joyeux, C., Tissot-Carayon, M.J., Pages, A., and Rochefort, H.,** Expression of the progestin-induced fatty acid synthetase in benign mastopathies and breast cancer as measured by RNA in situ hybridization, *J. Natl. Cancer Inst.,* 82, 602, 1990.
13. **Boyle, P.,** Epidemiology of breast cancer, *Bailliere's Clin. Oncol.,* 2, 1, 1988.
14. **MacMahon, B., Cole, P., and Brown, J.,** Etiology of human breast cancer: a review, *J. Natl. Cancer Inst.,* 50, 21, 1973.
15. **Trichopoulos, D., MacMahon, B., and Cole, P.,** Menopause and breast cancer risk, *J. Natl. Cancer Inst.,* 48, 605, 1972.
16. **Pike, M.C., Spicer, D.V., Dahmoush, L., and Press, M.F.,** Estrogens, progestogens, normal breast cell proliferation, and breast cancer risk, *Epidemiol. Rev.,* 15, 17, 1993.
17. **de Waard, F. and Baanders-van Halewijn, E.A.,** A prospective study in general practice on breast cancer risk in postmenopausal women, *Int. J. Cancer,* 14, 153, 1974.
18. **Vermeulen, A. and Verdonck, L.,** Sex hormone concentrations in post-menopausal women. Relation to obesity, fat mass, age and years post-menopause, *Clin. Endocrinol.,* 9, 59, 1978.
19. **James, V.H.T., Reed, M.J., and Folkerd, E.J.,** Studies of estrogen metabolism in postmenopausal women with cancer, *J. Steroid Biochem.,* 15, 235, 1981.
20. **Barber, H.R.K.,** Cancers of breast, uterus and ovary, in *Risk Factors and Multiple Cancer,* Stoll, B.A., Ed., John Wiley & Sons, New York, 1984, 315.
21. **De Waard, F., Cornelis, J.P., and Aoki, K.,** Breast cancer incidence according to weight and height in two cities in the Netherlands and in Aichi Prefecture, Japan, *Cancer,* 40, 1269, 1977.

22. **Pike, M.C., Krailo, M.D., Henderson, B.E., Casagrande, J.T., and Hoel, D.G.,** "Hormonal" risk factors, "breast tissue age" and the age-incidence of breast cancer, *Nature*, 303, 767, 1983.

23. **Colton, T., Greenberg, R., Noller, K., Resseguie, L., Van-Bennekom, C., Heeren, T., and Zhang, Y.,** Breast cancer in mothers prescribed diethylstilbestrol in pregnancy, *J. Am. Med. Assoc.*, 269, 2069, 1993.

24. **Weiss, N.S., Ure, C.L., Ballard, J.H., Williams, A.R., and Daling, J.R.,** Decreased risk of fractures of the hip and lower forearm with postmenopausal use of estrogen, *N. Engl. J. Med.*, 303, 1195, 1980.

25. **Henrich, J.B.,** The postmenopausal estrogen/breast cancer controversy, *J. Am. Med. Assoc.*, 268, 1900, 1992.

26. **Bluming, A.Z.,** Hormone replacement therapy: benefits and risks for the general postmenopausal female population and for women with a history of previously treated breast cancer, *Semin. Oncol.*, 20, 662, 1993.

27. **Armstrong, B.K.,** Estrogen therapy after the menopause: Boon or bane? *Med. J. Aust.*, 148, 213, 1988.

28. **Dupont, W.D. and Page, D.L.,** Menopausal estrogen replacement therapy and breast cancer, *Arch. Intern. Med.*, 151, 67, 1991.

29. **Steinberg, K., Thacker, S.B., Smith, J., Stroup, D.F., Zack, M.M., and Flanders, W.D.,** A meta-analysis of the effect of estrogen replacement therapy on the risk of breast cancer, *J. Am. Med. Assoc.*, 265, 1985, 1991.

30. **Colditz, G.A., Egan, K.M., and Stampfer, M.J.,** Hormone replacement therapy and risk of breast cancer: Results from epidemiologic studies, *Am. J. Obstet. Gynecol.*, 168, 1473, 1993.

31. **Pike, M.C., Henderson, B.E., Casagrande, J.T., Rosario, I., and Gray, G.E.,** Oral contraceptive use and early abortion as risk factors for breast cancer in young women, *Br. J. Cancer*, 43, 72, 1981.

32. **Pike, M.C., Henderson, B.E., Krailo, M.D., Duke, A., and Roy, S.,** Breast cancer in young women and use of oral contraceptives: possible modifying effect of formulation and age at use, *Lancet*, 2, 926, 1983.

33. **McPherson, K., Vessey, M.P., Neil, A., Doll, R., Jones, L., and Roberts, M.,** Early oral contraceptive use and breast cancer: results of another case-control study, *Br. J. Cancer*, 56, 653, 1987.

34. **Chilvers, C.E.D. and Smith, S.J.,** The effect of patterns of oral contraceptive use on breast cancer risk in young women, *Br. J. Cancer*, 67, 922, 1994.

35. **Fentiman, I.S.,** The endocrine dimension, in *Prevention of Breast Cancer*, RG Landes Company, Austin, 1993.

36. **Bulbrook, R.D., Hayward, J.L., and Spicer, C.C.,** Relation between urinary androgen and corticosteroid excretion and subsequent breast cancer, *Lancet*, 2, 395, 1971.

37. **De Waard, F. and Banders-van Halewun, E.A.,** A prospective study in general practice on breast cancer risk in postmenopausal women, *Int. J. Cancer*, 14, 153, 1974.

38. **Bulbrook, R.D., Hayward, J.L., Wang, D.Y., Thomas, B.S., Clark, G.M., Allen, D.S., and Moore, J.W.,** Identification of women with a high risk of breast cancer, *Breast Cancer Res. Treat.*, 7(Suppl.), 5, 1986.

39. **Bulbrook, R.D. and Thomas, B.S.,** Hormones are ambiguous risk factors for breast cancer, *Acta Pathol. Scand.*, 28, 841, 1989.

40. **Bulbrook, R.D., Moore, J.W., Clark, G.M.G., Wang, D.Y., Millis, R.R., and Hayward, J.L.,** Relation between risk of breast cancer and bioavailability of estradion in blood: prospective study in Guernsey, in *Endocrinology of the Breast: Basic and Clinical Aspects*, Angelli, A., Bradlow, H.L., and Dogliotto, L., Eds., *Ann. N.Y. Acad. Sci.*, 464, 373, 1986.

41. **Bulbrook, R.D., Moore, J.W., Allen, B.S., Thomas, B.S., Gravelle, I.H., Hayward, J.L., and Wang, D.Y.,** Sex-hormone-binding globulin and the natural history of breast cancer, *Ann. N.Y. Acad. Sci.*, 538, 248, 1988.

42. **Dao, T.L., Sinha, D.K., Nemoto, T., and Patel, J.,** Effect of estrogen and progesterone on cellular replication of human breast cancers, *Cancer Res.*, 42, 359, 1982.

43. **Conte, P., Pronento, P., and Rubagatti, A.,** Conventional versus cytokinetic polychemotherapy with estrogenic recruitment in metastatic breast cancer: results of a randomised co-operative trial, *J. Clin. Oncol.*, 53, 339, 1987.

44. **Bluming, A.Z.,** Hormone replacement therapy: Benefits and risks for the general postmenopausal female population and for women with a history of previously treated breast cancer, *Semin. Oncol.*, 20, 662, 1993.

45. **Soule, H.D., Vasquez, J., Long, A., Albert, S., and Brennan, M.,** Human cell line from a pleural effusion derived from a human breast carcinoma, *J. Natl. Cancer Inst.*, 51, 1409, 1973.

46. **Engle, L.W., Young, N.A., Tralka, T.S., Lippman, M.E., O'Brien, S.J., and Joyce, M.J.,** Establishment and characterisation of three new continuous cell lines derived from human breast carcinomas, *Cancer Res.*, 38, 3352, 1978.

47. **Keydar, I., Chen, L., Karyby, S., Weiss, F.R., DeLarea, J., Radu, M., Chaitcik, S., and Brenner, H.J.,** Establishment and characterisation of a cell line of human breast carcinoma origin, *Eur. J. Cancer*, 15, 659, 1979.

48. **Whitehead, R.H., Quirk, S.J., Vitali, A.A., Funder, J.W., Sutherland, R.L., and Murphy, L.C.,** A new human breast carcinoma cell line (PMC42) with stem cell characteristics. III. Hormone receptor status and responsiveness, *J. Natl. Cancer Inst.*, 73, 643, 1984.

49. **Leung, B.S., Quereshi, S., and Leung, J.S.,** Response to estrogen by the human mammary carcinoma cell line CAMA-1, *Cancer Res.*, 42, 5060, 1982.

50. **Reiner, G.C.A. and Katzenellenbogen, B.S.,** Characterization of estrogen and in MDA-MB-134 human breast cancer cells, *Cancer Res.*, 46, 1124, 1986.

51. **Engle, L.W. and Young, N.W.,** Human breast carcinoma cells in continuous culture: a review, *Cancer Res.*, 38, 4327, 1978.

52. **Soule, H.D. and McGrath, C.M.,** Estrogen responsive proliferation of clonal human breast carcinoma cells in athymic mice, *Cancer Lett.*, 10, 177, 1980.

53. **Lippman, M.E. and Dickson, R.B.,** Growth control of normal and malignant breast epithelium, *Proc. R. Soc. Edinburgh*, 95B, 89, 1989.

54. **Clarke, C.L. and Sutherland, R.L.,** Progestin regulation of cellular proleration, in *Endocrine Aspects of Cancer*, Horwitz, K.B., Ed., Endocrine Review Monographs, The Endocrine Society Press, Bethesda, 1993, 96.

55. **Birrell, S.N., Bentel, J.M., Hickey, T.E., Ricciardelli, C., Weger, M.A., Horsfall, D.J., and Tilley, W.D.,** Androgens induce divergent proliferative responses in human breast cancer cell lines, *J. Steroid Biochem. Mol. Biol.*, 52, 459, 1995.

56. **Zava, D.T. and McGuire, W.L.,** Human breast cancer: androgen action mediated by estrogen receptor, *Science*, 199, 787, 1978.

57. **Clarke, R., Brunner, N., Thompson, E.W., Glanz, P., Katz, D., Dickson, R.B., and Lippman, M.E.,** The inter-relationships between ovarian-independent growth, anti-estrogen resistance and invasiveness in the malignant progression of human breast cancer, *J. Endocrinol.*, 122, 331, 1989.

58. **Badwe, R.A., Gregory, W.M., Chaudrey, M.A., Richards, M.A., Bentley, A.E., Rubens, R.D., and Fentiman, I.S.,** Timing of surgery during the menstrual cycle and survival of premenopausal women with operable breast cancer, *Lancet*, 337, 1261, 1991.

59. **Badwe, R.A., Richards, M.A., Fentiman, I.S., Gregory, W., Saad, Z., Chaudary, M.A., Bentley, A., and Ruben, R.D.,** Surgical procedures, menstrual cycle phase and prognosis in operable breast cancer, *Lancet*, 338, 815, 1991.

60. **Blanco, G., Alavaikho, M., and Ojalo, A.,** Estrogen and progesterone receptors in breast cancers: relationships to tumor histopathology and survival of patients, *Anticancer Res.*, 4, 383, 1984.

61. **King, R.J.B.,** Progression from steroid sensitive to insensitive state in breast tumours, *Cancer Surv.*, 14, 131, 1992.

62. **Clarke, R., Skaar, T., Baumann, K., Leonessa, F., James, M., Lippman, J., Thompson, E.W., Freter, C., and Brunner, N.,** Hormonal carcinogenesis in breast cancer: cellular and molecular studies of malignant progression, *Breast Cancer Res. Treat.*, 31, 237, 1994.

63. **May, F.E. and Westley, B.R.,** Identification and characterisation of oestrogen regulated RNAs in human breast cancer cells, *J. Biol. Chem.*, 263, 12901, 1988.

64. **Dixon, R.B. and Lippmam, M.E.,** Estrogen regulation of growth and polypeptide growth factor secretion in human breast carcinoma, *Endocr. Rev.*, 8, 29, 1986.

65. **Dubik, D., Dembriniki, T.C., and Shiu, R.P.,** Stimulation of c-myc oncogene expression associated with estrogen-induced proliferation of human breast cancer cells, *Cancer Res.*, 47, 6517, 1987.

66. **Dati, C., Antoniotti, S., Taverna, D., Perroteau, I., and De Bortoli, M.,** Inhibition of c-erbB-2 oncogene expression by estrogens in human breast cancer cells, *Oncogene*, 5, 1001, 1990.

67. **Thompson, A.M., Steel, C.M., Foster, M.E., Kerr, D., Paterson, D., Deane, D., Hawkins, R.A., Carter, D.C., and Evans, H.J.,** Gene expression in oestrogen-dependent human breast cancer xenograft tumours, *Br. J. Cancer*, 62, 78, 1990.

68. **Morisset, M., Capony, F., and Rochefort, H.,** Processing and estrogen regulation of the 52 kDa protein inside MCF7 breast cancer cells, *Endocrinology*, 119, 2773, 1986.

69. **Sirbasku, D.A.,** Estrogen-induction of growth factors specific for hormone responsive mammary, pituitary and kidney tumor cells, *Proc. Natl. Acad. Sci. U.S.A.*, 75, 3786, 1978.

70. **Lippman, M.E., Dickson, R.B., Bates, S., Knabbe, C., Huff, K., Swain, S., McManaway, M., Bronzert, D., Kasid, A., and Gelmann, E.P.,** Autocrine and paracrine growth regulation of human breast cancer, *Breast Cancer Res. Treat.*, 7, 59, 1986.

71. **Dickson, R.B., McManaway, M., and Lippman, M.E.,** Estrogen induced factors of breast cancer cells partially replace estrogen to promote tumor growth, *Science*, 232, 1540, 1986.

72. **Bates, S.E., McManaway, M.E., and Lippman, M.E.,** Characterization of estrogen responsive transforming activity in human breast cancer cell lines, *Cancer Res.*, 46, 1707, 1986.

73. **Salomon, D.S., Kidwell, W.R., Kim, N., Ciardiello, F., Bates, S.E., Valverius, E., Lippman, M.E., Dickson, R.B., and Stampfer, M.,** Modulation by estrogen and growth factor of transforming growth factor alpha (TGF-α) expression in normal and malignant human mammary epithelial cells, *Recent Cancer Res.*, 113, 57, 1989.

74. **Huff, K.K., Knabbe, C., Lindsey, R., Kaufman, D., Bronzert, D., Lippman, M.E., and Dickson, R.B.,** Multi-hormonal regulation of insulin-like growth factor-I-related protein in MCF-7 human breast cancer cells, *Mol. Endicrinol.*, 2, 200, 1988.

75. **Brunner, N., Yee, D., and Kern, F.G.,** Effect of endocrine therapy on growth of T61 human breast cancer xenografts is directly correlated to a specific down-regulation of insulin-like growth factor-II (IGF-II), *Eur. J. Cancer*, 29, 562, 1993.

76. **Bronzert, D.A., Pantazis, P., Antoniades, H.N., Kasid, A., Davidson, N., Dickson, R.B., and Lippman, M.E.,** Synthesis and secretion of PDGF-like growth factor by human breast cancer cell lines, *Proc. Natl. Acad. Sci. U.S.A.*, 84, 5763, 1987.

77. **Yarden, Y., Escobedo, J.A., Kuang, W.J., Yang-Feng, T.L., Daniel, T.O., Tremble, P.M., Chen, E.Y., Ando, M.E., Harkins, R.N., and Francke, U.,** Structure of the receptor for platelet-derived growth factor helps define a family of closely related growth factors, *Nature*, 323, 226, 1986.

78. **Clemmons, D.R. and Shaw, D.S.,** Variables controlling somatomedin production by cultured human fibroblasts, *J. Cell. Physiol.*, 115, 137, 1983.

79. **Knabbe, C., Lippman, M.E., and Wakefield, L.M.,** Evidence that TGF-beta is a hormonally regulated growth factor in human breast cancer cells, *Cell*, 48, 417, 1987.

80. **Arteaga, C.L., Tandon, A.K., Von Hoff, D.D., and Osborne, C.K.,** Transforming growth factor β: potential autocrine growth inhibitor of estrogen receptor negative human breast cancer cells, *Cancer Res.*, 48, 3898, 1988.

81. **Rosenfield, R.J., Furlanetto, R., and Bock, D.,** Relationship of somatomedin C concentration to pubertal changes, *Pediatrics*, 103, 723, 1983.

82. **Laron, Z., Kanli, R., and Pertzelan, A.,** Clinical evidence on the role of oestrogens in the development of the breasts, *Proc. R. Soc. Edin.*, 95B, 13, 1988.

83. **Bates, S.E., Valverius, E.M., Ennis, B.W., Bronzert, D.A., Sheridan, J.P., Stampfer, M.R., Mendelsohn, S., Lippman, M.E., and Dickson, R.B.,** Expression of the transforming growth α/epidermal growth factor receptor pathway in normal human breast epithelial cells, *Endocrinology*, 126, 596, 1990.

84. **Perusinghe, N.P., Monaghan, P., O'Hare, M.J., Ashley, S., and Gusterson, B.A.,** Effects of growth factors on proliferation of basal and luminal cells in human breast epithelial explants in serum-free culture, *In Vitro Cell. Dev. Biol.*, 28A, 90, 1992.

85. **Anderson, T.J. and Miller, W.R.,** Morphological and biological observations relating to the development and progression of breast cancer, in *Mammary Tumorigenesis and Malignant Progression*, Dickson, R. and Lippman, M., Eds., Kluwer, 1994, 3.

86. **Bano, M., Worland, P., Kidwell, W.R., Lippman, M.E., and Dickson, R.B.,** Receptor-induced phosphorylation by mammary-derived growth factor 1 in mammary epithelial cell lines, *J. Biol. Chem.*, 267, 10389, 1991.

87. **Ness, J.C., Sedghinasab, M., Moe, R.E., and Tapper, D.,** Identification of multiple proliferative growth factors in breast cyst fluid, *Am. J. Surg.*, 166, 237, 1993.

88. **Smith, K., Miller, W.R., Fennelly, J.A., Matthews, J.N., Scott, W.N., and Harris, A.L.,** Quantification of epidermal growth factor in human breast cyst fluids: correlation with dehydroepiandrosterone-sulphate and electrolyte concentrations, *Int. J. Cancer*, 44, 229, 1989.

89. **Lai, L.C., Kadory, S., Siraj, A.K., and Lennard, T.W.,** Oncostatin M, interleukin 2, interleukin 6 and interleukin 8 in breast cyst fluid, *Int. J. Cancer*, 59, 369, 1994.

90. **Klijn, J.G.M., Berns, P.M.J.J., Schmitz, P.I.M., and Foekens, J.A.,** The clinical significance of epidermal growth factor receptor (EGF-R) in human breast cancer: a review on 5232 patients, *Endocr. Rev.*, 13, 3, 1992.

91. **Fox, S.B., Smith, K., Hollyer, J., Greenall, M., Hastrich, D., and Harris, A.L.,** The epidermal growth factor receptor as a prognostic marker: results of 370 patients and a review of 3009 patients, *Breast Cancer Res. Treat.*, 29, 41, 1994.

92. **Slamon, D.J., Clark, G.M., Wong, S.G., Levin, W.J., Ullrich, A., and McGuire, W.L.,** Human breast cancer: correlation of relapse and survival with amplification of the HER-2/neu oncogene, *Science*, 235, 177, 1987.

93. **Gusterson, B.A., Gelber, R.D., Goldhirsch, A., Price, K.N., Save-Soderborgh, J., Anbuzhagen, R., Styles, J., Rudenstam, C.M., Golouh, R., Reed, R., Martinez-Tello, F., Tiltman, A., Torhorst, J., Grigolato, P., Bettelheim, R., Neville, A.M., Burki, K., Castaglione, M., Collins, J., Lindtner, J., and Senn, H.-J.,** Prognostic importance of c-erbB-2 expression in breast cancer, *J. Clin. Oncol.*, 10, 1049, 1992.

94. **Gregory, H., Thomas, C.E., Willshire, I.R., Young, J.A., Anderson, H., Baildan, A., and Howell, A.,** Epidermal and transforming growth factor α in patients with breast tumours, *Br. J. Cancer*, 59, 605, 1989.

95. **Travers, M.R., Barrett-Lee, P.J., Berger, U., Luqmani, Y.A., Gazet, J.-C., Powles, T.J., and Coombes, R.C.,** Growth factor expression in normal, benign and malignant breast tissue, *Br. Med. J.*, 296, 1621, 1988.

96. **Pekonen, F., Partanen, S., Makinen, T., and Rutanen, E.M.,** Receptors for epidermal growth factor and insulin like grow factor I and their relation to steroid receptors in human breast cancer, *Cancer Res.*, 48, 1343, 1988.

97. **Jammes, H., Peyrat, J.P., Ban, E., Vilain, M.O., Haour, F., Djiane, J., and Bonneterre, J.,** Insulin-like growth factor receptors in human breast tumour — localization and quantification by histo-autoradiographic analysis, *Br. J. Cancer*, 66, 248, 1992.

98. **Bates, S.E., Davidson, N.E., Valverius, E., Freter, C., Dickson, R.B., Tam, J.D., Kudlow, J.E., Lippman, M.E., and Salomon, D.S.,** Expression of transforming growth factor alpha and its messenger ribonucleic acid in human breast cancer: its regulation by estrogen and its possible functional significance, *Mol. Endocrinol.*, 2, 543, 1988.

99. **Ennis, B.W., Valverius, E.M., Lippman, M.E., Bellot, F., Kris, R., Schlessinger, J., Masui, H., Goldberg, A., Mendelsohn, J., and Dickson, R.B.,** Anti-EGF receptor antibodies inhibit growth of MDA-MB-468 breast cancer cells, *Mol. Endocrinol.*, 3, 1830, 1989.

100. **Ciardiello, F., McGready, M., Kim, N., Basalo, F., Hynes, N., Langton, B.C., Yokozaki, H., Sucki, T., Elliot, J.W., Masui, H., Mendelsohn, J., Soule, H., Russo, J., and Salomon, D.,** TGF-α expression is enhanced in human mammary epithelial cells transformed by an activated c-Ha-ras but not the c-neu protooncogene and overexpression of the TGF-α cDNA leads to transformation, *Cell. Growth Differen.*, 1, 407, 1990.

101. **Clarke, R., Brunner, N., Katz, D., Glenz, P., Dickson, R.B., Lippman, M.E., and Kern, F.,** The effects of a constitutive production of TGF-α on the growth of MCF-7 human breast cancer cells *in vitro* and *in vivo*, *Mol. Endocrinol.*, 3, 372, 1989.

102. **Madsen, M.W., Lykkesfeldt, A.E., Laursen, I., Nielsen, K.V., and Brian, P.,** Altered gene expression of c-myc, epidermal growth factor receptor, transforming growth factor α, and c-erb-B2 in an immortalized human breast epithelial cell line, HMT-3522, is associated with decreased growth factor requirements, *Cancer Res.*, 52, 1210, 1992.

103. **Dickson, R.B. and Lippman, M.E.,** Growth factors in breast cancer, *Endocr. Rev.*, 16, 559, 1995.
104. **Ethier, S.P.,** Growth factor synthesis and human breast cancer progression, *J. Natl. Cancer Inst.*, 87, 964, 1995.
105. **Furlanetto, R.W. and DiCarlo, J.N.,** Somatomedin C receptors and growth effects in human breast cells maintained in long-term culture, *Cancer Res.*, 44, 2122, 1984.
106. **Baxter, R.C., Maitland, J.E., and Raisur, R.L.,** High molecular weight somatomedin C (IGF-I) from T47D human mammary carcinoma cells: immunoreactivity and bioactivity, in *Insulin-Like Growth-Factors/Somatomedins*, Spencer, E.M., Ed., Walter de Gruyter Co., Berlin, 1983, 615.
107. **Yee, D., Paik, S., Lebovic, G.S., Marcus, R.R., Favoni, R.E., Cullen, K.J., Lippman, M.E., and Rosen, N.,** Analysis of insulin-like growth factor-I gene expression in malignancy: evidence for a paracrine role in human breast cancer, *Mol. Endocrinol.*, 3, 509, 1989.
108. **Yee, D., Cullen, K.J., Paik, S., Perdue, J.F., Hampton, B., Schwartz, A., Lippman, M.E., and Rosen, N.,** Insulin-like growth factor II mRNA expression in human breast cancer, *Cancer Res.*, 48, 6691, 1988.
109. **Wakefield, L.M., Colletta, A.A., Maccune, B.K., and Sporn, M.B.,** Roles for transforming growth factors β in the genesis, prevention and treatment of breast cancer, in *Genes, Oncogenes and Hormones*, Dickson, R.B. and Lippman, M.E., Eds., Kluwer, Boston, 97, 1992.
110. **Knabbe, C. and Zugmaier, G.,** Expression of transforming growth factor-β in breast cancer, *Endocr.-Related Cancers*, 1, 5, 1994.
111. **Ervin, P.R., Kaminski, M., Cody, R.L., and Wicha, M.S.,** Production of mammostatin, a tissue-specific growth inhibitor, by normal human mammary cells, *Science*, 244, 1585, 1989.
112. **Walker, R.A. and Dearing, S.J.,** Transforming growth factor beta 1 in ductal carcinoma in situ and invasive carcinomas of the breast, *Eur. J. Cancer*, 28, 641, 1992.
113. **Goesch, S.M., Memoli, V.A., Stukel, T.A., Gold, L.I., and Arrick, B.A.,** Immunohistochemical staining for transforming growth factor β1 associates with disease progression in human breast cancer, *Cancer Res.*, 52, 6949, 1992.
114. **Dickson, R.B., Bates, S.E., McManaway, M.E., and Lippman, M.E.,** Characterisation of estrogen responsive transforming activity in human breast cancer cell lines, *Cancer Res.*, 46, 1707, 1986.
115. **King, R.J.B., Wang, D.Y., Daly, R.J., and Darbre, P.D.,** Approaches to studying the role of growth factors in the progression of breast tumours from the steroid sensitive to insensitive state, *J. Steroid Biochem.*, 34, 133, 1989.
116. **Thompson, A.M., Kerr, D.J., and Steel, C.M.,** Transforming growth factor beta 1 is implicated in the failure of tamoxifen therapy in human breast cancer, *Br. J. Cancer*, 63, 609, 1991.
117. **Nicholson, R.I., McClelland, R.A., Gee, J.M.W., Manning, D.L., Cannon, P., Robertson, J.F.R., Ellis, I.O., and Blamey, R.W.,** Transforming growth factor-α and endocrine sensitivity in breast cancer, *Cancer Res.*, 54, 1684, 1994.
118. **Fitzpatrick, S.L., Brightwell, J., Wittliff, J., Barrows, G.H., and Schultz, G.S.,** Epidermal growth factor binding by breast tumor biopsies and relationship to estrogen and progestin receptor levels, *Cancer Res.*, 44, 3448, 1984.
119. **Sainsbury, J.R.C., Farndon, J.R., Sherbert, G.V., and Harris, A.L.,** Epidermal growth factor receptors and oestrogen receptors in human breast cancers, *Lancet*, 1, 364, 1985.
120. **Palkowetz, K.H., Royer, C.L., Garofalo, R., Rudloff, H.E., Schmalstieg, F.C., and Goldman, A.S.,** Production of interleukin-6 and interleukin-8 by human mammary gland epithelial cells, *J. Reprod. Immunol.*, 26, 57, 1994.
121. **Basolo, F., Conaldi, P.G., Fiore, L., Calvo, S., and Toniolo, A.,** Normal breast epithelial cells produce interleukins 6 and 8 together with tumor necrosis factor: defective IL6 expression in mammary carcinoma, *Int. J. Cancer*, 55, 926, 1993.
122. **Pusztai, L., Clover, L.M., Cooper, K., Starkey, P.M., Lewis, C.E., and McGee, J.O.,** Expression of tumour necrosis factor alpha and its receptors in carcinoma of the breast, *Br. J. Cancer*, 70, 289, 1994.
123. **Eglinton, B.A., Roberton, D.M., and Cummins, A.G.,** Phenotype of T cells, their soluble receptor levels, and cytokine profile of human breast milk, *Immunol. Cell. Biol.*, 72, 306, 1994.
124. **Munoz, C., Endres, S., van der Meer, J., Schlesinger, L., Arevalo, M., and Dinarello, C.,** Interleukin-1 beta in human colostrum, *Res. Immunol.*, 141, 505, 1990.
125. **Saito, S., Mayuyama, M., Kato, Y., Moriyama, I., and Ichijo, M.,** Detection of IL-6 in human milk and its involvement in IgA production, *J. Reprod. Immunol.*, 20, 267, 1991.
126. **Paciotti, G.F. and Tamarkin, L.,** Interleukin-1 regulates hormone-dependent human breast cancer cell proliferation in vitro, *Mol. Endocrinol.*, 2, 459, 1988.
127. **Toi, M., Bicknell, R., and Harris, A.L.,** Inhibition of colon and breast carcinoma cell growth by interleukin-4, *Cancer Res.*, 52, 275, 1992.
128. **Chen, L., Shulman, L.M., and Revel, M.,** IL-6 receptors and sensitivity to growth inhibition by IL-6 in clones of human breast carcinoma cells, *J. Biol. Regul. Homeost. Agents*, 5, 125, 1991.
129. **Katano, M., Matsou, T., Morisaki, T., Naito, K., Nagumo, F., Kubota, E., Nakamura, M., Hisatsugu, T., and Tadano, J.,** Increased proliferation of a human breast carcinoma cell line by recombinant interleukin-2, *Cancer Immunol. Immunother.*, 39, 161, 1994.
130. **Pusztai, L. and Lewis, C.E., McGee, J.,** Growth arrest of the breast cancer cell line, T47D, by TNF alpha; cell cycle specificity and signal transduction, *Br. J. Cancer*, 67, 290, 1993.
131. **Coradini, D., Biffi, A., Pirronello, E., and Di Fronzo, G.,** The effect of alpha-, beta- and gamma-interferon on the growth of breast cancer cell lines, *Anticancer Res.*, 14, 1779, 1994.

132. **Verhasselt, B., Van Damme, J., van Larebeke, L., Put, W., Bracke, M., De Potter, C., and Mareel, M.,** Interleukin-1 is a motility factor for human breast carcinoma cells in vitro: additive effect with interleukin-6, *Eur. J. Cell. Biol.*, 59, 449, 1992.

133. **Macdiarmid, F., Wang, D., Duncan, L.J., Purohit, A., Ghilchick, M.W., and Reed, M.J.,** Stimulation of aromatase activity in breast fibroblasts by tumor necrosis factor alpha, *Mol. Cell. Endocrinol.*, 106, 17, 1994.

134. **Reed, M.J., Coldham, N.G., Patel, S.R., Ghilchik, M.W., and James, V.H.,** Interleukin 1 and interleukin-6 in breast cyst fluid: their role in regulating aromatase activity in breast cancer cells, *J. Endocrinol.*, 132, 5, 1992.

135. **Duncan, L.J., Coldham, N.G., and Reed, M.J.,** The interaction of cytokines in regulating oestradiol 17 beta-hydroxysteroid dehydrogenase activity in MCF-7 cells, *J. Steroid Biochem. Mol. Biol.*, 49, 63, 1994.

136. **Speirs, V., Birch, M.A., Boyle-Walsh, E., Green, A.R., Gallagher, J.A., and White, M.C.,** Interleukin-3; a putative protective factor against breast cancer which is secreted by male but not female breast fibroblasts, *Int. J. Cancer*, 61, 416, 1995.

137. **Danforth, D.N. and Sgagias, M.K.,** Interleukin 1 alpha blocks estradiol-stimulated growth and down-regulates the estrogen receptor in MCF-7 breast cancer cells in vitro, *Cancer Res.*, 51, 1488, 1991.

138. **Danforth, D.N. and Sgagias, M.K.,** Interleukin-1 alpha and interleukin-6 act additively to inhibit growth of MCF-7 breast cancer cells in vitro, *Cancer Res.*, 53, 1538, 1993.

139. **Danforth, D.N. and Sgagias, M.K.,** Tumour necrosis factor-alpha modulates oestradiol responsiveness of MCF-7 breast cancer cells in vitro, *J. Endocrinol.*, 138, 517, 1993.

140. **Pusztai, L., Lewis, C.E., and McGee, J.O.,** Epidermal growth factor, insulin-like growth factor-1 and basic fibroblast growth factor modulate the cytostatic effect of tumour necrosis factor-alpha on the breast cancer cell line, T47D, *Cytokine*, 5, 169, 1993.

141. **Budd, G.T., Green, S., Baker, L.H., Hersh, L.H., Weick, J.K., and Osborne, C.K.,** A Southwest Oncology Group phase II trial of recombinant tumor necrosis factor in metastatic breast cancer, *Cancer*, 68, 1694, 1991.

142. **Lissoni, P., Barni, S., Ardizzoia, A., Olivini, G., Brivio, F., Tisi, E., Tancini, G., Characiejus, D., and Kothari, L.,** Cancer immunotherapy with low-dose interleukin-2 subcutaneous administration: potential efficacy in most solid tumor histotypes by a concomitant treatment with the pineal hormone melatonin, *J. Biol. Regul. Homeostatic Agents*, 7, 121, 1993.

143. **Dillman, R.O., Church, C., Oldham, R.K., West, W.H., Schwartzberg, L., and Birch, R.,** Inpatient continuous infusion interleukin-2 in 788 patients with cancer. The National Biotherapy Study Group experience, *Cancer*, 71, 2358, 1993.

144. **Seymour, L. and Bezwoda, W.R.,** Interferon plus tamoxifen treatment for advanced breast cancer: in vivo biologic effects of two growth modulators, *Br. J. Cancer*, 68, 352, 1993.

Chapter 4

Oncogenes and Tumor-Suppressor Genes in Human Breast Cancer

Rosemary A. Walker

CONTENTS

1. INTRODUCTION

Breast carcinoma is the most common malignancy in women in North America and Western Europe. Despite advances in treatment, only modest improvements in survival have been achieved. There are probably several reasons for this, but an important factor is that breast cancers are heterogenous with respect to biological and clinical behavior. Programs have been introduced for the earlier detection of cancers when they are small and less likely to have metastasized. However, there is no clear understanding about how breast cancers develop and how they progress. Knowledge of these mechanisms could lead to more directed forms of screening and/or therapy. Molecular genetic studies of benign and malignant colonic tumors have identified an adenoma to carcinoma sequence with some genetic changes occurring preferentially at certain stages.[1] Similar studies should provide information about the natural history of breast cancer.

The first molecular studies of breast cancer required fresh, frozen tissue for extraction of DNA for use in Southern blotting techniques. This meant that carcinomas had to be of a certain size and could have gone through multiple genetic events. The introduction of the polymerase chain reaction (PCR), which can be applied to DNA extracted from formalin fixed, paraffin-embedded tissue, has markedly extended the range of lesions which can be studied e.g., atypical hyperplasia ("premalignant"), *in-situ* carcinomas, and small, impalpable invasive carcinomas. The use of microdissection techniques can extend analysis even further.

This chapter will consider how a knowledge of alterations to proto-oncogenes, tumor-suppressor genes and potential tumor-suppressor genes can give insight into: (1) the early genetic events resulting in the development of breast cancer and (2) how cancers will behave, i.e., prediction of prognosis. Familial breast cancer, DNA repair defects, and instability will also be discussed.

2. PROTO-ONCOGENES

The proteins encoded by the majority of proto-oncogenes are components of a cell's growth and other regulatory pathways. Alterations to certain of the genes appear to be important in a proportion of breast cancers (Table 1).

Table 1 Summary of Incidence of Alterations to Oncogenes and Tumor-Suppressor Genes in Breast Cancer

Gene	Alteration	Incidence
c-myc	Amplification/rearrangement	20–25%
ras	Mutation	Rare
c-erbB-2	Amplification	20–30%
int-2	Amplification	10–20%
retinoblastoma	Deletion/rearrangement	19–26%
p53	Deletion	20–60% (depending on marker)
	Mutation	18–45%

2.1 C-Myc

C-Myc proto-oncogene encodes a nuclear phosphoprotein which functions as a transcriptional regulator, controlling cell proliferation, differentiation, and apoptosis. The c-myc protein is a positive regulator of cell-cycle progression.[2]

The introduction of *myc* genes into reconstituted mouse mammary epithelium results in an increase in the number of ducts. The effects are local and may give a mammary cell a selective growth advantage, so promoting clonal expansion. This suggests that alterations to *myc* may be an important early event in the development of tumors.[3]

Studies of human breast carcinoma are less conclusive. Alterations to the *c-myc* gene, either amplification or rearrangement, have been found in about 25% of carcinomas.[4-8] Although Bonilla et al.[6] considered *c-myc* alterations to be associated with the development of breast cancer, other studies[4,5,7,8] have found c-myc alterations to correlate with aggressive features and/or poor prognosis. Whereas most tumors in which there is an amplified or rearranged *c-myc* show elevated expression of mRNA and protein, the converse does not apply.[9,10] Locker et al. found that high levels of p62 c-myc protein were associated with better differentiation and had no relationship to prognosis; that is, quite different from the findings for gene alteration.

2.2 Ras

Activation of the *ras* proto-oncogene family appears to play a role in the development and/or maintenance of certain animal and human neoplasms. Transforming capacity is acquired by single point mutations in codons 12 or 61 leading to expression of an aberrant protein, or by amplification and over-expression of the normal protein.[12]

An association between tumor development and point mutation of *H-ras* has been found in rat mammary carcinomas induced by N-nitroso-N'-methylurea, with the mutation always at codon 12.[13] The introduction of *v-H-ras* into mouse mammary epithelium made hyperplastic by the insertion of *v-myc* results in the fairly rapid development of tumors.[14]

However, the role of ras genes in human breast carcinomas is less clear. Mutations in *H-ras* and *K-ras* have been identified in breast cancer cell lines.[14,15] Theillet et al.[16] did not detect any point mutations in either *H-ras* or *K-ras* in 32 and 64 breast carcinomas respectively, and Rochlitz et al.[17] found a mutation in *K-ras* in only one of 40 primary carcinomas assessed. No amplification or rearrangement was found in 104 carcinomas, but loss of one *H-ras-1* allele correlated with aggressive features.[18]

There are conflicting results regarding ras protein expression, which is further compounded by the subsequent finding of lack of specificity of one antibody used by several groups, RAP-5.[19] Some studies have found greater expression in carcinomas,[20,21] while others have not.[22] Going et al.[23] found an increase in staining from normal to *in situ* carcinomas, but noted that in normal breast myoepithelial cells showed greater staining, which seems rather surprising.

The main interest for ras in breast cancer comes from the tight linkage of H-ras-1 to a mini satellite locus which consists of four common alleles and several rare alleles. There is a highly significant association of rare H-ras-1 alleles with cancer and as many as 1 in 11 breast cancers might be attributed to this.[24]

2.3 C-erbB-2

C-*erb*B-2 (also called neu or HER2) encodes a 185 kDa transmembrane glycoprotein that has extensive homology with epidermal growth-factor receptor and is a putative growth-factor receptor.[25] A candidate ligand, heregulin, was identified and then found to exist as multiple isoforms.[26,27] Although capable of stimulating tyrosine phosphorylation of c-erbB-2 in some mammary cell lines,[28] it has since been discovered that both erbB-3 and erbB-4 are receptors for heregulin. The response of c-erbB-2 to heregulin is mediated through either of these class I receptors by heterodimerization, which stimulates tyrosine phosphorylation.[29] Transgenic mice bearing an activated *neu* oncogene rapidly develop mammary tumors, apparently as a result of "one hit,"[30] although the level of expression of *neu* mRNA appears to be critical for tumorigenesis.[31] Carcinomas *in situ* develop, followed by clonal tumors. This mouse model appears to recapitulate events in human breast cancer.

C-*erb*B-2 has been of particular interest in human breast carcinomas. Amplification of the gene is found in 20 to 30% of invasive carcinomas[5,32-36] and a number of groups have found a correlation between amplification and aggressive features and poor short term prognosis.[5,32,34,35] Others have failed to find a relationship with prognosis.[8,36,37] The reason for this is unclear but the situation becomes more complex when other factors are considered.

The majority of studies have used Southern blotting techniques, but some recent work has employed differential polymerase chain reaction, since it can be applied to fixed, embedded tissue.[38,39] Liu et al.[38] identified amplification in 21% of Stage II carcinomas, but Hubbard et al.[39] found it in much higher numbers.

Unlike c-*myc* there is generally a good concordance between c-*erb*B-2 gene amplification and overexpression of mRNA and protein.[10,33,35,41] In some tumors there is overexpression in the absence of amplification. C-erbB-2 protein can be detected readily in fixed, paraffin-embedded tissue using immunohistochemistry, and this has resulted in a wealth of antibodies and publications. Some have not found a relationship between overexpression and prognosis,[37,42,43] or only in node positive cases,[44,45] but many other studies have found c-erbB-2 overexpression to be an independent predictor of poorer disease-free interval and survival.[46-51] A detailed analysis of the different antibodies used suggests that their ability to detect overexpression is variable and this may account for some of the differences relating to the relationship between c-erbB-2 and prognosis.[52]

The prognostic data would suggest that c-erbB-2 is involved in the progression of the disease. However, it was soon noted in immunohistochemical studies that the protein could be detected in ductal carcinoma *in situ*,[42,53] which indicates that it has a role in development, and in this respect is similar to the findings from transgenic mice studies.[30] More extensive studies of ductal carcinoma *in situ*[54-56] have found c-erbB-2 expression in 40 to 60% of cases, with an association between staining and comedo, high grade types. Liu et al.[38] found amplification in 48% of cases using differential polymerase chain reaction and suggested that alterations to c-*erb*B-2 gene occur at an early stage in the development of breast cancer. No expression has been detected in atypical ductal hyperplasia though.[42,57] Allred et al.[57] found expression in 56% of cases of pure ductal carcinoma *in situ*, 22% of infiltrating carcinomas with a ductal carcinoma *in situ* component and 11% of infiltrating carcinomas without. They proposed that either overexpression of c-*erb*B-2 decreases as carcinomas evolve from *in situ* to invasive or that many invasive carcinomas arise *de novo* by mechanisms not involving c-*erb*B-2. The most likely situation is that there are several molecular pathways by which breast carcinomas arise and alterations to c-*erb*B-2 is just one of them.

The third area of interest with regard to c-erbB-2 relates to therapy. There is clear data that c-erbB-2 positive tumors show a poor response to endocrine therapy,[58-60] but there are differences in the data relating to chemotherapy.

2.4 Int-2/CCND1

Evidence for the involvement of genes on chromosome 11q13 in human breast cancer comes from retroviral studies in mice. There is a high incidence of mammary tumors due to the milkborne transmission of mouse mammary tumor virus (MMTV). The oncogenic effect of MMTV is due to transcriptional activation of one or more cellular proto-oncogenes by the nearby integration of viral DNA. *Int-1* and *Int-2* (now referred to as *Wnt*-1 and *FGF*-3) are the best examples of insertion sites corresponding to cellular genes.[61,62] The virus activates expression of these genes, both of which encode growth factors, which would normally be silent. Transgenic mice expressing either *Wnt*-1 and/or *Fgf*-3 under the control of MMTV develop hyperplasia and neoplasms.[63]

There is evidence that the *int*-2/*FGF*-3 gene is amplified in 10 to 20% of human breast carcinomas.[64-68] An association between the presence of an estrogen receptor and amplification has been found,[66] although some groups have claimed that amplification is associated with poor prognosis.[65,67,68] *int*-2/*FGF*-3 has been localized to chromosome 11q13 and analysis shows that there is amplification of this region. Despite efforts to find FGF-3 transcripts in human tumors it is now clear that amplification of 11q13 does not result in overexpression of FGF-3,[66] and it is just a useful indicator of amplification of another gene in that region.

CCND1 is on 11q13 and encodes cyclin D1[69] which controls cell-cycle progression in G1, when complexed with its associated cyclin-dependent kinase, by phosphorylating retinoblastoma protein.[70,71] The gene was first identified as a rearranged DNA fragment in a parathyroid adenoma (PRAD1).[72] Overexpression of cyclin D1 in mammary cells in transgenic mice results in abnormal proliferation including the development of adenocarcinomas.[73] Mice lacking cyclin D1 fail to undergo pregnancy-associated proliferation of mammary epithelium despite normal ovarian steroid hormone levels, suggesting that steroid-induced proliferation may be driven through cyclin D1.[74]

Another gene mapping to 11q13 is EMS1, which encodes a p80/85 protein which is a major substrate for phosphorylation by the src oncogene. Transformation by src or overexpression due to gene amplification causes the protein to accumulate between the cell and substratum,[75] which may contribute to abnormal cell-cell interactions.

Unlike FGF-3, CCND1 and EMS1 are expressed at elevated levels in breast carcinomas in which there is amplification of 11q13.[76] Buckley et al.[77] identified overexpression of cyclin D1 in both the presence and absence of amplification in breast cancer cell lines, and considered dysregulation of expression of cyclin D1 a potential factor in the pathogenesis of breast cancer. Overexpression of a short cyclin D1 mRNA, associated with a truncated CCND1, has been found in MDA-MB-453 cells.[78] Antibodies have been generated against recombinant human cyclin D1, and these have identified breast cancers with overexpression of cyclin D1, both associated with amplification and with no amplification.[79,80] Patients whose carcinomas co-express cyclin D1 with epidermal growth factor receptors were found in an immunohistochemical study to have a poorer prognosis.[81] Aberrant cyclin D1 expression appear to be important, at least in a subset of breast cancers.

3. TUMOR-SUPPRESSOR GENES

3.1 Retinoblastoma Gene

The retinoblastoma gene (RB1) is the classical example of a tumor-suppressor gene. Both familial and sporadic forms of retinoblastoma arise due to changes in the one gene altered by a "two-hit" mechanism.[82]

The first hit may be passed through the germline and may be due to a point mutation or small intragenic deletion. The second hit is a chromosomal mechanism and occurs somatically. It can be due to a variety of changes resulting in loss of heterozygosity (LOH) at or around the RB1 gene.[83]

The RB1 gene has been located on chromosome 13q14,[84] and encodes a Mr 105,000 protein which is involved in normal cell-growth regulation. The unphosphorylated form restricts cell cycle progression in G1 by interacting with E2F transcription factor.[85] Alterations to the gene are found in all retinoblastomas and these result in either loss of expression, or expression of an aberrant protein.

Alterations to chromosome 13q14 have been found in breast carcinomas and breast cancer cell lines. Lundberg et al.[86] demonstrated LOH at several loci on 13q. Rearrangement in RB1 and loss of Rb proteins were reported to have occurred in 2 of 9 cell lines.[87] In another study, 4 of 16 breast cancer cell lines and 3 of 41 primary breast tumors were found to have structural aberrations, including internal or total deletions and duplication of a part of the RB1 gene.[88] Varley et al.[89] found allele loss or structural rearrangement in 19% of primary breast carcinomas studied. Some degree of loss of RB protein expression was found in all but one carcinoma, which had a rearranged gene, while a small number of cases showed some loss of protein by immunohistochemistry but no alteration to RB1. Borg et al.[90] identified LOH in 26% of informative breast carcinomas, with low or absent levels of Rb protein in 15% of immunoblot-analyzed cancers. However, most of these retained heterozygosity and those cases with LOH often showed high Rb protein expression. In the former study, alterations to Rb occurred more frequently in advanced cases, and in the latter there was a correlation with aneuploidy and high S phase fraction.

These observations, along with the finding that sequences at 13q14 have been excluded as the site of the primary lesions in breast cancer,[91] suggesting that structural alteration to Rb is not an initiating event in the development of breast cancer, but is associated with progression of the disease and is an event occurring in an unstable genome.

3.2 p53

Unlike the retinoblastoma gene there is substantial evidence that p53 is involved in both the development and progression of breast cancer.

For many years, the p53 gene was assumed to be an oncogene, but it was subsequently shown that a variety of mutations within the gene cause a change in function and confer oncogenic features.[92] It has now been clearly demonstrated that the wild-type protein has a tumor-suppressor function. The steady-state level of p53 protein is post-transcriptionally elevated in response to both intrinsic and environmental DNA damage. Accumulation of the wild-type p53 protein results in either cell cycle arrest or apoptosis,[93] and hence tumor suppression. The cell cycle arrest occurs at G_1-S and is due to the ability of p53 to regulate cell cycle check point related genes e.g., mdm-2,[94] GADD45,[95] and p21 (WAF1/GP1).[96] Mutations to the p53 gene result in an alteration to these functions, and further defects in regulatory control can occur if there is loss of the wild-type allele.[97]

The p53 gene maps to the 13.1 region of the short arm of chromosome 17.[98] The gene consists of 11 exons with 5 conserved regions within which most of the p53 mutations identified in tumors have been found.

Germline p53 mutations have been found in families with Li-Fraumeni Syndrome,[99] a rare syndrome in which there is young-onset sarcoma associated with breast cancer, primary brain tumor, or leukemia in a first degree relative < 45 years. The mutations were initially found in a region in exon 7, but subsequently they have been identified in other areas in the conserved regions. However, only about half of the Li-Fraumeni families have p53 mutations.[100] Other analyses of patients with early-onset breast cancer and those with a strong family history have rarely detected germline p53 mutations.[101-103]

Alterations to chromosome 17p have been found in about 60% of breast cancers, with LOH for markers (YNZ22.1) mapping near the tip of 17p at 13.3.[104-107] A lower percentage is found when markers for the p53 locus are used.[108] Deletions from chromosome 17p in the vicinity of the p53 gene have been found in ductal carcinoma *in situ*.[109] suggesting that at least one locus on 17p is involved early in the development of breast cancer.

Mutations in the evolutionarily conserved codons of the p53 gene are common in a wide variety of tumors and analysis can provide clues to etiology.[11] The reported incidence of mutations in breast carcinoma ranges from 17.6% to 45%.[111-120] More than one mutation can occur within a tumor.[111] The pattern of mutations does vary between populations. G:C to T:A transversions were prominent in a Scottish population,[113] while G:C to A:T transitions predominated in U.S. whites from the midwest,[114] but A:T to G:C transitions were in excess from an American black cohort with a high mortality.[117] These different patterns of mutation may reflect different endogenous or environmental factors.

Irrespective of the type of mutation, there is an association between the presence of mutations and aggressive features within breast carcinomas e.g., lack of estrogen receptor,[112,116,118] high S phase index.[115] Andersen et al.[116] found a significant association between p53 mutations, disease-free, and overall survival. Bergh et al.[120] identified mutations throughout the whole coding sequence, but the sites were partly different for node-positive and node-negative cases. Mutations in conserved regions II and V were associated with significantly worse prognosis and adjuvant therapy, particularly tamoxifen, was of less value in p53 mutation, lymph node positive cases.

In view of the roles of p53 in the mechanisms of cells to repair DNA damage in normal cells, Eyfjord et al.[119] examined the relationship between p53 mutations and genetic instability of breast carcinomas, as determined by gene amplification, allele loss, karyotype analysis, and fluorescent *in situ* hybridization. They concluded that p53 abnormalities lead to increased genetic instability which could be important in both development and progression of breast cancer.

Many antibodies have been developed against p53 protein, with the emphasis more recently on the generation of ones which will work on fixed, embedded tissue.[121-123] Wild type p53 protein has a short half life and is not detected by immunohistochemical methods. Initially, it was thought that all p53 detected by immunohistochemistry was mutant, but there is now clear evidence that this is not the case. The incidence of mutations and immunoreactive protein in the same group of carcinomas is different,[107,124] although Thor et al.[125] found a correlation between p53 protein accumulation and gene mutation. An increase in p53 expression can occur in response to DNA damage and can be detected by immunohistochemistry.[126] Therefore, caution should be taken in interpreting the significance of p53 staining. However, the findings from several studies are similar to those found when tumor characteristics are related to the presence or absence of mutation. Prominent reactivity for p53 is associated with the lack of estrogen receptor, poor differentiation, high proliferation rates, and the presence of epidermal growth-factor.[124,125,127-130] It has also been shown to be an independent marker of prognosis.[125,131,132]

Analysis of ductal carcinoma *in situ* by immunohistochemistry has shown that p53 can be detected, again emphasizing that it has a role in both development and progression. Reactivity is seen predominantly in high grade, comedo type cases,[107,125,128,133] which have the same characteristics (e.g., in relation to receptor status) as the invasive carcinomas in which p53 protein is detected.

The overall evidence is that p53 has an important role in the development and progression of breast carcinoma. Analysis can help in our understanding of the different pathways of breast cancer development, possibly give clues to the environmental factors involved, can provide information about prognosis and aid in selection of appropriate therapy.

3.3 Other Tumor-Suppressor Genes

Two tumor-suppressor loci which are important in sporadic colorectal cancer, the adenomatous polyposis coli (APC) gene on 5q21-22 and Deleted in Colon Cancer (DCC) gene on 18q21, have been studied to a limited extent in breast carcinoma. Loss of heterozygosity has been found in 28% (5q21) and 31% (18q21) of cases, with expression of DCC mRNA in 42% of carcinomas. An abnormal DCC mRNA was found in 5 cases with LOH.[134] No rearrangements of DCC have been found.[134,135] Accumulation of abnormalities at 5q21, 17p13, and 18q21 may confer a growth advantage to the breast cancers.

4. LOSS OF HETEROZYGOSITY

As already referred to in relation to RB1 and p53, frequent LOH at a certain chromosomal locus in tumor tissue highlights the presence of a tumor-suppressor gene. This has provided the basis of many studies which have examined a variety of chromosomes for LOH in series of breast cancers. Specific allelic losses have been reported for chromosome 1p, 1q, 3p, 6q, 7q, 8p, 9q, 11q, 13q, 14q, 15q, 16q, 17p, 17q, 18q, 22q, and Xp, as reviewed by Devilee et al.[136] Frequent alterations are shown in Table 2. Certain of the allelic losses or imbalances appear to be late events (1p, 3p, 3q, 6p, 16p, 18p, 18q, and 22q, and possibly 6q and 11p) since alterations to these areas have been observed with very low frequency at the in situ stage, while changes affecting 7p, 16q, 17p and 17q appear to be early abnormalities since they have been found in 25 to 30% of ductal carcinoma *in situ* cases.[137] Studies that use many markers per chromosome arm frequently show that there are tumors with multiple alternating regions of retention and LOH. Multiple genes may be targets for LOH on a single chromosome but possibly at a different stage of the disease.

The introduction of microdissection techniques will make interpretation of LOH data easier due to exclusion of nonmalignant tissue, and will allow analysis of much earlier stages of the disease.[109-137]

5. DNA REPLICATION ERRORS

Studies of both sporadic and familial colon cancer have shown instability in short, repeated DNA sequences (microsatellites)[138] which is due to defects in DNA mismatch repair genes e.g., MSH2 on chromosome 2p22-21.[139] Loeb has proposed that mutations in genetic stability genes may be an early event in tumorigenesis. Relaxed genomic stability could be initiated by primary alterations in genes involved in DNA replication, DNA repair or chromosomal segregation.[140]

Instability of predominantly tri- and tetranucleotide microsatellite repeats has been found in 10.6% of breast cancers by Wooster et al.[141] Yee et al.[142] detected instability in 4 of 20 breast cancers and suggested that microsatellite instability (MSI) is an early event in mammary tumorigenesis, while LOH may occur at a later stage. Differences in the incidence of MSI between invasive ductal and lobular carcinomas has been found, with a higher frequency in the latter.[137] We have examined early, mammographically detected invasive carcinomas using 10 markers and found MSI in 4 of 30 cases, with two showing instability at 9 of 10 loci.[143] No instability has been found in tubular carcinomas. The data suggest that MSI is an early event in the genesis of some sporadic breast cancers.

6. INHERITANCE AND BREAST CANCER

The great majority of breast cancers are due to acquired mutations. Only 5 to 8% of breast cancer patients have a strong family history indicating that they have inherited mutations leading to the disease.

Inherited early-onset breast cancer is linked to two genes, BRCA1 and BRCA2. BRCA1 was mapped to chromosome 17q21[144] and has now been cloned.[145] There is evidence that BRCA1 is a tumor-suppressor gene.[146] It is particularly important in inherited breast and ovarian cancer. The genomic structure of BRCA1 is complex and the rapid detection of mutations for screening of high risk cases will be complex.

BRCA2 has been localized to chromosome 13q12-13.[147] It confers a high risk in the susceptible group, but not of ovarian cancer.

As already described, germline mutations in the p53 gene have been found in a proportion of Li-Fraumeni syndrome families.[99,100]

The other major area of interest relating to inherited breast cancer is Ataxia Telangiectasia. This rare hereditary disorder causes ataxia, depressed immunity, a high risk of leukemia and an extreme sensitivity to X-rays. Female carriers of the gene have a fivefold higher incidence of breast cancer,[148] and their cells show an increased susceptibility to DNA damage after exposure to X-rays.[149] This radiation sensitivity is important to recognize since it may be that carriers will not be suitable for mammographic screening.

7. CONCLUSIONS

The introduction of techniques which can be applied to early breast cancers and "at-risk" lesions means that an understanding of the disease in its early stages should improve. More thorough analysis of these oncogenes and tumor-suppressor genes already recognized has meant that they are becoming increasingly valuable as tools for predicting prognosis and helping selection of therapy. Although familial breast cancer represents a small proportion of cases, linkage analysis has resulted in the identification of genes clearly important in breast cancer. As our understanding of genomic instability and DNA repair increases, along with defining the interrelationship with genes such as p53, we will markedly extend our knowledge about the development and progression of breast cancer and be able to link basic science to clinical outcome.

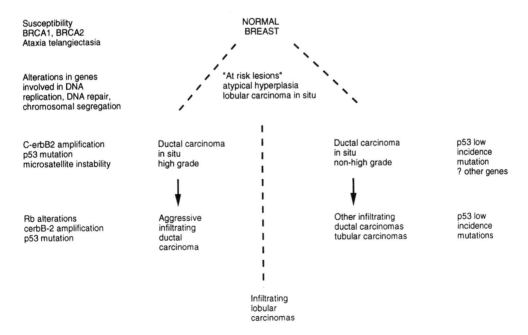

Figure 1 Possible sequence of events in the development and progression of breast cancer.

Table 2 Loss of Heterozygosity in Breast Cancer

Chromosome Arm	Cases LOH (%)	Range (%)	Example	Cases LOH (%)
1p	26.5	3–47	1p 31	28%
1q	33.0	0–50	1q 21–31	65%
3p	33.6	9–47	3p 14.2–14.3	38%
6q	36.4	9–52	6q 25–27	35%
7q	23.7	0–41	7q 23	41%
8p	38.7	27–50	8p 12–21	50%
9q	24.0	9–36	9q 34	20%
11p	27.1	8–41	11p 15.5	35%
11q	53.0	42–66	11q 23	42%
13q	27.7	0–33	13q 12–13	33%
16q	52.3	40–62	16q 24.2qter	52%
17p	57.0	37–75	17p 13.3	48%
17q	38.6	5–75	17q 11.2q12	59%
18q	20	3–36	18q 21	31%

ACKNOWLEDGMENTS

I am grateful to Mrs. Beverley Richardson for typing the manuscript, to Drs. Jenny Varley and Jacqui Shaw for fruitful discussions over the years, and to Steve Chappell and Tom Walsh, Ph.D. students.

REFERENCES

1. **Fearon, E.R.,** Genetic alterations underlying colorectal carcinogenesis, *Cancer Surv.*, 12, 119, 1992.
2. **Evan, G.I., Wyllie, A.H., Gilbert, C.S., Littlewood, T.D., Land, H., Brooks, M., Waters, C.M., Penn, L.Z., and Hancock, D.C.,** Induction of apoptosis in fibroblasts by c-*myc* protein, *Cell*, 69, 119, 1992.
3. **Edwards, P.A.W., Ward, J.L., and Bradbury, J.M.,** Alteration of morphogenesis by the v-*myc* oncogene in transplants of mouse mammary gland, *Oncogene*, 2, 407, 1988.
4. **Escot, C., Theillet, C., Lidereau, R., Spyratos, F., Champene, M.-H., Gest, J., and Callahan, R.,** Genetic alterations of the c-*myc* proto-oncogene (MYC) in human breast carcinomas, *Proc. Natl. Acad. Sci. U.S.A.*, 83, 4834, 1986.
5. **Varley, J.M., Swallow, J.E., Brammar, W.J., Whittaker, J.L., and Walker, R.A.,** Alterations to either c-*erb*B-2 (neu) or c-*myc* proto-oncogenes in breast carcinomas correlate with poor short-term prognosis, *Oncogene*, 1, 423, 1987.
6. **Bonilla, M., Ramirez, M., Lopez-Cueto, J., and Gariglio, P.,** In vivo amplification and rearrangements of c-myc oncogene in human breast tumors, *J. Natl. Cancer Inst.*, 80, 665, 1988.
7. **Garcia, I., Dietrich, P.-Y., Aapiro, M., Vauthier, G., Vadas, L., and Engel, E.,** Genetic alterations of c-myc, c-erB-2 and c-Ha-ras proto-oncogenes and clinical associations in human breast carcinomas, *Cancer Res.*, 49, 6675, 1989.
8. **Berns, P.M.J.J., Krijn, J.G.M., van Putten, W.L.J., van Steveren, I.L., Portengen, H., and Foekens, J.A.,** C-myc amplification is a better prognostic factor than HER2/neu amplification in primary breast cancer, *Cancer Res.*, 52, 1107, 1992.
9. **Mariani-Costantini, R., Escot, C., Theillet, C., Gentile, A., Merlo, G., Lidereau, R., and Callahan, R.,** In situ c-myc expression and genomic status of the c-myc locus in infiltrating ductal carcinomas of the breast, *Cancer Res.*, 48, 199, 1988.
10. **Walker, R.A., Senior, P.V., Jones, J.L., Critchley, D.R., and Varley, J.M.,** An immunohistochemical and in situ hybridisation study of c-myc and c-erbB-2 expression in primary breast tumours, *J. Pathol.*, 165, 203, 1991.
11. **Locker, A.P., Dowle, C.S., Ellis, I.O., Elston, C.W., Blamey, R.W., Sikora, K., Evan, G., and Robins, R.A.,** C-myc oncogene product exression and prognosis in operable breast cancer, *Br. J. Cancer*, 60, 669, 1989.
12. **Barbacid, M.,** *Ras* genes, *Annu. Rev. Biochem.*, 56, 779, 1987.
13. **Sukumar, S., Notario, V., Martin-Zanca, D., and Barbacid, M.,** Induction of mammary carcinomas in rats by nitroso-methyl-urea involves the malignant activation of the H-*ras*-1 locus by single point mutations, *Nature*, 306, 658, 1983.
14. **Bradbury, J.M., Sykes, H., and Edwards, P.A.W.,** Induction of mouse mammary tumours in a transplantation system by sequential introduction of the c-*myc* and c-*ras* oncogenes, *Int. J. Cancer*, 48, 908, 1991.

15. **Kraus, M.H., Yuasa, Y., and Aaronson, S.A.,** A position 12-activated H-ras oncogene in all Hs578T mammary carcinosarcoma cells but not normal mammary cells of the same patient, *Proc. Natl. Acad. Science U.S.A.*, 81, 5384, 1984.

16. **Kozma, S.C., Bogaard, M.E., Buser, K., Saurer, S.M., Bos, J.L., Groner, B., and Hynes, N.E.,** The human c-Kirsten *ras* gene is activated by a novel mutation in codon 13 in the breast carcinoma cell line MDA-MB-231, *Nucleic Acid Res.*, 15, 5963, 1987.

17. **Thiellet, C., Lidereau, R., Escot, C., Brunet, M., Gest, J., Schlom, J., and Callahan, R.,** Loss of a c-H-ras-1 allele and aggressive human primary breast carcinomas, *Cancer Res.*, 46, 4776, 1986.

18. **Rochlitz, C.F., Scott, G.K., Dodson, J.M., Liu, E., Dollbaum, C., Smith, H.S., and Benz, C.C.,** Incidence of activating ras oncogene mutations associated with primary and metastatic breast cancer, *Cancer Res.*, 49, 357, 1989.

19. **Robinson, A., Williams, A.R.W., Piris, J., Spandidos, D.A., and Wyllie, A.H.,** Evaluation of a monoclonal antibody to ras peptide, RAP-5, claimed to bind preferentially to cells of infiltrating carcinomas, *Br. J. Cancer*, 54, 877, 1986.

20. **Ohuchi, N., Thor, A., Pge, D.L., Horan Hand, P., Halter, S., and Schlom, J.,** Expression of the 21,000 molecular weight ras protein with a molecular weight ras protein in a spectrum of benign and malignant human mammary tissues, *Cancer Res.*, 47, 5290, 1986.

21. **Clair, T., Miller, W.R., and Cho-Chung, Y.S.,** Prognostic significance of the expression of a ras protein with a molecular weight of 21,000 by human breast cancer, *Cancer Res.*, 47, 5290, 1986.

22. **Walker, R.A. and Wilkinson, N.,** p21 ras expression in benign and malignant human breast, *J. Pathol.*, 156, 147, 1988.

23. **Going, J.J., Anderson, T.J., and Wyllie, A.H.,** Ras p21 in breast tissue: associations with pathology and cellular localisation, *Br. J. Cancer*, 65, 45, 1992.

24. **Krontiris, T.G., Devlin, B., Karp, D.D., Robert, J.J., and Risch, N.,** An association between the risk of cancer and mutations in the Hras1 mini-satellite locus, *N. Engl. J. Med.*, 329, 517, 1993.

25. **Coussens, L., Yang-Feng, T.L., Chen, Y.-C., Gray, A., Mcgrath, J., Seeburg, P.H., Libermann, T.A., Schlessinger, J., Francke, U., Levinson, A., and Ullrich, A.,** Tyrosine kinase receptor with extensive homology to EGF receptor shares chromosomal location with *neu* oncogene, *Science*, 230, 1132, 1985.

26. **Holmes, W.E., Sliwkowski, M.X., Akita, R.W., Henzel, W.J., Lee, J., Park, J.W., Yansura, D., Abadi, N., Raab, H., Lewis, G.D., Shepard, H.M., Kuang, W.-J., Wood, W.I., Goeddel, D.V., and Vanden, R.L.,** Identification of heregulin, a specific activator of p185erbB2, *Science*, 256, 1205, 1992.

27. **Peles, E., Bacus, S.S., Koski, R.A., Lu, H.S., Wen, D., Ogden, S.G., Ben-Levy, R., and Yarden, Y.,** Isolation of the neu/HER-2 stimulatory ligand: a 44 kD glycoprotein that induces differentiation of mammary tumour cells, *Cell*, 69, 205, 1992.

28. **Peles, E., Ben-Levy, R., Tzahar, E., Liu, N., Wen, D., and Yarden, Y.,** Cell type specific interactions of Neu differentiation factor (NDF/heregulin) with Neu/HER-2 suggests complex ligand-receptor relationships, *EMBO Journal*, 2, 961, 1993.

29. **Carraway, III, K.L. and Cantley, L.C.,** A neu acquaintance for ErbB3 and ErbB4: a role for receptor dimerization in growth signalling, *Cell*, 78, 5, 1994.

30. **Muller, W.J., Sinn, E., Pattengale, P.K., Wallace, R., and Leder, P.,** Single-step induction of mammary adenocarcinoma in transgenic mice bearing the activated c-*neu* oncogene, *Cell*, 54, 105, 1988.

31. **Bouchard, L., Lamarre, L., Tremblay, P.J., and Jolicouer, P.,** Stochastic appearance of mammary tumours in transgenic mice carrying the activated c-neu oncogene, *Cell*, 57, 931, 1989.

32. **Slamon, D.J., Clark, G.M., Wong, S.G., Levin, W.J., Ullrich, A., and McGuire, W.L.,** Human breast cancer: corelation of relapse and survival with amplificaton of the HER-2/neu oncogene, *Science*, 235, 177, 1987.

33. **Van der Vijver, M., van de Bersselaar, R., Devilee, P., Cornelisse, C., Peterse, J., and Nusse, R.,** Amplification of the neu (c-erbB-2) oncogene in human mammary tumours is relatively frequent and is often accompanied by amplification of the linked c-erbA oncogene, *Mol. Cell. Biol.*, 7, 2019, 1987.

34. **Zhou, D., Battifora, H., Yokata, J., Yamamoto, T., and Cline, M.J.,** Association of multiple copies of the c-erbB-2 oncogenre with the spread of breast cancer, *Cancer Res.*, 47, 6123, 1987.

35. **Borresen, A.-L., Ottestad, L., Gaustad, A., Andersen, T.I., Heikkila, R., Jahnsen, T., Tveit, K.M., and Nesland, J.M.,** Amplification and protein overexpression of the neu/HER-2/c-erbB-2 proto-oncogene in human breast carcinomas: relationship to loss of gene sequences on chromosome 17, family history and prognosis, *Br. J. Cancer*, 62, 585, 1990.

36. **Clark, G.M. and McGuire, W.L.,** Follow-up study of HER-2/neu amplification in primary breast cancer, *Cancer Res.*, 51, 944, 1991.

37. **Zhou, D.-J., Ahuja, H., and Cline, M.J.,** Proto-oncogene abnormalities in human breast cancer: c-erbB-2 amplification does not correlate with recurrence of disease, *Oncogene*, 4, 105, 1989.

38. **Liu, E., Thor, A., He, M., Barcos, M., Ljung, B.-M., and Benz, C.,** The HER-2 (c-erbB-2) oncogene is frequently amplified in in situ carcinoma of the breast, *Oncogene*, 7, 1027, 1992.

39. **Hubbard, A.L., Doris, C.P., Thompson, A.M., Chetty, T.J., and Anderson, T.J.,** Critical determination of the frequency of c-erbB-2 amplification in breast cancer, *Br. J. Cancer*, 70, 434, 1994.

40. **Ciocca, D.R., Fujimura, F.K., Tandon, A.K., Clark, G.M., Mark, C., Lee-Chen, G.-J., Pounds, G.W., Vendeley, P., Owens, M.A., Pandian, M.R., and McGuire, W.L.,** Correlation of HER-2/neu amplification with expression and with other prognostic factors in 1103 breast cancers, *J. Natl. Cancer Inst.*, 84, 1279, 1992.

41. **Venter, D.J., Tuzi, N.L., Kumar, S., and Gullick, W.J.,** Overexpression of the c-erbB-2 oncoprotein in human breast carcinomas: immunological assessment correlates with gene amplification, *Lancet*, 2, 69, 1987.

42. **Gusterson, B.A., Machin, L.G., Gullick, W.J., Gibbs, N.M., Powles, T.J., Elliot, C., Ashley, S., Monaghan, P., and Harrison, S.,** c-erbB-2 expression in benign and malignant breast disease, *Br. J. Cancer*, 58, 453, 1988.

43. **Barnes, D.M., Lammie, G.A., Millis, R.R., Gullick, W.J., Allen, D.S., and Altman, D.G.,** An immunohistochemical evaluation of c-erbB-2 expression in human breast carcinoma, *Br. J. Cancer*, 58, 448, 1988.

44. **Tandon, A.K., Clark, G.M., Chamness, G.C., Ullrich, A., and McGuire, W.L.,** HER-2/neu oncogene expression and prognosis in breast cancer, *J. Clin. Oncol.*, 7, 1120, 1989.

45. **O'Reilly, S.M., Barnes, D.M., Camplejohn, R.A., Bartkova, J., Gregory, W.M., and Richards, M.A.,** The relationship between c-erbB-2 expression, S-phase fraction and prognosis in breast cancer, *Br. J. Cancer*, 63, 444, 1991.

46. **Wright, C., Angus, B., Nicholson, S., Sainsbury, J.R.C., Cairns, J., Gullick, W.J., Kelly, P., Harris, A.L., and Horne, C.H.W.,** Expression of c-erbB-2 oncoprotein: a prognostic indicator in human breast cancer, *Cancer Res.*, 49, 2087, 1989.

47. **Walker, R., Gullick, W.J., and Varley, J.M.,** An evaluation of immunoreactivity for c-erbB-2 protein as a marker of short term prognosis in breast cancer, *Br. J. Cancer*, 60, 426, 1989.

48. **Gullick, W.J., Love, S.B., Wright, C., Barnes, D.M., Gusterson, B., Harris, A.L., and Altman, D.G.,** C-erbB-2 protein overexpression in breast cancer is a risk factor in patients with involved and uninvolved lymph nodes, *Br. J. Cancer*, 63, 434, 1991.

49. **Lovekin, C., Ellis, I.O., Locker, A., Robertson, J.F.R., Bell, J., Nicholson, R., Gullick, W.J., Elston, C.W., and Blamey, R.W.,** C-erbB-2 oncoprotein expression and prognosis in operable breast cancer, *Br. J. Cancer*, 63, 439, 1991.

50. **Winstanley, J., Cooke, T., Muray, G.D., Platt-Higgins, A., George, W.D., Holt, S., Mystov, M., Spedding, A., Barraclough, B.R., and Rudland, P.S.,** The long-term prognostic significance of c-erbB-2 in primary breast cancer, *Br. J. Cancer*, 63, 447, 1991.

51. **Press, M.F., Pike, M.C., Chazin, V.R., Hung, G., Udove, J.J.A., Markowicz, M., Danyluk, J., Godolphin, W., Sliwkowski, M., Akita, R., Paterson, M.C., and Slamon, D.J.,** Her-2/neu expression in node negative breast cancer: direct tissue quantitation by computerised image analysis and association of overexpression with increased risk of recurrent disease, *Cancer Res.*, 53, 4960, 1993.

52. **Press, M.F., Hung, G., Godolphin, W., and Slamon, D.J.,** Sensitivity of HER-2/neu antibodies in archival tissue samples: potential source of error in immunohistochemical studies of oncogene expression, *Cancer Res.*, 54, 2771, 1994.

53. **Van der Vijver, M.J., Peterse, J.L., Mooi, W.J., Wisman, P., Lomans, J., Dalesio, O., and Nusse, R.,** Neu-protein overexpression in breast cancer: association with comedo-type ductal carcinoma in situ and limited prognostic value in Stage II breast cancer, *N. Engl. J. Med.*, 319, 1239, 1988.

54. **Bartkova, J., Barnes, D.M., Millis, R.R., and Gullick, W.J.,** Immunohistochemical demonstration of c-erbB-2 protein in mammary ductal carcinoma in situ, *Hum. Pathol.*, 21, 1164, 1990.

55. **Ramachandra, S., Machin, L., Ashley, S., Monaghan, P., and Gusterson, B.A.,** Immunohistochemical distribution of c-erbB-2 in in situ breast carcinoma, — a detailed morphological analysis, *J. Pathol.*, 161, 7, 1990.

56. **Lodata, R.J.F. and Maguire, H.C., Jr., Greene, M.I., Weiner, D.B., and LiVolsi, V.A.,** Immunohistochemical evaluation of c-erbB-2 oncogene expression in ductal carcinoma in situ and atypical ductal hyperplasia of the breast, *Mod. Pathol.*, 3, 449, 1990.

57. **Allred, D.C., Clark, G.M.R., Tandon, A.K., Schmitt, S.J., Gilchrist, K.W., Osborne, C.K., Tormey, D.C., and McGuire, W.L.,** Overexpression of HER-2/neu and its relationship with other prognostic factors change during the progression of in situ to invasive breast cancer, *Hum. Pathol.*, 23, 974, 1992.

58. **Wright, C., Nicholson, S., Angus, B., Sainsbury, J.R.C., Farndon, J., Cairns, J., Harris, A.L., and Horne, C.H.W.,** Relationship between c-erbB-2 protein product expression and response to endocrine therapy in advanced breast cancer, *Br. J. Cancer*, 65, 118, 1992.

59. **Nicholson, R.I., McClelland, R.A., Finlay, P., Eaton, C.L., Gullick, W.J., Dixon, A.R., Robertson, J.F.R., Ellis, I.O., and Blamey, R.W.,** Relationship between EGF-R, c-erbB-2 protein expression and Ki67 immunostaining in breast cancer and hormone sensitivity, *Eur. J. Cancer*, 29A, 1018, 1993.

60. **Klijn, J.G.M., Berns, E.M.M.J., and Foekens, J.A.,** Prognostic factors and response to therapy in breast cancer, *Cancer Surv.*, 18, 165, 1993.

61. **Nusse, R. and Varmus, H.E.,** Wnt genes, *Cell*, 69, 1073, 1992.

62. **Dickson, C., Smith, R., Brookes, S., and Peters, G.,** Tumorigenesis by mouse mammary tumor virus: proviral activation of a cellular gene in the common integration region int-2, *Cell*, 37, 529, 1984.

63. **Muller, W.J., Lee, F.S., Dickson, G., Peters, G., Pattengale, P., and Leder, P.,** The int-2 gene product acts as an epithelial growth factor in transgenic mice, *EMBO J.*, 9, 907, 1990.

64. **Varley, J.M., Walker, R.A., Casey, G., and Brammar, W.J.,** A common alteration to the int-2 proto-oncogene in DNA from primary breast carcinomas, *Oncogene*, 3, 87, 1988.

65. **Lidereau, R., Callahan, R., Dickson, C., Peters, G., Escot, C., and Ali, I.,** Amplification of the int-2 gene in primary breast tumors, *Oncogene Res.*, 2, 285, 1988.

66. **Fantl, V., Richards, M.A., Smith, R., Lammie, G.J.A., Johnstone, G., Allen, D., Gregory, W., Peters, G., Dickson, C., and Barnes, D.M.,** Gene amplification on chromosome band 11q13 and oestrogen receptor status in breast cancer, *Eur. J. Cancer*, 26, 423, 1990.

67. **Borg, A., Sigurdsson, H., Clark, G., Ferno, M., Fuqua, S.A.W., Olsson, H., Killander, D., and McGuire, W.L.,** Association of INT2/HST1 coamplification in primary breast cancer with hormone-dependent phenotype and poor prognosis, *Br. J. Cancer*, 63, 136, 1991.

68. **Henry, J.A., Hennessy, C., Levett, D.L., Lennard, T.W.J., Westley, B.R., and May, F.E.B.,** Int-2 amplification in breast cancer: association with decreased survival and relationship to amplification of c-erbB-2 and c-myc, *Int. J. Cancer*, 53, 774, 1993.

69. **Xiong, Y., Menninger, J., Beach, D., and Ward, D.C.,** Molecular cloning and chromosomal mapping of CCND genes encoding human D-type cyclins, *Genomics*, 13, 575, 1992.

70. **Scherr, C.J.,** G1 phase progression: cycling on cue, *Cell*, 79, 551, 1994.

71. **Weinberg, R.A.,** The retinoblastoma protein and cell cycle control, *Cell*, 81, 323, 1995.

72. **Arniold, A., Kim, H.G., Gaz, R.D., Eddy, R.L., Fukushima, Y., Byers, M.G., Shows, T.B., and Kronenberg, H.M.,** Molecular cloning and chromosomal mapping of DNA rearranged with the parathyroid hormone gene in a parathyroid adenoma, *J. Clin. Invest.*, 83, 2034, 1989.

73. **Wang, T.C., Cardiff, R.D., Zukerberg, L., Lees, E., Arnold, A., and Schmidt, E.V.,** Mammary hyperplasia and carcinoma in MMTV-cyclin D1 transgenic mice, *Nature*, 369, 669, 1994.

74. **Sicinski, P., Donaher, J.L., Parker, S.B., Li, T., Fazeli, A., Gardner, H., Haslam, S.Z., Bronson, R.T., Elledge, S.J., and Weinberg, R.A.,** Cyclin D1 provides a link between development and oncogenesis in the retina and breast, *Cell*, 82, 621, 1995.

75. **Schuuring, E., Verhoeven, E., Mooi, W.J., and Michalides, R.J.A.M.,** Identification and cloning of two overexpressed genes U21B31/PRAD1 and EMS1 within the amplified chromosome 11q13 region in human carcinomas, *Oncogene*, 7, 355, 1992.

76. **Lammie, G.A., Fantl, V., Smith, R., Schuuring, E., Brookes, S., Michalides, R., Dickson, C., Arnold, A., and Peters, G.,** D11 S287, a putative oncogene on chromosomal 11q13 is amplified and expressed in squamous cell and mammary carcinomas and linked to BCL-1, *Oncogene*, 7, 2381, 1991.

77. **Buckley, M.F., Sweeney, K.J.E., Hamilton, J.A., Sini, R.L., Manning, D.L., Nicholson, R.I., de Fazio, A., Watts, C.K.W., Musgrove, E.A., and Sutherland, R.L.,** Expression and amplification of cyclin genes in human breast cancer, *Oncogene*, 8, 2127, 1993.

78. **Lebwohl, D.E., Muise-Helmericks, R., Sepp-Lorenzino, L., Serve, S., Timaul, M., Bol, R., Borgen, P., and Rosen, N.,** A truncated cyclin D1 gene encodes a stable mRNA in a human breast cancer cell line, *Oncogene*, 9, 1925, 1994.

79. **Bartkova, J., Lukas, J., Muller, H., Lutzhoft, D., Strauss, M., and Bartek, J.,** Cyclin D1 protein expression and function in human breast cancer, *Int. J. Cancer*, 57, 353, 1994.

80. **Gillet, C., Fantl, V., Smith, R., Fisher, C., Bartek, J., Dickson, C., Barnes, D.M., and Peters, G.,** Amplification and overexpression of cyclin D1 in breast cancer detected by immunohistochemical staining, *Cancer Res.*, 54, 1812, 1994.

81. **McIntosh, G.G., Anderson, J.J., Milton, I., Steward, M., Parr, A.H., Thomas, M.D., Henry, J.A., Angus, B., Lennard, T.W.J., and Horne, C.H.W.,** Determination of the prognostic value of cyclin D1 overexpression in breast cancer, *Oncogene*, 11, 885, 1995.

82. **Knudson, A.G.,** Mutation and cancer: a statistical study of retinoblastoma, *Proc. Natl. Acad. Sci. U.S.A.*, 68, 820, 1971.

83. **Cavanee, W.K., Dryja, T.P., Phillips, R.A., Benedict, W.F., Godbou, R., Gallie, B.L., Murphree, A.L., Strong, L.C., and White, R.L.,** Expression of recessive alleles by chromosomal mechanisms in retinoblastoma, *Nature*, 305, 779, 1983.

84. **Friend, S.H., Bernards, R., Rogelj, S., Weinberg, R.A., Rapaport, J.M., Albert, D.M., and Dryja, T.P.,** A human DNA segment with properties of the gene that predisposes to retinoblastoma and osteosarcoma, *Nature*, 323, 643, 1986.

85. **Chellappan, S.P., Hierbert, S., Mudryj, M., Horowitz, J.M., and Nevins, J.R.,** The E2F transcription factor is a cellular target for the RB protein, *Cell*, 65, 1053, 1991.

86. **Lundberg, C., Skoog, L., Cavanee, W.K., and Nordenskjold, M.,** Loss of heterozygosity in human ductal breast tumors indicates a recessive mutation on chromosome 13, *Proc. Natl. Acad. Sci. U.S.A.*, 84, 2372, 1987.

87. **Lee, E.Y.-H.P., To, H., Shew, J.-Y., Bookstein, R., Scully, P., and Lee, W.-H.,** Inactivation of the retinoblastoma susceptibility gene in human breast cancers, *Science*, 241, 218, 1988.

88. **Tang, A., Varley, J.M., Chakraborty, S., Murphree, A.L., and Fung, Y.-K.T.,** Structural rearrangement of the retinoblastoma gene in human breast carcinoma, *Science*, 242, 263, 1988.

89. **Varley, J.M., Armour, J., Swallow, J.E., Jeffreys, A.J., Ponder, B.A.J., Tang, A., Fung, Y.-K.T., Brammar, W.J., and Walker, R.A.,** The retinoblastoma gene is frequently altered leading to loss of expression in primary breast tumours, *Oncogene*, 4, 725, 1989.

90. **Borg, A., Zhang, Q.-X., Alm, P., Olsson, H., and Sellberg, G.,** The retinoblastoma gene in breast cancer: allele loss is not correlated with loss of gene protein expression, *Cancer Res.*, 52, 2991, 1992.

91. **Bowcock, A.M., Hall, J.M., Hebert, J.M., and King, M.-C.,** Exclusion of the retinoblastoma gene and chromosome 13q as the site of a primary lesion for human breast cancer, *Am. J. Hum. Genet.*, 46, 12, 1990.

92. **Levine, A.J.,** The p53 tumour suppressor gene and product, *Cancer Surv.*, 12, 59, 1992.

93. **Prives, C.,** Doing the right thing: feedback control and p53, *Curr. Opin. Cell. Biol.*, 5, 214, 1993.

94. **Wu, X., Boyle, J.H., Olson, D., and Levine, A.J.,** The p53-mdm2 autoregulatory feedback loop, *Genes Dev.*, 7, 1126, 1993.

95. **Kastan, M.B., Zhan, Q., El-Deiry, W.S., Carrier, R., Jacks, T., Walsh, W.V., Plunkett, B.S., Vogelstein, B., and Fornace, A.J.,** A mammalian cell cycle checkpoint pathway utilizing p53 and GADD45 is defective in ataxia-telangiectasia, *Cell*, 71, 587, 1992.

96. **El-Diery, W.S., Tokino, T., Welculescu, V.E., Levy, D.B., Parsons, R., Trent, J.M., Lin, D., Mercer, W.E., Linzer, K.W., and Vogelstein, B.,** WAF-1, a potential mediator of p53 tumour suppression, *Cell*, 75, 817, 1993.

97. **Baker, S.J., Fearon, E.R., Nigro, J.M., Hamilton, S.R., Preisinger, A.C., Jessup, J.M., Varitulinen, P., Ledbetter, D.H., and Barker, D.F.,** Chromosome 17 deletions and p53 gene mutations in colorectal carcinomas, *Science*, 244, 217, 1989.

98. **Miller, C., Mohandas, T., Wolf, D., Prokocimer, M., Rotter, V., and Koeffler, H.P.,** Human p53 gene localized to short arm of chromosome 17, *Nature*, 319, 873, 1986.

99. **Malkin, D., Li, F.P., Strong, L.C., Fraumeni, J.F., Nelson, C.E., Kim, D.H., Kassel, J., Gryka, M.A., Bischoff, F.Z., Tainsky, M.A., and Friend, S.H.,** Germ line p53 mutations in a familial syndrome of breast cancer, sarcomas and other neoplasms, *Science*, 250, 1233, 1990.

100. **Santbanez-Koref, M.F., Birch, J.M., and Hartley, A.L.,** Norris Jones, P.H., Craft, A.H. Eden, T., Crowther, D., Kelsey, A.M., and Harris, M., p53 germline mutations in Li Fraumeni syndrome, *Lancet*, 338, 1490, 1991.

101. **Prosser, J., Elder, P.A., Candie, A., MacFayden, I., Steel, C.M., and Evan, H.J.,** Mutations in p53 do not account for heritable breast cancer: a study in five affected families, *Br. J. Cancer*, 63, 181, 1991.

102. **Sidransky, D., Tokins, T., Helzisouer, K., Rauch, G., Zehnbauer, B., Shelton, B., Prestiglacomo, L., Vogelstein, B., and Davidson, N.,** Inherited p53 gene mutations in breast cancer, *Cancer Res.*, 52, 2984, 1992.

103. **Warren, W., Eeles, R.A., Ponder, B.A.J., Easton, D.F., Averill, D., Ponder, M.A., Anderson, K., Evans, A.M., De Mars, R., and Love, R.,** No evidence for germline mutations in exons 59 of the p53 gene in 25 breast cancer families, *Oncogene*, 7, 1043, 1992.

104. **Mackay, J., Elder, P.A., Steel, C.M., Forrest, A.P.M., and Evans, H.J.,** Allele loss on short arm of chromosome 17 in breast cancers, *Lancet*, 2, 1384, 1988.

105. **Deville, P., Van den Brock, M., Kuipers-Dijkshoorn, N., Kolluri, R., Meera Khan, P., Pearson, P.L., and Cornelisse, C.J.,** At least four different chromosomal regions are involved in loss of heterozygosity in human breast carcinoma, *Genomics*, 5, 554, 1989.

106. **Thompson, A.M., Steel, C.M., Chetty, U., Hawkins, R.A., Miller, W.R., Carter, D.C., Forrest, A.P.M., and Evans, H.J.,** p53 gene mRNA expression and chromosome 17p allele loss in breast cancer, *Br. J. Cancer*, 61, 74, 1990.

107. **Varley, J.M., Brammar, W.J., Lane, D.P., Swallow, J.E., Dolan, C., and Walker, R.A.,** Loss of chromosome 17p13 sequences and mutations of p53 in human breast carcinomas, *Oncogene*, 6, 413, 1991.

108. **Thompson, A., M., Steel, C.M., Chetty, U., and Carter, D.C.,** Evidence for the multistep theory of carcinogenesis in human breast cancer, *The Breast*, 1, 29, 1992.

109. **Radford, D.M., Fair, K., Thompson, A.M., Ritter, J.H., Holt, M., Steinbrueck, T., Wallace, M., Wells, S.A., and Donis-Keller, H.R.,** Allelic loss on chromosome 17 in ductal carcinoma in situ of the breast, *Cancer Res.*, 53, 2947, 1993.

110. **Hollstein, M., Sidransky, D., Vogelstein, B., and Harris, C.C.,** p53 mutations in human cancers, *Science*, 253, 49, 1991.

111. **Osborne, R.J., Merlo, G.R., Mitsudomi, T., Venesio, T., Liscia, D.S., Cappa, A.P.M., Chiba, I., Takahashi, T., Nau, M.M., Callahan, R., and Minna, J.D.,** Mutations in the p53 gene in primary human breast cancers, *Cancer Res.*, 51, 6194, 1991.

112. **Mazars, R. and Spinardi, L., Ben Cheikh, M., Simony-Lafontaine, J., Jeanteur, P., and Theillet, C.,** p53 mutations occur in aggressive breast cancer, *Cancer Res.*, 52, 3918, 1992.

113. **Coles, C., Condie, A., Chetty, U., Steel, C.M., Evans, H.J., and Prosser, J.,** p53 mutations in breast cancer, *Cancer Res.*, 52, 5291, 1992.

114. **Sommer, S.S., Cunningham, J., McGovern, R.M., Saitoh, S., Schroeder, J.J., Wold, L., and Kovach, J.S.,** Pattern of p53 gene mutations in breast cancers of women in the Midwestern United States, *J. Natl. Cancer Inst.*, 84, 246, 1992.

115. **Merlo, G.R., Bernardi, A., Diella, F., Venesio, T., Cappa, A.P.M., Callahan, R., and Liscia, D.S.,** In primary human breast carcinomas mutations in exons 5 and 6 of the p53 gene are associated with high S-phase index, *Int. J. Cancer*, 53, 531, 1993.

116. **Andersen, T.I., Holm, R., Nesland, J.M., Heimdel, K.R., Ottestad, L., and Borresen, A.-L.,** Prognostic significance of TP53 alterations in breast carcinoma, *Br. J. Cancer*, 68, 540, 1993.

117. **Blaszyk, H., Vaughn, C.B., Hartmann, A., McGovern, R.M., Schroeder, J.J., Cunningham, J., Schaid, D., Sommer, S.S., and Kovach, J.S.,** Novel pattern of p53 mutations in an American black cohort with high mortality from breast cancer, *Lancet*, 343, 1195, 1994.

118. **Caleffi, M., Teague, M.W., Jensen, R.A., Vnencak-Jones, C.L., Dupont, W.D., and Parl, F.F.,** p53 gene mutations and steroid receptor status in breast cancer, *Cancer*, 73, 1247, 1994.

119. **Eyfjord, J.E., Thorlacius, S., Steinarsdottir, M., Valgardsdottir, R., Ogmundsdottir, H.M., and Anamthawat-Jonsson, K.,** p53 abnormalities and genomic instability in primary human breast carcinomas, *Cancer Res.*, 55, 646, 1995.

120. **Bergh, J., Norbert, T., Sjogren, S., Lindgren, A., and Holmberg, L.,** Complete sequencing of the p53 gene provides prognostic information in breast cancer patients, particularly in relation to adjuvant systemic therapy and radiotherapy, *Nat. Med.*, 1, 1029, 1995.

121. **Midgeley, C.A., Fisher, C.J., Bartek, J., Vojtesek, B., Lane, D., and Barnes, D.M.,** Analysis of p53 expression in human tumours: an antibody raised against human p53 expressed in *Escherichia coli, J. Cell Sci.*, 101, 183, 1992.

122. **Vojkesek, B., Bartek, J., Midgley, C.A., and Lane, D.P.,** An immunohistochemical analysis of the human nuclear phosphoprotein p53: new monoclonal antibodies and epitope mapping using recombinant p53, *J. Immunol. Methods*, 151, 237, 1992.

123. **Bartek, J., Bartkova, J., Lukas, J., Staskova, Z., Vojtesck, B., and Lane, D.P.,** Immunohistochemical analysis of the p53 oncoprotein on paraffin sections using a series of novel monoclonal antibodies, *J. Pathol.*, 169, 27, 1993.

124. **Jacqumier, J., Moles, J.P., Penault-Llorca, F., Adelaide, J., Torrente, M., Viens, P., Birnbaum, D., and Theillet, C.,** p53 immunohistochemical analysis in breast cancer with four monoclonal antibodies: comparison of staining and PCR: SSCP results, *Br. J. Cancer*, 69, 846, 1994.

125. **Thor, A.D. and Moore, II, D.H., Edgerton, S.M., Kawaski, E.S., Reihsaus, E., Lynch, H.T., Marcus, J.N., Schwartz, L., Chen, L-C., Mayall, B.H., and Smith, S.,** Accumulation of p53 tumor-suppressor gene protein: an independent marker of prognosis in breast cancers, *J. Natl. Cancer Inst.*, 84, 845, 1992.

126. **Hall, P.A., McKee, P.H., Menage, H., Du, P., Dover, R., and Lane, D.P.,** High levels of p53 protein in UV-irradiated normal human skin, *Oncogene*, 8, 203, 1993.

127. **Cattoretti, G., Andreola, S., Clemente, C., D'Amato, L., and Rilke, F.,** Vimentin and p53 expression in epidermal growth factor receptor-positive oestrogen receptor-negative breast carcinomas, *Br. J. Cancer*, 57, 353, 1988.

128. **Walker, R.A., Dearing, S.J., Lane, D.P., and Varley, J.M.,** Expression of p53 protein in infiltrating and in situ breast carcinomas, *J. Pathol.*, 165, 203, 1991.

129. **Poller, D.N., Hutchings, C.E., Galea, M., Bell, J.A., Nicholson, R.A., Elston, C.W., Blamey, R.W., and Ellis, I.O.,** p53 protein expression in human breast carcinoma: relationship to expression of epidermal growth factor receptor, c-erbB-2 protein overexpression, and oestrogen receptor, *Br. J. Cancer*, 66, 583, 1992.

130. **Fisher, C.J., Gillet, C.E., Vojtesek, B., Barnes, D.M., and Millis, R.,** Problems with p53 immunohistochemical staining: the effect of fixation and variation in the methods of evaluation, *Br. J. Cancer*, 69, 26, 1994.

131. **Barnes, D.M., Dublin, E.H., Fisher, C.J., Levison, D.A., and Millis, R.R.,** Immunohistochemical detection of p53 protein in mammary carcinoma: an important new independent indicator of prognosis? *Hum. Pathol.*, 24, 469, 1993.

132. **Lipponen, P., Ji, H., Aaltomaa, S., Syrjanen, S., and Syrjanen, K.,** p53 protein expression in breast cancer as related to histopathological characteristics and prognosis, *Int. J. Cancer*, 55, 51, 1993.

133. **Bobrow, L.G., Happerfield, L.C., Gregory, W.M., and Millis, R.R.,** Ductal carcinoma in situ: assessment of necrosis and nuclear morphology and their association with biological markers, *J. Pathol.*, 176, 333, 1995.

134. **Thompson, A.M., Morris, R.G., Wallace, M., Wyllie, A.H., Steel, C.M., and Carter, D.C.,** Allele loss from 5q21 (APC/MCC) and 18q21 (DCC) and DCC mRNA expression in breast cancer, *Br. J. Cancer*, 68, 64, 1993.

135. **Sato, T., Akiyama, F., Sakamoto, G., Kasumi, F., and Nakamura, Y.,** Accumulation of genetic alterations and progression of primary breast cancer, *Cancer Res.*, 51, 5794, 1991.

136. **Devilee, P. and Cornelisse, C.J.,** Somatic genetic changes in human breast cancer, *Biochim. Biophys. Acta*, 1198, 113, 1994.

137. **Aldaz, C.M., Chen, T., Sahin, S., Cunningham, J., and Bondy, M.,** Comparative allelotype of in situ and invasive human breast cancer: high frequency of microsatellite instability in lobular breast carcinomas, *Cancer Res.*, 55, 397, 1995.

138. **Thibodeau, S.N., Bren, G., and Schaid, D.,** Microsatellite instability in cancer of the proximal colon, *Science*, 260, 816, 1993.

139. **Fishel, R., Lescoe, M.K., Rao, M.R.S., Copeland, N.G., Jenkins, N.A., Garber, J., Kane, M., and Kolodner, R.,** The human mutator gene homolog MSH2 and its association with hereditary nonpolyposis colorectal cancer, *Cell*, 75, 1027, 1993.

140. **Loeb, L.A.,** Microsatellite instability: marker of a mutator phenotype in cancer, *Cancer Res.*, 54, 5059, 1994.

141. **Wooster, R., Cleton-Jansen, A.-M., Collins, N., Mangion, J., Cornelis, R.S., Cooper, C.S., Gusterson, B.A., Ponder, B.A.J., von Deimhing, A., Wiestler, O.D., Cornelisse, C.J., Devilee, P., and Stratton, M.R.,** Instability of short tandem repeats (microsatellites) in human cancers, *Nat. Genet.*, 6, 152, 1994.

142. **Yee, C.J., Roodi, N., Verrier, C.S., and Parl, F.P.,** Microsatellite instability and loss of heterozygosity in breast cancer, *Cancer Res.*, 54, 1641, 1994.

143. **Shaw, J.A., Walsh, T., Chappell, S.A., Carey, N., Johnson, K., and Walker, R.A.,** Microsatellite instability in early breast cancer, *Br. J. Cancer*, in press, 1996.

144. **Hall, J.M., Lee, M.K., Newman, B., Morron, J.E., Anderson, L.A., Heney, B., and King, M.-C.,** Linkage of early onset familial breast cancer to chromosome 17q21, *Science*, 250, 1648, 1990.

145. **Miki, Y., Swensen, J., Shattuck-Eidens, J.D., Futreal, P.A., Harshman, K., Tavtigian, S., Liu, Q., Cochran, C., Bennett, L.M., Ding, W., Bell, R., Rosenthall, J., Hussey, C., Tran, T., McClure, M., Frye, C., Hattier, T., Phelps, R., Hangen-Strauo, A., Kalchen, H., Yakumo, K., Gholami, Z., Shaffer, D., Stone, S., Bayer, S., Wray, C., Bogden, R., Dayananth, P., Ward, J., Tonin, P., Narod, S., Bristow, P.K., Norris, F.H., Helvering, L., Morrison, P., Rosteck, P., Lai, M., Barrett, J.C., Lewis, C., Neuhausen, S., Cannon-Albright, L., Goldgar, D., Wiseman, R., Kamb, A., and Skolnick, M.H.,** A strong candidate for the breast and ovarian susceptibility gene BRCA1, *Science*, 266, 66, 1994.

146. **Friedman, L.S., Ostermeyer, E.A., Lynch, E.D., Szabo, C.I., Anderson, L.A., Dowd, P., Lee, H.K., Rowell, S.E., Boyd, J., and King, M.-C.,** The search for BRCA1, *Cancer Res.*, 54, 6374, 1994.

147. **Wooster, R., Neuhausen, S.L., Mangion, J., Quirk, Y., Ford, D., Collins, N., Nguyen, K., Seal, S., Tran, T., Averill, D., Fields, P., Marshall, G., Narod, S., Lenoir, G.M., Lynch, H., Feunteun, J., Devilee, P., Conelisse, C.J., Menko, F.H., Daly, P.A., Ormiston, W., McManus, J.R., Pye, C., Lewis, C.M., Cannon-Albright, L.A., Peto, J., Ponder, B.A.J., Skolnick, M.H., and Easton, D.F., Goldgar, D.B., and Stratton, M.R.,** Localization of a breast cancer susceptibility gene, BRCA2 to chromosome 13q 12-13, *Science*, 265, 2088, 1994.

148. **Swift, M., Morrell, R., Massey, C.L., and Chase, C.L.,** Incidence of cancer in 161 families affected by ataxia-telangiectasia, *N. Engl. J. Med.*, 325, 1831, 1991.

149. **Friedberg, E.C., Walker, G.C., and Siede, W.,** *DNA Repair and Mutagenesis*, American Society for Microbiology, Washington D.C., 1995, 662.

Chapter 5

Biological Basis for Therapy in Breast Cancer

David A. Cameron and Robert C.F. Leonard

CONTENTS

1. INTRODUCTION

Breast cancer is the commonest cancer in women in the Western world, as well as the leading cause of death from cancer. In the U.S. there are approximately 180,000 new cases per annum, with 46,000 deaths, whereas for the U.K., there are approximately 25,000 cases per annum, and 15,000 deaths, respectively. This translates to a life-time risk of developing this cancer of between 8 and 12% for the individual woman. The incidence has been slowly increasing throughout the second half of this century, and the etiological factors implicated in this epidemic have been discussed (see Chapter 2), with a dominance of hormone-related events. The exploitation of alterations in the hormonal milieu in a woman with breast cancer is an example of one of the earliest effective systemic therapies for any malignant disease,[1] and remains a mainstay in the current therapeutic armamentarium. The adversarial metaphor is deliberate — despite extensive laboratory and clinical experience, the death rate from breast cancer has remained almost unchanged for most of this century, although there are data to suggest that it might be beginning to fall as we approach its end.

Since Beatson's experiment,[1] there has been a major change in emphasis in the management of breast cancer; from local to systemic treatment. The theoretical basis for Halstead's radical mastectomy, namely, that the disease spreads through predictable anatomical routes, dominated the clinician's approach to a woman presenting with a malignant breast lump.[2] However, over the past 25 years this concept has been successfully challenged, and the current paradigm is that the disease is, at presentation, already systemic in the majority of women. The ability of systemic postoperative adjuvant therapy to improve survival has vindicated this assumption, and surgery is now seen essentially as a method of obtaining local control of the disease and providing prognostic information on which to base systemic treatment. A recent trend towards preoperative therapy may result in even fewer women requiring surgery, although this is not yet proven nor standard therapy. The current paradigm is not without its dissenters, whose objections are based as much on the inability of current practice to alter the overall death rate, as on the definite proposal of a scientifically-validated alternative model. Devitt's reminder[3] that the survival of patients can be very different with apparently similar tumors demands that we review the role a patients's physiology and/or immunity may play in determining the long-term outcome. In contrast, Craig Henderson is less controversial, merely questioning whether adjuvant systemic therapy can cure or only retard the inevitable outcome.[4] But viewed from an historical perspective, these objections to a new model are not sufficient to reject it.

Despite the near universality of the current paradigm, that the disease is essentially systemic in most women, there remains controversy concerning the most effective approach to eradicating metastatic disease. Prognostic factors, such as the number of involved axillary nodes, define the probability of a patient relapsing from her disease,[5-7] but not always the most appropriate therapy; equally a predictive factor for response to treatment, such as the estrogen-receptor (ER) status,[8] does not determine the necessity for such systemic therapy. Finally, the management of women with overt metastatic disease remains largely palliative; yet radical treatment is the basis of the approach in early breast cancer, when the majority of women also have metastases, albeit undetectable by conventional techniques.

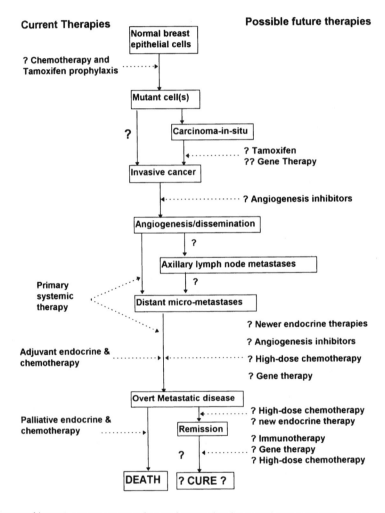

Figure 1 Stages of breast cancer progression and strategies for actual and potential therapeutic intervention.

2. PRINCIPLES

2.1 Loco-Regional Treatment

It is helpful when considering the principles behind the treatment of breast cancer to return briefly to the Halsteadian paradigm. It must not be forgotten that women have been, and will continue to be, cured of breast cancer by surgery alone,[9] since there is no evidence that breast cancer originates outside the breast. The definition of what is adequate local treatment for different tumors remains controversial, but it has now been clearly demonstrated by several studies that when excision of the tumor alone is followed by radical radiotherapy to the breast, the recurrence rate is no different from that achieved with a mastectomy.[10] This demonstrates that radiotherapy is capable of sterilizing malignant cells that remain in the breast after lumpectomy, and that the normal breast tissue does not all need to be excised, as would happen with a mastectomy. This has been shown to be the case for preinvasive cancer such as ductal carcinoma in situ (DCIS);[11,12] though a recent review highlights the fact that there are no direct comparisons with mastectomy of this conservation approach in noninvasive carcinoma.[13]

Current practice is to administer radiotherapy after loco-regional surgery, partly because historically the use of radiotherapy in the primary management of breast cancer started with its adjuvant administration after radical mastectomy. Although it undoubtedly reduces the local relapse rate,[14] it became apparent that there was no clear survival benefit, particularly with older radiotherapy techniques.[15,16] Thus, when there was a move towards less radical surgery, the radiotherapy continued to be offered after surgery. However there are biological reasons for this sequence; there is a sigmoid dose-response curve for the effect and side-effects of radiotherapy,[17] such that it is the tolerance of the normal tissues that limit the dose of radiation that can be administered. Hence the dose of radiotherapy is chosen so that it is on the threshold of toxicity, but on the steep, middle part of the dose-response curve. Furthermore, there is good evidence, particularly in locally advanced or inoperable breast cancer,[18] that the efficacy of radiotherapy in achieving local control depends on the volume of tumor present.

2.2 Systemic Treatment

Whether there are any tumors that can be safely identified as having a no metastatic potential is still open to debate, but most authorities would agree that one can identify a group of tumors for whom adjuvant systemic therapy is not warranted. They are small (e.g., less than 1 cm in size) tumors that have not spread to the axillary glands and are of either low grade or a special type (tubular, medullary).[19] Fifteen-year survivals for such tumors with no systemic therapy are of the order of 80%;[20] since the Lancet overview demonstrated that systemic adjuvant therapy improves survival at 5 years by 25 to 30%,[21] such treatment in these "good prognosis" tumors would need to be administered to over 200 women for each life that might be saved. It is usually felt that the toxicity and cost of such adjuvant therapy would not justify its use in this setting. Exactly what are the underlying biological characteristics of these good prognostic tumors is unclear; the role of the S-phase fraction is under scrutiny,[22] as are various oncogenes including p53[23] and c-erb-B2.[24] This is an important area of research, and there is a real need to follow definite guidelines in assessing prognostic factors[25,26] since the long-term survival of women with such tumors is based on historical series,[5,7] and with the advent of breast-screening, an increasing number of smaller tumors are identified, whose natural history may not be the same as small tumors that presented in the past.

However, the majority of early breast cancer is felt to be potentially or actually systemic at presentation, and adjuvant therapy is given to eliminate this burden of micrometastatic disease. The evidence that the disease is metastatic at presentation, rather than as a consequence of the primary surgery or later dissemination of residual local disease, is first the demonstration of malignant epithelial cells in the bone marrow of women undergoing surgery for operable breast cancer,[27,28] and second, some evidence that these are more commonly found in those women who are more likely to relapse from their breast cancer.[29] Furthermore, there is the clear demonstration in a number of trials that adjuvant systemic therapy improves survival over effective loco-regional treatment alone.[21] It must be stated, however, that a recent analysis of the adjuvant trials conducted by the Ludwig consortium have challenged this latter view, in that they observed that the biggest effect of adjuvant therapy was in reducing loco-regional relapse, and there was less effect on the prevention of metastatic disease.[30]

The first ever randomized trial in the treatment of any cancer was of radiation ovarian ablation versus no further treatment after mastectomy for women with operable breast cancer.[31] The key to this approach is that over 80% of a premenopausal woman's circulatory estrogen is synthesized in her ovaries, and a proportion of breast cancers demonstrate not only a dependence on estrogen for growth, but respond to estrogen deprivation.[32,33] The exact mechanism for this response was unclear, but recent work in cell lines strongly suggests that it may be as a consequence of apoptosis[34] which occurs naturally in the breast epithelium,[35] as well as the loss of a mitogenic stimulus. Although not every trial has shown a significant advantage for ovarian ablation, its benefit was upheld in the *Lancet* overview,[21] and the Scottish-Guy's trial showed that it was women with estrogen-receptor positive tumors who stood to gain — indeed, in that trial, their survival was better than women with estrogen-receptor-poor tumors given adjuvant CMF chemotherapy.[36]

The estrogen-dependence of many breast cancers was the stimulus to the use of tamoxifen in the treatment of breast cancer. Originally developed as an oral contraceptive, it emerged at the end of the 1950's as an effective antiestrogen. Relative success in the treatment of metastatic disease led to its use in the adjuvant setting.[37] During the 1970's, several large multicenter, randomized trials of adjuvant tamoxifen were begun.[38,39] Most of the larger ones, particularly if employing tamoxifen for 2 years or more, have shown a significant survival advantage for its use in the adjuvant setting as compared with observation alone. The *Lancet* overview confirmed this, and showed that tamoxifen was an effective treatment for premenopausal as well as postmenopausal women; although with less statistical certainty, since this latter conclusion was based on the pooled data of only 9,500 women, as compared with 21,000 postmenopausal women.[21] The exact mechanism of action remains unclear; tamoxifen certainly binds to the estrogen receptor-[40] (ER), causing, in model systems, an increase in the cell-loss factor,[41] necrosis[42] (perhaps by inhibiting angiogenesis[43]) and probable G_0/G_1 arrest,[44] and reduced S-phase fraction.[45,46] It may also cause a rise in apoptosis,[34] and has effects independent of the estrogen receptor, possibly mediated via TGF-β.[47] The actual cellular effects *in vivo* remain unclear, particularly since it may be of benefit in ER-negative tumors, although a reduction in the growth rate is likely.[48] Since in postmenopausal women peripheral conversion of androgens to estrogens results in a significant contribution to the overall level of circulating estrogens,[49] inhibition of the aromatase enzyme responsible has offered an alternative therapeutic strategy. Such inhibitors have not yet found a role in adjuvant therapy, but with the advent of new potent and less toxic oral aromatase inhibitors, there are plans to test the addition of an aromatase inhibitor to tamoxifen in the adjuvant setting, or even possibly to replace it.

Breast cancer has always been considered relatively sensitive to chemotherapy, and in the 1970s several trials of adjuvant chemotherapy were begun. Superficially, the rationale for this approach would seem similar to the use of the endocrine therapies discussed above; administering drugs in order to eliminate the (possible) micrometastatic disease present in a woman presenting with operable breast cancer. However, the underlying mechanisms are rather more complicated. For a metastatic lesion to be radiographically detectable, there need to be 10^9 to 10^{10} cells within it, although not all may, in fact, be viable tumor cells. However, with a tumor burden of this order of magnitude, cure with conventional chemotherapy is essentially impossible. There may be several reasons for this, but two that have bearing on the justification for using chemotherapy in the adjuvant setting must be considered. First, cancer cells can develop resistance to cytotoxics, thus rendering them less effective.[50] In some cases the mechanism for this resistance occurs as a result of a genetic mutation within the cells,[50] and if the tumor burden is smaller, for example around 10^6 to 10^8 cells, there is less chance of this having occurred.[51] Second, there are data to suggest that tumors have a higher growth rate when at a smaller volume, demonstrating a growth pattern that can be described by the Gompertzian function (*vide infra*).[52] Cytotoxics will kill any rapidly dividing cells in the patient, and in general malignant cells may be more sensitive than the normal tissues. Thus, the higher the proliferating fraction of the malignant cells, the better the chance of eliminating them with chemotherapy. Repeating the treatment every few weeks allows the normal tissues to recover, but hopefully the malignant cells, even if they have regrown somewhat, will not have reached their previous volume, and thus through the use of successive courses they will eventually be reduced to a level that may permit host immune cell killing to eradicate the remaining cells.

However, it was soon realized that more than 6 months' therapy produced little increment in cure,[53] and in view of the large proportion of women who are rendered menopausal by cytotoxic therapy, it was questioned whether the chemotherapy was really killing the breast cancer cells or acting via ovarian ablation. This latter issue is still unresolved, but the Scottish-Guy's trial[36] has shown that even if the two treatments are equivalent in some cases, they are not identical in their effect on all premenopausal women. Whether or not part of the effect of adjuvant chemotherapy is mediated by its effect on the host, rather than directly on the tumors cells, the current adjuvant chemotherapy regimens may not be of the best design. The administration of a fixed number of intermittent cycles of the same doses of the same drugs has never been shown to be the most efficacious; the concept has more to do with minimizing toxicity than maximizing the efficacy against an individual tumor. Indeed, in one of the few trials to seriously address the issue of schedule alone, Bonadonna has clearly shown that the order of drug administration is important, with significantly better survival for patients given doxorubicin before CMF, rather than in an alternating regimen.[54]

The issue of tumor growth rates can be examined further, and has therapeutic implications. *In vivo* studies of tumor growth rates are difficult to perform, but several approaches have been tried. The simplest is to monitor tumor volume over time, and deduce tumor-doubling times from these data. However, it is clearly unethical to leave a tumor untreated, and so many of these studies are based on historical series, or on comparison of previous mammogram volumes, once a tumor has subsequently been diagnosed. Some groups have looked at *in vivo* labeling of the S-phase fraction with ^{3}H-thymidine, or measurement of the same fraction with flow cytometry. These can only give a measure of instantaneous growth, and take no account of cell loss factors. Therefore, a considerable volume of data has been generated from animal studies wherein human breast cancers have been implanted into immunodeficient animals. There are excellent data to support the idea of decelerating growth for tumor xenografts, and the Gompertzian model describes this very well. This mathematical function describes exponential growth at all times, but the rate of growth decays exponentially with time, resulting in an infinite doubling time at the maximum volume.[55] Studies of several tumor types in different hosts have clearly shown that the maximum volume is host-specific rather than tumor-specific,[56] and thus it is relatively easy to imagine a similar model applying to clinical tumors, with the maximum volume taken as around 1 litre or 10^{12} cells.[52] Certainly it is uncommon to find many patients surviving with a measurable clinical tumor burden of greater than this! There is clear evidence in myeloma[57] and metastatic testicular cancer[58] that Gompertzian growth occurs in human tumors. For breast cancer, controversy remains as to the precise nature of the growth function that best describes the data. There is one well-documented historical series of untreated breast cancer,[59] and two separate groups have drawn different conclusions from the same data. Norton has fitted the data with a Gompertzian function.[52] However, Retsky et al. have fitted a more complicated growth pattern, where they permit all growth to be Gompertzian, but tumors can have long periods of dormancy.[60] This idea has been given further credence by an analysis of time to local recurrence in women who had a mastectomy.[61] Spratt showed that data on mammographic measurements of primary breast cancer could be fitted with any one of a number of decelerating equations, with the Gompertzian function not in fact giving the best fit.[62] Furthermore, Brown showed that simple exponential growth (which is in essence a simplified form of Gompertzian growth with an infinite maximal volume) gives as good fit as either Gompertzian or logistic functions to historical data on tumor size at presentation.[63] Clearly there is still much controversy about actual clinical tumor growth rates, but the important conclusion from these studies is that if the growth rate is faster in smaller tumors, it may be more difficult to treat disease as it regresses, such as occurs during effective chemotherapy.[64] Norton, a major advocate of Gompertzian growth, has therefore argued that the dose of chemotherapy should be increased during therapy[65] — although this conclusion depends heavily on a dose-response relationship which even if well established *in vitro*, remains unproven in clinical breast cancer. Several on-going studies, including the CALGB-SWOG and ECOG-SWOG studies in the U.S., and the Anglo-Celtic study in Western Europe, seek to test this by examining the benefit of myelo-ablative therapy after several cycles of anthracycline-based adjuvant chemotherapy; unfortunately, there has been no direct test of this concept.

3. PRACTICE

It is not within the scope of this chapter to detail many of the different regimens used in the treatment of breast cancer. However, to clarify the issues in the remainder of the discussion, it is appropriate to comment on the treatments in more detail.

3.1 Tamoxifen

Tamoxifen is increasingly used for the adjuvant treatment of breast cancer and, in most centers, 20 mg daily is the usual dose. However, it should be remembered that this is an empirical dose — and although others have been tried, particularly in Scandinavia,[65] there is little evidence to indicate what is the correct dose. Daily doses of more than 20 mg, particularly for more than 2 years, seem to be associated with an increased risk of second malignancy,[65] but there has been no attempt to define doses that are tailored to individuals, unlike chemotherapy when drugs are commonly given corrected for the patient's (estimated) surface area. Furthermore, although the ideal duration of treatment remains unknown, a minimum of 2 years and maximum of 5 years is the usual practice. However, as the role of adjuvant tamoxifen is extended from single-agent use in postmenopausal women to premenopausal women and in combination with chemotherapy, there are few data as to optimum schedule and duration — even the issue of whether or not tamoxifen should be started with or after the chemotherapy is unclear, with no definite evidence for an antagonistic effect,[67] but theoretical considerations suggest concurrent use may not be the ideal — tamoxifen reduces the tumor cell proliferation,[68] and thus may make the tumor less sensitive to the chemotherapy. There is a small study suggesting increased expression of P-glycoprotein in tumors that fail to respond to tamoxifen[69] (the P-glycoprotein efflux pump is one of the mechanisms by which cells become resistant to cytotoxics drugs[50]). If confirmed, this would have profound implications for concurrent chemo-endocrine therapy, since it could render hormone-resistant tumors less sensitive to what is currently the only other effective adjuvant strategy. Phase III trials of sequential versus concurrent adjuvant chemo-endocrine therapy are under way, and results are awaited with interest. However, the EORTC have reported a randomized trial in locally advanced breast cancer in which disease-free intervals were statistically significantly longer for those women given combined chemotherapy and endocrine therapy (which included tamoxifen).[70] An NSABP trial also showed a survival benefit for the combination of chemotherapy and endocrine therapy, but largely in ER-positive postmenopausal women, and the chemotherapy (melphalan and 5-fluorouracil) may have been suboptimal.[71]

3.2 Ovarian Ablation

Since the benefit of this mode of adjuvant therapy has been reconfirmed,[21] there are now more ways of inducing a postmenopausal environment for a tumor. There has always been the option of a radiation-induced menopause, as used in the original Manchester trial.[31] Surgical bilateral oophorectomy is much easier and safer than it was in the past, largely due to a combination of improved anesthetic technique, and the use of laparoscopic or "key-hole" surgery. Finally the LHRH agonists, delivered as a subcutaneous implant every 4 weeks, will result in reversible ovarian suppression. Thus, if the woman is uncertain as to whether she would tolerate an iatrogenic menopause as part of her treatment, it can be achieved temporarily every 4 weeks by an injection, and later rendered permanent by surgery if appropriate. Equally, when treating disease preoperatively[72] such a strategy can again be used to ensure that there is sufficient therapeutic benefit to warrant permanent ovarian ablation with its concomitant symptoms and long-term consequences for the woman; and for metastatic disease, ovarian ablation can be maintained in this manner only for as long as it is effective.[73,74]

3.3 Chemotherapy

There are innumerable chemotherapy regimens in current use, both adjuvantly, and in particular for metastatic disease. However, there remain a few commonly used combinations that are often considered to be "standard" — at least in the sense of providing data with which newer regimens can be compared. In the adjuvant setting there remains one unresolved issue, which is whether anthracyclines are superior. An interim analysis of a large Canadian multicenter trial comparing epirubicin with methotrexate in combination with cyclophosphamide and 5-fluorouracil has shown improved relapse-free survival for the use of epirubicin, but with increased toxicity and as yet no survival benefit.[75] However, other trials suggest only marginal benefit.[76,77] The higher-risk patients, such as those with more than four nodes involved, may benefit, particularly if doxorubicin is administered alone for the first four cycles.[54] However, there is concern about both the short-term toxicity and long-term cardiac side-effects of anthracyclines. Thus, many centers still use CMF as standard adjuvant chemotherapy — either in the "classical" regimen of cyclophosphamide 100 mg daily for the first 14 days, and methotrexate 40 mg/m^2 and 5-fluorouracil 600 mg/m^2 on days 1 and 8 of a 4-week cycle; or the intravenous version, where 600 to 750 mg/m^2 of cyclophosphamide are given together with 50 mg/m^2 of methotrexate and 600 mg/m^2 every 3 weeks, usually for six cycles. For anthracycline-based regimens, there are perhaps three standards — AC (doxorubicin 60 mg/m^2 and cyclophosphamide 600 mg/m^2) every 3 weeks for four cycles, CAF (cyclophosphamide 600 mg/m^2, doxorubicin 40 mg/m^2 and 5-fluorouracil 600 mg/m^2), and FEC (5-fluorouracil 600 mg/m^2, epirubicin 60 mg/m^2 and cyclophosphamide 600 mg/m^2); both these latter usually given for six cycles at three weekly intervals.

3.4 Inoperable Disease

A large part of the above discussion has centered on operable, and thus potentially curable breast cancer. However, for various reasons, many women still present with disease that is not operable. In the past some of these women were treated with a "toilet" mastectomy, but the local relapse rate was high, and the long-term survival very poor. Over the past few decades, there has been a strong move toward preoperative treatment, with systemic drug treatment and/or radiotherapy. The best results for local control of the disease appear to be with the use of preoperative chemotherapy and radiotherapy;[78,79] some controversy remains about the necessity for surgery.[80] However, the initial optimism that the earlier administration of systemic therapy might improve the overall survival has not stood the test of time. Thus, many centers now concentrate more on obtaining effective local control, and where possible, the down-staging of the tumor such that a mastectomy (or even wide-local excision) becomes feasible. To this end, similar therapeutic options exist as for systemic therapy as for adjuvant treatment.

Chemotherapy, often with an anthracycline-based regimen, is the mainstay of the drug treatment, although some choose to offer endocrine therapy to the woman with an ER-positive tumor — ovarian ablation for the premenopausal and tamoxifen for the postmenopausal patient. In general, down-staging occurs in the majority of patients with less side-effects than chemotherapy — where the tumor fails to respond, chemotherapy can be administered. Since there are few data to suggest that the preoperative therapy alters long-term outcome, the use of an endocrine maneuver is unlikely to have any detrimental effect on the woman's survival.

3.5 Metastatic Disease

Although overtly metastatic breast cancer remains sensitive to the same treatment modalities used for earlier stage disease, the practice is somewhat different since at least with conventional regimens there is no realistic hope of cure. In the majority of cases, the emphasis is to treat the symptoms due to disease by reducing tumor size. Thus, wherever possible, an endocrine maneuver is used, and for those women who have relapsed despite tamoxifen given adjuvantly, or in whom the disease has progressed after initial tamoxifen therapy, hormone approaches are possible. ER-positive breast cancer cells often express progesterone receptors, and may respond to treatment with progestogens perhaps by undergoing further differentiation, in a manner analogous to endometrial cells in the luteal phase of the menstrual cycle.[81] Although progestogens are not cytotoxic, the approach is effective in treating advanced disease, and megestrol acetate (megace) is commonly used as a second-line agent for metastatic disease, at a dose of 160 mg daily. However, progestogens are associated with significant side-effects, of which fluid retention is probably the most troublesome, and thus the newer aromatase inhibitors are currently undergoing phase III trials in direct comparison with megace. The aromatase inhibitor aminoglutethimide (AMG) is still widely used, both as second- and third-line treatment for advanced disease. In an analysis of hormone treatments for metastatic breast cancer, Rose and Mourisden reported 30% cross-sensitivity between tamoxifen and AMG with surprisingly low cross-resistance of about 10%.[82] This led the authors to recommend the use of AMG rather than progestogens as second-line hormone therapy in view of the side-effect profile of megace.

Estrogen-receptor-negative disease has a low response rate to tamoxifen treatment (of the order of 10%),[82] and thus there is an important role for chemotherapy in the treatment of metastatic disease. Furthermore disease that was initially hormone-sensitive may become resistant to these maneuvers, and chemotherapy can then be used, although the response rate is often lower than when treating disease that has not been given prior hormonal therapy.[83] Visceral metastases, in particular liver involvement, are associated with a particularly poor prognosis,[84] and many clinicians will treat all patients with visceral disease with cytotoxics. Primarily, this is because the pace of disease becomes important; and the higher and faster response rates in general seen with cytotoxic therapy dictate that this is usually the treatment of choice; and if there is significant disturbance of liver function this may place constraints on the use of certain drugs due to altered metabolism.

For the treatment of metastatic or inoperable disease the same regimens as used in adjuvant therapy (or variations upon) are often used, but because of the overall poor long-term results, this is a fertile ground for the testing of newer regimens, drugs (e.g., the taxanes[85]), and routes of administration (e.g., with continuous 5-fluorouracil[86]). Of particular interest at present is the use of increased dose-intensity, including so-called "high-dose chemotherapy," with the patients' own bone-marrow progenitor cells re-infused after the chemotherapy in order to permit the bone marrow to reconstitute. In the past bone-marrow progenitor cells were obtained by withdrawing some of a patients' own bone marrow under a general anaesthetic. However, there is increasing use of peripheral blood progenitor cells which are removed from the patients' peripheral blood using a cell-separator after a suitable priming regimen, usually consisting of a cycle of chemotherapy and G-CSF. Phase II studies in North America suggest that up to 25% of patients may remain in durable complete responses;[87,88] but whether any of these women have been cured is yet to be determined. Randomized trials are under way in North America and Europe.

4. RECENT DEVELOPMENTS

Adjuvant therapy for breast cancer is well-established, but current data suggest that it may cure at most 20 to 30% of women with operable breast cancer. Despite the almost universal application of systemic treatment to women with metastatic disease, none are cured by this approach. It is clear therefore that current treatments are inadequate, and an understanding of the biology of the disease may allow more effective treatments to be designed and tested.

4.1 Alternative Endocrine Strategies

Mention has been made earlier of the use of progestogens in the treatment of metastatic disease, which are thought to act by promoting differentiation of the cells, thus reducing the proliferative fraction. It has often been noted that tamoxifen treatment causes a fall in the measured estrogen receptor level;[89] yet at least in the early stages of treatment it may promote progestogen receptor synthesis, as a consequence of its partial agonist activity. Thus, sequential cyclical administration of tamoxifen followed by a progestogen might allow replenishment of the progestogen receptor level and hence increase the efficacy of the progestogen administered. A small study testing this approach had been undertaken in Scotland,[90] in a randomized study comparing continuous tamoxifen 20 mg daily against a repeated six-week cycle of tamoxifen for two weeks, followed by megace 160 mg daily for three weeks, followed by one week off treatment. Interestingly there was no effect on response, but a significant survival benefit of 134 weeks as compared with 90 weeks in favour of the sequential arm. Statistically the only benefit appeared to be delayed progression in the nonresponders, but in fact there was a trend to more complete and partial responses as well. The authors' explanation for this benefit appearing largely in non-responders is that tumors that are borderline-endocrine responders may be the only one to benefit from this approach — those that will respond to tamoxifen may well do so with or without the addition of progestogen. What is needed then is a further study to test the efficacy of this approach in disease that appears stable on tamoxifen; where the comparison might be against standard second-line therapy with continuous megace.

Another area of current development is in the testing of more specific, and less toxic aromatase inhibitors. This has already been noted above in brief, but it is pertinent to discuss their potential role further. Local estrogen synthesis occurs within the breast.[91] Furthermore, there is evidence that breast cancers display aromatase activity which converts androgens into estrogens.[92] Thus aromatase inhibition might be a more logical method of estrogen deprivation, particularly in the postmenopausal woman, since it can potentially inhibit synthesis of estrogen both within and beyond the breast. It remains uncertain how important local estrogen synthesis is in the maintenance of a breast cancer, but this is clearly an area for further work, and potential therapeutic developments. Indeed if this approach proves successful, one could speculate on the possible benefits of a more complete estrogen blockade that might be achievable in premenopausal women by the co-administration of an LHRH agonist together with an aromatase inhibitor.

4.2 Tamoxifen Resistance

As tamoxifen is increasingly used in the adjuvant setting, there will be a higher number of women who will present with disease that has relapsed while on tamoxifen. The reasons for the development of such tamoxifen-resistant disease will be multifarious, including the fact that some cases will have been primarily resistant to tamoxifen. For these women various secondline hormone strategies are already available, but they all rely on the presence of on-going endocrine sensitivity. It has been demonstrated *in vitro* that tamoxifen resistance can develop to the extent that the agent actually stimulates growth[93] — and reports that about 10% of patients demonstrate a response to tamoxifen withdrawal alone suggests that this is a real phenomenon.[94] This has profound implications for treatment of relapsed disease, because until the mechanism is clearly understood, one cannot be sure that a switch to an alternative endocrine maneuver will be effective.

One solution to this alteration in response to tamoxifen might be the newer pure antiestrogens, and it has been shown that the parenterally-administered agent ICI 182 780 can inhibit the stimulatory effect of both tamoxifen and estrogen in this setting.[95] Furthermore, this drug has also been shown to be active in tamoxifen-resistant advanced disease.[96] As far as can be determined, alterations in tamoxifen metabolism[97] and mutations in the estrogen receptor[95] are not the main mechanisms, although both are known to occur.

A word of caution needs to be expressed when considering future developments in endocrine treatment of breast cancer. Tamoxifen is not universally effective, even in ER-rich tumors. This might in part be because it is a partial-estrogen agonist. But as a partial agonist it has directly beneficial effects, both in the postmenopausal woman as well as potentially in a premenopausal woman who is rendered post-menopausal as a consequence of her treatment. There is no doubt that it can prevent bone loss,[98] and in some studies it has been shown to reduce cardiovascular mortality in postmenopausal women.[99] Any increased survival benefit from the use of a pure antiestrogen in the adjuvant setting could be counterbalanced by an increased mortality and morbidity from these nonmalignant pathologies. Other by-products of the search for an effective antiestrogen may have a role in the secondary prevention of these problems, and Raloxifene (LY139481 Hcl) is a candidate drug. It has been shown that this drug can reduce bone loss and lower serum cholesterol in ovariectomized rats without demonstrating any proestrogenic effect on the uterus.[100]

4.3 Oncogene-Directed Therapy

Any progress in the development of better endocrine therapies for breast cancer is unlikely to impact on the treatment of ER-poor disease. These tumors manifest little or no estrogen receptor, and as such are unlikely to benefit from alterations in the level of circulating estrogen, or from receptor inhibition. A "gene therapy" approach has been considered for these tumors — if it were possible with a vector to induce expression of a functioning receptor, or possibly even reactivate the dormant genes naturally responsible for the expression of the estrogen receptor (and post-receptor apparatus). Certainly both mutant and wild-type ER can be transfected into an ER-negative breast cancer cell line — although paradoxically it was found that estradiol inhibited growth in this system!

An alternative therapeutic window for "gene therapy" relates to the oncogene c-erb-B2. Known to be overexpressed in around one third of invasive breast cancers, it tends to be associated with ER negative, poorer prognosis tumors.[101] If the growth advantage following expression of this oncogene at the message level could be nullified by the use of an antisense DNA sequence, to which the mRNA would bind, then the tumor might be rendered more susceptible to alternative therapeutic agents. A trial of locally administered c-erb-B2 antisense oligonucleotides is underway at the Hammersmith Hospital London (K. Sikora, personal communication). Even if this approach is active in the setting under consideration, which is locally relapsed skin disease, there is still a long way to go before it might help the majority of women with c-erb-B2 positive breast cancer.

However, the expression of c-erb-B2 offers another avenue for therapy. A monoclonal antibody has been synthesized, and it has been clearly shown that co-administration with cisplatinum results in regression of breast cancer xenografts.[102] Pilot studies in patients have shown some clinical activity,[103] and therefore multicenter, randomized trials are being planned, combining standard chemotherapy for advanced disease with (or without) a monoclonal antibody to c-erb-B2 in patients whose tumors express this oncogene.

Increasing numbers of oncogenes are being recognized as functionally important in breast cancer, and similar approaches may be possible for other genes. However, if overexpression of an oncogene is the primary problem, it might be helpful to target therapy to the source of this overexpression, which in some circumstances is in extra-chromosomal DNA or "double minute" (DM) chromosomes.[104] These can occur in breast cancers, and preliminary data from San Antonio suggests that in another female malignancy, ovarian cancer, low-dose hydroxyurea can reduce the number of such DM chromosomes, at least temporarily (von Hoff, personal communication).

4.4 Chemotherapy

There are some new drugs on the horizon of clinical practice. Phase I and II studies with the taxanes (taxotere and taxol) have demonstrated their efficacy in breast cancer,[85] even when previously exposed to anthracyclines. This is important because there are situations where an anthracycline is used in the adjuvant setting. Vinorelbine, a synthetic vinca-alkaloid, appears to be active even as an oral agent, but without the neurotoxicity of its predecessors in the class.[105] Whether these drugs will ever replace those currently used in the adjuvant setting is uncertain.

What is perhaps more important is the continued exploration of the best method of using the current pharmacopoeia. As mentioned above, the use of myeloablative therapy is being tested in randomized trials both for metastatic and high-risk operable breast cancer. Support for such therapy by peripherally harvested bone marrow progenitor cells should allow a reduction in its morbidity and mortality, so that any therapeutic gains can be translated into a survival advantage. Alternative ways of using drugs are being explored, and Lokich has perhaps more than any other clinician pioneered the continuous intravenous administration of cytotoxics,[106] when as for 5-fluorouracil, the pharmacology suggests that this may be a more logical method of administration.[107] However, it is perhaps in understanding the biology of the disease that there may be more scope for development. Since cytotoxics, under the current paradigm, are meant to act on actively dividing cells, increasing the growth rate by the prior administration of estrogen might improve their efficacy. That this is possible has been suggested in inoperable disease, but not on a wider scale. A SWOG phase II study confirmed that the administration of estradiol before each cycle of chemotherapy could increase the s-phase fraction of the tumor, but there was no evidence of an enhanced response to the anthracycline-based chemotherapy.[108] However, it is now appreciated that many cytotoxics cause sufficient damage to the DNA of a cell that its own surveillance system determines that it is beyond repair and the cell undergoes programmed cell death or apoptosis.[109] Mutant p53 appears to disable this cellular self-destruct pathway,[110] and an increased understanding of the signals that determine whether or not a cell undergoes apoptosis might permit therapeutic strategies that can increase the cell-kill delivered by the drugs we currently have, but with little or no increase in dose (and thus toxicity).

4.5 Angiogenesis

In order for a tumor to successfully grow beyond a small size, it must develop a nexus of blood vessels within itself that are connected to the host circulation. Angiogenesis, the process of developing these new vessels, has been the focus of research over the past 30 years. Many different biochemical and genetic mechanisms are involved, both from the tumor and the host.[111] The clinical importance of angiogenesis has become apparent from a number of lines of research. Several studies of the degree of vascularization of primary breast cancer have shown this to be an independent prognostic factor for survival,[112-114] but this does not prove that angiogenesis is the important factor — it may be a surrogate marker for another biological difference in some tumors. A large number of factors inhibitory to angiogenesis have been identified, including IFN-α and β,[115] and more importantly a 38 kDa protein called angiostatin, which is produced by some primary tumors, and disappears from the circulation within days of surgical excision.[112] In a xenograft system the rate of apoptosis in metastases significantly fell when the primary tumor was removed, an effect which could be prevented by the angiogenesis inhibitor TNP-470.[117] This change in the cell-loss rate was accompanied by an increase in metastatic volume, as measured by lung weight. The implications for this in the clinical setting are enormous; if surgical excision of the primary really does confer a growth advantage on micro-metastatic disease, as demonstrated in xenografts by Fisher,[118] and this is due to changes in angiogenesis, as suggested by the effect of the angiogenesis-inhibitor, then the timing and/or type of adjuvant systemic therapy that is given will need to be rethought. However, it must be borne in mind that there are clinical data which suggest that the solution is not necessarily that simple. Despite the survival benefit shown by a single perioperative cycle of cyclophosphamide in the classic trial reported by Nissen-Meyer,[119] several other trials of immediate postoperative chemotherapy have failed to show an advantage over more conventionally timed adjuvant treatment, and the recent interest in preoperative chemotherapy for operable disease has yet to show a convincing survival advantage. Perhaps more specific antiangiogenesis therapy is required, and several inhibitors including antibodies to endothelium[111] are undergoing phase I and II studies. However, when considering any application of these therapies to patients before radiotherapy, it needs to be recalled that hypoxic tumors are more radioresistant, and thus adjuvant antiangiogenesis treatment could have unexpected consequences for the efficacy of standard radiotherapy.

5. CONCLUSIONS

The appalling death rate from breast cancer is still a major health care problem in the Western world, and despite the newer therapeutic strategies that have resulted from our increased understanding of its biology, we are a long way from having a cure for this disease. In the past, a "magic bullet" was sought for cancers; the very least that the biology teaches us is that we are more likely to improve the outcome if we consider the heterogeneity of the disease as necessitating a variety of different treatments for different women.

REFERENCES

1. **Beatson, G.T.,** On the treatment of inoperable cases of carcinoma of the mamma: suggestions for a new method of treatment, with illustrative cases, *Lancet*, 2, 104, 1896.
2. **Halstead, W.S.,** The results of operations for the cure of cancer of the breast performed at the Johns Hopkins Hospital from June 1889 to January 1894, *Johns Hopkins Hosp. Rep.*, 4, 297, 1894.
3. **Devitt, J.E.,** Breast cancer: have we missed the forest because of the trees?, *Lancet*, 344, 734, 1994.
4. **Henderson, I.C.,** Paradigmatic shifts in the management of breast cancer, *N. Engl. J. Med.*, 332, 951, 1995.

5. **Schottenfeld, D., Nash, A.G., Robbins, G.F., and Beattie, E.J.,** Ten-year results of the treatment of primary operable breast cancer, *Cancer*, 38, 1001, 1976.

6. **Valagussa, P., Bonadonna, G., and Veronesi, U.,** Patterns of relapse and survival following radical mastectomy: analysis of 716 consecutive patients, *Cancer*, 41, 1170, 1978.

7. **Fisher, B., Slack, N., and Bross, I.D.,** Cancer of the breast: size of neoplasm and prognosis, *Cancer*, 24, 1071, 1969.

8. **Osborne, C.K., Yochmowitz, M.G., Knight, W.A., and McGuire, W.L.,** The value of estrogen and progesterone receptors in the treatment of breast cancer, *Cancer*, 46, 2884, 1980.

9. **Adair, F., Berg, J., Joubert, L., and Robbins, G.F.,** Long-term follow-up of breast cancer patients: the 30 year report, *Cancer*, 33, 1145, 1974.

10. **Harris, J.R., Morrow, M., and Bonadonna, G.,** Cancer of the Breast, in *Cancer: Principles and Practice of Oncology*, 3rd ed., DeVita, V.T., Jr., Hellman, S., and Rosenburg, S.A., Eds., J.B. Lippincott Co., Philadelphia, 1993, 1284.

11. **Silverstein, M.J., Cohlan, B.F., Gierson, E.D., Furmanski, M., Gamagami, P., Colburn, W.J., Lewinsky, B.S., and Waisman, J.R.,** Duct carcinoma in situ: 227 cases without microinvasion, *Eur. J. Cancer*, 28, 630, 1992.

12. **Fisher, B., Costantino, J., Redmond, C., Fisher, E.R., Margolese, R., Dimitrov, N., Wolmark, N., Wickerham, D.L., Deutsch, M., Ore, L., Mamounas, E., Poller, W., and Kavanah, M.,** Lumpectomy compared with lumpectomy and radiation therapy for the treatment of intraductal breast cancer, *N. Engl. J. Med.*, 328, 1581, 1993.

13. **Frykberg, E.R. and Bland, K.I.,** Overview of the biology and management of ductal carcinoma in situ of the breast, *Cancer*, 74, 350, 1994.

14. **Harris, J.R., Morrow, M., and Bonadonna, G.,** Cancer of the breast, in *Cancer: Principles and Practice of Oncology*, 3rd ed., DeVita, V.T., Jr., Hellman, S., and Rosenburg, S.A., Eds., J.B. Lippincott Co., Philadelphia, 1993, 1289.

15. **Jones, J. and Ribeiro, G.,** Mortality patterns over 34 years of breast cancer patients in a clinical trial of postoperative radiotherapy, *Clin. Radiol.*, 40, 204, 1989.

16. **Høst, H., Brennhovd, I.O., and Loeb, M.,** Post-operative radiotherapy in breast cancer-long-term results from the Oslo study, *Int. J. Radiat. Oncol. Biol. Phys.*, 12, 727, 1986.

17. **Hellman, S.,** Principles of radiation therapy, in *Cancer: Principles and Practice of Oncology*, 3rd ed., DeVita, V.T., Jr., Hellamn, S., and Rosenburg, S.A., Eds., J.B. Lippincott Co., Philadelphia, 1993, 263.

18. **Fletcher, G. and Montague, E.,** Carcinoma of the breast: criteria of operability, *Am. J. Roentgenol. Radium Ther. Nucl. Med.*, 93, 573, 1965.

19. **McGuire, W.L. and Clark, G.M.,** Prognostic factors and treatment decisions in axillary-node-negative breast cancer, *N. Engl. J. Med.*, 326, 1756, 1992.

20. **Galea, M.H., Blamey, R.W., Elston, C.E., and Ellis, I.O.,** The Nottingham Prognostic Index in primary breast cancer, *Breast Cancer Res. Treat.*, 22, 207, 1992.

21. Early Breast Cancer Trialists Collaborative Group, Systemic treatment of early breast cancer by hormonal, systemic or immune therapy: 133 randomised trials involving 31,000 recurrences and 24,000 deaths among 75,000 women, *Lancet*, 339, 18, 71, 1992.

22. **Clark, G.M., Mathieu, M.-C., Owens, M.A., Dressler, L.G., Eudey, L., Tormey, D.C., Osborne, C.K., Gilchrist, K.W., Mansour, E.G., Abeloff, M.D., and McGuire, W.L.,** Prognostic significance of S-phase fraction in Good-risk, Node negative breast cancer patients, *J. Clin. Oncol.*, 10, 428, 1992.

23. **Allred, D.C., Clark, G.M., Elledge, R., Fuqua, S.A.W., Brown, R.W., Chamness, G.C., Osborne, C.K., and McGuire, W.L.,** Association of p53 Protein expression with tumour cell proliferation rate and clinical outcome in Node-negative breast cancer, *J. Natl. Cancer Inst.*, 85, 200, 1993.

24. **Gasparini, G.,** Dalla Parma, P., Maluta, S., Caffo, P., Leonardi, E., Meli, S., Cazzavillan, S., Pozza, F., and Bevelacqua, S., Expression of the c-erbB-3 and cerbB-2 Gene products lacks prognostic value i node-negative breast carcinoma (NNBC), *Proc. A.S.C.O.*, 12, 65, 1993.

25. **McGuire, W.L.,** Breast cancer prognostic factors: evaluation guidelines, *J. Natl. Cancer Inst.*, 83, 154, 1991.

26. **Leonard, R.C.F.,** Tumour markers of prognosis, *J. Clin. Oncol.*, 9, 1102, 1991.

27. **Ménard, S., Squicciarini, P., Luini, A.J., Sacchini, V., Rovini, D., Tagliabue, E., Veronesi, P., Salvadori, B., Veronesi, U., and Colnaghi, M.I.,** Immunodetection of bone marrow micrometastases in breast carcinoma patients and its correlation with primary tumour prognostic features, *Br. J. Cancer*, 699, 1126, 1994.

28. **Mansi, J.L., Berger, U., McDonnell, T., Pople, A., Rayter, Z., Gazet, J.C., and Coombes, R.C.,** The fate of bone marrow micrometastases in patients with primary breast cancer, *J. Clin. Oncol.*, 7, 445, 1989.

29. **Ünal, E., Çamlibel, M., Coskun, F., Cengiz, Ö., Ruacan, S., and Dinçtürk, C.,** The malignant cells in sternal bone marrow of breast cancer patients and staging. A prospective study, *J. Exp. Clin. Cancer Res.*, 13, 165, 1994.

30. **Goldhirsch, A., Gelber, R.D., Price, K.N., Castiglione, M., Coates, A.S., Rudenstam, C.M., Collins, J., Lindtner, J., Hacking, A., Marini, G., Byrne, M., Cortés-Funes, H., Schnürch, G., Brunner, K.W., Tatersall, M.H.N., Forbes, J., and Senn, H.-J.,** Effect of systemic adjuvant treatment on first sites of breast cancer relapse, *Lancet*, 343, 377, 1994.

31. **Cole, M.,** A clinical trial of an artificial menopause in carcinoma of the breast, *Horm. Breast Cancer INSERM*, 55, 143, 1975.

32. **Buchanan, R.B., Blamey, R.W., Durrant, K.R., Howell, A., Paterson, A.G., Preece, P.E., Smith, D.C., Williams, C.J., and Wilson, R.G.,** A randomised comparison of tamoxifen with surgical oophorectomy in premenopausal patients with advanced breast cancer, *J. Clin. Oncol.*, 4, 1326, 1986.

33. **Anderson, E.D.C., Forrest, A.P.M., Levack, P.A., Chetty, U., and Hawkins, R.A.,** Response to endocrine manipulation and oestrogen receptor concentration in large operable primary breast cancer, *Br. J. Cancer*, 60, 223, 1989.

34. **Warri, A.M., Huovinen, R.L., Martikainen, P.M., and Härkönen, P.L.,** Apoptosis in Toremifene-induced growth inhibition of human breast cancer cells in vivo and in vitro, *J. Natl. Cancer Inst.*, 85, 1412, 1993.

35. **Ferguson, D.J.P. and Anderson, T.J.,** Ultrastructural observations on cell death by apoptosis in the "resting" human breast, *Virchows Arch. [Pathol. Anat.]*, 393, 193, 1981.

36. Scottish Cancer Trials Breast Group and ICRF Breast Unit, Guy's Hospital, London, Adjuvant ovarian ablation versus CMF chemotherapy in premenopausal women with pathological stage II breast carcinoma: the Scottish trial, *Lancet*, 341, 1293, 1993.

37. **Cole, M.P., Jones, C.T., and Todd, I.D.,** A new anti-oestrogenic agent in late breast cancer. An early clinical appraisal of ICI 46,474, *Br. J. Cancer*, 25, 270, 1971.

38. Breast Cancer Trials Committee: Scottish Cancer Trials Office (MRC) Adjuvant tamoxifen in the management of operable breast cancer: the Scottish trial, *Lancet*, 2, 171, 1987.

39. Nolvadex Adjuvant Trial Organisation (NATO), Controlled trial of tamoxifen as a single adjuvant agent in the management of early breast cancer: analysis at 8 years, *Br. J. Cancer*, 57, 608, 1988.

40. **Horwitz, K.B. and McGuire, W.L.,** Nuclear mechanisms of estrogen action: effects of oestradiol and anti-estrogens on estrogen receptors and nuclear receptor processing, *J. Biol. Chem.*, 253, 8185, 1978.

41. **Brunner, N., Bronzert, D., Vindeløv, L.L., Rygaard, K., Spang-Thomsen, M., and Lippman, M.E.,** Effect on growth and cell cycle kinetics of estradiol and tamoxifen on MCF-7 human breast cancer cells grown in vitro and in nude mice, *Cancer Res.*, 497, 1515, 1989.

42. **Haran, E.F., Maretzek, A.F., Goldberg, I., and Degani, H.,** Tamoxifen enhances cell death in implanted MCF7 breast cancer by inhibiting endothelium growth, *Cancer Res.*, 549, 5511, 1994.

43. **Gagliardi, A. and Collins, D.C.,** Inhibition of angiogenesis by antoestrogens, *Cancer Res.*, 530, 533, 1993.

44. **Tominaga, T., Yoshida, Y., Matsumoto, A., Hayashi, K., and Kosaki, G.,** Effects of tamoxifen and the derivative (tat) on cell-cycle of MCF-7 in-vitro, *Anticancer Res.*, 13, 661, 1993.

45. **Osborne, C.K., Boldt, C.H., Clark, G.M., and Trent, J.M.,** Effects of tamoxifen on human breast cancer cell cycle kinetics: accumulation of cells in early G1 phase, *Cancer Res.*, 43, 3583, 1983.

46. **Lykkesfeldt, A.E., Larsen, J.K., Christensen, I.J., and Briand, P.,** Effects of the antioestrogen tamoxifen on the cell cyle kinetics of the human breast cancer cell line, MCF-7, *Br. J. Cancer*, 49, 717, 1984.

47. **Colleta, A.A., Benson, J.R., and Baum, M.,** Alternative mechanisms of action of anti-oestrogens, *Breast Cancer Res. Treat.*, 31, 5, 1994.

48. **Clarke, R.B., Laidlaw, I.J., Jones, L.J., Howell, A., and Anderson, E.,** Effect of tamoxifen on Ki67 labelling index in human breast tumours and its relationship to oestrogen and progesterone receptor status, *Br. J. Cancer*, 67, 606, 1993.

49. **Siiteri, P.K. and MacDonald, P.C.,** Role of extraglandular estrogen in human endocrinology, in *Handbook of Physiology*, Vol. II, Part 1, Geiger, S.R., Astwood, E.B., and Greep, R.O., Eds., American Physiological Society, Bethesda, Maryland, 1973, 615.

50. **Harris, A.L. and Hochhauser, D.,** Mechanisms of multidrug resistance in cancer treatment, *Acta Oncol.*, 31, 205, 1992.

51. **Jaffrezou, J.P., Chen, G., Duran, G.E., Kuhl, J.S., and Sikic, B.I.,** Mutation-rates and mechanisms of resistance to etoposide determined from fluctuation analysis, *J. Natl. Cancer Inst.*, 86, 1152, 1994.

52. **Norton, L.,** A Gompertzian model of human breast cancer growth, *Cancer Res.*, 48, 7067, 1988.

53. **Bonadonna, G., Valagussa, P., and Rossi, A.,** Ten-year results with CMF-based adjuvant chemotherapy in resectable breast cancer, *Breast Cancer Res. Treat.*, 5, 95, 1985.

54. **Buzzoni, R., Bonadonna, G., Valagussa, P., and Zambetti, M.,** Adjuvant chemotherapy with doxorubicin plus cyclophosphamide, methotrexate, and fluorouracil in the treatment of resectable breast cancer with more than three positive axillary nodes, *J. Clin. Oncol.*, 9, 2134, 1991.

55. **Laird, A.K.,** Dynamics of tumor growth, *Br. J. Cancer*, 18, 490, 1964.

56. **Brunton, G.F. and Wheldon, T.E.,** Characteristic species dependent growth patterns of mammalian neoplasms, *Cell Tissue Kinet.*, 11, 161, 1978.

57. **Sullivan, P.W. and Salmon, S.E.,** Kinetics of tumor growth and regression in IgG multiple myeloma, *J. Clin. Invest.*, 51, 1697, 1972.

58. **Demicheli, R.,** Growth of testicular neoplasm lung metastases: tumor-specific relation between two gompertzian parameters, *Eur. J. Cancer*, 16, 1603, 1980.

59. **Bloom, H., Richardson, M., and Harries, B.,** Natural history of untreated breast cancer (1804-1933); comparison of untreated and treated cases according to histological grade of malignancy, *Br. Med. J.*, 2, 213, 1962.

60. **Speer, J.F., Petrosky, V.E., Retsky, M.W., and Wardwell, R.H.,** A stochastic numerical model of breast cancer growth that simulates clinical data, *Cancer Res.*, 44, 4124, 1984.

61. **Demicheli, R., Terenziani, M., Valagussa, P., Moliterni, A., Zambetti, M., and Bonadonna, G.,** Local recurrences following mastectomy: Support for the concept of tumor dormancy, *J. Natl. Cancer Inst.*, 86, 45, 1994.

62. **Spratt, J.A., von Fournier, D., Spratt, J.S., and Weber, E.E.,** Decelerating growth and human breast cancer, *Cancer*, 71, 2013, 1993.

63. **Brown, B.W., Atkinson, E.N., Bartoszynski, R., Thompson, J.R., and Montague, E.D.,** Estimation of human tumour growth rate from distribution of tumor size at detection, *J. Natl. Cancer Inst.*, 72, 31, 1984.

64. **Norton, L. and Simon, R.,** Tumor size, sensitivity to therapy, and design of treatment schedules, *Cancer Treat. Rep.*, 61, 1307, 1977.

65. **Norton, L. and Simon, R.,** The Norton-Simon hypothesis revisited, *Cancer Treat. Rep.*, 70, 163, 1986.

66. **Fornander, T., Rutqvist, L.E., and Cedermark, B.,** Adjuvant tamoxifen in early breast cancer: Occurrence of new primary cancers, *Lancet*, 1, 117, 1989.

67. **Boccardo, F., Rubagotti, A., Amoroso, D., Sismondi, P., Genta, F., Nenci, I., Piffanelli, A., Farris, A., Castagnetta, L., Traina, A., Cappellini, M., Pacini, P., Sassi, M., Malacarne, P., Donati, D., Mustacchi, G., Galletto, L., Schieppati, G., Villa, E., Bolognesi, A., Gallo, L., and** other participants in the GROCTA chemotherapy versus tamoxifen versus chemotherapy plus tamoxifen in node-positive, oestrogen-receptor positive breast cancer patients. An update at 7 years of the GROCTA (Breast Cancer Adjuvant Chemo-hormone Therapy Cooperative Group) Trial, *Eur. J. Cancer*, 28, 673, 1992.

68. **Sarkaria, J.N., Gibson, D.F.C., Jordan, V.C., Fowler, J.F., Lindstrom, M.J., and Mulcahy, R.T.,** Tamoxifen-induced increase in the potential doubling time of MCF-7 xenografts as determined by bromodeoxyuridine labelling and flow cytometry, *Cancer Res.*, 530, 4413, 1993.

69. **Keen, J., Miller, E.P., Bellamy, C., Dixon, J.M., and Miller, W.R.,** P-glycoprotein and resistance to tamoxifen, *Lancet*, 343, 1047, 1994.

70. **Rubens, R.D., Bartelink, H., Engelsman, E., Hayward, J.L., Rotmensz, N., Sylvester, R., van der Schukren, E., Papadiamantis, J., Vassilaros, S.D., Wildiers, J., and Winter, P.J.,** Locally advanced breast cancer. The contribution of cytotoxic and endocrine treatment to radiotherapy, *Eur. J. Cancer Clin. Oncol.*, 25, 667, 1989.

71. **Fisher, B., Redmond, C., and Brown, A.,** Adjuvant chemotherapy with and without tamoxifen in the treatment of primary breast cancer: 5 year results from the NSABP trial, *J. Clin. Oncol.*, 4, 459, 1986.

72. **Anderson, E.D.C., Forrest, A.P.M., Hawkins, R.A., Anderson, T.J., and Leonard, R.C.F.,** Primary systemic therapy for operable breast cancer, *Br. J. Cancer*, 63, 561, 1991.

73. **Blamey, R.W., Jonat, W., Kaufmann, M., Bianco, A.R., and Namer, M.,** Goserelin depot in the treatment of premenopausal advanced breast cancer, *Eur. J. Cancer*, 28A, 810, 1992.

74. **Buzzoni, R., Biganzoli, L., Bajetta, E., Celio, L., Fornasiero, A., Mariani, L., Zilembo, N., Di Bartolomeo, M., Di Leo, A., Arcangeli, G., Aitini, E., Farina, G., Schieppati, G., Galluzzo, D., and Martinetti, A.,** Combination goserelin and tamoxifen therapy in premenopausal advanced breast cancer: a multicentre study by the ITMO group, *Br. J. Cancer*, 71, 1111, 1995.

75. **Levine, M., Bramwell, V., Bowman, D., Norris, B., Findlay, B., Warr, D., Pritchard, K.I., MacKenzie, R., Robert, J., Arnold, A., Tonkin, K., Shepherd, L., Ottaway, J., and Myles, J.,** A clinical trial of intensive CEF versus CMF in premenopausal women with node positive breast cancer, *Proc. ASCO*, 14, 103, 1995.

76. **Moliterni, A., Bonadonna, G., Valagussa, P., Ferrari, L., and Zambetti, M.,** Cyclophosphamide, methorexate, and fluorouracil with and without doxorubicin in the adjuvant treatment of resectable breast cancer with one to three positive axillary nodes, *J. Clin. Oncol.*, 9, 1124, 1991.

77. **Fisher, B., Brown, A.M., Dimitrov, N.V., Poisson, R., Redmond, C., Margolese, R., Bowman, D., Wolmark, N., Wickerham, D.L., Kardinal, C.G., Shibata, H., Paterson, A.H.G., Sutherland, C.M., Robert, N.J., Ager, P.J., Levy, L., Wolter, J., Wozniak, T., Fisher, E.R., and Deutsch, M.,** Two months of doxorubicin-cyclophosphamide with and without interval re-induction therapy compared with 6 months of cyclophosphamide, methotrexate, and fluorouracil in positive-node breast cancer patients with tamoxifen-nonresponsive tumours: results from the National Surgical Adjuvant Breast and Bowel Project B-15, *J. Clin. Oncol.*, 8, 1483, 1990.

78. **Perez, C.A., Fields, J.N., Fracasso, P.M., Philpott, G., Soares, R.L., Jr., Taylor, M.E., Lockett, M.A., and Rush, C.,** Management of Locally advanced carcinoma of the breast. II Inflammatory Carcinoma, *Cancer*, 74, 466, 1994.

79. **Perez, C.A., Graham, M.L., Taylor, M.E., Levy, J.F., Mortimer, J.E., Philpott, G.W., and Kucik, N.A.,** Management of locally advanced carcinoma of the breast: I. Noninflammatory, *Cancer*, 74, 453, 1994.

80. **Ahern, V., Barraclough, B., Bosch, C., Langlands, A., and Boyages, J.,** Locally advanced breast cancer: defining an optimum treatment regimen, *Int. J. Radiat. Oncol. Biol. Phys.*, 28, 867, 1994.

81. **Vecchietti, G., Gerzeli, G., Zanoio, L., Novelli, G.G., Patton, R., and Barni, S.,** Cytohistochemical observations of the endometrial adenocarcinoma before and after treatment with 6-methyl-17-hydroxyprogesterone acetate (MPA), in *Role of Medroxyprogesterone in Endocrine Related Tumours*, Iacobelli, S. and Di Marco, A., Eds., Raven Press, New York, 1980, 107.

82. **Rose, C. and Mouridsen, H.,** Endocrine therapy of advanced breast cancer, *Acta Oncol.*, 27A, 721, 1988.

83. **Swenerton, K.D., Legha, S.S., Smith, T., Hortobagyi, G.N., Gehan, E.A., Yap, H., Gutterman, J.U., and Blumenschein, G.R.,** Prognostic factors in metastatic breast cancer treated with combination chemotherapy, *Cancer Res.*, 39, 1552, 1979.

84. **Gregory, W.M., Smith, P., Richards, M.A., Twelves, C.J., Knight, R.K., and Rubens, R.D.,** Chemotherapy of advanced breast cancer: outcome and prognostic factors, *Br. J. Cancer*, 68, 988, 1993.

85. **O'Shaughenessy, J.A. and Cowan, K.H.,** Current status of paclitaxel in the treatment of breast cancer, *Breast Cancer Res. Treat.*, 33, 27, 1995.

86. **Cameron, D.A., Gabra, H., and Leonard, R.C.F.,** Continuous 5-fluorouracil in the treatment of breast cancer, *Br. J. Cancer*, 70, 120, 1994.

87. **Dunphy, F.R. and Spitzer, G.,** Rossiter Fornoff, J.E., Yau, J.C., Huan, S.A., Dicke, K.A., Buzdar, A.U., and Hortobagyi, G.N., Factors predicting long-term survival for metastatic breast cancer patients treated with High-Dose chemotherapy and bone marrow support, *Cancer*, 73, 2157, 1994.

88. **Antman, K., Ayash, L., Elias, A., Wheeler, C., Hunt, M., Eder, J.P., Teicher, B.A., Critchlow, J., Bibbo, J., Schnipper, L.E., and Frei, E.,** III, A Phase II study of high-dose cyclophosphamide, thiotepa, and carboplatin with autologous marrow support in women with measurable advanced breast cancer responding to standard-dose therapy, *J. Clin. Oncol.*, 10, 102, 1992.

89. **Crawford, D.J., Cowan, S., Fitch, R., Smith, D.C., and Leake, R.E.,** Stability of oestrogen receptor status in sequential biopsies from patients with breast cancer, *Br. J. Cancer*, 56, 137, 1987.

90. **Crawford, D.J., George, W.D., Smith, D.C., Stewart, M., Paul, J., and Leake, R.E.,** Comparison of cyclical tamoxifen and megestrol acetate with tamoxifen alone in advanced breast cancer, *Breast*, 14, 35, 1992.

91. **James, V.H.T., Reed, M.J., Adams, E.F., Ghilchick, M., Lai, L.C., Coldham, N.G., Newton, C.J., Purohit, A., Owen, A.M., Singh, A., and Islam, S.,** Oestrogen uptake and metabolism in vivo, *Proc. R. Soc. Edinburgh*, Sec B, 95, 185, 1989.

92. **Miller, W.R. and O'Neill, J.S.,** The relevance of local oestrogen metabolism within the breast, *Proc. R. Soc. Edinburgh*, Sec B, 95, 203, 1989.

93. **Gottardis, M.M. and Jordan, V.C.,** Development of tamoxifen-stimulated growth of MCF-7 tumours in athymic mice after long-term antiestrogen stimulation, *Cancer Res.*, 48, 5183, 1988.

94. **Howell, A., Dodwell, D.J., Anderson, H., and Redford, J.,** Response after withdrawal of tamoxifen and progestogens in advanced breast cancer, *Ann. Oncol.*, 302, 611, 1992.

95. **Osborne, C.K. and Fuqua, S.A.W.,** Mechanisms of tamoxifen resistance, *Breast Cancer Res. Treat.*, 32, 49, 1994.

96. **Howell, A., Defriend, D., Robertson, J., Blamey, R.A., and Walton, P.,** Response to a specific antiestrogen (ici-182780) in tamoxifen-resistant breast-cancer, *Lancet*, 345, 29, 1995.

97. **Wolf, D.M., Langan-Fahey, S.M., Parker, C.J., McCague, R., and Jordan, V.C.,** Investigation of the mechansim of tamoxifen-stimulated growth with non-isomerizable analogues of tamoxifen and metabolites, *J. Natl. Cancer Inst.*, 85, 806, 1993.

98. **Love, R.R., Mazess, R.B., Barden, H.S., Epstein, S., Newcomb, P.A., Jordan, V.C., Carbone, P.P., and DeMets, D.L.,** Effects of tamoxifen on bone mineral density in postmenopausal women with breast cancer, *N. Engl. J. Med.*, 326, 852, 1992.

99. **MacDonald, C.C. and Stewart, H.J.,** Fatal myocardial infarction in the Scottish adjuvant tamoxifen trial. The Scottish Breast Cancer Committee, *Br. Med. J.*, 303, 435, 1991.

100. **Black, L.J., Sato, M., Rowley, E.R., Magee, D.E., Bekele, A., Williams, D.C., Cullinan, G.J., Bendele, R., Kauffman, R.F., Bensch, W.R., Frolik, C.A., Termine, J.D., and Bryant, H.U.,** Raloxifene (LY 139481 HCl) prevents bone loss and reduces serum cholesterol without causing uterine hypertrophy in ovariectomized rats, *J. Clin. Invest.*, 93, 63, 1994.

101. **Têtu, B. and Brisson, J.,** Prognostic significance of HER-2/neu oncoprotein expression in node-positive breast cancer, *Cancer*, 73, 2359, 1994.

102. **Baselga, J., Norton, L., Shalaby, R., and Mendelsohn, J.,** Anti Her2 humanised monoclonal antibody (MAb) alone and in combination with chemotherapy against human breast carcinoma xenografts, *Proc. A.S.C.O.*, 13, 63, 1994.

103. **Baselga, J., Tripathy, D., Mendelsohn, J., Benz, C., Dantis, L., Moore, J., Rosen, P.P., Henderson, I.C., Baughman, S., Twaddell, T., and Norton, L.,** Phase II study of recombinant human anti-HER2 monoclonal antibody (rhuMAb HER2) in stage IV breast cancer (BC): HER2-shedding dependant pharmacokinetics and antitumor activity, *Proc. ASCO*, 14, 103, 1995.

104. **Nielsen, J.L., Walsh, J.T., Degen, D.R., Drabek, S.M., McGill, J.R., and Von Hoff, D.D.,** Evidence of gene amplification in the form of double minute chromosomes is frequently observed in lung cancer, *Cancer Genet. Cytogenet.*, 65A, 120, 1993.

105. **Spicer, D., McCaskill-Stevens, W., Oke, M., Abrams, J., Hines, J., Cooper, B., Craig, J., Bertsch, L., and Hohneker, J.,** Oral Navelbine (NVB) in women with previously treated advanced breast cancer (ABC): a US multicenter phase II trial, *Proc. A.S.C.O.*, 13, 76, 1994.

106. **Lokich, J.J. and Anderson, N.,** Infusional chemotherapy for breast cancer: the Cancer Centre of Boston, *J. Infus. Chemother.*, 3, 9, 1993.

107. **Vogelzang, N.J.,** Continuous infusion chemotherapy: a critical review, *J. Clin. Oncol.*, 2, 1289, 1984.

108. **Fabian, C.J., Kimler, B.F., McKittrick, R., Park, C.H., Lin, F., Krichnan, L., Jewell, W.R.J., Osborne, C.K., Martino, S., Hutchins, L.F., Leong, L.A., and Green, S.,** Recruitment with high physiological doses of estradiol preceding chemotherapy: flow cytometric and therapeutic results in women with locally advanced breast cancer — a Southwest Oncology Group study, *Cancer Res.*, 54, 5357, 1994.

109. **Lane, D.P.,** A death in the life of p53, *Nature*, 362, 786, 1993.

110. **Wang, Y., Ramqvist, T., Szekely, L., Axelson, H., Klein, G., and Wiman, K.G.,** Reconstitution of wild-type p53 expression triggers apoptosis in a p53-negative v-myc retrovirus-induced T-cell lymphoma line, *Cell Growth Differ.*, 4, 467, 1993.

111. **Gasparini, G. and Harris, A.L.,** Clinical importance of the determination of tumor angiogenesis in breast carcinoma: much more than a new prognostic tool, *J. Clin. Oncol.*, 13, 765, 1995.

112. **Fox, S.B., Leek, R.D., Smith, K., Hollyer, J., Greenall, M., and Harris, A.L.,** Tumour angiogenesis in node-negative breast carcinomas — relationship with epidermal growth factor receptor, estrogen receptor, and survival, *Breast Cancer Res. Treat.*, 29, 109, 1994.

113. **Weidner, N., Folkman, J., Pozza, F., Bevilacqua, P., Allred, E.N., Moore, D.H., Meli, S., and Gasparini, G.,** Tumor angiogenesis: a new significant and independent prognostic factor in early-stage breast carcinoma, *J. Natl. Cancer Inst.*, 84, 1875, 1992.

114. **Bosari, S., Lee, A.K.C., DeLellis, R.A., Wiley, B.D., Heatley, G.J., and Silverman, M.L.,** Microvessel quantitation and prognosis in invasive breast carcinoma, *Hum. Pathol.*, 23, 755, 1992.

115. **Fidler, I.J. and Ellis, L.M.,** The implications of angiogenesis for the biology and therapy of cancer metastasis, *Cell*, 79, 185, 1994.

116. **Folkman, J.,** Angiogenesis in cancer, vascular, rheumatoid and other disease, *Nature Med.*, 1, 27, 1995.

117. **Holmgren, L., O'Reilly, M.S., and Folkman, J.,** Dormancy of micrometastases: Balanced proliferation and apoptosis in the presence of angiogenesis suppression, *Nature Med.*, 1, 149, 1995.

118. **Fisher, B., Gunduz, N., Coyle, J., Rudock, C., and Saffer, E.,** Presence of a growth-stimulating factor in serum following primary tumor removal in mice, *Cancer Res.*, 49, 1996, 1989.

119. **Nissen-Meyer, R., Kjellgren, K., Malmio, K., Månsson, B., and Norin, T.,** Surgical adjuvant chemotherapy: results with one short course with cyclophosphamide after mastectomy for breast cancer, *Cancer*, 41, 2088, 1978.

Part III.
Ovarian Cancer

Chapter 6

Ovarian Cancer — An Introduction

Hani Gabra and John F. Smyth

CONTENTS

1. INTRODUCTION

Ovarian cancer is the most common cause of death from gynecological malignancy, and the fifth most common cause of cancer death in women after breast, lung, colon, and (in the U.K.) stomach cancer. About 24,000 and 5,000 cases occur annually in the U.S. and the U.K., respectively, with approximately 13,500 and 3,500 deaths in the same period. This high mortality rate is due to presentation at an advanced stage of the disease, as symptoms of the disease are insidious in onset and nonspecific in nature.

2. EPIDEMIOLOGY

Approximately 1 woman in 70 will develop ovarian cancer, and 1 woman in 100 will die from it. Incidence is age-related, peaking at 55 per 100,000 per year in the 7th and 8th decades, with the disease being uncommon in those under 45 (less than 15 cases per 100,000 per year). The overall 5-year survival rate is poor; about 30%, primarily due to the late presenting nature of the disease. There is a particularly high incidence in Scandinavia, followed by Northern Europe and North America. Despite being industrialized, Japan has a particularly low incidence of the disease; however, the incidence increases sharply in Japanese migrants to the U.S. and their daughters (a phenomenon seen in common with other cancers), suggesting profound environmental component(s) to etiology.[1] Demographically, there are wide mortality rate differences; Denmark having a mortality rate six times that of Japan. These may be due to genetic differences between populations.[2] There has been a slight improvement in ovarian cancer survival over the last two decades, almost certainly due to improved multimodality therapies. This improvement has been more pronounced in younger patients (less than 50 years old). The age-specific mortality rate curve in those with ovarian cancer becomes parallel to that of the general age-specific mortality rate at 15 to 20 years post diagnosis, indicating those who are probably "cured" of their disease by that time.

3. ETIOLOGY

Ninety-five percent of all patients with ovarian cancer have no family history, suggesting that almost all cases can be attributed to spontaneous or environmentally induced carcinogenesis. Many individual environmental factors have been examined. The epidemiological studies that are published for these potential factors are diverse in their methodologies and power. What is clear is that for many factors, the epidemiological data is (1) conflicting from study to study and (2) small in magnitude of effect, often at the limits of what is legitimately detectable by the study designs. Putting these points together with that of publication bias gives an idea of how opaque this area continues to be. Potential candidate carcinogens such as inhaled (tobacco[3]), ascending (talc,[4-6] asbestos), and some dietary (alcohol[7-8]) factors have been examined and do not seem to confer excess risk.

The low incidence of the disease in Japan does suggest that the causes are not simply due to environmental carcinogens as a result of industrialization; however, the rising incidence in Japanese migrants to the U.S. suggests environmentally related carcinogenesis. Whether this is due to carcinogenesis from, say, dietary or other sources in the U.S., or whether there are dietary or other protectants that prevent environmentally associated carcinogenesis in Japan is unknown.

3.1 Diet

Coffee drinking has been variously ascribed with slightly increased risk or no risk depending on the epidemiological overview or opinion, and there may be some consensus that a slightly increased risk of ovarian cancer is associated with coffee ingestion.[3,9,10] Japanese women with ovarian cancer ate significantly more fish and drank significantly less milk than controls.[11] Initial evidence that consumption of animal fat is associated with an increased risk of ovarian cancer,[10] was followed with a large case/control study into diet and ovarian cancer risk. This study suggested that every 10 g of ingested saturated fat per day increased ovarian cancer risk by 20%, whereas the same ingested weight of vegetable fiber reduced ovarian cancer risk by 37%.[12] The extent of independence of these factors is unclear. However, this seemingly unambiguous study is virtually unique, and since there are very few other confirmatory studies, its findings should be treated cautiously. Although no relationship has been shown between the exogenous ingestion of estrogen and the incidence of ovarian cancer,[13] re-examination of this area may be warranted with these new observations because of the role of estrogens and other steroid hormones often present in saturated animal fat,[12] the role of phytoestrogens acting to inhibit endogenous estrogen production,[14-16] and the role of vegetable fiber in binding estrogen in the gut.[17,18]

During the late 1980s, a case/control study was performed which showed an association between higher consumption of lactose/galactose (in the form of yogurt and cottage cheese) and lower levels of the enzyme galactose-1-phosphate uridyl transferase (GPUT) in erythrocytes in ovarian cancer cases compared with controls.[19] Deficiency of this enzyme may be a genetic risk factor for early menopause. Early menopause may be associated with hypergonadotrophic hypogonadism (also associated with decreased GPUT), and such hormonal perturbation may be involved in the pathoetiology of ovarian cancer.[20] This work awaits confirmation by follow-up studies.

3.2 Molecular Epidemiology

The molecular study of genes involved in carcinogenesis has opened new avenues into epidemiology. It is possible by analysis of genes thought to be important in the etiology of a particular tumor type to make inferences about causation, the relative balance of inherited to somatic events, and whether the somatic events are due to environmental or spontaneous processes. For instance, molecular epidemiology of the p53 gene in hepatocellular carcinoma where aflatoxin has been the underlying carcinogen reveals a high frequency of transversions (mutation where purines are replaced by pyrimidines or vice versa) of the guanine to thymine (G to T) type at a specific site of the gene's sequence (codon 249).[21] The presence of transversions at such sites represent events related to exogenous carcinogens, whereas transitions (conversion of pyrimidine to pyrimidine, or purine to purine) are more likely to be due to errors of fidelity during DNA replication, which may be a spontaneous process.

The p53 gene is frequently mutated in epithelial ovarian cancer; however similar analysis of the p53 gene in ovarian cancer revealed no such transversion "hot-spots"; mutation of transition type were predominant (72% of all mutation events detected) and dispersed throughout the gene (predominantly exons 5 to 8).[22] This predominance of transition mutations suggests that the impact of environmental carcinogens to the development of ovarian cancer may be minimal. Although these data are generally supported by the epidemiological evidence available, it is important to remember that environmental carcinogens may mediate their action on other genes involved at other stages of ovarian cancer development.

3.3 Hormonal Factors and Incessant Ovulation

Although no formal evidence exists to suggest that exogenous estrogens or other steroid hormones have a role to play in the initiation of ovarian cancer in humans, there is quite a body of literature that would suggest that ovarian cancer is hormone sensitive. The ovary is the main site of synthesis of estrogen and progesterone and is also a target organ for these hormones. The actions of these hormones are mediated by specific intracellular receptors which function as hormone-inducible nuclear transcription factors with context-specific, often conflicting effects on proliferation and differentiation of target tissues.[23]

Progesterone receptor (PR) is regulated by estrogen via the estrogen receptor (ER);[24] estradiol induced PR expression has been demonstrated in ovarian cancer cell lines which express ER.[25] Ovarian cancer has been reported to respond to antiestrogens in about 10 to 20% of cases, and to progestins with an average of 36%, within a range between 0 and 60%.[26,27] That histopathological subtypes of epithelial ovarian cancer (EOC) differ in their tumoral PR expression is suggested by at least six reports that endometrioid ovarian carcinomas contain relatively more PR than other histologic types,[26] and that PR positivity is associated with well differentiated ovarian tumors in premenopausal women.[28] Tumoral PR content has prognostic significance in several studies,[26-29] but the associations are not unequivocal.[30,31] Estrogen receptors have been measured in ovarian cancer in many studies; the evidence for an association of tumoral ER content with prognosis is conflicting.[30-33] Thus, although the data are currently inconclusive, it is likely that the endocrine, paracrine, and autocrine milieux are involved at various stages of ovarian cancer development. Although exogenous estrogens do not appear to be associated with the development of human ovarian carcinoma (despite evidence that chronic estrogen and progestin administration in animal studies results in ovarian cancer[34,35]), hormonal perturbations with consequent effects on ovulation appear to profoundly affect the development of ovarian cancer, as predicted by the hypothesis of "incessant ovulation." To this extent, the etiology of ovarian cancer may further be considered to be endocrine related. In 1971, Fathalla proposed that "incessant ovulation" results in repeated proliferative repair cycles of the ovarian surface epithelium.[36] These epithelial cells can acquire genetic damage during proliferative repair, and this cumulative genetic damage due to repeated ovulatory cycles predisposes to the development of ovarian cancer. This hypothesis is perfectly understandable if one considers that in a panmycitic natural state, a female might only ovulate 20-40 times in her life due to recurrent childbearing, whereas in late twentieth-century economically developed societies, this figure may be an order of magnitude higher. Thus, any factor which reduces the number of times a woman ovulates (late menarche, oral contraceptive use, multigravidity, prolonged lactation, early menopause) can be predicted to reduce ovarian cancer risk. Nulliparity and a low mean number of pregnancies have been associated with increased risk.[37] Conversely, childbirth confers a protective effect, and the magnitude of this protective effect relates to parity with an estimated 30-60% reduction in risk for those women with two or more pregnancies. Increased total pregnancy and lactation time also confer a protective effect.[38]

Ovarian cancer risk is decreased 30 to 60% by the use of oral contraceptives; in one study, ingestion of oral contraceptive over a 5 year period was associated with 37% reduction in ovarian cancer risk.[39]

This effect is believed to persist for many years after the cessation of the oral contraceptive. The high gonadotrophic milieu of the postmenopausal state may confer additional risk.[37] Fertility drugs (exogenous gonadotrophins and stimulants of pituitary gonadotrophins which act by ovarian hyperstimulation) have been suggested to confer increased risk (as might be expected from the Fathalla hypothesis); however, the evidence is not unequivocal. A small cohort study showed no apparent increase in risk of ovarian cancer, although there was an increase in endometrial cancer.[40] A subsequent review of 12 American case/control studies suggested a threefold increase risk of ovarian cancer in women who had used fertility drugs.[41,42] However, it has been pointed out that this was a small study with wide confidence intervals; in addition, controls lacked a history of infertility, and this presents a problem as nulliparous women are intrinsically at increased ovarian cancer risk.[43] Recently, another case/control study in nearly 4000 women reported that women treated with infertility drugs had a 2.5 times increased risk of ovarian cancer compared to the general population.[44] Women undergoing fertility treatment, therefore, need to know that they may be at increased risk of ovarian cancer, although this information requires careful presentation management given its equivocacy.

3.4 Inherited Predisposition

Clearly defined predisposing/hereditary factors account for only about 5% of patients with ovarian cancer, as judged by familial clustering. In most of these familial cases, the etiology is probably multifactorial, consisting of both environmental and genetic elements. Women with one first-degree relative affected run a 5% lifetime ovarian cancer risk (1 in 20). The risk associated with having two or more affected first-degree relatives is 7% (1 in 15). Those with definable autosomal dominant syndromes run a much higher risk (40 to 50%) of developing ovarian cancer, but these individuals constitute only about 3% of women with two affected first degree relatives.[45]

3.4.1 Hereditary Syndromes

Three hereditary syndromes have been defined, and patients with these tend to be younger (usually 10 or more years less than the median) and have bilateral and multifocal tumors more commonly. Major international efforts are in progress to identify the spectrum of genes involved in these syndromes.

3.4.2 Hereditary Site-Specific Ovarian Cancer

Hereditary site-specific ovarian cancer is rare and in the majority linked to BRCA1 (see Chapter 8),[46] a putative tumor-suppressor gene located on chromosome 17q. This gene is very large, dispersed over about 100Kb, with 1836 amino-acids in 22 coding exons. Such proportions make the task of classifying all mutations with functional consequences somewhat Herculean. The BRCA1 protein contains an amino-acid sequence known as a zinc-finger motif which suggests that at least one of its functions is as a transcriptional factor.[47]

3.4.3 Hereditary Breast/Ovarian Cancer

Clustering of breast and ovarian cancer occurs in this syndrome, and hereditary ovarian cancer is most likely to occur as part of this syndrome. Again, most of these are BRCA1 linked.[48] BRCA1 mutation carriers appear to have a variable lifetime risk of ovarian cancer, although estimates are as high as 63% by age 70.[49] Although BRCA1 mutation is likely to underlie most hereditary ovarian cancer, by definition this means that other genes must account for the non-BRCA1 linked cases. A second locus has been mapped by linkage studies, to chromosome 13q proximal to RB1, the retinoblastoma tumor-suppressor gene.[49] This gene has been called BRCA2 (see Chapter 8). It does not account for all other heritable cases and it is likely that there are other genes involved in the residuum of inherited ovarian cancers.

3.4.4 Lynch Syndrome II

Lynch syndrome II is the integrated syndrome which incorporates hereditary nonpolyposis colorectal cancer (Lynch syndrome I) with gastrointestinal (gastric, small intestinal, pancreatic), gynecological (endometrial and ovarian), urological, or breast cancer.[50-52] This syndrome is linked to a family of DNA mismatch repair genes, disruption of which result in "promutator" effects on the genome, often observed as a high incidence of microsatellite instability. These genes have been designated hMSH2,[53] hMLH1,[54] hPMS1, and hPMS2.[55]

4. BIOLOGY AND PATHOPHYSIOLOGY

4.1 Developmental Histogenesis

The bewildering array of different neoplasms arising from the ovary can be understood if one considers the embryological development of the female reproductive system, the range of epithelial tissue types that it generates, and the physiology of the ovary.[56] As early as the 4th week after fertilization, the developing gonad commences as the genital ridge in the embryo by a proliferation of coelomic epithelium (a cell layer of mesodermal origin which goes on to form the peritoneal epithelial lining). Primordial germ cells appear at week 6 and migrate from the wall of the yolk sac to the genital ridge. As the germ cells invade the genital ridge, the overlying coelomic epithelium proliferates and invades the underlying mesenchyme where it forms intimate relationships, surrounding the germ cells in structures called sex-cords. As the ovary develops, the sex-cords form irregular clusters and the central part of the ovary loses the cords. These are replaced by vascular stroma and become the ovarian medulla. The cords proliferate in the cortical region and form oogonia while the surrounding epithelial cells form the enveloping follicular cells. The mesonephric (Wölffian) duct that goes on to connect the gonads to the urethra in the adult male has no way of connecting with the developing ovarian cortex, and so germ cells in the adult female must be shed from the ovarian epithelium rather than transported in the same way as spermatozoa. At week 6, a second duct forms from the coelomic epithelium dorsal to the Wölffian (mesonephric) duct. This is the Müllerian (paramesonephric) duct which at the cranial end opens into the coelomic cavity (forming the future fallopian tubes), and at the caudal end fuses with the contralateral Müllerian duct to form the uterine canal. In the female, the Müllerian duct continues to develop into the main genital duct while the Wölffian duct atrophies almost completely. As the ovary develops and migrates caudally, so the Müllerian duct goes on to form the fallopian tubes, the uterine body and the uterine cervix, each with distinctive lining epithelia. Although in continuity with these structures, the vagina develops separately, arising from the urogenital sinus located caudally. Common epithelial tumors of the ovary (including epithelial ovarian cancer) therefore arise from the ovarian surface epithelium (derived from the primitive coelomic mesothelium), and exhibit a range of histological structures which have appearances similar to other Müllerian derived structures; i.e., serous differentiation (similar to fallopian tube epithelium), endometrioid differentiation (uterine body endometrium), and mucinous differentiation (endocervical epithelium). Structures within the ovary not derived from this epithelium produce much rarer tumor types. The mesoderm underlying the primitive coelomic mesothelium generates theca, granulosa, Leydig and Sertoli cells, and all these cell types can give rise to tumors. The primordial germ cells go on to form oocytes and from these can arise a range of germ cell tumors, including some that are identical to mucinous epithelial tumors. Dysgerminoma, teratoma (often mature, benign and of parthenogenetic origin in females), embryonal carcinoma, and choriocarcinoma may occur. They will not be discussed further here (Figure1).

4.2 Pathology of the Common Epithelial Tumors

Epithelial ovarian carcinomas represent over 90% of all ovarian malignant neoplasms. Epithelial neoplasms of the ovary comprise benign tumors, borderline tumors (tumors with low malignant potential), and frank epithelial carcinoma. These three categories of neoplasm usually exhibit Müllerian histogenic differentiation as described above.

4.2.1 Benign Tumors

These tumors have an excellent prognosis. The essential cytological features of benign tumors are a single layer of columnar cells, lack of cellular atypia, a normal nucleocytoplasmic ratio, few mitoses, and no evidence of either microinvasion or frank invasion into the underlying stroma. Benign tumors usually exhibit either serous or mucinous differentiation, are usually cystic, and can grow to enormous size. Half of all serous tumors are benign serous cystadenomas. Although sometimes bilateral, they are generally unilocular with a few intracystic papillae which may be either pedunculated with fronds or sessile. The fluid within the cyst is watery (serous). The serous epithelium is well differentiated, and forms a regular single layer of cuboidal cells with centrally placed nuclei lying on loose stroma. Mucinous cystadenomas constitute 80% of all mucinous tumors and are almost always unilateral. The cysts are multilocular containing a gelatinous mucinous secretion. The epithelium consists of tall columnar epithelial cells with basal nuclei. Sometimes this epithelium is similar to endocervical epithelium consistent with a Müllerian origin, however, more commonly the epithelium is reminiscent of colonic

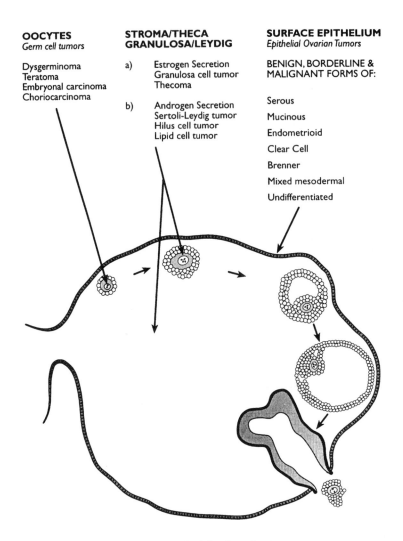

OOCYTES
Germ cell tumors

Dysgerminoma
Teratoma
Embryonal carcinoma
Choriocarcinoma

STROMA/THECA
GRANULOSA/LEYDIG

a)   Estrogen Secretion
     Granulosa cell tumor
     Thecoma

b)   Androgen Secretion
     Sertoli-Leydig tumor
     Hilus cell tumor
     Lipid cell tumor

SURFACE EPITHELIUM
Epithelial Ovarian Tumors

BENIGN, BORDERLINE &
MALIGNANT FORMS OF:

Serous

Mucinous

Endometrioid

Clear Cell

Brenner

Mixed mesodermal

Undifferentiated

Figure 1   Sites of origin of ovarian tumors.

mucinous glandular epithelium with the presence of goblet or Paneth cells, suggesting a tumor of non-Müllerian origin. Some have hypothesized that these "intestinal" type tumors are monodermal benign ovarian teratomas with complete colonic epithelium differentiation, but this is unlikely, and an alternative explanation is awaited. Occasionally spillage of mucin-secreting epithelial cell clusters into the peritoneal cavity can result in pseudomyxoma peritonei, a chronic and often ultimately fatal condition with mucinous secretion filling the peritoneum and causing bowel obstruction without destructive invasion. Brenner tumors are usually benign ovarian tumors in which there are cords of urothelium embedded in a dense stroma. This tumor is thought to arise from remnants of Wölffian origin (see above), part of the primitive urothelial system. Endometrioid and clear cell tumors are rarely benign.

4.2.2 Borderline Tumors (Tumors of Low Malignant Potential, LMP)

These tumors have a good prognosis generally. Borderline tumors are characterized histologically by the presence of pseudostratification of the cells forming the tumor's epithelial lining on a fibrovascular underlying stroma. Nuclear and cellular atypia, increased mitotic activity and detachment, and reimplantation of neoplastic cell clusters are all features. However, the pathological diagnosis also requires the absence of neoplastic cell invasion of the underlying stroma. This last feature differentiates borderline tumors from malignant invasive tumors. Serous LMP tumors, the commonest histological type of borderline tumor, in addition to the above, may also contain psammoma bodies (which confer a favourable prognosis on serous tumors), and may be multicentric (the clonality of which is currently unclear).

Mucinous LMP tumors are less common than serous LMPs and may be difficult to distinguish from malignant tumors. Again, the majority show intestinal rather than Müllerian differentiation, and pseudomyxoma peritonei may be a complication. Endometrioid LMPs constitute a fifth of all endometrioid tumors and may be associated with endometriosis (as can clear cell LMPs).

4.2.3 Malignant Tumors (Epithelial Ovarian Cancer)

Frank ovarian adenocarcinomas carry an extremely poor prognosis if disseminated from the primary site (which they do in the majority of cases). The primary tumors frequently contain solid regions of malignant tissue which can grow rapidly and often exhibit hemorrhagic and necrotic regions. The tumor tends to invade destructively and may breach the ovarian capsule, inducing adhesions locally and spreading into the peritoneal cavity. The histological hallmarks of malignant tissue are the same as for borderline tumors, with cellular atypia, a high nucleocytoplasmic ratio, and frequent mitoses. There is, however, destructive invasion of the underlying stroma by the malignant adenocarcinoma cells. As part of histopathological diagnosis, an attempt at estimating the degree of differentiation of ovarian adenocarcinomas should be made.[57] Well-differentiated carcinomas demonstrate little cellular atypia and relatively few mitoses (grade 1). Moderately differentiated carcinomas exhibit retention of the histological differentiation but there is increasing cellular atypia and more frequent mitoses (grade 2). Poorly differentiated carcinomas are more aggressive still, with abundant mitoses and sheets of poorly or undifferentiated cells with often very few features providing clues to the histological category (grade 3). Tumors can entirely lose their histogenic features leaving an undifferentiated adenocarcinoma, which is rapidly growing and behaves aggressively. One third of all serous tumors are malignant, and serous adenocarcinomas account for just under half of all epithelial ovarian cancers (EOCs). They are frequently bilateral and are often cystic with solid regions within. Endometrioid tumors are usually malignant and account for about 15% of all EOCs. They are usually less cystic than serous and mucinous adenocarcinomas. In a quarter of cases, endometrioid adenocarcinomas may co-exist with endometrial carcinoma, and have identical histology. They can also co-exist with, or arise on, a previous site of endometriosis.[58] Mucinous adenocarcinomas account for one fifth of all mucinous tumors, and represent only about 12% of EOCs. They are frequently cystic and multilocular, may be bilateral, and are often well differentiated when recognized. Clear cell carcinomas are uncommon (6% of EOCs) and so-called because of the appearance of the cytoplasm after removal of the abundant cytoplasmic glycogen during the specimen preparation process. Like endometrioid carcinoma, they may co-exist with endometriosis. Undifferentiated EOCs account for about 17% of the total, and tend to behave in an aggressive fashion.

4.3 Cytogenetic and Molecular Lesions in Sporadic Disease

As would be expected from a disorder primarily affecting the genome, gross chromosomal abnormalities are evident in ovarian cancer. The methodological difficulties resulting in having to resort to malignant cells from ascites specimens have generally skewed the data towards more advanced ovarian malignancies, with evidence of the bizarre end-products of extreme genomic instability. A problem associated with this approach is how to ascribe causality (via genetic lesions underlying the chromosomal aberrations) to the observed cytogenetic lesion, and one can try to describe frequent nonrandom rearrangements such as those described for hematological tumors. Often, however, the complexity of the rearrangements precludes such definitive comment and therefore one has to use molecular genetics, using both structural and functional studies to attempt to answer this question. In-situ hybridization methods are yet to be fully assessed in solid tumors and it may be that this technology will provide more specific observations at the cytogenetic level.[59] Among those cytogenetic lesions that are frequently reported are those involving chromosomes 1, 3, 6, and 11.[60-62] Molecular analysis of sporadic ovarian cancer has absorbed a large research effort over the last 10 years. A detailed account is beyond the scope of this chapter (see Chapter 8), but in principle the approach has been twofold. First, attempts have been made to define the role of previously identified "candidate" molecules implicated in the neoplastic process with studies concentrating on expression at protein and RNA levels, genomic alterations, and associated clinicopathological features. Second, a genome-mapping approach using polymorphic DNA probes has been used to identify regions of possible deletion (often based on prior cytogenetic evidence of specific frequent chromosomal disruption). This approach is based on the hypothesis proposed by Knudson[63] extrapolated from classical genetics observations in hereditary retinoblastoma, suggesting that a process whereby both alleles for a gene involved in maintaining the nonneoplastic state are knocked out by any of several possible mechanisms resulting in loss of function. The loss of function in turn uncovers the neoplastic state. This "two-hit" hypothesis has become a paradigm for the characterization of tumor-suppressor genes, and such "allele loss" studies presume that one "hit" is due to a deletion involving one allele, the other hit being a more subtle event such as a point mutation. Allele loss does not provide direct proof of a tumor-suppressor, however, specific chromosomal regions demonstrating very high levels of allele loss flanked by regions of much lower loss provide attractive candidate regions for a positional cloning approach (obviously in these cases, reasonably large populations of tumor patients are required) . The finding of statistical associations with clinicopathological features may suggest possible functions of such regions.[64] One major drawback of this type of analysis is that many random allele losses may occur in advanced sporadic tumors which may have no causal relationship with the neoplastic process, but merely represent consequences of primary genome instability associated with cancer. Many chromosomal regions have been implicated by allele loss analysis, and a detailed discussion of these is beyond the scope of this chapter (see Chapter 8).

BRCA 1, the familial ovarian/breast cancer gene recently cloned, was initially expected to exhibit profound effects in the sporadic counterparts of the disease. It came as some surprise, therefore, when it was reported that in an apparently sporadic series of 12 ovarian and 32 breast cancers with LOH found using tightly linked and intragenic microsatellites, only four had mutations of the gene, and in these four, the mutations were found in the germline — suggesting previously undetected familial cases.[65] Subsequently, a further report[66] demonstrated that of 47 ovarian cancers with LOH at one or both of two polymorphic CA repeat microsatellites intragenic to the BRCA 1 gene, 4 exhibited evidence of somatic mutation by single strand conformational polymorphism analysis (SSCP). However, this still leaves 90% of the tumors with LOH with no obvious BRCA 1 mutation. Some speculate that all these LOH examples will ultimately be accounted for by more thorough analysis of the BRCA 1 region including flanking regulatory regions. Others suggest that this discrepancy points to a second gene involved in sporadic cancers of ovary and breast in this region. Time will tell.

In parallel, the second familial locus for breast and ovarian cancer, BRCA 2, remains uncloned but has been localized to chromosome 13q just proximal to Rb-1.[49] The retinoblastoma gene locus may also be significant in ovarian cancer. Several reports have indicated a high level of allele loss in sporadic ovarian cancer in the region of Rb 1, but in at least two studies,[67,68] Rb 1 expression was normal in the tumors with LOH, suggesting that another gene was responsible.

4.4 The Nature of the Early Histological Lesion and Relationships between Benign, Borderline, and Malignant Disease

The FIGO staging system[69] (Table 1) and the anatomical view of epithelial tumor spread suggests an orderly progression from normal cell through benign tumor to early malignant tumor followed by the breaching of an underlying serosal layer with vascular and lymphatic tumor dissemination, and finally, metastases. However, although such a pathological multistep model may be valid for colorectal cancer,[70] the evidence does not favor this for ovarian cancer. Relatively little work has been done on the features and the nature of the earliest histological lesion in ovarian cancer. In patients who had oophorectomy where ovarian carcinoma was not recognized preoperatively, and who were found to have *de novo* ovarian carcinoma on review, tumors were small, ranging from microscopic foci to 7 mm diameter. The tumors were all unilateral, and some were multifocal. They arose from the surface epithelium or from the crypts/inclusion bodies associated. The noncarcinomatous surface epithelium and inclusion bodies exhibited "severe atypia" in a fifth of cases, suggesting a preneoplastic lesion. Several women went on to develop peritoneal carcinomatosis, suggesting stage III disease essentially from the outset, with fatal outcome. A significant proportion had normal serum CA125 levels (see below) and their tumors would not have been detected radiologically. It is therefore clear that some ovarian cancers are capable of progressing from microscopic ovarian malignant disease straight to potentially fatal stage III disease with no prospect of screening-related intervention.[71] The presence of multifocality, atypia and inclusion cysts in these very early malignant lesions suggests the possibility of field changes affecting the ovarian epithelium. A case/control study examining in detail the contralateral ovaries from patients with unilateral ovarian carcinoma compared to age-matched controls showed a significantly higher number of inclusion cysts in the unaffected ovaries of ovarian cancer patients with frequent evidence of serous metaplasia within these inclusion cysts. In the ovarian cancer patient group, an age related increase in the number of inclusion cysts was seen, which was not observed in the control group.[72] These features suggest a histologically abbreviated route to malignant ovarian cancer.

Over 90% of specimens of mucinous epithelial ovarian cancer contain surrounding benign neoplastic epithelium. In a quarter of cases one can observe a region of transition from benign to malignant tissue.[73] Such benign epithelia are seen frequently in mucinous carcinomas but relatively infrequently in serous

Table 1 FIGO Staging Classification for Ovarian Cancer

Stage 1	**Tumor Limited to Ovaries**
	Ia: involving only one ovary, no ascites, capsule intact, no tumor on external surface
	Ib: involves both ovaries, no ascites, capsule intact, no tumor on external surface
	Ic: stage Ia or Ib with tumor on ovarian surface, rupture of capsule, malignant ascites, or positive peritoneal washings
Stage II	**Tumor Involves One or Both Ovaries with Pelvic Extension**
	IIa: extension to uterus and/or fallopian tubes
	IIb: extension to other pelvic tissues
	IIc: stage IIa or IIb with tumor on the surface of one or both ovaries, rupture of capsule, malignant ascites, or positive peritoneal washings
Stage III	**Tumor Involves One or Both Ovaries with Peritoneal Implants Outside the Pelvis and/or Positive Retroperitoneal or Inguinal Lymph Nodes; Superficial Liver Metastases; or Tumor Limited to the True Pelvis, but There Is Histologically Proven Extension to the Small Bowel or Omentum**
	IIIa: grossly limited to the true pelvis with negative lymph nodes, but microscopic seeding of the abdominal peritoneal surface
	IIIb: tumor involves one or both ovaries with histologically proven implants of the abdominal peritoneal surfaces, none >2 cm
	IIIc: abdominal implants >2 cm diameter and/or positive retroperitoneal lymph nodes
Stage IV	**Tumor Involves One or Both Ovaries Plus Distant Metastases; Pleural Effusion Must Be Cytologically Positive for Stage IV; or Presence of Parenchymal Liver Metastases**

carcinomas, with a suggestion therefore that serous carcinomas may arise more frequently *de novo*. A recent analysis[74] demonstrated that approximately 50% of ovarian cystadenocarcinomas contained p53 mutations, whereas there was no evidence of p53 mutation in 37 benign and borderline tumors. These researchers next identified malignant tumors with contiguous regions of apparently benign neoplasia and demonstrated that benign-looking regions contained the same mutations in p53 as were present in the adjacent malignant regions. These findings suggest that benign neoplasia in the context of cystadeno-carcinoma represents a genetically and biologically distinct entity from the solitary benign cystadenoma, and at a molecular level is clonally contiguous with the adjacent malignant tumor. Whether these benign regions represent precursors that progress to frank adenocarcinoma or are products of a differentiating effect such as that seen in teratoma is unclear and hinges on further analysis of these regions to demonstrate whether or not they share the full complement of genetic abnormalities found in the adjacent adenocarcinoma. The alternative, that the benign lesions may be precursors to advanced disease is also a possibility. First- and second-degree relatives of ovarian cancer patients have five times the incidence of benign tumors compared to those without a family history.[75]

Endometriosis may well be a premalignant condition. It co-exists with endometrioid cancer in nearly a third of cases; and with clear cell carcinoma in a half of cases, and there are documented cases of endometriosis progressing to epithelial ovarian cancer in some patients taking exogenous estrogens.[58] In contrast, true serous borderline tumors (without evidence of invasion) undergo malignant transformation in only 0.7% of cases,[76] suggesting that the pathways to borderline disease do not form part of the multistep pathway from normal surface epithelium to epithelial ovarian cancer (Figure 2). These phenotypic lesions do not preclude molecular lesions in phenotypically normal ovarian surface epithelium, and there is evidence that mutant or overexpressed normal p53 is detectable in apparently normal epithelia adjacent to ovarian neoplastic tissue, suggesting the possibility of phenotypically normal precursor lesions. To what extent this represents a molecular field change is unknown.

4.5 Staging and Primary Surgical Treatment

The most powerful determinant of prognosis for epithelial ovarian cancer is the extent to which the malignant tumor has disseminated from its primary site at the time of diagnosis. In order to stage the patient adequately, thorough surgical evaluation by exploratory laparotomy needs to be undertaken by a surgeon experienced in gynecological oncology. Painstaking and careful removal of all tumor within the pelvis and peritoneal cavity offers the patient the best chance of ultimate control of her disease. Thus, definitive surgery is employed with abdominal exploration in a procedure that constitutes primary

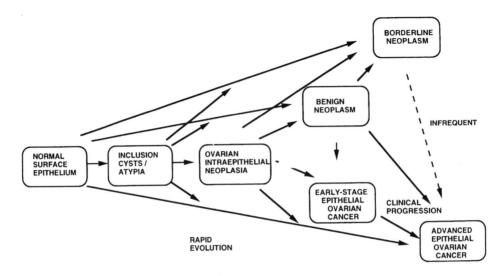

Figure 2 Multistep pathway from normal ovarian epithelium to epithelial ovarian cancer.

treatment as well as indicating the patient's prognosis. In order to properly evaluate the extent of spread, adequate access to the pelvis and peritoneum is facilitated by a vertical midline incision. Ascites should be aspirated, or washings from the sites where peritoneal fluid circulates (both paracolic gutters and under both hemidiaphragms) should be performed and sent for cytological analysis. The tumor should be removed intact, as should the uterus (hysterectomy), both fallopian tubes, and the contralateral ovary (bilateral salpingo-oophorectomy). Removal of the omentum and sampling of para-aortic and pelvic lymph nodes should be routinely performed. In those without overt evidence of tumor dissemination random biopsies from the peritoneum and peritoneal surface of the diaphragm should be undertaken. This rigorous approach to surgical evaluation has resulted in the widespread adoption of the FIGO (1988) staging system for ovarian cancer, where ascending stage correlates with worsening prognosis (Table 1). Accurate staging of the patient is crucial as it determines subsequent therapy, which will contribute to outcome. If there is obvious evidence of tumor spread, debulking of residual tumor masses (up to a maximum diameter of 10 cm) to under 2 cm maximum diameter should be undertaken as there is evidence that this results in better response and survival with subsequent chemo- or radio-therapy. This extent of debulking is achievable in most patients. Those patients with tumor masses < 2 cm postoperatively have a median survival of 45 months, while for those with masses > 2 cm the figure is 16 months.[77] There is additional evidence that those with residual masses < 0.5 cm diameter have a better median survival than those with residual masses between 0.5 and 2 cm.[78] These survival figures are the same irrespective of the preoperative size of the largest masses (up to a maximum diameter of 10 cm).[79] Additionally, there is some recent evidence that the number of masses rather than just their size may be important prognostically.

4.6 Other Prognostic Factors

Many features of ovarian cancer: surgical, pathological, and biological may be associated with survival. It is of importance whether these so-called prognostic factors are truly independent in themselves or simply reflect associations with other prognostic factors. Multivariate statistical methods have been used to answer this question. Such analyses suggest that there are general independent prognostic factors such as younger patient age and good performance status. Lower FIGO stages, well differentiated tumors (lower tumor grade), low residual tumor volume, and low preoperative tumor volume are also favorable independent prognostic factors.[80] Serum CA125 (see below) measured 3 months after the start of chemotherapy correlated well with survival of patients.[81] Patients with clear cell histology have worse prognosis, specifically in early stage disease.[82,83] Flow cytometric analysis of tumoral DNA content can indicate the presence of aneuploid tumor populations and is a significant adverse independent prognostic factor.[84,85] However, it is not enough to simply derive prognostic indices from multivariate analysis, these factors must then be applied in prospective studies to test their usefulness in practice. The use of molecular biological analysis will hopefully yield tumoral markers which have prognostic utility. Tumoral estrogen and progesterone receptor content, erbB-2 (HER2/neu) and epidermal growth factor receptor levels are promising candidates. The data examining the role and independence of p53 abnormality as a prognostic factor in ovarian cancer is conflicting, with some investigations showing no association with p53 abnormalities,[86-89] and others showing that p53 status functions as an independent prognostic factor.[90-93] Mutation of p53 is reported to be significantly associated with serous histology.[87, 94] Recent data suggest that those ovarian cancers which have a normal p53 gene by sequence analysis are more likely to respond to chemotherapy than those with p53 mutation.[95]

The use of some of these markers in combination (PR, DNA ploidy, S-phase fraction, and Ki67, a marker of cell cycling) has been shown to have similar (though prognostically independent) power to FIGO stage in correlation with progression-free interval and overall survival of ovarian cancer patients.[96]

4.7 CA-125

CA-125 is a high molecular weight mucin glycoprotein. It is normally found on serosal surfaces in the fetus. In adults it is expressed on the surface of mesothelium lining the pleura, the pericardium, and the peritoneal cavity. It is also produced by Müllerian-derived epithelia (fallopian tubes, endometrium, and endocervix).[97] CA-125 may be expressed on the surface of both benign and malignant ovarian tumors.[98] Cell death and turnover releases CA-125 from its anatomical site and it circulates in the bloodstream with a half-life of 4.5 days. Its level is elevated in 80% of ovarian cancers.[99] CA-125 is not specific for ovarian cancer; it may be elevated in cancers of the breast, colon, pancreas, lung, lymphomas, and other gynecological cancers.[100] It can also be elevated in liver disease and during inflammatory processes involving serosal surfaces. The main applications of the measurement of CA-125 levels at the present time are in monitoring response to therapy and the detection of relapse. The fall in CA-125 with treatment correlates well with tumor response to therapy, so much so that it is being considered for incorporation as part of objective response criteria. This is especially important given the difficulties of tumor imaging and radiological criteria for response in ovarian cancer.

CA-125 measurement has prognostic utility (see above), and in those who have normalized levels, it provides a sensitive index for subclinical relapse, preceding disease recurrence by 3 months on average.[101] Attempts are being made to boost the specificity of tumor-marker diagnosis of ovarian cancer for screening purposes by combining CA-125 with newer ovarian cancer markers (see below).

5. TREATMENT: BIOLOGIC PRINCIPLES

5.1 Surgery/Controversies in Surgical Management

As outlined above, the importance of optimal surgery cannot be overstated in the context of current practice. In early stage disease with good prognostic factors, i.e., Stage Ia/b well-differentiated ovarian cancer, surgery can be considered definitive and adequate, without need for further adjuvant therapy.[102] In those with such early disease who have not completed their families, unilateral salpingo-oophorectomy may be performed to preserve fertility.[103] In all other patients who achieve optimal debulking regardless of presenting tumor burden (see above), adjuvant therapy (i.e., further anticancer therapy given "prophylactically" to eradicate potential micrometastatic disease which could proliferate and manifest as "recurrence") is mandatory to maximize the opportunity for long term control of the disease.

It has been suggested that those patients who are optimally debulked, are operable because of intrinsic biological factors of their disease, and that optimal surgery simply highlights those who will intrinsically do well. This hypothesis was recently tested in an elegant and influential study by Van der Berg et al.[104] involving the procedure of interval debulking. Patients with advanced EOC (stages III and IV) who had been suboptimally debulked (had residual disease ≥ 1 cm left after primary surgery) were given three cycles of chemotherapy. Those who did not have progressive disease at this stage were then randomized to either receive or not a second debulking procedure where a second attempt was made to remove all residual disease and all randomized patients went on to receive a further three cycles of chemotherapy. Those who had this second procedure had significantly increased survival compared to those who did not. Furthermore, this benefit was primarily observed in those who had their disease successfully reduced to ≤ 1 cm residual disease after the second procedure. This study confirms that surgery has therapeutic rather than merely prognostic value in the management of ovarian cancer.

5.2 Role of Radiotherapy

Radiotherapy can be employed as an adjuvant treatment in certain stage/grade criteria. In general, this option is not frequently exercised because it is not as utilitarian as chemotherapy in this disease, and because there is a defined incidence of serious complications including bowel obstruction, adhesions, and radiation enteritis. As adjuvant therapy, the whole abdomen is irradiated, with the pelvis treated to a high dose and the rest of the peritoneal cavity fully encompassed as far up as both hemidiaphragms. This approach may be indicated in early stage disease without good prognostic features, i.e., stage Ic, stage I with moderately/poorly differentiated carcinoma, stage II disease, and possibly very early stage III disease.

5.3 Chemotherapy

Patients who do not have early-stage ovarian cancer with good prognostic features should receive chemotherapy as first-line adjuvant therapy (stage Ic, stage I with moderately/poorly differentiated carcinoma, stage II-IV disease, and clear cell carcinoma irrespective of stage). "Advanced disease" can be considered according to the consequences of laparotomy: whether or not debulking was optimal. The definition of "optimal" depends on whether one has a European or North American perspective. Optimal debulking (GOG) is reduction of residual diameter to lesions of ≤ 1 cm. Its European equivalent (GCCG), "small volume disease" is defined as maximum residual diameter ≤ 2 cm. Above these diameters patients are regarded as having "suboptimally debulked" or "large volume" disease. A reanalysis of the American data by Hoskins[105] reveals that a significant advantage for debulking to < 2 cm exists for the GOG cohorts also.

Standard chemotherapy generally consists of platinum containing regimens, usually with an alkylating agent, delivered as six cycles typically three-weekly, and these regimens appear to be as effective as more complex and toxic regimens containing anthracyclines and hexamethylmelamine. The best results suggest that using multimodality therapy, the 5-year survival of advanced (stage III-IV) epithelial ovarian cancer can be improved from approximately 12% to approximately 30%. The improvement in survival is mostly but probably not completely attributable to combination platinum chemotherapy.[106] Taxanes are an important new development and are likely to find a mainstream role in the management of ovarian cancer.

5.3.1 Platinum

The current evidence is that (1) immediate platinum-based therapy is better than a non-platinum regimen; (2) platinum in combination is better than single-agent platinum when used at the same dose (although the studies used a lower dose of cisplatinum in the single agent arm), and (3) carboplatin and cisplatin are equally effective (although there is evidence that this may not be the case in small volume disease). Platinum analogues function as cytotoxic drugs by binding to amino or hydroxyl groups of nucleoside bases and forming DNA intrastrand and interstrand crosslinks. This DNA crosslinking results in interference with replication/repair mechanisms. They are currently the most important cytotoxic agents used in ovarian cancer. The choice of platinum agent has been the subject of much research and speculation. Superficially, carboplatin appears to be tolerated better with reduction in gastrointestinal, neurological, and renal toxicity, while cisplatinum exhibits relatively little myelosuppression.

5.3.2 Evidence for Platinum-Associated Survival Benefit

Neijt showed data from two large randomized studies,[107] demonstrating that large-residual-volume disease patients (>2 cm) had no difference in survival comparing platinum with nonplatinum chemotherapy. However, in those with small volume disease, a significant survival advantage was seen for the platinum arm. Thus, a subgroup of advanced EOC can be defined who will do badly with standard therapy, and a group can be defined who benefit from platinum-based therapy.

A large meta-analysis of chemotherapy in advanced ovarian cancer[108] did detect a survival advantage for platinum combinations over combinations not containing platinum and for platinum combinations over single agent platinum. However, no difference was demonstrated between carboplatin and cisplatin based therapy, as several randomized trials have demonstrated,[109,110] and generally the median survival differences between the two agents are non-significant, although there remains some evidence that cisplatin-based combinations may be more effective than carboplatin combinations in terms of survival.[111]

5.3.3 Is Platinum Given at Optimum Doses?

Perhaps of more relevance to the compromise of long-term control of optimally debulked disease is the contribution of platinum toxicity to reducing the platinum dose actually delivered to any one patient. The issues are first, how to ensure the delivery of planned dose intensity; second, is there a dose-response curve, and if so, what are its characteristics, and how does one exploit it to improve patient outcome?

Several approaches have been explored to attempt to improve the pharmacodynamic effect of cis-platinum on the tumor whilst minimizing toxicity (improving the "therapeutic index"). Prehydration and the use of osmotic diuresis to minimize renal toxicity and the use of increasingly sophisticated antiemetics to minimize vomiting, both previously major toxicities, have revolutionized the acceptability of the drug. Abrogation of platinum drug resistance *in vitro* has been observed when tamoxifen is added concurrently. The mechanism underlying this observation is obscure, and is under intensive investigation.[112,113] However, clinical studies do not support the use of a platinum/tamoxifen at the moment.[114] The combination of platinum with interferon, based on *in-vitro* evidence for synergy between platinum and interferon in ovarian cancer[115] to improve response rates, has produced only modest clinical benefit.[116,117] Results from a randomized study attempting to reduce the toxicity of cisplatinum using glutathione (an intracellular free-radical scavenger) suggest that more full-dose cycles can be given on time with significant reduction in nephrotoxicity. The incidence of nausea, vomiting, and neuropathy were significantly reduced in the glutathione group. The response rate was not reduced, suggesting that glutathione improved the therapeutic index of cisplatinum.[118]

5.3.4 Platinum Dose Intensity

The issue of the correct platinum dose is both controversial and confusing. At its heart lies Levin's and Hryniuk's concept of dose intensity and their hypothesis that as the dose intensity of a drug is increased, response and survival rates increase. This hypothesis took its empirical evidence from retrospective meta-analyses of largely randomized studies in ovarian cancer.[119,120] Thereafter, the hypothesis was tested in a series of prospective randomized trials comparing different dose intensities of platinum.[121]

"Dose intensity" (DI) was a concept developed to try and compare the impact of differing dosing practices. This is the amount of drug administered in milligrams per meter squared body surface area per week, and can be calculated in practice for an individual's chemotherapy course. "Relative dose intensity" (RDI) is the amount of drug delivered per unit time relative to an arbitrarily chosen standard. "Average dose intensity" (ADI) takes the RDIs of different components of a regimen and averages them with respect to the chosen standard regimen. Levin and Hryniuk made assumptions in their analysis that all drugs contribute equally to the regimen, and that the schedule and route of administration are irrelevant. They considered prospectively randomized studies of patients with advanced ovarian cancer receiving first-line therapy where patients received some or all components of a standard test regimen (CHAP). Drug analogs were converted to "equivalents" of the drugs used in the CHAP regimen. RDI showed an association with response rates for single agent and combination platinum, single agent cyclophosphamide, and doxorubicin in combination regimens. Additionally, platinum RDI was associated with improved median survival time.

Thus, platinum DI is an important determinant of outcome, but in these studies, its upper limit was not defined. The dose/response curve for platinum was found to be steep only at RDI < 1.5; above 1.5, it is less steep but did not plateau. The curve reflected a suboptimal range of platinum doses (<36 mg/m^2, three weekly).

Levin and Hryniuk's work therefore suggests that dropping the platinum dose below optimal would predict a steep decline in response and survival, but no comment could be made about platinum RDIs > 1.5.[119,120]

Prospective studies comparing low and high DIs of platinum were subsequently performed to test this hypothesis, and two types of study can be generally identified which tried to answer these issues. The first type of study compared two doses of platinum with a twofold increase in DI between the arms. They showed higher response rates and longer median survival times in patients receiving higher DIs.[122,123] These studies suffered from several criticisms. First, the studies compared different total doses not just intensities since both arms received the same number of cycles. Second, the DI of cisplatin used was generally of standard magnitude in the "dose-intense" arm versus what would be regarded as a suboptimal dose in the other. Third, the staging was unclear in so far as the definition of "advanced" was skewed to better prognosis subgroups as a result of entry criteria. In contrast, the second type of study, which strictly compared two DIs while keeping total dose the same and varying overall treatment time showed no significant survival differences.[124] Hence, total dose is felt to be at least as important as DI. The relative importance of these two concepts in practice is not yet resolved.

As introduced above, detailed curve-fitting of the retrospective data suggests that the platinum dose-response curve is sigmoid with a plateau of response at 25 mg/m^2/week (cisplatin combinations) — 30 mg/m^2/week (single agent cisplatin). This data is supported by the findings of another retrospective study which showed that as the pharmacokinetic dose term "area under the curve" (AUC) for carboplatin increased above 5 mg/ml × min, there was no significant increase in tumor response.[125] This clinical evidence for a plateau reflects the fact that only modest increases in platinum dose are possible because of a steep toxicity curve.

5.3.5 High-Dose Chemotherapy

Observations on chemotherapy dose in murine cancer models by Skipper which showed that a 20% drop in dose did not alter the complete response rate but diminished the cure rate by 50%, have stimulated interest in chemotherapy dose escalation in solid tumors.

In vitro data with ovarian cancer cells suggest that increasing platinum dose five- to tenfold can overcome both acquired resistance[126] and therefore potentially the intrinsic "shoulder" of the sigmoid dose-response curve observed over conventional clinical dose escalation, suggesting that if this could be achieved clinically without excessive morbidity, it might produce a clear increment in outcome. Furthermore, there is evidence that synergistic effects occur between platinum and alkylating agents *in vitro*.[127]

Levin and Hryniuk's retrospective observation of a relationship between platinum dose intensity and outcome parameters over a conventional dose-range were reflected in two prospective studies demonstrating survival advantage for those receiving higher platinum dose intensity/total dose (see above); but from the above discussion, it is apparent that these studies have not addressed the issue of raising DI to overcome such relative resistance.

The problem with many high-dose chemotherapy studies (rescued by either autologous bone marrow, or peripheral blood stem-cells) are that the patients usually have large tumor burdens consisting of refractory pretreated ovarian cancer. In addition, these patients, having been heavily pretreated, and in general having higher tumor burdens, are poor candidates for highly toxic chemotherapy, and therefore exhibit florid toxicity. As expected, these studies show a high overall response rate, a low pathological complete response rate (where it has been assessed) and only a modest increment (if at all) in survival, usually in comparison to historical controls.

The only substantial published series in which high-dose chemotherapy with PBSC rescue was performed early in the course of disease, is more promising. Shinozuka treated 42 patients after primary surgical debulking with two cycles of high dose cyclophosphamide, doxorubicin, and cisplatinum with PBSC rescue.[128] In those optimally debulked patients (n = 23), 70% were alive after 4 years. Further work is required to identify which groups (if any) of ovarian cancer patients might benefit from such treatment.

Research protocols are now accumulating patients whose stem cells are *ex-vivo* transfected with the multidrug resistance gene (MDR 1) which results in the transferred stem cells post chemotherapy being drug resistant, thus reducing the patient's myelosuppresion with subsequent cycles of high-dose chemotherapy. However, it is not simply dose which is of importance. Although not considered by Levin and Hryniuk, the number of cycles of the regimen is also important. No randomized clinical trials of single versus multiple courses of the same chemotherapy in curable cancers exist, however, reducing the number of cycles of cisplatin/etoposide from four to three produces inferior results in testicular cancer,[129] and anecdotally, testicular cancer chemotherapy defaulters who default after one cycle have poor outcomes.

Thus, not only should platinum RDI be increased to about five, but one should test multiple cycles rather than the current practice of a single cycle of high dose chemotherapy as late intensification. The toxicity profile of cisplatin precludes its use at RDIs > 2.5, but carboplatin possesses excellent characteristics for dose escalation, and indeed, feasibility studies[130] have suggested that three cycles of PBSC supported carboplatin (1200 mg/m^2) can be delivered safely with the main limitation being ototoxicity.

5.3.6 Taxanes

Taxanes are natural (taxol) and semisynthetic (taxotere) diterpenoids. Taxol (Paclitaxel) was discovered in 1971 when it was isolated from the bark of the western yew tree (Taxus brevifolia). Taxoids induce tubulin polymerization, and also prevent the depolymerization of microtubules, resulting in stable, nonfunctional microtubules disrupting cellular organization and producing effective mitotic arrest. Phase II studies of taxol and taxotere demonstrated a 30 to 35% objective response rate in patients with ovarian cancer refractory to other drugs.[131-133] The American gynecologic oncology group (GOG protocol 111) have recently reported mature data from a randomized study of suboptimally debulked ovarian cancer patients, comparing standard first line chemotherapy cisplatin/cyclophosphamide) (CP) with cisplatin/taxol (TP) . Both the complete and partial response rates were significantly increased in the taxol/cisplatin arm and median survival was increased from 2 years to 3 years in this arm.[134] Although severe adverse effects were noted for the TP arm, more patients completed the 6 cycles than in the other arm. This study is being repeated in a large European/Canadian joint study, and if confirmed, will increase the acceptance of platinum/taxane as the gold standard (as it now is in the U.S.) in Europe and Canada in this subgroup of patients. The GOG has several other protocols currently underway testing taxol in other situations. GOG 114 is essentially the same as GOG 111 except that it tests taxol in optimally debulked disease. The CP arm was dropped after the mature data from GOG 111, and it now tests TP × 6 against moderate dose carboplatin initially followed by TP × 6. GOG 132 tests the relative contributions of taxol and platinum in a three-arm study in suboptimally debulked disease; T alone, P alone, and TP. A crossover design assesses the impact and ordering of sequential monotherapies. It is the first trial to test T alone in previously untreated disease. GOG 34 is another taxol protocol that tests 3 DIs of taxol in platinum refractory, taxol naive patients. Unfortunately, this protocol is accruing poorly because of the difficulty in finding patients who have not received previous taxol.

5.3.7 Drug Resistance

It is generally accepted that chemotherapy fails to eradicate cancer due to a combination of intrinsic and acquired drug resistance. Physiological host factors may alter drug disposition within the body resulting in a reduction of the amount of active drug getting to the tumor (reduced activation, enhanced excretion, and breakdown), but resistance is determined more by tumor cellular factors. In terms of ontogeny, tumor cells represent extremely old human cells (having ceaselessly divided over a long period of time), and have acquired many genetic lesions that allow them to escape normal growth regulation. They can evoke mechanisms to circumvent attempts to control their growth. One manifestation of this survival pathway redundancy is drug resistance. Tumor cells may limit drug accumulation by influx pump inhibition or by efflux pump augmentation (multidrug resistance, p-glycoprotein). Intracellular inactivators of drugs may be upregulated by tumor cells. Such inactivators include scavenger antioxidants such as glutathione for example, which binds intracellular platinum among other drugs and may prevent DNA adduct formation. Damage to genes which maintain the stability and integrity of the genome allow the cells to deregulate their tight control over the genome with rapid and often bizarre changes mediated by selective pressure from a hostile environment. p53 abnormality and mutation of mismatch repair genes resulting in the "promutator phenotype" may contribute to a phenotype tolerant to the adverse environment created by the cytotoxic milieu. This may manifest in the ability to circumvent apoptosis by mutations acquired in proteins involved in the proapoptotic pathway. Additionally, tumor cells may be able to develop DNA repair mechanisms that are more efficient than normal, counteracting the genotoxic effects of cytotoxic chemotherapy, and hence contributing to resistance to DNA damaging drugs. An overview of DNA repair mechanisms is presented in Chapter 9: many enzymes are involved in the process of damage recognition, nucleotide excision, resynthesis of the damaged region with fidelity, and ligation to reestablish the continuity of chromatin. Drug resistance may also be associated with alteration of the topoisomerase pathway (either by mutation or altered expression). This enzyme is involved in cleavage of DNA, passage of a DNA strand through the cleaved segment and resealing of the DNA. Type I topoisomerases cleave only one strand; type II cleave both strands.

6. SCREENING: PROSPECTS FOR PREVENTION

The development by Bast et al. of CA 125 as a serum tumor marker in the early 1980s[99] opened up the possibility of early detection of ovarian cancer by serial blood tests in populations. The underlying philosophy for this is that ovarian cancer progresses through the FIGO stages and therefore detection of the disease while still limited to stages I and II could dramatically increase the cure rate of the disease. While the evidence suggests that this is certainly the situation in many cases, the preceding discussion about the early lesion raises the possibility that clinical early stage ovarian cancer may be a different disease from the much commoner advanced ovarian cancer. Detailed examination of ovaries showing microscopic or very small primary lesions in association with extensive intra-peritoneal disease have been demonstrated, which may render the art of the achievable in screening more modest. Having said that, progress towards effective screening is being made, although its positive predictive value is not yet sufficient to enable widespread implementation. CA 125 is elevated in only 60% of early stage EOCs.[135] Studies of stored serum have demonstrated that up to 5 years previously, CA 125 is elevated in up to a quarter of those in a population destined to develop EOC, suggesting that at least some ovarian cancers indeed progress over a long time-frame. Eighteen months before developing the disease, 50% of the population had elevated CA 125.[136] Two prospective interventional studies have indicated the spectrum of disease likely to be picked up. In a large study by Jacobs et al.[137] 22,000 postmenopausal women were screened with CA 125; 340 women had elevated CA 125. These patients were then screened further by transvaginal sonography (TVS) and 40 abnormalities were detected. Ovarian cancer was detected in 11, but a further 18 ovarian cancers were missed by the CA 125 screen in this study. In another study by Einhorn et al.[138] 5,550 "normal" women were screened. Of those older than 50 years, 2% had elevated CA 125. Those with elevated CA 125 and an equal number of CA 125 negative controls were followed up with 3 monthly CA 125s, 6 monthly gynecological examinations, and 6 monthly transabdominal ultrasonographies. Six cases of EOC were detected, four were early stage. There were three cases (i.e., one third of the total) that were completely undetected by screening. CA 125 levels alone are therefore not sensitive enough and an approach of complementary serum markers has been investigated. Using CA 125, OVX 1, and M-CSF a retrospective study in 48 patients with stage I disease demonstrated a sensitivity of 98%.[139] The 18 patients without elevated CA 125 studied by Jacobs et al. (above) who went on to develop ovarian cancer were assessed using these two other markers and 8 (44%) demonstrated elevation of the other two markers. The overall specificity of these markers was 89%. A study in postmenopausal women is now underway using CA 125 and OVX 1 as a screen followed by TVS, and laparotomy in those with abnormal sonograms. The lowest target specificity required to achieve a positive predictive value of 10% (i.e., one cancer picked up for every ten operations) is 99.6%, and it remains to be seen if the OVX 1/CA 125 study will achieve this. Until then, population screening for ovarian cancer cannot be recommended.

REFERENCES

1. **Herrinton, L.J., Stanford, J.L., Schwartz, S.M., and Weiss, N.S.,** Ovarian cancer incidence among Asian migrants to the United States and their descendants, *J. Natl. Cancer Inst.,* 86, 1336, 1994.

2. **Parazzini, F., Franceschi, S., La-Vecchia, C., and Fasoli, M.,** The epidemiology of ovarian cancer, *Gynecol. Oncol.,* 43, 9, 1991.

3. **Whittemore, A.S., Wu, M.L., Paffenbarger, R., Jr., Sarles, D. L., Kampert, J. B., Grosser, S., Jung, D. L., Ballon, S., and Hendrickson, M.,** Personal and environmental characteristics related to epithelial ovarian cancer. II. Exposures to talcum powder, tobacco, alcohol, and coffee, *Am. J. Epidemiol.,* 128, 1228, 1988.

4. **Cramer, D.W., Welch, W.R., Scully, R.E., and Wojciechowski, C.A.,** Ovarian cancer and talc: a case-control study, *Cancer,* 50, 372, 1982.

5. **Harlow, B.L., Cramer, D.W., Bell, D.A., and Welch, W.R.,** Perineal exposure to talc and ovarian cancer risk, *Obstet. Gynecol.,* 80, 19, 1992.

6. **Wehner, A.P.,** Biological effects of cosmetic talc, *Food Chem. Toxicol.,* 32, 1173, 1994.

7. **Gwinn, M.L., Webster, L.A., Lee, N.C., Layde, P.M., and Rubin, G.L.,** Alcohol consumption and ovarian cancer risk, *Am. J. Epidemiol.,* 123, 759, 1986.

8. **Kato, I., Tominaga, S., and Terao, C.,** Alcohol consumption and cancers of hormone-related organs in females, *Jpn. J. Clin. Oncol.,* 19, 202, 1989.

9. **Byers, T., Marshall, J., Graham, S., Mettlin, C., and Swanson, M.,** A case-control study of dietary and nondietary factors in ovarian cancer, *J. Natl. Cancer Inst.,* 71, 681, 1983.

10. **Cramer, D.W., Welch, W.R., Hutchison, G.B., Willett, W., and Scully, R.E.,** Dietary animal fat in relation to ovarian cancer risk, *Obstet. Gynecol.,* 63, 833, 1984.

11. **Mori, M., Harabuchi, I., Miyake, H., Casagrande, J.T., Henderson, B.E., and Ross, R.K.,** Reproductive, genetic, and dietary risk factors for ovarian cancer, *Am. J. Epidemiol.,* 128, 771, 1988.

12. **Risch, H.A., Jain, M., Marrett, L.D., and Howe, G.R.,** Dietary fat intake and risk of epithelial ovarian cancer, *J. Natl. Cancer Inst.,* 86, 1409, 1994.

13. **Kaufman, D.W., Kelly, J.P., Welch, W.R., Rosenberg, L., Stolley, P.D., Warshauer, M.E., Lewis, J., Woodruff, J., and Shapiro, S.,** Noncontraceptive estrogen use and epithelial ovarian cancer, *Am. J. Epidemiol.,* 130, 1142, 1989.

14. **Adlercreutz, H., Mousavi, Y., Clark, J., Hockerstedt, K., Hamalainen, E., Wahala, K., Makela, T., and Hase, T.,** Dietary phytoestrogens and cancer: *in vitro* and *in vivo* studies, *J. Ster. Biochem. Mol. Biol.,* 41, 331, 1992.

15. **Adlercreutz, H., Bannwart, C., Wahala, K., Makela, T., Brunow, G., Hase, T., Arosemena, P.J., Kellis, J., Jr., and Vickery, L.E.,** Inhibition of human aromatase by mammalian lignins and isoflavonoid phytoestrogens, *J. Ster. Biochem. Mol. Biol.,* 44, 147, 1993.

16. **Mousavi, Y. and Adlercreutz, H.,** Genistein is an effective stimulator of sex hormone-binding globulin production in hepatocarcinoma human liver cancer cells and suppresses proliferation of these cells in culture, *Steroids,* 58, 301, 1993.

17. **Rose, D.P., Goldman, M., Connolly, J.M., and Strong, L.E.,** High-fiber diet reduces serum estrogen concentrations in premenopausal women, *Am. J. Clin. Nutr.,* 54, 520-525, 1991.

18. **Goldin, B.R., Woods, M.N., Spiegelman, D.L., Longcope, C., Morrill-LaBrode, A., Dwyer, J.T., Gualtieri, L.J., Hertzmark, E., and Gorbach, S.L.,** The effect of dietary fat and fiber on serum estrogen concentrations in premenopausal women under controlled dietary conditions, *Cancer,* 74, 1125, 1994.

19. **Cramer, D.W., Harlow, B.L., Willett, W.C., Welch, W.R., Bell, D.A., Scully, R.E., Ng, W.G., and Knapp, R.C.,** Galactose consumption and metabolism in relation to the risk of ovarian cancer, *Lancet,* 2, 66, 1989.

20. **Cramer, D.W.,** Epidemiologic aspects of early menopause and ovarian cancer, *Ann. N.Y. Acad. Sci.,* 592, 363, 1990.

21. **Bressac, B., Kew, M., Wands, J., and Ozturk, M.,** Selective G-mutation to T-mutation of the p53 gene in hepatocellular carcinoma in Southern Africa, *Nature,* 350, 429, 1991.

22. **Kohler, M.F., Marks, J.R., Wiseman, R.W., Jacobs, I.J., Davidoff, A.M., Clarke-Pearson, D.L., Soper, J.T., Bast, R., Jr., and Berchuck, A.,** Spectrum of mutation and frequency of allelic deletion of the p53 gene in ovarian cancer, *J. Natl. Cancer Inst.,* 85, 1513, 1993.

23. **O'Malley, B.W.,** The steroid receptor superfamily: more excitement predicted for the future, *Mol. Endocrinol.,* 4, 363, 1990.

24. **Horwitz, K.B. and McGuire, W.L.,** Estrogen control of progesterone receptor in human breast cancer: correlation with nuclear processing of estrogen receptor, *J. Biol. Chem.,* 253, 2223, 1978.

25. **Langdon, S.P., Hirst, G.L., Miller, E.P., Hawkins, R.A., Tesdale, A.L., Smyth, J.F., and Miller, W.R.,** The regulation of growth and protein expression by estrogen in vitro: a study of 8 human ovarian carcinoma cell lines, *J. Ster. Biochem. Mol. Biol.,* 50, 131, 1994.

26. **Slotman, B.J. and Rao, B.R.,** Ovarian cancer (review), *Anticancer Res.,* 8, 417, 1988.

27. **Bonte, J.,** Developments in endocrine related therapy of endometrial and ovarian cancer, *Rev. Endocr. Rel. Cancer,* 3, 11, 1979.

28. **Friedman, M., Lagios, M., Markowwitz, A., Jones, H., Resser, K., and Hoffman, P.,** Estradiol (ER) and Progesterone receptors (PR) in ovarian cancer — clinical and pathological correlation, *Clin. Res.,* 27, 385A, 1979.

29. **Sevelda, P., Denison, U., Schemper, M., Spona, J., Vavra, N., and Salzer, H.,** Oestrogen and progesterone receptor content as a prognostic factor in advanced epithelial ovarian carcinoma, *Br. J. Obstet. Gynaecol.,* 97, 706, 1990.

30. **Rose, P.G., Reale, F.R., Longcope, C., and Hunter, R.E.,** Prognostic significance of estrogen and progesterone receptors in epithelial ovarian cancer, *Obstet. Gynecol.,* 76, 258, 1990.

31. **Masood, S., Heitmann, J., Nuss, R.C., and Benrubi, G.I.,** Clinical correlation of hormone receptor status in epithelial ovarian cancer, *Gynecol. Oncol.,* 34, 57, 1989.

32. **Kieback, D.G., McCamant, S.K., Press, M.F., Atkinson, E.N., Gallager, H.S., Edwards, C.L., Hajek, R.A., and Jones, L.A.,** Improved prediction of survival in advanced adenocarcinoma of the ovary by immunocytochemical analysis and the composition adjusted receptor level of the estrogen receptor, *Cancer Res.,* 53, 5188, 1993.

33. **Kieback, D.G., Press, M.F., Atkinson, E.N., Edwards, G.L., Mobus, V.J., Runnebaum, I.B., Kreienberg, R., and Jones, L.A.,** Prognostic significance of estrogen receptor expression in ovarian cancer. Immunoreactive Score (IRS) vs. Composition Adjusted Receptor Level (CARL), *Anticancer Res.,* 13, 2489, 1993.

34. **Gardner, W.U.,** Further studies on experimental ovarian tumorigenesis, *Proc. Am. Assoc. Cancer Res.,* 2, 300, 1958.

35. **Jabara, A.G.,** Induction of canine ovarian tumors by diethylstilbestrol and progesterone, *Aust. J. Exp. Biol.,* 40, 139, 1962.

36. **Fathalla, M.F.,** Incessant ovulation — a factor in ovarian neoplasia?, *Lancet,* 2, 163, 1971.

37. **Cramer, D.W., Hutchison, G.B., Welch, W.R., Scully, R.E., and Ryan, K.J.,** Determinants of ovarian cancer risk I. Reproductive experiences and family history, *J. Natl. Cancer Inst.,* 71, 711, 1983.

38. **Weiss, N.S., Lyon, J.L., Liff, J.M., Vollmer, W.M., and Daling, J.R.,** Incidence of ovarian cancer in relation to the use of oral contraceptives, *Int. J. Cancer,* 28, 669, 1981.

39. **Cramer, D.W., Hutchison, G.B., Welch, W.R., Scully, R.E., and Knapp, R.C.,** Factors affecting the association of oral contraceptives and ovarian cancer, *N. Engl. J. Med.,* 307, 1047, 1982.

40. **Ron, E., Lunenfeld, B., Menczer, J., Blumstein, T., Katz, L., Oelsner, G., and Serr, D.,** Cancer incidence in a cohort of infertile women, *Am. J. Epidemiol.,* 125, 780, 1987.

41. **Whittemore, A.S., Harris, R., and Itnyre, J.,** Characteristics relating to ovarian cancer risk: collaborative analysis of 12 US case-control studies. II. Invasive epithelial ovarian cancers in white women. Collaborative Ovarian Cancer Group [see comments], *Am. J. Epidemiol.,* 136, 1184 , 1992.

42. **Whittemore, A.S.,** Fertility drugs and risk of ovarian cancer, *Hum. Reprod.,* 8, 999, 1993.

43. **Spirtas, R., Kaufman, S.C., and Alexander, N.J.,** Fertility drugs and ovarian cancer: red alert or red herring? [see comments], *Fertil. Steril.,* 59, 291, 1993.

44. **Rossing, M.A., Daling, J.R., Weiss, N.S., Moore, D.E., and Self, S.G.,** Ovarian tumors in a cohort of infertile women [see comments], *N. Engl. J. Med.,* 331, 771, 1994.

45. **Narod, S.A.,** Genetics of breast and ovarian cancer, *Br. Med. Bull.,* 50, 656, 1994.

46. **Steichen-Gersdorf, E., Gallion, H.H., Ford, D., Girodet, C., Easton, D.F., DiCioccio, R.A., Evans, G., Ponder, M.A., Pye, C., and Mazoyer, S., et al.,** Familial site-specific ovarian cancer is linked to BRCA1 on 17q12-21, *Am. J. Hum. Genet.,* 55, 870, 1994.

47. **Miki, Y., Swensen, J., Shattuck-Eidens, D., Futreal, P.A., Harshman, K., Tavtigian, S., Liu, Q., Cochran, C., Bennett, L.M., Ding, W., et al.,** A strong candidate for the breast and ovarian cancer susceptibility gene BRCA1, *Science,* 266, 66, 1994.

48. **Easton, D.F., Bishop, D.T., Ford, D., Crockford, G.P., and the Breast Cancer Consortium**, Genetic linkage analysis in familial breast and ovarian cancer: results from 214 families, *Am. J. Hum. Genet.*, 52, 678, 1993.

49. **Wooster, R., Neuhausen, S.L., Mangion, J., Quirk, Y., Ford, D., Collins, N., Nguyen, K., Seal, S., Tran, T., Averill, D., Fields, P., Marshall, G., Narod, S., Lenoir, G., Lynch, H., Feunteun, J., Devilee, P., Cornelisse, C., Menko, F., Daly, P., Ormiston, W., McManus, R., Pye, C., Lewis, C., Cannon-Albright, L., Peto, J., Ponder, B., Skolnick, M., Easton, D., Goldgar, D., and Stratton, M.**, Localization of a breast cancer susceptibility gene, BRCA2, to chromosome 13q12-13, *Science*, 265, 2088, 1994.

50. **Lynch, H.T., Schuelke, G.S., Kimberling, W.J., Albano, W.A., Lynch, J.F., Biscone, K.A., Lipkin, M.L., Deschner, E.E., Mikol, Y.B., Sandberg, A.A., et al.,** Hereditary nonpolyposis colorectal cancer (Lynch syndromes I and II). II. Biomarker studies, *Cancer*, 56, 939, 1985.

51. **Lynch, H.T., Kimberling, W., Albano, W.A., Lynch, J.F., Biscone, K., Schuelke, G.S., Sandberg, A.A., Lipkin, M., Deschner, E.E., Mikol, Y.B., et al.,** Hereditary nonpolyposis colorectal cancer (Lynch syndromes I and II). I. Clinical description of resource, *Cancer*, 56, 934, 1985.

52. **Lynch, H.T. and Lynch, J.F.,** The Lynch syndromes, *Curr. Opin. Oncol.*, 5, 687, 1993.

53. **Fishel, R., Lescoe, M.K., Rao, M.R., Copeland, N.G., Jenkins, N.A., Garber, J., Kane, M., and Kolodner, R.,** The human mutator gene homolog MSH2 and its association with hereditary nonpolyposis colon cancer [published erratum appears in Cell 1994 Apr 8;77(1):167], *Cell*, 75, 1027, 1993.

54. **Kolodner, R.D., Hall, N.R., Lipford, J., Kane, M.F., Morrison, P.T., Finan, P.J., Burn, J., Chapman, P., Earabino, C., Merchant, E., et al.,** Structure of the human MLH1 locus and analysis of a large hereditary nonpolyposis colorectal carcinoma kindred for mlh1 mutations, *Cancer Res.*, 55, 242, 1995.

55. **Nicolaides, N.C., Papadopoulos, N., Liu, B., Wei, Y.F., Carter, K.C., Ruben, S.M., Rosen, C.A., Haseltine, W.A., Fleischmann, R.D., Fraser, C.M., et al.,** Mutations of two PMS homologues in hereditary nonpolyposis colon cancer, *Nature*, 371, 75, 1994.

56. **Young, R.H. and Scully, R.E.,** Pathology of epithelial tumors, *Hematol./Oncol. Clin. N. Am.*, 6, 739, 1992.

57. **Baak, J.P., Chan, K.K., Stolk, J.G., and Kenemans, P.,** Prognostic factors in borderline and invasive ovarian tumors of the common epithelial type, *Pathol. Res. Pract.*, 182, 755, 1987.

58. **Heaps, J.M., Nieberg, R.K., and Berek, J.S.,** Malignant neoplasms arising in endometriosis, *Obstet. Gynecol.*, 75, 1023, 1990.

59. **Iwabuchi, H., Sakunaga, H., Sakamoto, M., Yang-Feng, T.L., Pinkel, D., and Gray, J.W.,** Genetic progression model in ovarian cancer with comparative genomic hybridisation (meeting abstract), *Proc. Am. Assoc. Cancer Res.*, 36, 226, 1995.

60. **Jenkins, R.B., Bartelt, D., Stalboerger, P., Persons, D., Dahl, R.J., Podratz, K., and Keeney, G.,** Cytogenetic studies of epithelial ovarian-carcinoma, *Cancer Genet. Cytogenet.*, 71, 76, 1993.

61. **Bello, M.J. and Rey, J.A.,** Chromosome aberrations in metastatic ovarian cancer: relationship with abnormalities in primary tumours, *Int. J. Cancer*, 45, 50, 1990.

62. **Pejovic, T., Heim, S., Mandhal, N., Baldetorp, B., Elmfors, B., Floderus, U.-M., Furgyik, S., Helm, G., Himmelman, A., Willen, H., and Mitelman, F.,** Chromosome aberration in 35 primary ovarian carcinomas, *Genes Chromosom. Cancer*, 4, 58, 1992.

63. **Knudson, A.G.,** Mutation and cancer: statistical study of retinoblastoma, *Proc. Natl. Acad. Sci. U.S.A.*, 68, 820, 1971.

64. **Gabra, H., Taylor, L., Cohen, B.B., Lessels, A., Eccles, D.M., Leonard, R.C.F., Smyth, J.F., and Steel, C.M.,** Chromosome 11 allele imbalance and clinicopathological correlates in ovarian tumours, *Br. J. Cancer*, in press, 1995.

65. **Futreal, P.A., Liu, Q., Shattuck-Eidens, D., Cochran, C., Harshman, K., Tavtigian, S., Bennett, L.M., Haugen-Strano, A., Swensen, J., Miki, Y., et al.,** BRCA1 mutations in primary breast and ovarian carcinomas, *Science*, 266, 120, 1994.

66. **Merajver, S.D., Pham, T.M., Caduff, R.F., Chen, M., Poy, E.L., Cooney, K.A., Weber, B.L., Collins, F.S., Johnston, C., and Frank, T.S.,** Somatic mutations in the BRCA1 gene in sporadic ovarian tumours, *Nature Genet.*, 9, 439, 1995.

67. **Dodson, M.K., Cliby, W.A., Xu, H.J., DeLacey, K.A., Hu, S.X., Keeney, G.L., Li, J., Podratz, K.C., Jenkins, R.B., and Benedict, W.F.,** Evidence of functional RB protein in epithelial ovarian carcinomas despite loss of heterozygosity at the RB locus, *Cancer Res.*, 54, 610, 1994.

68. **Kim, T.M., Benedict, W.F., Xu, H.J., Hu, S.X., Gosewehr, J., Velicescu, M., Yin, E., Zheng, J., D'Ablaing, G., and Dubeau, L.,** Loss of heterozygosity on chromosome 13 is common only in the biologically more aggressive subtypes of ovarian epithelial tumors and is associated with normal retinoblastoma gene expression, *Cancer Res.*, 54, 605, 1994.

69. **Committee, F.C.,** Staging announcement, *Gynecol. Oncol.*, 25, 383, 1986.

70. **Vogelstein, B. and Kinzler, K.W.,** The multistep nature of cancer, *Trends Genet.*, 9, 138, 1993.

71. **Bell, D.A. and Scully, R.E.,** Early de-novo ovarian-carcinoma — a study of 14 cases, *Cancer*, 73, 1859, 1994.

72. **Mittal, K.R., Zeleniuch-Jacquotte, A., Cooper, J.L., and Demopoulos, R.I.,** Contralateral ovary in unilateral ovarian carcinoma: A search for preneoplastic lesions, *Int. J. Gynecol. Pathol.*, 12, 59, 1993.

73. **Puls, L.E., Powell, D.E., DePriest, P.D., Gallion, H.H., Hunter, J.E., Kryscio, R.J., van, N.J., Jr.,** Transition from benign to malignant epithelium in mucinous and serous ovarian cystadenocarcinoma, *Gynecol. Oncol.*, 47, 53, 1992.

74. **Zheng, J., Benedict, W.F., Xu, H.-J., Hu, S.-X., Kim, T.M., Velicescu, M., Wan, M., Cofer, K.F., and Dubeau, L.,** Genetic disparity between morphologically benign cysts contiguous to ovarian carcinomas and solitary cystadenomas, *J. Natl. Cancer Inst.*, 87, 1146, 1995.

75. **Bourne, T.H., Whitehead, M.I., Campbell, S., Royston, P., Bhan, V., and Collins, W.P.,** Ultrasound screening for familial ovarian cancer, *Gynecol. Oncol.*, 43, 92, 1991.

76. **Kurman, R.J. and Trimble, C.L.,** The behavior of serous tumors of low malignant potential: Are they ever malignant?, *Int. J. Gynecol. Pathol.*, 12, 120, 1993.

77. **Delgado, G., Oram, D.H., and Petrilli, E.S.,** Stage III epithelial ovarian cancer: the role of maximal surgical reduction, *Gynecol. Oncol.*, 18, 293, 1984.

78. **Griffiths, C.T.,** Surgical resection of tumor bulk in the primary treatment of ovarian carcinoma, *Natl. Cancer Inst. Monogr.*, 42, 101, 1975.

79. **Hacker, N.F., Berek, J.S., Lagasse, L.D., Nieberg, R.K., and Elashoff, R.M.,** Primary cytoreductive surgery for epithelial ovarian cancer, *Obstet. Gynecol.*, 61, 413, 1983.

80. **de-Souza, P.L. and Friedlander, M.L.,** Prognostic factors in ovarian cancer, *Hematol. Oncol. Clin. N. Am.*, 6, 761, 1992.

81. **Sevelda, P., Schemper, M., and Spona, J.,** CA 125 as an independent prognostic factor for survival in patients with epithelial ovarian cancer, *Am. J. Obstet. Gynecol.*, 161, 1213, 1989.

82. **Guthrie, D., Davy, M.L., and Philips, P.R.,** A study of 656 patients with 'early' ovarian cancer, *Gynecol. Oncol.*, 17, 363, 1984.

83. **O'Brien, M.E., Schofield, J.B., Tan, S., Fryatt, I., Fisher, C., and Wiltshaw, E.,** Clear cell epithelial ovarian cancer (mesonephroid): Bad prognosis only in early stages, *Gynecol. Oncol.*, 49, 250, 1993.

84. **Friedlander, M.L., Hedley, D.W., Taylor, I.W., Russell, P., Coates, A.S., and Tattersall, M.H.,** Influence of cellular DNA content on survival in advanced ovarian cancer, *Cancer Res.*, 44, 397, 1984.

85. **Friedlander, M.L., Hedley, D.W., Swanson, C., and Russell, P.,** Prediction of long-term survival by flow cytometric analysis of cellular DNA content in patients with advanced ovarian cancer, *J. Clin. Oncol.*, 6, 282, 1988.

86. **Kohler, M.F., Kerns, B.J.M., Humphrey, P.A., Marks, J.R., Bast, R.C., and Berchuck, A.,** Mutation and overexpression of p53 in early-stage epithelial ovarian-cancer, *Obstet. Gynecol.*, 81, 643, 1993.

87. **Niwa, K., Itoh, M., Murase, T., Morishita, S., Itoh, N., Mori, H., and Tamaya, T.,** Alteration of p53 gene in ovarian-carcinoma — clinicopathological correlation and prognostic-significance, *Br. J. Cancer*, 70, 1191, 1994.

88. **Sheridan, E., Silcocks, P., Smith, J., Hancock, B.W., and Goyns, M.H.,** P53 mutation in a series of epithelial ovarian cancers from the uk, and its prognostic-significance, *Eur. J. Cancer*, 11, 1701, 1994.

89. **Frank, T.S., Bartos, R.E., Haefner, H.K., Roberts, J.A., Wilson, M.D., and Hubbell, G.P.,** Loss of heterozygosity and overexpression of the p53 gene in ovarian-carcinoma, *Mod. Pathol.*, 7, 3, 1994.

90. **Henriksen, R., Strang, P., Wilander, E., Backstrom, T., Tribukait, B., and Oberg, K.,** P53 expression in epithelial ovarian neoplasms — relationship to clinical and pathological parameters, ki-67 expression and flow-cytometry, *Gynecol. Oncol.*, 53, 301, 1994.

91. **Klemi, P.J., Takahashi, S., Joensuu, H., Kiilholma, P., Narimatsu, E., and Mori, M.,** Immunohistochemical detection of p53 protein in borderline and malignant serous ovarian-tumors, *Int. J. Gynecol. Pathol.*, 13, 228, 1994.

92. **Levesque, M.A., Katsaros, D., Yu, H., Zola, P., Sismondi, P., Giardina, G., and Diamandis, E.P.,** Mutant p53 protein overexpression is associated with poor outcome in patients with well or moderately differentiated ovarian-carcinoma, *Cancer*, 75, 1327, 1995.

93. **Vanderzee, A.G.J., Hollema, H., Suurmeijer, A.J.H., Krans, M., Sluiter, W.J., Willemse, P.H.B., Aalders, J.G., and Devries, E.G.E.,** Value of p-glycoprotein, glutathione-s-transferase-pi, c-erbB-2, and p53 as prognostic factors in ovarian carcinomas, *J. Clin. Oncol.*, 13, 70, 1995.

94. **Eccles, D.M., Russell, S., Haites, N.E., Atkinson, R., Bell, D.W., Gruber, L., Hickey, I., Kelly, K., Kitchener, H., and Leonard, R.,** Early loss of heterozygosity on 17q in ovarian-cancer, *Oncogene*, 7, 2069, 1992.

95. **Al-Azraqi, A., Chapman, C., Challen, C., Aswad, S., Sinha, D., Calvert, A.H., and Lunec, J.,** P53 mutations in primary human ovarian cancer as a determinant of resistance to carboplatin (meeting abstract), *Proc. Am. Assoc. Cancer Res.*, 36, 228, 1995.

96. **Kaufmann, M., Vonminckwitz, G., Kuhn, W., Schmid, H., Costa, S., Goerttler, K., and Bastert, G.,** Combination of new biologic parameters as a prognostic index in epithelial ovarian-carcinoma, *Int. J. Gynecol. Cancer*, 5, 49, 1995.

97. **Kabawat, S.E. Bast, R., Jr., Bhan, A.K., Welch, W.R., Knapp, R.C., and Colvin, R.B.,** Tissue distribution of a coelomic-epithelium-related antigen recognized by the monoclonal antibody OC125, *Int. J. Gynecol. Pathol.*, 2, 275, 1983.

98. **Niloff, J.M., Knapp, R.C., Schaetzl, E., Reynolds, C., and Bast, R., Jr.,** CA125 antigen levels in obstetric and gynecologic patients, *Obstet. Gynecol.*, 64, 703, 1984.

99. **Bast, R., Jr., Klug, T.L., St-John, E., Jenison, E., Niloff, J.M., Lazarus, H., Berkowitz, R.S., Leavitt, T., Griffiths, C.T., Parker, L., Zurawski, V., Jr., and Knapp, R.C.,** A radioimmunoassay using a monoclonal antibody to monitor the course of epithelial ovarian cancer, *N. Engl. J. Med.*, 309, 883, 1983.

100. **Niloff, J.M., Klug, T.L., Schaetzl, E., Zurawski, V., Jr., Knapp, R.C., and Bast, R., Jr.,** Elevation of serum CA125 in carcinomas of the fallopian tube, endometrium, and endocervix, *Am. J. Obstet. Gynecol.*, 148, 1057, 1984.

101. **Niloff, J.M., Knapp, R.C., Lavin, P.T., Malkasian, G.D., Berek, J.S., Mortel, R., Whitney, C., Zurawski, V., Jr., and Bast, R., Jr.,** The CA 125 assay as a predictor of clinical recurrence in epithelial ovarian cancer, *Am. J. Obstet. Gynecol.,* 155, 56, 1986.

102. **Young, R.C., Walton, L.A., Ellenberg, S.S., Homesley, H.D., Wilbanks, G.D., Decker, D.G., Miller, A., Park, R., and Major, F., Jr.,** Adjuvant therapy in stage I and stage II epithelial ovarian cancer. Results of two prospective randomized trials [see comments], *N. Engl. J. Med.,* 322, 1021, 1990.

103. **Miyazaki, T., Tomoda, Y., Ohta, M., Kano, T., Mizuno, K., and Sakakibara, K.,** Preservation of ovarian function and reproductive ability in patients with malignant ovarian tumors, *Gynecol. Oncol.,* 30, 329, 1988.

104. **Van der Burg, M.E., van Lent, M., Buyse, M., Kobierska, A., Colombo, N., Favalli, G., Lacave, A.J., Nardi, M., Renard, J., and Pecorelli, S.,** The effect of debulking surgery after induction chemotherapy on the prognosis in advanced epithelial ovarian cancer, *N. Engl. J. Med.,* 332, 675, 1995.

105. **Hoskins, W.J., McGuire, W.P., Brady, M.F., Homesley, H.D., Creasman, W.T., Berman, M., Ball, H., and Berek, J.S.,** The effect of diameter of largest residual disease on survival after primary cytoreductive surgery in patients with suboptimal residual epithelial ovarian carcinoma, *Am. J. Obstet. Gynecol.,* 170, 974, 1994.

106. **Omura, G., Blessing, J.A., Ehrlich, C.E., Miller, A., Yordan, E., Creasman, W.T., and Homesley, H.D.,** A randomized trial of cyclophosphamide and doxorubicin with or without cisplatin in advanced ovarian carcinoma. A Gynecologic Oncology Group Study, *Cancer,* 57, 1725, 1986.

107. **Neijt, J.,** Ovarian cancer: rethinking prognostic factors and chemotherapy, *Proc. Am. Soc. Clin. Oncol.,* 214, 1986.

108. Chemotherapy in advanced ovarian cancer: an overview of randomised clinical trials. Advanced Ovarian Cancer Trialists Group [see comments], *Br. Med. J.,* 303, 884, 1991.

109. **Alberts, D.S., Green, S., Hannigan, E.V., O'Toole, R., Stock-Novack, D., Anderson, P., Surwit, E.A., Malvlya, V.K., Nahhas, W.A., and Jolles, C.J.,** Improved therapeutic index of carboplatin plus cyclophosphamide versus cisplatin plus cyclophosphamide: final report by the Southwest Oncology Group of a phase III randomized trial in stages III and IV ovarian cancer [published erratum appears in *J. Clin. Oncol.* 1992 Sep;10(9):1505], *J. Clin. Oncol.,* 10, 706, 1992.

110. **Swenerton, K., Jeffrey, J., Stuart, G., Roy, M., Krepart, G., Carmichael, J., Drouin, P., Stanimir, R., O'Connell, G., MacLean, G., et al.,** Cisplatin-cyclophosphamide versus carboplatin-cyclophosphamide in advanced ovarian cancer: a randomized phase III study of the National Cancer Institute of Canada Clinical Trials Group [see comments], *J. Clin. Oncol.,* 10, 718, 1992.

111. **Vermorken, J.B., ten-Bokkel-Huinink, W. W., Eisenhauer, E. A., Favalli, G., Belpomme, D., Conte, P. F., and Kaye, S. B.,** Advanced ovarian cancer. Carboplatin versus cisplatin, *Ann. Oncol.,* 4, S41, 1993.

112. **McClay, E.F., Albright, K.D., Jones, J.A., Eastman, A., Christen, R., and Howell, S.B.,** Modulation of cisplatin resistance in human malignant melanoma cells, *Cancer Res.,* 52, 6790, 1992.

113. **Nakata, B., Albright, K.D., Barton, R.M., Howell, S.B., and Los, G.,** Synergistic interaction between cisplatin and tamoxifen delays the emergence of cisplatin resistance in head and neck cancer cell lines, *Cancer Chemother. Pharmacol.,* 35, 511, 1995.

114. **Schwartz, P.E., Chambers, J.T., Kohorn, E.I., Chambers, S.K., Weitzman, H., Voynick, I.M., MacLusky, N., and Naftolin, F.,** Tamoxifen in combination with cytotoxic chemotherapy in advanced epithelial ovarian cancer. A prospective randomized trial, *Cancer,* 63, 1074, 1989.

115. **Nehme, A., Julia, A.M., Jozan, S., Chevreau, C., Bugat, R., and Canal, P.,** Modulation of cisplatin cytotoxicity by human recombinant interferon-gamma in human ovarian cancer cell lines, *Eur. J. Cancer.,* 30a, 4, 520, 1994.

116. **Markman, M., Berek, J.S., Blessing, J.A., McGuire, W.P., Bell, J., and Homesley, H.D.,** Characteristics of patients with small-volume residual ovarian cancer unresponsive to cisplatin-based ip chemotherapy: Lessons learned from a Gynecologic Oncology Group phase II trial of ip cisplatin and recombinant alpha-interferon, *Gynecol. Oncol.,* 45, 3, 1992.

117. **Ferrari, E., Maffeo, D.A., Graziano, R., Gallo, M.S., Pignata, S., De-Rosa, L., Montella, M., and Pergola, M.,** Intraperitoneal chemotherapy with carboplatin and recombinant interferon alpha in ovarian cancer, *Eur. J. Gynaecol. Oncol.,* 15, 437, 1994.

118. **Smyth, J., Bowman, A., Perren, T., Wilkinson, P., Prescott, R., Quinn, K., and Tedeschi, M.,** Glutathione improves the therapeutic index of cisplatin and quality of life for patients with ovarian cancer, *Proc. Annu. Meet. Am. Soc. Clin. Oncol.,* 14, 273, 1995.

119. **Levin, L. and Hryniuk, W.M.,** Dose intensity analysis of chemotherapy regimens in ovarian carcinoma, *J. Clin. Oncol.,* 5, 756, 1987.

120. **Levin, L., Simon, R., and Hryniuk, W.,** Importance of multiagent chemotherapy regimens in ovarian carcinoma: Dose intensity analysis, *J. Natl. Cancer Inst.,* 85, 1732, 1993.

121. **Ozols, R.F., Thigpen, J.T., Dauplat, J., Colombo, N., Piccart, M.J., Bertelsen, K., Levin, L., and Lund, B.,** Advanced ovarian cancer. Dose intensity, *Ann. Oncol.,* 4, S49-56, 1993.

122. **Kaye, S.B., Lewis, C.R., Paul, J., Duncan, I.D., Gordon, H.K., Kitchener, H.C., Cruickshank, D.J., Atkinson, R.J., Soukop, M., Rankin, E.M., Cassidy, J., Davis, J.A., Reed, N.S., Crawford, S.M., MacLean, A., Swapp, G.A., Sarkar, T.K., Kennedy, J.H., and Symonds, R.P.,** Randomised study of two doses of cisplatin with cyclophosphamide in epithelial ovarian cancer, *Lancet,* 340, 329, 1992.

123. **Ngan, H.Y.S., Choo, Y.C., Cheung, M., Wong, L.C., Ma, H.K., Collins, R., Fung, C., Ng, C.S., Wong, V., Ho, H.C., Leung, P., Wong, R., Chan, E., Simon, M.T.P., Ho, L.C., and Chan, Y.F.,** A randomized study of high-dose versus low-dose cis-platinum combined with cyclophosphamide in the treatment of advanced ovarian cancer, *Chemotherapy,* 35, 221, 1989.

124. **McGuire, W.P., Hoskins, W.J., Brady, M.F., Homesley, H.D., and Clarke-Pearson, D.L.,** A phase III trial of dose-intense (di) vs standard-dose (ds) cisplatin (cddp) and cytoxan (ctx) in advanced ovarian cancer (aoc), *Proc. Annu. Meet. Am. Soc. Clin. Oncol.,* 11, 1992.

125. **Jodrell, D.I., Egorin, M.J., Canetta, R.M., Langenberg, P., Goldbloom, E.P., Burroughs, J.N., Goodlow, J.L., Tan, S., and Wiltshaw, E.,** Relationships between carboplatin exposure and tumor response and toxicity in patients with ovarian cancer [see comments], *J. Clin. Oncol.,* 10, 520, 1992.

126. **Behrens, B.C., Hamilton, T.C., Masuda, H., Grotzinger, K.R., Whang-Peng, J., Louie, K.G., Knutsen, T., McKoy, W.M., Young, R.C., and Ozols, R.F.,** Characterization of a cis-diamminedichloroplatinum (II)-resistant human ovarian cancer cell line and its use in evaluation of platinum analogues, *Cancer Res.,* 47, 414, 1987.

127. **Lidor, Y.J., Shpall, E.J., Peters, W.P., and Bast, R., Jr.,** Synergistic cytotoxicity of different alkylating agents for epithelial ovarian cancer, *Int. J. Cancer,* 49, 704, 1991.

128. **Shinozuka, T., Murakami, M., and Miyamoto, T., Fujii, A., Tokuda, Y., Tajima, T., Mitomi, T., Watanabe, K., and Tsuji, K.,** High-dose chemotherapy (hdc) with autologous bone marrow transplantation (abmt) in ovarian cancer (meeting abstract), *Proc. Annu. Meet. Am. Soc. Clin. Oncol.,* 10, 1991.

129. **Williams, S.D., Loehrer, P.J., Nichols, C.R., and Einhorn, L.N.,** Chemotherapy of male and female germ cell tumors, *Semin. Oncol.,* 19, 19, 1992.

130. **Shea, T.C., Mason, J.R., Storniolo, A.M., Newton, B., Breslin, M., Mullen, M., Ward, D.M., Miller, L., Christian, M., and Taetle, R.,** Sequential cycles of high-dose carboplatin administered with recombinant human granulocyte-macrophage colony-stimulating factor and repeated infusions of autologous peripheral-blood progenitor cells — a novel and effective method for delivering multiple courses of dose-intensive therapy, *J. Clin. Oncol.,* 10, 464, 1992.

131. **Einzig, A.I., Wiernik, P.H., Sasloff, J., Runowicz, C.D., and Goldberg, G.L.,** Phase II study and long-term follow-up of patients treated with taxol for advanced ovarian adenocarcinoma, *J. Clin. Oncol.,* 10, 1748, 1992.

132. **Kavanagh, J.J., Kudelka, A.P., Freedman, R.S., Edwards, C.L., Pazdur, R., Bellet, R., Bayssas, M., Finnegan, M.B., and Newman, B.M.,** A Phase II trial of Taxotere (RP56976) in ovarian cancer patients (pts) refractory to cisplatin/carboplatin therapy, *Proc. Annu. Meet. Am. Soc. Clin. Oncol.,* 12, 1993.

133. **Thigpen, J.T., Blessing, J.A., Ball, H., Hummel, S.J., and Barrett, R.J.,** Phase II trial of paclitaxel in patients with progressive ovarian carcinoma after platinum-based chemotherapy: A Gynecologic Oncology Group study, *J. Clin. Oncol.,* 12, 1748, 1994.

134. **McGuire, W.P., Hoskins, W.J., Brady, M.F., Kucera, P.R., Partridge, E.E., Look, K.Y., and Davidson, M.,** Taxol and cisplatin (TP) improves outcome in advanced ovarian cancer (AOC) as compared to cytoxan and cisplatin (CP) (meeting abstract), *Proc. Annu. Meet. Am. Soc. Clin. Oncol.,* 14, 275, 1995.

135. **Jacobs, I. and Bast, R., Jr.,** The CA 125 tumour-associated antigen: A review of the literature, *Hum. Reprod.,* 4, 1, 1989.

136. **Zurawski, V., Jr., Orjaseter, H., Andersen, A., and Jellum, E.,** Elevated serum CA 125 levels prior to diagnosis of ovarian neoplasia: Relevance for early detection of ovarian cancer, *Int. J. Cancer,* 42, 677, 1988.

137. **Jacobs, I., Davies, A.P., Bridges, J., Stabile, I., Fay, T., Lower, A., Grudzinskas, J.G., and Oram, D.,** Prevalence screening for ovarian cancer in postmenopausal women by CA 125 measurement and ultrasonography, *Bmj,* 306, 1030, 1993.

138. **Einhorn, N., Sjovall, K., Knapp, R.C., Hall, P., Scully, R.E., Bast, R., Jr., and Zurawski, V., Jr.,** Prospective evaluation of serum CA 125 levels for early detection of ovarian cancer, *Obstet. Gynecol.,* 80, 14, 1992.

139. **Woolas, R.P., Xu, F.J., Jacobs, I.J., Yu, Y.H., Daly, L., Berchuck, A., Soper, J.T., Clarke-Pearson, D.L., Oram, D.H., and Bast, R., Jr.,** Elevation of multiple serum markers in patients with stage I ovarian cancer, *J. Natl. Cancer Inst.,* 85, 1748, 1993.

Chapter 7

Hormonal, Growth Factor and Cytokine Control of Ovarian Cancer

Simon P. Langdon, Barbara J.B. Simpson, John M.S. Bartlett, and William R. Miller

CONTENTS

1. INTRODUCTION

The two major roles of the ovary, namely, the controlled release of ova and the production of steroid hormones, involve the co-ordinated interactions of a variety of different cell types, regulated by hormones, growth factors, and cytokines. The malignant tumors that arise from these cell types develop under the same regulatory processes which may become aberrant during the pathogenesis and progression of the disease. This chapter will review the evidence that these regulatory molecules influence both the incidence of ovarian cancer and the growth of established tumors.

2. HORMONES

2.1 Role of Hormones in the Etiology of Ovarian Cancer

The ovary is the major site of synthesis of the sex steroid hormones and the target organ for the gonadotrophins. There is evidence to implicate both these groups of hormones in the development of ovarian cancer and this includes observations obtained from animal studies where administration of estrogens,[1] androgens,[2] progestins,[3] and gonadotrophins[4] all produced ovarian tumors (Figure 1). Epidemiological data support the role of a hormonal component in so far as parity, length of reproductive life, and the use of oral contraceptives all have an effect on the incidence of this disease. Compared to nulliparous females, women with 1 or 2 children have one half the risk, women with 3 or 4 children, one third the risk, and women with 5 or more children, one quarter the risk of developing ovarian cancer.[5,6] Variable findings have been obtained with respect to age of menopause with some studies finding that late age increases the risk,[7,8] although other studies have disputed this.[9] Oral contraceptive use is also associated with decreased risk.[10-13]

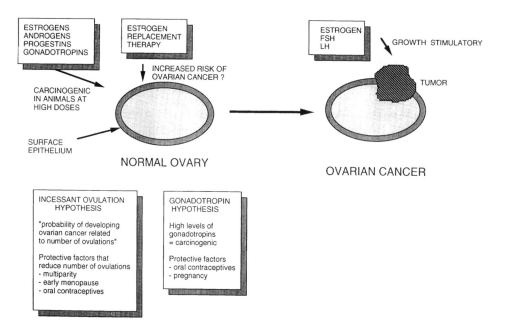

Figure 1   Involvement of steroid hormones and gonadotropins in the development and growth of ovarian cancer.

Two hypotheses have been put forward to explain the protective effects of increased parity and oral contraceptive use — the "gonadotropin hypothesis"[14] and the "incessant ovulation" hypothesis.[15,16] Both pregnancy and the use of oral contraceptive agents lower gonadotropin levels and suppress ovulation.[14] At menopause, the levels of gonadotropins are high and then decline, a pattern reflected in the age-incidence of ovarian cancer. If gonadotropins are associated with the incidence of ovarian cancer, their effects may be either direct or mediated via their ability to stimulate steroidogenesis. The alternative view, that of "incessant ovulation," is based on observations that there is a high frequency of ovarian tumors arising from the surface epithelium in women compared to other mammals.[15,16] Thus, in women, ovulatory cycles occur almost continuously from puberty to the menopause while other mammals are more economical with their ova.[15] Each ovulation involves minor trauma to the surface epithelium and repair is associated with rapid proliferation which may enhance the likelihood of malignancy. Experimental evidence in support of this proposal comes from the observation that when domestic hens ovulate continuously (by the use of artificial lighting), they have an almost 100% incidence of ovarian cancer, compared with hens kept under normal lighting condition which ovulate seasonally and in which ovarian cancer does not appear.[17] Hormonal factors which reduce the number of ovulations, such as increasing parity, early menopause and oral contraceptive use, would therefore be expected to reduce the risk of ovarian cancer. Direct evidence indicating that the surface epithelium of the normal ovary will undergo malignant transformation under conditions in which proliferation has been induced has been demonstrated in a rat model.[18] Further, indirect clinical support of this hypothesis is provided by the report of the 2.5-fold increased incidence of ovarian cancer in women whose ovaries were hyperstimulated to produce multiple ovulations as a result of treatment for infertility.[19] The effect of estrogen replacement therapy (ERT) on the incidence of ovarian cancer is controversial. ERT in postmenopausal women has been suggested to increase the risk of ovarian cancer with relative risks varying from 1.6 — 5.7,[5,7,20,21] although other studies report no greater risk[22-25] and at least one study has claimed a protective effect.[26] In several of the studies reporting an increased risk, this increased with extended use (>5 to 6 years).[5,21] Among the various histological forms of ovarian cancer, increased incidence was more likely to be associated with the endometrioid subtype.[5,27] A possible explanation as to why estrogens should result in increased risk has been proposed by Cramer and Welch who have suggested that the first stage of ovarian malignancy involves the formation of inclusion cysts by the entrapment of surface epithelium into the stroma; estrogen then promotes proliferation and increases the possibility of malignant transformation.[14]

2.2. Sensitivity of Established Tumors to Hormones

While ovarian cancer has not been generally regarded as an endocrine-sensitive disease, there are several lines of evidence to indicate that a percentage of this disease is hormonally sensitive. This view is based on the following: (1) the presence of high numbers of hormone receptors in some primary ovarian tumors, (2) the association of receptors and prognosis, (3) evidence of sensitivity to hormones in experimental models of ovarian cancer, and (4) the clinical responsiveness of certain ovarian tumors to endocrine therapies.

A large number of studies have examined the prevalence of hormone receptors in ovarian cancer. A review of 40 studies investigating the presence of steroid receptors in over 1600 patients reported the incidence of estrogen receptors in 63%, progesterone receptors in 48% and androgen receptors in 69% of ovarian carcinomas although it should be appreciated that criteria of positivity varied between individual reports.[28] Serous tumors were most frequently positive for the estrogen receptor and endometrioid tumors for the progesterone receptor with only a low percentage of mucinous and clear cell histologies containing either type of receptor.[28] Luteinizing hormone releasing hormone (LHRH) receptors are reported to be present in 80% of ovarian cancers.[29,30]

In several studies, the presence of estrogen and progesterone receptors in ovarian tumors has been associated with survival. Estrogen receptor as measured by ligand binding methods[31-33] or immunohistochemistry[34] has been shown to relate to improved survival although this has not been a universal finding.[35-41] Possible explanations for the inconsistencies between these studies include differences in methodology and cut-off points. Recent studies have accounted for the cellular heterogeneity within tumor samples by adjusting for the composition of the sample when using ligand binding technology — the CARL value (Composition Adjusted Receptor Level)[33] or taking into account both the percentage of cells positive and the intensity of staining to give an immunoreactive score (IRS) when using immunohistochemistry.[34] These refinements are likely to give more accurate assessments of ER status and both methods demonstrate an association between ER positivity and improved survival.[33,34]

The presence of the progesterone receptor has similarly been related to improved survival in at least four studies.[39,42-44] Again, both ligand binding[39,42,43] and immunohistochemical assessments[44] have shown an association with survival and in at least one of these studies,[43] the patient subgroup with the most favorable survival was the ER-positive/PR-positive group. More direct evidence for endocrine-sensitivity has been obtained in experimental systems and in clinical trials of hormonal therapies and these studies are elaborated below.

2.3 Estrogen Regulation of Growth and Protein Expression in Ovarian Cancer Models

Despite the majority of primary ovarian tumors being ER-positive and many containing high concentrations of estrogen receptors, it is only recently that the corresponding model systems have been derived to allow study of the role of estrogen regulation in ovarian cancer.[45-56] Estrogen-stimulated growth has been demonstrated in several cell lines and these are characterized by possessing an ER content greater than 20 to 30 fmol/mg protein.[45-53,56] Cell lines with lower ER concentrations appear unresponsive.[45,47] In breast and endometrial cancer, a number of proteins are induced by estrogen; several of these have been investigated in ER-positive ovarian cancer models (Figure 2). The interest in these estrogen-regulated proteins is twofold. First, they may help to identify tumors under estrogen-regulation and therefore their use might help to predict those ER-positive tumors most likely to respond to antiestrogen therapy. Second, the proteins may be the mediators of the mitogenic action of estrogen and as such may be new targets for therapy.

On exposure to 17 β-estradiol (E_2), the ER content of ER-positive cell lines is down-regulated[47] and the PR content is increased both in culture and *in vivo*.[46,47,54] Production of secreted proteins, such as procathepsin D, and a 120 kDa glycoprotein are also increased by E_2.[48,52] These effects are similar to those obtained in ER-positive breast cancer cell lines. In breast cancer, the mitogenic effects of estrogen have been proposed to act via mediation of growth factors, including transforming growth factor-α (TGF-α) and insulin-like growth factor-I (IGF-I).[57] Modulation by E_2 of several of these factors has been observed in ER-positive ovarian cancer models. For example, E_2 increases levels of TGF-α mRNA as detected by northern blot analysis[58] or RNAse protection assay and this is reflected in increased TGF-α protein secretion (unpublished data). Furthermore, concentrations of epidermal growth factor (EGF) receptors are reduced after treatment by E_2.[59] Higher concentrations of TGF-α are found in ER-positive/PR-positive primary ovarian tumors, consistent with possible estrogen regulation in tumors.[60]

Although IGF-1 levels appear to be unchanged by E_2, several IGF binding proteins (IGF-BPs) are modulated.[61] Therefore concentrations of IGF-BP-3 are decreased by E_2 while those of IGF-BP5 are

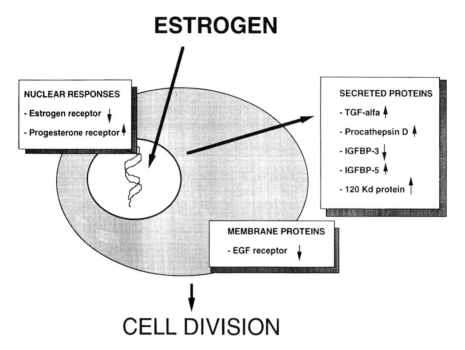

ESTROGEN

NUCLEAR RESPONSES

- Estrogen receptor ↓

- Progesterone receptor ↑

SECRETED PROTEINS

- TGF-alfa ↑

- Procathepsin D ↑

- IGFBP-3 ↓

- IGFBP-5 ↑

- 120 Kd protein ↑

MEMBRANE PROTEINS

- EGF receptor ↓

CELL DIVISION

Figure 2 Estrogen-regulated responses in ER-positive ovarian cancer cells. Estrogen can increase or decrease protein expression as shown. See text for details.

increased; other IGF-BPs are unaffected.[61] The effects of these growth factors are discussed further in Section 3.

Estrogen-stimulated growth in ER-positive cell lines and xenografts is inhibited by antiestrogens, a finding which supports the occasional clinical responsiveness of primary tumors to tamoxifen (see below).[45,49,62] Experimental support for tamoxifen-sensitivity in ER-positive tumors emanated from a study of tamoxifen against primary ovarian tumor cells and demonstrated that tamoxifen decreased colony formation by > 50% in primary tumor samples with an ER content ≥ 30 fmol/mg protein, whereas no effects were evident in tumors with an ER content < 30 fmol/mg protein.[63] This cut-off value is consistent with that obtained for E_2 stimulation of growth in cell lines.

2.4 Gonadotropin Regulation of Growth in Experimental Systems

Experimental models have indicated a role for gonadotropins in the growth regulation of ovarian cancer.[64] FSH (follicle stimulating hormone) and LH stimulate growth of ovarian cancer cells both in culture and *in vivo*.[65-67] Suppression of LH and FSH levels by the LHRH agonist leuprolide is associated with growth inhibition in an ovarian carcinoma xenograft[68] and the agonist has direct inhibitory effects on cultured ovarian cancer cells.[69] Another agonist, [D-Trp[6]]-LHRH, has also produced growth inhibition of ovarian cancer cells *in vitro*[64] and *in vivo*.[70]

2.5 Clinical Studies of Endocrine Therapy in Ovarian Cancer

A number of clinical studies have evaluated the potential of targeting steroid hormone or gonadotropin receptors in ovarian cancers. Many of these trials have produced response rates in excess of 10%.[71-75]

The first report of the use of antiestrogen therapy was in 1981 and described two patients who responded to tamoxifen.[71] Subsequently, several small investigations throughout the 1980s had a cumulative response rate of approximately 10%. Two of the most recent trials are worthy of mention. In the largest study to date, consisting of 105 patients with stage III/IV disease who were considered chemoresistant, a 10% complete and 7% partial response rate was reported after treatment with tamoxifen.[74] In another recent study, of 29 patients with stage III/IV chemoresistant disease, an 8% complete and 10% partial response rate was reported.[75] Although these studies suggest an overall response rate is only obtained in about 1 in 6 patients, this could be clinically useful if responsive patients could first be identified. The use of the aforementioned markers (Section 2.3) may help define these estrogen-sensitive tumors.

Estrogen can induce the progesterone receptor and the administration of progestins have been investigated in a number of trials, Overall, these studies have yielded a response rate similar to the rate achieved with antiestrogens (10 to 20%).[76-81] The most impressive response rate in these studies was a 55% response (18 of 33 patients) achieved by medroxyprogesterone acetate in a study of advanced endometrioid ovarian cancer.[80] Sequentially administered hormonal therapy of estrogen treatment (to induce PR) followed by a progestin has produced a 14 to 17% response rate in two trials.[82,83]

The positive experimental data obtained with LHRH agonists has led to several small clinical studies. These studies have produced response and disease stabilization rates of 10 to 50%.[84-87] A large international multicenter study, aiming to recruit 150 to 200 patients, is currently underway to evaluate [D-Trp^6]-LHRH after cisplatin-based chemotherapy in advanced stage tumors.[64]

Overall, these clinical studies suggest that a small percentage of patients can benefit from endocrine-based therapies. The challenge at present is to identify this group of patients reliably.

3. PEPTIDE GROWTH FACTORS

Several families of peptide growth factors that regulate the growth of ovarian cancer cells have been identified; these may act on the same cell from which they are secreted (autocrine control), on neighboring cells (paracrine regulation) or on distant cells (endocrine control). Autocrine regulation by growth factors has been suggested to be be a key mechanism whereby malignant cells become autonomous[88] and evidence has been obtained to indicate such control may operate in ovarian cancer cells.

The EGF family, the insulin-like growth factor family and the TGF-β family may all have important roles in ovarian cancer and these are discussed below.

3.1 Epidermal Growth Factor Family

Of the growth factor families, the EGF-related peptides have been the most extensively studied in ovarian cancer. EGF and the related factors TGF-α, amphiregulin and cripto all bind to, and activate, the EGF receptor. All have been identified in ovarian tumors and cultured ovarian carcinoma cells. TGF-α is reported to be present in 50 to 100% of tumors,[89-93] EGF in 28% to 71%,[89,93] amphiregulin in 18%[93] and cripto in 53%.[93] In experimental systems, TGF-α and EGF stimulated the growth of ovarian cancer cell lines *in vitro*.[94-98] Antibodies directed against either TGF-α or the EGF receptor can inhibit the proliferation of ovarian cancer cells which both produce TGF-α and possess the EGF receptor, an effect consistent with autocrine growth stimulation.[98-100] The EGF receptor, a 170 kDa glycosylated membrane-spanning protein is present in between 33 and 75% of primary ovarian tumors and has been detected by both ligand binding,[98,101-104] and immunohistochemical methods.[98,104-106] Levels of EGF receptor appear to be higher in malignant than benign tumors or normal ovary suggesting a possible biological role in malignant progression,[107,108] and rat ovarian epithelial cells which have gained the ability to grow in anchorage-independent assays show an increased responsiveness to EGF. Perhaps the strongest indication of a role, however, for this receptor derives from reports that its presence relates to poor prognosis in malignant tumors[101,102,105,109] although this has not been a universal finding.[110,111]

The EGF receptor (c-erbB-1) is a member of the type I tyrosine kinase growth factor receptor family and shares structural similarities with c-erbB-2, c-erbB-3 and c-erbB-4. The c-erbB-2 (HER-2/neu) protein is a 185 kDa transmembrane protein that is overexpressed in 20 to 30% of ovarian tumors, primarily as a result of gene amplification.[112,113] Like the EGF receptor, increased expression of c-erbB-2 is associated with poor survival.[112,113]

The c-erbB-3 receptor is present in the majority of ovarian tumors with 89% of malignant, 100% of borderline and 61% of benign tumors reported positive by immunohistochemical staining.[114] Overexpression seems to be more strongly associated with borderline and early invasive lesions.[114]

Ligand-induced activation of the c-erbB receptors encourages receptor dimerization which in turn initiates a signaling cascade via the ras/MAP kinase pathway resulting in transcriptional activation. Not only do receptors of the same type produce dimers (homodimerization) but different members of the type I kinase family can interact (heterodimerization), for example experimental evidence demonstrating interactions between the EGF receptor and c-erbB-2 has been obtained in ovarian cancer lines.[115] Multiple expression of c-erbB receptors is significantly higher in malignant than in borderline or benign ovarian tumors (Table 1) [116] and the formation of ligand-induced c-erbB heterodimers may confer a selective advantage to cells expressing more than one receptor.[116]

The family of type I kinase receptors clearly represent possible targets for therapy and a number of approaches are currently under consideration. One strategy is the use of the combination of cisplatin

Table 1 c-Erb-B Receptor Family Co-Expression in Ovarian Tumors

	% Tumors Positive		
	Malignant	**Borderline**	**Benign**
EGF receptor	65	38	39
c-erb-B2	76	25	31
c-erb-B3	89	100	62
EGF-r + c-erb-B2	50	13	23
EGF-r + c-erb-B3	63	38	39
c-erb-B2 + c-erb-B3	74	25	23
EGF-r + c-erb-B2 + c-erb-B3	50	13	23

Data obtained from Reference 115.

and antibodies targetting c-erbB-2. Co-administration of antibody with drug markedly enhanced both the *in vitro* and *in vivo* cytotoxicity of cisplatin against ovarian cancer models that overexpress this receptor. In addition to the antibody blocking a potentially mitogenic signalling pathway, it also appears to enhance the effects of cisplatin in resistant cells, indicating a possible application in chemo-resistant disease.[117,118]

3.2 Transforming Growth Factor-Beta Superfamily

The transforming growth factor-β family of polypeptide growth factors are involved in cell growth regulation, tissue remodeling, angiogenesis, and immune suppression.[119] Three forms of TGF-β have been identified in human systems, namely TGF-β1, TGF-β2 and TGF-β3, and these exist as homodimeric chains of between 111 and 113 amino acid, with molecular weights of 25 kDa. These growth factors interact with cell surface serine-threonine kinase linked receptors which mediate their regulatory effects.[120,121] The TGF-β isoforms bind directly to the TGF-βII receptor, whereupon the type I receptor is recruited into the complex, becomes phosphorylated and in turn propogates the signal to downstream substrates.[121]

TGF-β peptides have been shown to exert negative growth control in normal epithelial ovarian cultures and in most (95%) ovarian cancer cultures obtained from ascites.[122] The growth of approximately 50% of immortalized ovarian carcinoma cell lines are also inhibited by TGF-β.[123-127] It has been proposed that TGF-β may be an important regulator of normal ovarian epithelium and autocrine growth inhibition may be lost in many ovarian cancer cell lines, perhaps as an early step in the development of some ovarian cancers. Certain ovarian cancers that are growth inhibited by TGF-β are also more prone to undergo apoptosis than normal ovarian epithelial cells.[128] All three isoforms of TGF-β can be expressed in ovarian cancer cells and both TGF-β1 and TGF-β2 have been shown to be growth inhibitory to cultured malignant cells.[123]

In primary ovarian cancer, mRNA for the three isoforms, TGF-β1, TGF-β2, and TGF-β3 has been detected in 46, 66, and 66% of malignant tumors, respectively; the predominant pattern of expression being either dual or triple co-expression.[129] The TGF-β II receptor was present in over 90% of samples. Patterns of expression were similar between malignant, borderline, and benign tumors. TGF-β 3 was associated with advanced stage and reduced survival suggesting that perhaps the influence of this factor on angiogenesis and other features of tumor progression are more significant than direct inhibitory effects on growth.[129]

Two other members of the TGF-β superfamily, inhibin and Müllerian inhibiting substance (MIS) have also been studied in ovarian cancer. Inhibin is a polypeptide hormone produced by the granulosa cells of the ovary; its function is to inhibit FSH secretion by the pituitary gland. Inhibin is produced by all granulosa cell tumors and a positive serum level has been proposed as a marker for this subtype of ovarian cancer in postmenopausal women.[130] Inhibin has also been investigated in two studies of epithelial ovarian cancer.[131,132] In the first, sera levels of inhibin were elevated in 9 of 29 cases and in the second, in 14 of 24 cases. In the latter study, the survival time of the women with elevated levels of inhibin was five times longer than that for women not producing inhibin;[132] FSH levels were also significantly lower in the inhibin-producing patients. These data would be consistent with inhibin acting as a physiological defense mechanism to reduce elevated gonadotropin levels.

Like inhibin, MIS shares homology with TGF-β at the C-terminal domain.[133] MIS, in the male embryo, causes regression of Müllerian duct tissues that would otherwise develop into the fallopian tubes, the uterus, and upper vagina. Given its normal physiological role, it has been investigated for antitumor efficacy in ovarian tumor models. A limited degree of activity has been demonstrated against ovarian cancer cells grown both in culture[134,135] and *in vivo*.[136]

3.3 Insulin-Like Growth Factors

The insulin-like growth factors, IGF-I and IGF-II, are an important pair of mitogenic growth factors which show close structural similarity to insulin.[137] IGF-II is considered the major IGF mitogen in fetal growth, while IGF-I is the more important from birth onwards. The structures of the IGFs are sufficiently similar to insulin that they can influence metabolic activity via the insulin receptor and exert their mitogenic activities via IGF receptors, the IGF type-I receptor being the major mediator of IGF activities. These receptors belong to the type II receptor tyrosine kinase class. The IGFs bind to specific carrier proteins, the IGF binding proteins (IGFBPs), when circulating in extracellular fluids.[138]

The IGFs have important roles in the normal ovary and exert intraovarian control in the replication and differentiation processes of folliculogenesis.[139,140] In these processes they synergize with gonadotropins and interact with both thecal and granulosa cells in autocrine and paracrine pathways.

The IGFs, their receptors (insulin, type I, and type II receptors) and members of the IGFBP family (IGFBP-2,-3,-4,-5,-6) have been identified in a number of ovarian cancer cell lines[141-145] and in ovarian tumors.[146-150] IGF-I and insulin, when added to these cell lines, stimulate growth and since both the peptide and receptors are co-expressed, this provides the potential for autocrine control. Consistent with this, a DNA antisense oligonucleotide targeted to the mRNA for the IGF-I receptor (leading to its degradation) resulted in growth inhibition of an ovarian cancer cell line.[145]

3.4 Other Growth Factors

While factors such as the EGF and IGF families, and to a lesser degree the TGF-β family, tend to be have their major effects on epithelial cells, factors such as platelet derived growth factor (PDGF) and fibroblast growth factors (FGFs), are produced by ovarian epithelial cells but are more likely to influence surrounding stromal cells in a paracrine manner.

Platelet-derived growth factor (PDGF) is expressed by many ovarian cancer cell lines and is found in primary ovarian tumors.[151-153] The presence of PDGF receptor (a type III receptor tyrosine kinase) expression (found in 34% tumors) was correlated with a shorter survival.[153] Since ovarian cancer cell lines in culture are unresponsive to the factor, its action is more likely to be paracrine than autocrine.

The FGF family consists of seven FGF peptides and five receptors (members of the type IV receptor tyrosine kinase family) which possess varying affinities for each ligand. Basic fibroblast growth factor (bFGF) and its receptor are both expressed in ovarian cancer cells while addition of bFGF to cultured cells produces growth stimulation suggesting that the factor can act in both an autocrine or paracrine manner.[154,155] Amplification of several FGF receptors including FGFR1 (the flg oncogene), FGFR3 and FGFR4 have been demonstrated in ovarian tumors as has one of the FGF ligands, the oncogene int-2.

Other factors, produced by ovarian cancer cells, such as platelet-derived endothelial cell growth factor (PDECGF) and vascular endothelial growth factor (VEGF) are important angiogenic factors. PDECGF is produced by ovarian cancer cells and a recent study has indicated that increased expression of the factor is associated with areas of high blood velocity in malignant tumors.[156] VEGF, has also been shown to be overexpressed in ovarian carcinomas and is co-expressed with its receptor.[157]

4. CYTOKINES

The cytokines are low molecular weight (<80 kDa) peptides that regulate cell growth and immune function. Their effects are complex and dependent on the context into which they are placed; consequently, a cytokine may either stimulate or inhibit growth depending on its concentration, the cell type with which it interacts and the presence of other cytokines. Several cytokines are involved in the pathogenesis of ovarian cancer and the growth of established disease. Indeed, epithelial ovarian cancer has been suggested to be a "cytokine-propelled" disease.[158] Other cytokines, such as the interferons, are growth inhibitory and their therapeutic potential has been investigated. The roles of the major cytokine families in this disease are described below.

4.1 Interleukins

The interleukins are a family of low molecular weight cytokines initially identified as regulators of the proliferation and differentiation of lymphatic and myeloid cells but are now realized to be capable of influencing the growth of epithelial cells also. The normal ovarian surface epithelium secretes several interleukins including IL-1 and IL-6 but not IL-2, IL-3, or IL-4.

Interleukin 1-α (IL-1α), and IL-1β are produced by ovarian cancer cells in culture or when isolated from ascites and concentrations of IL-1α in the ascitic fluid of ovarian cancer patients are reported at 2 to 14 pg/ml.[159] This cytokine has been reported to both stimulate and inhibit proliferation of ovarian cancer cell lines and increases expression of TNF-α (tumor necrosis factor) mRNA and protein.[160] Production of TNF is thought to mediate the growth proliferative effects of IL-1, since antibodies targeting either TNF-α or its receptor block the effects of IL-1. Furthermore, transfecting ovarian cancer cells with TNF-α antisense mRNA abolishes the proliferative response to IL-1.[161]

Interleukin 6 (IL-6) is a 26 kDa glycoprotein produced by ovarian cancer cells.[103] The concentrations expressed in ascitic fluids and sera of women with ovarian cancer are elevated compared to non-cancer controls.[162,163] Sera concentrations > 0.2 units/ml were found in 76% of ovarian cancer patients with macroscopic disease, compared to 13% of patients with microscopic disease and 17% of normal controls.[163] Increased levels of IL-6 are also associated with increased levels of acute phase proteins (C′-reactive protein, haptoglobin and α2 macroglobulin) and IgA in sera and ascites of ovarian cancer patients, consistent with IL-6 inducing the production of acute phase proteins and inducing the differentiation of activated B cells to Ig secreting cells.[164] In addition to ovarian carcinoma cells, activated human monocytes also produce IL-6 providing a paracrine source of the cytokine.[165] Expression of IL-6 is upregulated by other cytokines, including IFN-γ, IL-1 and TNF.[162] The cytokine interacts with the IL-6 receptor, an 80 kDa binding protein, which may also be present in ovarian cancer cells indicating the potential for autocrine regulation.[164] To test whether autocrine control involving IL-6 operates, the loop has been inhibited by antisense oligonucleotide treatment which decreased the growth of three ovarian cancer cell lines in which it was tested.[166] Interleukin 10 (IL-10), or cytokine synthesis inhibitory factor, is a 35- to 40-kDa protein which is present in high levels in ascitic fluids of ovarian cancer patients.[167] In contrast to IL-6, IL-10 appears to not be produced by ovarian cancer cells, its source is most likely to be T cells, B cells or monocytes.[168] IL-10 inhibits production of a wide variety of cytokines and may be responsible for the peritoneal immunosuppression characteristically seen in ovarian cancer.[164] IL-4 is another T cell-derived cytokine which can interact with ovarian cancer cells and inhibit their growth. High affinity receptors for this cytokine have been demonstrated on ovarian cancer cells.

The hematopoietic growth factor, IL-3, has been administered to ovarian cancer patients to ameliorate chemotherapy-induced myelotoxicity. Its effects are dose- and schedule-dependent but it appears to reduce the thrombocytopenia associated with aggressive treatment with cyclophosphamide and carboplatin.

4.2 Tumor Necrosis Factor

Tumor necrosis factor (TNF) was originally identified as a factor that was cytotoxic to tumor cells.[169] However, it is now recognized to produce a variety of effects in differing cell types; and in ovarian cancer models, its effects are primarily associated with growth stimulation and disease progression.[170,171] TNF is expressed in two thirds (45 of 63) of ovarian tumors[171] and *in situ* hybridization has shown that up to 8% of cells in the epithelial areas expressed TNF mRNA.[172] This contrasts with studies of colorectal, breast, and cervical carcinomas wherein less than 0.1% of cells are positive for TNF mRNA.[172] The TNF protein as detected by immunohistochemistry is found predominantly in the stroma and at the epithelial/stromal border consistent with the distribution of macrophages.[171] The p55 TNF receptor was detected in all cancers studied and confined to epithelial areas while the p75 TNF receptor was found at the tumor-stromal interface, again consistent with macrophage distribution.[171] Addition of recombinant TNF-α to ovarian cell lines stimulated growth and increased expression of TNF mRNA.[159] The growth stimulation could be blocked with antibodies targeted to TNF-α or its soluble receptor.[160] In xenograft models of ovarian cancer, the factor promoted tumor progression by causing ascitic ovarian cancer cells to aggregate and form solid deposits on the surface of the peritoneum.[170] A further indication of positive effects on invasion is the observation that cells transfected with the TNF gene show an enhanced ability to invade peritoneal surfaces and metastasize in nude mice.[173] A neutralizing antibody to TNF blocked the metastatic activity of these cells.[173]

From the above observations, a hypothesis has emerged that proposes an interaction between tumor epithelial cells and macrophages.[161] Ovarian tumors produce chemotactic factors such as M-CSF and MCP-1 and these attract macrophages which generate TNF, IL-1 and other cytokines. These cytokines in turn stimulate production of TNF in the tumor cells causing cell proliferation.

4.3 Colony-Stimulating Factors

The hematopoietic regulators, macrophage colony stimulating factor (M-CSF), granulocyte colony stimulating factor (G-CSF) and granulocyte-macrophage colony stimulating factor (GM-CSF) are all secreted by surface ovarian epithelia cells when placed into short-term culture. Since several of these factors are being given to ovarian cancer patients to help achieve greater dose intensity with cytotoxic drugs, their direct effects on tumor cells are important.

M-CSF (CSF-1) is a homodimeric 90 kDa protein which has been demonstrated to be produced by cultured ovarian cancer cells.[174,175] Its receptor (FMS), a 150 kDa transmembrane tyrosine kinase, is also found within ovarian cancer cells, suggesting the potential for autocrine control.[174,175] Parallel studies in primary tumors have demonstrated the presence of cytokine and receptor and increased levels (approximately threefold) were found in the sera of over 80% of patients with ovarian cancer.[176] The co-expression of M-CSF and its receptor FMS has been suggested to enhance invasiveness and metastatic spread, as shown experimentally by the stimulatory effect of M-CSF on invasiveness of FMS-positive cell lines *in vitro*,[177] and by the simultaneous expression of factor and receptor transcripts in metastases, while expression was negative in the corresponding primary tumors.[175] Its chemotactic effect, mentioned above, may also contribute to ovarian tumor growth.

G-CSF and GM-CSF are used to treat or prevent neutropenia caused by marrow-toxic anticancer therapy. In a study investigating these factors on ovarian cancer cell lines and primary cultures, G-CSF produced an increase in growth in one of five primary cultures while GM-CSF had no significant effect.[178] In contrast, in an earlier study, GM-CSF stimulated colony formation in four human ovarian cancer cell lines.[179] These results suggest that CSFs may act as growth factors in some but not all ovarian cancers.

4.4 Interferons

The interferons have cytotoxic effects on a wide variety of cancers and have been shown to produce growth inhibitory effects on ovarian cancer cells *in vitro* and *in vivo*.[180] These observations have lead to clinical studies and all three interferons (α-IFN, β-IFN and γ-IFN) have been evaluated for the treatment of ovarian cancer. While these cytokines have minimal effects when used against advanced disease, activity has been reported for all three against minimal residual disease when given via the intraperitoneal route.[181-187] Nonrandomized studies combining α-IFN initially with cytotoxic agents such as mitoxantrone and cisplatin and later with carboplatin have produced pathological complete response rates of > 50% and up to 90% in some studies. These impressive response rates require confirmation in randomized trials to evaluate their future clinical potential.[188-194]

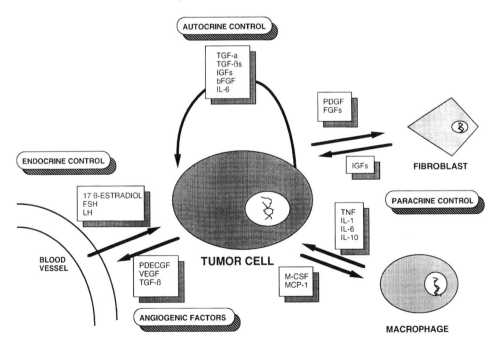

Figure 3 The network of interacting hormonal, growth factor and cytokine influences acting on an ovarian carcinoma cell *in vivo*. See text for details.

5. CONCLUSION

It is clear from the above that ovarian cancer is driven by a complex network of interacting hormonal, growth factor and cytokine pathways with the balance of factors varying in individual cancers (Figure 3). Since ovarian cancers clearly produce and respond to many of these factors, it is likely that alterations in expression of these substances or their receptors have roles in the pathogenesis and progression of this disease. An improved understanding of these pathways should help identify the best targets for therapy.

REFERENCES

1. **Jabara, A.G.,** Induction of canine ovarian tumors by diethylstilbestrol and progesterone, *Aust. J. Exp. Biol.,* 40, 139, 1962.
2. **Gardner, W.U.,** Further studies on experimental ovarian tumorigenesis, *Proc. Am. Assoc. Cancer Res.,* 2, 300, 1958.
3. **Horning, E.S.,** Carcinogenic action of androgens, *Br. J. Cancer,* 12, 414, 1958.
4. **Biskind, M.S. and Bisking, G.R.,** Development of tumors in the rat ovary after transplantation into spleen, *Proc. Soc. Exp. Biol. Med.,* 55, 176, 1944.
5. **Cramer, D.W., Hutchison, G.B., Welch, W.R., Scully, R.E., and Ryan, K.J.,** Determinants of ovarian cancer risk. I. Reproductive experiences and family history, *J. Natl. Cancer. Inst.,* 71, 711, 1983.
6. **Casagrande, J.T., Louie, E.W., Pike, M.C., Roy, S., Ross, R.K., and Henderson, B.E.,** Incessant ovulation and ovarian cancer, *Lancet,* 2, 170, 1979.
7. **Polychronopoulou, A., Tzonou, A., Hsieh, C., Kaprinis, G., Rebelakos, A., Toupadaki, N., and Trichopoulous, D.,** Reproductive variables, tobacco, ethanol, coffee and somatometry as risk factors for ovarian cancer, *Int. J. Cancer,* 55, 402, 1993.
8. **Franceschi, S., La Vecchia, C., Booth, M., Tzonou, A., Negri, E., Parazzini, F., Trichopoulos, D., and Beral, V.,** Pooled analysis of 3 European case-control studies of ovarian cancer. II. Age at menarche and at menopause, *Int. J. Cancer,* 49, 57, 1991.
9. **Whittemore, A.S., Harris, R., Itnyre, J., Halpern, J., and The Collaborative Ovarian Cancer Group**, Characteristics relating to ovarian cancer risk: collaborative analysis of 12 US case-control studies. II. Invasive epithelial cancers in white women, *Am. J. Epidemiol.,* 136, 1184, 1992.
10. **Weiss, N.S., Lyon, J.L., Liff, J.M., Vollmer, W.M., and Daling, J.R.,** Incidence of ovarian cancer in relation to the use of oral contraceptives, *Int. J. Cancer,* 28, 669, 1981.
11. **Cramer, D.W., Hutchison, G.B., Welch, W.R., Scully, R.E., and Knapp, R.C.,** Factors affecting the association of oral contraceptives and ovarian cancer, *N. Engl. J. Med.,* 307, 1047, 1982.
12. Centers for Disease Control Cancer and the National Institute of Child Health and Human Development Cancer and Steroid Hormone Study: the reduction in risk of ovarian cancer associated with oral contraceptive use, *N. Engl. J. Med.,* 316, 650, 1987.
13. **Rosenblatt, K.A., Thomas, D.B., and Noonan, E.A.,** High-dose and low-dose combined oral contraceptives: protection against epithelial ovarian cancer and the length of the protective effect, *Eur. J. Cancer,* 28A, 1872, 1992.
14. **Cramer, D.W. and Welch, W.R.,** Determinants of ovarian cancer risk. II. Interferences regarding pathogenesis, *J. Natl. Cancer Inst.,* 71, 717, 1983.
15. **Fathalla, M.F.,** Incessant ovulation — a factor in ovarian neoplasia?, *Lancet,* 2, 163, 1971.
16. **Fathalla, M.F.,** Factors in the causation and incidence of ovarian cancer, *Obstet. Gynecol. Surv.,* 27, 751, 1972.
17. **Wilson, J.E.,** Adenocarcinomata in hens kept in a constant environment, *Poult. Sci.,* 37, 1253, 1958.
18. **Godwin, A.K., Testa, J.R., Handel, L.M., Liu, Z., Vanderveer, L., and Tracey, P.A.,** Spontaneous transformation of rat ovarian surface epithelial cells: association with cytogenetic changes and implications of repeated ovulation in the etiology of ovarian cancer, *J. Natl. Cancer Inst.,* 84, 592, 1992.
19. **Whittemore, A.S.,** Fertility drugs and risk of ovarian cancer, *Hum. Reprod.,* 8, 999, 1993.
20. **Parazzini, F., La Vecchia, C., Negri, E., and Villa, A.,** Estrogen replacement therapy and ovarian cancer risk, *Int. J. Cancer,* 57, 135, 1994.
21. **Rodriguez, C., Calle, E.E., Coates, R.J., Miracle-McMahill, H.L., Thun, M.J., and Heath, C.W.,** Estrogen replacement therapy and fatal ovarian cancer, *Am. J. Epidemiol.,* 141, 828, 1995.
22. **La Vecchia, C., Liberati, A., Franceschi, S., and Decarli, A.,** Non-contraceptive estrogen use and occurrence of ovarian cancer, *J. Natl. Cancer Inst.,* 69, 1207, 1982.
23. **Kaufman, D.W., Kelly, J.P., Welch, W.R., Rosenberg, L., Stolley, P.D., Warshauer, M.E., Lewis, J., Woodruff, J., and Shapiro, S.,** Noncontraceptive estrogen use and epithelial ovarian cancer, *Am. J. Epidemiol.,* 130, 1142, 1989.
24. **Annegers, J.F., O'Fallon, W., and Kurland, L.T.,** Exogenous oestrogens and ovarian cancer, *Lancet,* 2, 869, 1977.
25. **Wynder, E.L., Dodo, H., and Barber, H.R.K.,** Epidemiology of cancer of the ovary, *Cancer,* 23, 352, 1969.
26. **Hartge, P., Hoover, R., McGowan, L., Lesher, L., and Norris, H.J.,** Menopause and ovarian cancer, *Am. J. Epidemiol.,* 127, 990, 1988.
27. **Weiss, N.S., Lyon, J.L., Krishnamurthy, S., Dietert, S.E., Liff, J.M., and Daling, J.R.,** Noncontraceptive estrogen use and the occurrence of ovarian cancer, *J. Natl. Cancer Inst.,* 68, 95, 1982.

28. **Slotman, B.J. and Rao, B.R.,** Ovarian cancer (review): etiology, diagnosis, prognosis, surgery, radiotherapy, chemotherapy and endocrine therapy, *Anticancer Res.*, 8, 417, 1988.

29. **Emons, G. and Schally, A.V.,** The use of luteinizing hormone releasing hormone agonists and antagonists in gynaecological cancers, *Hum. Reprod. Update*, 9, 1364, 1994.

30. **Emons, G., Pahwa, G.S., Brack, C., Sturm, R., Oberheuser, F., and Knuppen, R.,** Gonadotropin releasing hormone binding sites in human epithelial ovarian carcinomata, *Eur. J. Cancer Clin. Oncol.*, 25, 215, 1989.

31. **Creasman, W.T., Sasso, R.A., Weed, J.C., and McCarty, K.,** S., Ovarian carcinoma: histologic and clinical correlation of cytoplasmic estrogen and progesterone binding, *Gynecol. Oncol.*, 12, 319, 1981.

32. **Bizzi, A., Codegoni, A.M., Landoni, F., Marelli, G., Marsoni, S., Spina, A.M., Torri, W., and Mangioni, C.,** Steroid receptors in epithelial ovarian carcinoma: relation to clinical parameters and survival, *Cancer Res.*, 48, 6222, 1988.

33. **Kieback, D.G., McCamant, S.K., Press, M.F., Atkinson, E.N., Gallager, H.S., Edwards, C.L., Hjek, R.A., and Jones, L.A.,** Improved prediction of survival in advanced adenocarcinoma of the ovary by immunocytochemical analysis and the composition adjusted receptor level of the estrogen receptor, *Cancer Res.*, 53, 5188, 1993.

34. **Kieback, D.G., Press, M.F., Atkinson, E.N., Edwards, G.L., Mobus, V.J., Runnebaum, I.B., Kreienberg, R., and Jones, L.A.,** Prognostic significance of estrogen receptor expression in ovarian cancer. Immunoreactive score (IRS) vs. composition adjusted receptor level (CARL), *Anticancer Res.*, 13, 2489, 1993.

35. **Schwartz, P.E., LiVolsi, A., Hildrith, N., MacLusky, N.J., Naftolin, F.N., and Eisenfeld, A.J.,** Estrogen receptors in ovarian epithelial carcinoma, *Obstet. Gynecol.*, 59, 229, 1982.

36. **Quinn, M.A.,** Is endocrine status relevant to treatment planning in patients with gynecological cancer, *Aust. N.Z.J. Obstet. Gynecol.*, 24, 153, 1984.

37. **Slotman, B.J., Kuhnel, R., Rao, B.R., Dukhuizen, G.H., DeGraff, J., and Stotk, J.G.,** Importance of steroid receptors and aromatase activity in the prognosis of ovarian cancer: high tumor progesterone receptor levels correlate with longer survival, *Gynecol. Oncol.*, 33, 76, 1989.

38. **Schwartz, P.E., MacLusky, N., Merino, M.J., LiVolsi, V.A., Kohorn, E.I., and Eisenfeld, A.,** Are cytosol estrogen and progestin receptors of prognostic significance in the management of epithelial ovarian cancers, *Obstet. Gynecol.*, 68, 751, 1986.

39. **Iversen, O.E., Skaarland, E., and Utaaker, E.,** Steroid receptor content in human ovarian tumors: survival of patients with ovarian carcinoma related to steroid receptor content, *Gynecol. Oncol.,* 23, 65, 1986.

40. **Masood, S., Heitmann, J., Nuss, R.C., and Benrubi, G.I.,** Clinical correlation of hormone receptor status in epithelial ovarian cancer, *Gynecol. Oncol.*, 34, 57, 1989.

41. **Rose, P.G., Reale, F.R., Longcope, C., and Hunter, R.E.,** Prognostic significance of estrogen and progesterone receptors in epithelial ovarian cancer, *Obstet. Gynecol.*, 76, 258, 1990.

42. **Sevelda, P., Denison, U., Schemper, M., Spona, J., Vavra, N., and Salzer, H.,** Oestrogen and progesterone receptor content as a prognostic factor in advanced epithelial ovarian carcinoma, *Br. J. Obstet. Gynaecol.*, 97, 706, 1990.

43. **Harding, M., Cowan, S., Hole, D., Cassidy, L., Kitchener, H., Davis, J., and Leake, R.,** Estrogen and progesterone receptors in ovarian cancer, *Cancer*, 65, 486, 1990.

44. **Kommoss, F., Pfisterer, J., Thome, M., Schafer, W., Sauerbrei, W., and Pfleiderer, A.,** Steroid receptors in ovarian carcinoma: immunohistochemical determination may lead to new aspects, *Gynecol. Oncol.*, 47, 317, 1992.

45. **Langdon, S.P., Hawkes, M.M., Lawrie, S.S., Hawkins, R.A., Tesdale, A.L., Crew, A.J., Miller, W.R., and Smyth, J.F.,** Oestrogen receptor expression and the effects of oestrogen and tamoxifen on the growth of human ovarian carcinoma cell lines, *Br. J. Cancer*, 62, 213, 1990.

46. **Langdon, S.P., Ritchie, A., Young, K., Crew, A.J., Sweeting, V., Bramley, T., Hillier, S., Hawkins, R.A., Tesdale, A.L., Smyth, J.F., and Miller, W.R.,** Contrasting effects of 17 β-estradiol on the growth of human ovarian carcinoma cells in vitro and in vivo, *Int. J. Cancer*, 55, 459, 1993.

47. **Langdon, S.P., Hirst, G.L., Miller, E.P., Hawkins, R.A., Tesdale, A.L., Smyth, J.F., and Miller, W.R.,** The regulation of growth and protein expression by estrogen in vitro: a study of 8 human ovarian carcinoma cell lines, *J. Ster. Biochem. Mol. Biol.,* 50, 131, 1994.

48. **Rowlands, C., Krishnan, V., Wang, X., Santostefano, M., Safe, S., Miller, W.R., and Langdon, S.,** Characterization of the aryl hydrocarbon receptor and aryl hydrocarbon responsiveness in human ovarian carcinoma cell lines, *Cancer Res.*, 53, 1802, 1993.

49. **Nash, J.D., Ozols, R.F., Smyth, J.F., and Hamilton, T.C.,** Estrogen and anti-estrogen effects on the growth of human epithelial ovarian cancer in vitro, *Obstet. Gynecol.,* 73, 1009, 1989.

50. **Geisenger, K.R., Kute, T.E., Pettenati, M.J., Welander, C.E., Dennard, Y., Collins, L.A., and Berens, M.E.,** Characterization of a human ovarian carcinoma cell line with estrogen and progesterone receptors, *Cancer*, 63, 280, 1989.

51. **Geisenger, K.R., Berens, M.E., Duckett, Y., Morgan, T.M., Kute, T.E., and Welander, C.E.,** The effects of estrogen, progesterone, and tamoxifen alone and in combination with cytotoxic agents against human ovarian carcinoma in vitro, *Cancer*, 65, 1055, 1990.

52. **Galtier-Dereure, F., Capony, F., Maudelonde, T., and Rochefort, H.,** Estradiol stimulates cell growth and secretion of procathepsin D and a 120-kilodalton protein in the human ovarian cancer cell line BG-1, *J. Clin. Endocrin. Metab.*, 75, 1497, 1992.

53. **Sawada, M., Terada, N., Wada, A., Mori, Y., Yamasaki, M., Saga, T., and Endo, K.,** Estrogen and androgen-responsive growth of human ovarian adenocarcinoma heterotransplanted into nude mice, *Int. J. Cancer*, 45, 359, 1990.

54. **Hamilton, T.C., Behrens, B.C., Louie, K.G., and Ozols, R.F.,** Induction of progesterone receptor in human ovarian cancer, *J. Clin. Endocrinol. Metab.*, 59, 561, 1984.

55. **Hamilton, T.C., Young, R.C., McKoy, W.M., Grotzinge, K.R., Green, J.A., Chu, E.W., Whang-Peng, J., Rogan, A.M., Green, W.R., and Ozols, R.F.,** Characterization of a human ovarian carcinoma cell line (NIH:OVCAR-3) with androgen and estrogen receptors, *Cancer Res.*, 43, 5379, 1983.

56. **Chien, C.H., Wang, F.F., and Hamilton, T.C.,** Transcriptional activation of c-myc proto-oncogene by estrogen in human ovarian cancer cells, *Mol. Cell. Endocrinol.*, 99, 11, 1994.

57. **Lippman, M.E., Dickson, R.B., Gelmann, E.P., Rosen, N., Knabbe, C., Bates, S., Bronzert, D., Huff, K., and Kasid, A.,** Growth regulation of human breast carcinoma occurs through regulated growth factor secretion, *J. Cell. Biochem.*, 35, 1, 1987.

58. **Nash, J., Hall, L., Ozols, R., Young, R., Smyth, J., and Hamilton, T.,** Estrogenic regulation and growth factor expression in human ovarian cancer in vitro, *Proc. Am. Assoc. Cancer Res.*, 30, 1189, 1989.

59. **Crew, A.J., Bartlett, J.M.S., Scott, W.N., Miller, E.P., Rabiasz, G.J., Langdon, S.P., and Miller, W.R.,** Autocrine/paracrine regulation by EGF and TGF-α in human ovarian adenocarcinoma cell lines, *Br. J. Cancer*, 65, 979, 1992.

60. **Leake, R., Barber, A., Owens, O., Langdon, S., and Miller, B.,** Growth factors and receptors in ovarian cancer, in *Ovarian Cancer: 3,* Sharp, F., Mason, P., Blackett, T., and Berek, J., Eds., Chapman and Hall, 1994, 99.

61. **Krywicki, R.F., Figueroa, J.A., Jacjson, J.G., Wozelsky, T.W., Shimasaki, S., Von Hoff, D.D., and Yee, D.,** Regulation of insulin-like growth factor binding proteins in ovarian cancer cells by oestrogen, *Eur. J. Cancer*, 29A, 2015, 1993.

62. **Langdon, S.P., Crew, A.J., Ritchie, A.A., Muir, M., Wakeling, A., Smyth, J.F., and Miller, W.R.,** Growth inhibition of oestrogen receptor-positive human ovarian carcinoma by anti-oestrogens in vitro and in a xenograft model, *Eur. J. Cancer*, 30A, 682, 1994.

63. **Lazo, J.S., Schwartz, P.E., MacLusky, N.J., Labaree, D.C., and Eisenfeld, A.J.,** Antiproliferative actions of tamoxifen to human ovarian carcinomas in vitro, *Cancer Res.*, 44, 2266, 1984.

64. **Emons, G. and Schally, A.V.,** The use of luteinizing hormone releasing hormone agonists and antagonists in gynaecological cancers, *Hum. Reprod. Update,* 9, 1364, 1994.

65. **Simon, W.E. and Holzel, F.,** Hormone sensitivity of gynecological tumor cells in tissue cultures, *J. Cancer Res. Clin. Oncol.*, 94, 307, 1979.

66. **Simons, W.E., Albrecht, M., Hansel, M., Dietl, M., and Holzel, F.,** Cell line derived from human ovarian carcinomas: growth stimulation by gonadotropic and steroid hormones, *J. Natl. Cancer Inst.*, 70, 839, 1983.

67. **Ohtani, K.,** Effects of gonadotropin on the growth of malignant ovarian neoplasms assessed by subrenal capsule assay, *Nippon Sanka Fujinka Gakki Zasshi*, 42, 579, 1990.

68. **Peterson, C.M. and Zimniski, S.J.,** A long acting gonadotropin-releasing hormone agonist inhibits the growth of a human epithelial ovarian carcinoma (BG-1) heterotransplanted in the nude mouse, *Obstet. Gynecol.*, 76, 264, 1990.

69. **Thompson, M.A., Adelson, M.D., and Kaufman, L.M.,** Lupron retards proliferation of ovarian epithelial tumor cells, cultured in serum free medium, *J. Clin. Endocrinol. Metab.*, 72, 1036, 1991.

70. **Mortel, R., Satyaswaroop, P.G., Schally, A.V., Hamilton, T., and Ozols, R.,** Inhibitory effects of GnRH superagonist on the growth of human ovarian carcinoma NIH:OVCAR 3 in the nude mouse, *Gynecol. Oncol.*, 23, 254, 1986.

71. **Myers, A.M., Moore, G.E., and Major, F.J.,** Advanced ovarian carcinoma: response to antiestrogen therapy, *Cancer*, 48, 2368, 1981.

72. **Weiner, S.A., Alberts, D.S., Surwit, E.A., Davis, J., and Grosso, D.,** Tamoxifen therapy in recurrent epithelial ovarian carcinoma, *Gynecol. Oncol.*, 27, 208, 1987.

73. **Campbell, J.J., Rome, R.M., Quinn, M.A., Pepperell, J., and Morgan, W.J.,** Tamoxifen for recurrent progressive epithelial ovarian tumours, *Proc. XI Clin. Oncol. Soc. Aust.*, 73, 1984.

74. **Hatch, K.D., Beecham, J.B., Blessing, J.A., and Creasman, W.T.,** Responsiveness of patients with advanced ovarian carcinoma to tamoxifen. A Gynecologic Oncology Group study of second-line therapy in 105 patients, *Cancer*, 68, 269, 1991.

75. **Ahlgren, J.D., Ellison, N., Gottlieb, R.J., Laluna, F., Lokich, J.L., Sinclair, P.R., Ueno, W., Wampler, G.L., Yeung, K.-Y., Alt, D., and Fryer, J.G.,** Hormonal palliation of chemoresistant ovarian cancer: three phase II trials of the mid-atlantic oncology program, *J. Clin. Oncol.*, 11, 1957, 1993.

76. **Ward, H.W.C.,** Progestogen therapy for ovarian carcinoma, *Br. J. Obstet. Gynaecol.*, 79, 555, 1972.

77. **Timothy, I.,** Progestogen therapy for ovarian cancer, *Br. J. Obstet. Gynaecol.*, 89, 561, 1982.

78. **Malkasian, G.D., Dekker, D.G., Jorgensen, E.O., and Webb, M.J.,** Evaluation of 6, 17 α-dimethyl-6-dehydro-progesterone for treatment of recurrent and metastatic gynecologic malignancy, *Am. J. Obstet. Gynecol.,* 118, 461, 1974.

79. **Mangioni, C., Franceschi, S., La Vecchia, C., and D'Incalci, M.,** High-dose medroxyprogesterone acetate (MPA) in advanced epithelial ovarian cancer resistant to first — or second — line chemotherapy, *Gynecol. Oncol.,* 12, 314, 1981.

80. **Rendina, G.M., Donadio, C., and Giovannini, M.,** Steroid receptors and progestinic therapy in ovarian endometrioid carcinoma, *Eur. J. Gynaecol. Oncol.,* 3, 241, 1982.

81. **Geisler, H.E.,** The use of high-dose megestrol acetate in the treatment of ovarian adenocarcinoma, *Semin. Oncol.,* 12 (suppl), 20, 1985.

82. **Freedman, R.S., Saul, P.B., and Edwards, C.L.,** Ethinyl estradiol and medroxyprogesterone acetate in patients with epithelial ovarian carcinoma: a phase II study, *Cancer Treat. Rep.,* 70, 369, 1986.

83. **Fromm, G.L., Freedman, R.S., Fritsche, H.A., Atkinson, E.N., and Scott, W.,** Sequentially administered ethinyl estradiol and medroxyprogesterone acetate in the treatment of refractory epithelial ovarian carcinoma in patients with positive estrogen receptors, *Cancer,* 68, 1885, 1991.

84. **Ro, I.G., Wigler, N., Merimsky, O., Inbar, M.J., and Chaitchik, S.,** A phase II trial of D-Trp-6-LHRH (decapeptyl) in pretreated patients with advanced epithelial ovarian cancer, *Cancer Invest.,* 13, 272, 1995.

85. **Emons, G., Pahwa, G.S., Ortmann, O., Knuppen, R., Oberheuser, F., and Schulz, K.D.,** LHRH-receptors and LHRH-agonist treatment in ovarian cancer: an overview, *J. Ster. Biochem. Mol. Biol.,* 37, 1003, 1990.

86. **Savino, L., Baldini, B., Susini, T., Pulli, F., Antignani, L., and Massi, G.B.,** GnRH analogs in gynecological oncology: a review, *J. Chemother.,* 4, 312, 1992.

87. **Emons, G., Ortmann, O., Pahwa, G.S., Hackenberg, R., Oberheuser, F., and Schulz, K.D.,** Intracellular actions of gonadotropic and peptide hormones and the therapeutic value of GnRH-agonists in ovarian cancer, *Acta Obstet. Gynecol. Scand.,* 155, 31, 1992.

88. **Sporn, M.B. and Roberts, A.B.,** Autocrine growth factors and cancer, *Nature,* 313, 745, 1985.

89. **Owens, O.J., Stewart, C., and Leake, R.E.,** Growth factors in ovarian cancer, *Br. J. Cancer,* 64, 1177, 1991.

90. **Morishige, K., Kurachi, H., Amemiya, K., Fujita, Y., Yamamoto, T., Miyake, A., and Tanizawa, O.,** Evidence for the involvement of transforming growth factor α and epidermal growth factor receptor autocrine growth mechanism in primary human ovarian cancers in vitro, *Cancer Res.,* 51, 5322, 1991.

91. **Kommoss, F., Wintzer, H.O., von Kleist, S., Kohler, M., Walker, R., Langton, B., van Tran, K., Pfleiderer, A., and Bauknecht, T.,** In situ distribution of transforming growth factor α in normal human tissues and in malignant tumours of the ovary, *J. Pathol.,* 162, 223, 1990.

92. **Kohler, M., Bauknecht, T., Grimm, M., Birmelin, G., Kommoss, F., and Wagner, E.,** Epidermal growth factor receptor and transforming growth factor alpha expression in human ovarian carcinomas, *Eur. J. Cancer,* 28A, 1432, 1992.

93. **Stromberg, K., Johnson, G.R., O'Connor, D.M., Sorensen, C.M., Gullick, W.J., and Kannan, B.,** Frequent immunohistochemical detection of EGF supergene family members in ovarian carcinogenesis, *Int. J. Gynecol. Pathol.,* 13, 342, 1994.

94. **Crew, A.J., Langdon, S.P., Miller, E.P., and Miller, W.R.,** Mitogenic effects of epidermal growth factor and transforming growth factor-α on EGF-receptor positive human ovarian carcinoma cell lines, *Eur. J. Cancer,* 28, 337-341.

95. **Scambia, G., Benedetti-Panici, P., Battaglia, F., Ferrandina, G., Gaggini, C., and Mancuso, S.,** Presence of epidermal growth factor (EGF) receptor and proliferative response to EGF in six human ovarian carcinoma cell lines, *Int. J. Gynecol. Cancer,* 1, 253, 1991.

96. **Rodriguez, G.C., Berchuck, A., Whitaker, R.S., Schlosman, D., Clarke-Pearson, D.L., and Bast, R.,** Epidermal growth factor receptor expression in normal ovarian epithelium and ovarian cancer. 2. Relationship betwen receptor expression and response to epidermal growth factor, *Am. J. Obstet. Gynecol.,* 164, 745, 1991.

97. **Zhou, L.I. and Leung, B.S.,** Growth regulation of ovarian cancer cells by epidermal growth factor and transforming growth factor-α and β1, *Biochem. Biophys. Acta,* 1080, 130, 1992.

98. **Morishige, K., Kurachi, H., Amemiya, K., Fujita, Y., Yamamoto, T., Miyake, A., and Tanizawa, O.,** Evidence for the involvement of transforming growth factor-α and epidermal growth factor receptor autocrine growth mechanism in primary human ovarian cancers in vitro, *Cancer Res.,* 51, 5322, 1991.

99. **Kurachi, H., Morishige, K., Amemiya, K., Adachi, H., Hirota, K., Miyake, A., and Tanizawa, O.,** Importance of transforming growth factor α/epidermal growth factor receptor autocrine mechanism in an ovarian cancer cell line *in vivo, Cancer Res.,* 51, 5956, 1991.

100. **Jindal, S.K., Snoey, D.M., Lobb, D.K., and Dorrington, J.H.,** Transforming growth factor alpha localization and role in surface epithelium of normal human ovaries and in ovarian carcinoma cells, *Gynecol. Oncol.,* 53, 17, 1994.

101. **Bauknecht, T., Runge, M., Schwall, M., and Pfleiderer, A.,** Occurrence of epidermal growth factor receptors in human adnexal tumors and their prognostic value in advanced ovarian carcinomas, *Gynecol. Oncol.,* 29, 147, 1988.

102. **Battaglia, F., Scambia, G., and Beneditti Panici, P.,** Epidermal growth factor receptors in gynecologic malignancies, *Gynecol. Obstet. Invest.,* 27, 42, 1989.

103. **Owens, O.J., Stewart, C., Brown, I., and Leake, R.E.,** Epidermal growth factor receptors (EGFr) in human ovarian cancer, *Br. J. Cancer,* 64, 907, 1991.

104. **Henzen-Logmans, S.C., Berns, E.M.J.J., Klijn, J.G.M., van der Burg, M.E.L., and Foekens, J.A.,** Epidermal growth factor receptor in ovarian tumors: correlation of immunohistochemistry with ligand binding assay, *Br. J. Cancer,* 66, 1015, 1992.

105. **Berchuck, A., Rodriguez, G.C., Kamel, A., Dodge, R.K., Soper, J.T., Clarke-Pearson, D.L., and Bast, R.C.,** Epidermal growth factor expression in normal epithelium and ovarian cancer, *Am. J. Obstet. Gynecol.,* 164, 669, 1991.

106. **Owens, O.J., Stewart, C., Leake, R.E., and McNichol, A.M.,** A comparison of biochemical and immunohistochemical assessment of EGFr expression in ovarian cancer, *Anticancer Res.,* 12, 1455, 1992.

107. **Owens, O.J. and Leake, R.E.,** Epidermal growth factor receptor expression in malignant ovary, benign ovarian tumours and normal ovary: a comparison, *Int. J. Oncol.*, 2, 321, 1993.

108. **Berns, W.M.J.J., Klijn, J.G.M., Henzen-Logmans, S.C., Rodenburg, C.J., and van der Burg M.E.L.,** Receptors for hormones and growth factors and (onco)-gene amplification in human ovarian cancer, *Int. J. Cancer*, 52, 218, 1992.

109. **Scambia, G., Benedetti-Panici, P., Battaglia, F., Ferrandina, G., Baiocchi, G., Greggi, S., De Vincenzo, R., and Mancuso, S.,** Significance of epidermal growth factor receptor in advanced ovarian cancer, *J. Clin. Oncol.*, 10, 529, 1992.

110. Van der Burg, M.E.L., Henzen-Logmans, S.C., Foekens, J.A., Berns, E.M.J.J., Rodenburg, C.J., van Putten, W.L.J., and Klijn, J.G.M., The prognostic value of epidermal growth factor receptors, determined by both immunohistochemistry and ligand binding assays, in primary epithelial ovarian cancer: a pilot study, *Eur. J. Cancer*, 29A, 1951, 1993.

111. **Bauknecht, T., Birmelin, G., and Kommoss, F.,** Clinical significance of oncogenes and growth factors in ovarian carcinomas, *J. Ster. Biochem. Mol. Biol.*, 37, 855, 1990.

112. **Slamon, D.J., Godolphin, W., Jones, L.A., Holt, J., Wong, S.G., Keitch, D.E., Levin, W.J., Stuart, S.G., Udove, J., Ullrich, A., and Press, M.F.,** Studies of the HER-2/neu proto-oncogene in human breast and ovarian cancer, *Science*, 244, 707, 1989.

113. **Berchuck, A., Kamel, A., Whitaker, R., Kerns, B., Olt, G., Kinney, R., Soper, J.T., Dodge, R., Clarke-Pearson, D.L., Marks, P., McKenzie, S., Yin, S., and Bast, R.C.,** Overexpression of HER-2/neu is associated with poor survival in advanced ovarian cancer, *Cancer Res.*, 50, 4087, 1990.

114. **Simpson, B.J.B., Weatherill, J., Miller, E.P., Lessels, A.M., Langdon, S.P., and Miller, W.R.,** c-erbB-3 protein expression in ovarian tumours, *Br. J. Cancer*, 71, 758, 1995.

115. **Simpson, B.J.B., Phillips, H.A., Lessels, A.M., Langdon, S.P., and Miller, W.R.,** c-erbB growth factor receptors in ovarian tumours, *Int. J. Cancer*, 64, 202, 1995.

116. **Marth, C., Lang, T., Cronauer, M.V., Doppler, W., Zeimet, A.G., Bachmair, F., Ullrich, A., and Daxenbichler, G.,** Epidermal growth factor reduces HER-2 protein level in human ovarian carcinoma cells, *Int. J. Cancer*, 52, 311, 1992.

117. **Pietras, R.J., Fendley, B.M., Chazin, V.R., Pegram, M.D., Howell, S.B., and Slamon, D.J.,** Antibody to HER-2/neu receptor blocks DNA repair after cisplatin in human breast and ovarian cancer cells, *Oncogene*, 9, 1829, 1994.

118. **Langton-Webster, B.C., Xuan, J.-A., Brink, J.R., and Saloman, D.S.,** Development of resistance to cisplatin is associated with decreased expression of the gp185 c-erbB-2 protein and alterations in growth properties and response to therapy in an ovarian tumor cell line, *Cell Growth Different.*, 5, 1367, 1994.

119. **Roberts, A.B. and Sporn, M.B.,** The transforming growth factor-betas, in *Peptide Growth Factors and Their Receptors*, Sporn, M. B. and Roberts, A. B., Eds., Springer Verlag, Heidelberg, 1990, 419.

120. **Wrana, J.L., Attisano, L., Wieser, R., Ventura, F., and Massague, J.,** Mechanism of activation of the TGF-β receptor, *Nature*, 370, 341, 1994.

121. **Massague, J.,** Receptors for the TGF-β family, *Cell*, 69, 1067, 1992.

122. **Hurteau, J.A., Rodriguez, G.C., Whitaker, R.S., Shah, S., Mills, G., Bast, R.C., and Berchuck, A.,** Transforming growth factor-β inhibits proliferation of human ovarian cancer cells obtained from ascites, *Cancer*, 74, 93, 1994.

123. **Bartlett, J.M.S., Rabaiasz, G.J., Scott, W.N., Langdon, S.P., Smyth, J.F., and Miller, W.R.,** Transforming growth factor-β mRNA expression in growth control of human ovarian carcinoma cells, *Br. J. Cancer*, 65, 655, 1992.

124. **Berchuck, A., Olt, G.J., Everitt, L., Soisson, A.P., Bast, R.C., and Boyer, C.M.,** The role of peptide growth factors in epithelial ovarian cancer, *Obstet. Gynecol.*, 75, 255, 1990.

125. **Berchuck, A., Rodriguez, G.C., Olt, G.J., Boente, M.P., Whitaker, R.S., Arrick, B., Clarke-Pearson, D.L., and Bast, R.C.,** Regulation of growth of normal ovarian epithelial cells and ovarian cancer cell lines by transforming growth factor-β, *Am. J. Obstet. Gynecol.*, 166, 676, 1992.

126. **Marth, C., Lang, T., Koza, A., Mayer, I., and Daxenblicher, G.,** Transforming growth factor-β and ovarian carcinoma cells: regulation of proliferation and surface antigen expression, *Cancer Lett.*, 51, 221, 1990.

127. **Jozan, S., Guerrin, M., Mazars, P., Dutaur, M., Monsarrat, B., Cheutin, F., Bugat, R., Martel, P., and Valette, A.,** Transforming growth factor β1 (TGF-β1) inhibits growth of a human ovarian carcinoma cell line (OVCCR1) and is expressed in human ovarian tumors, *Int. J. Cancer*, 52, 766, 1992.

128. **Havrilesky, L.J., Hurteau, A., Whitaker, R.S., Elbendary, A., Wu, S., Rodriguez, G.C., Bast, R.C., and Berchuck, A.,** Regulation of apoptosis in normal and malignant ovarian epithelial cells by transforming growth factor β1, *Cancer Res.*, 55, 944, 1995.

129. **Bartlett, J.M.S., Langdon, S.P., Rabiasz, G.J., Scott, W.N., Love, S., Hawkins, R.A., Katsaros, D., Smyth, J.F., and Miller, W.R.,** Expression of transforming growth factor beta isoforms in human ovarian tumors and their relationships to clinicopathological factors, submitted.

130. **Lappohn, R.E., Burger, H.G., Bouma, J., Bangah, M., Krans, M., and De Bruijn, H.W.A.,** Inhibin as a marker for granulosa cell tumours, *N. Engl. J. Med.*, 321, 790, 1989.

131. **Cooke, I., O'Brien, M., Charnock, F.M., Groome, N., and Ganesan, T.S.,** Inhibin as a marker for ovarian cancer, *Br. J. Cancer*, 71, 1046, 1995.

132. **Blaakaer, J., Micic, S., Morris, I.D., Hording, U., Bennett, P., Toftager-Larsen, K., Djursing, H., and Bock, J.E.,** Immunoreactive inhibin-production in post menopausal women with malignant epithelial ovarian tumors, *Eur. J. Obstet. Gynecol.*, 52, 105, 1993.

133. **Cate, R.L., Mattaliano, R.J., Hession, C., Tizard, R., Farber, N.M., Cheung, A., Ninfa, E.G., Frey, A.Z., Gash, D.J., Chow, E.P., Fisher, R.A., Bertonis, J.M., Torres, G., Wallner, B.P., Ramachandran, K.L., Ragin, R.C., Manganaro, T.F., MacLaughlin, D.T., and Donahoe, P.K.,** Isolation of the bovine and human genes for mullerian inhibiting substance and expression of the human gene in animal cells, *Cell*, 45, 685, 1986.

134. **Fuller, A.F., Guy, S., Budzik, G.P., and Donahoe, P.K.,** Mullerian inhibiting substance inhibits colony growth of a human ovarian carcinoma cell line, *J. Clin. Endocrin. Metab.*, 54, 1051, 1982.

135. **Wallen, J.W., Cate, R.L., Kiefer, D.M., Riemen, M.W., Martinez, D., Hoffman, R.M., Donahoe, P.K., Von Hoff, D. D., Pepinsky, B., and Oliff, A.,** Minimal antiproliferative effect of recombinant mullerian inhibiting substance on gynecological tumor cell lines and tumor explants, *Cancer Res.,* 49, 2005, 1989.

136. **Donahoe, P.K., Fuller, A.F., Scully, R.E., Guy, S.R., and Budzik, G.P.,** Mullerian inhibiting substance inhibits growth of a human ovarian cancer in nude mice, *Ann. Surg.*, 194, 472, 1981.

137. **Barreca, A. and Minuto, F.,** Somatomedins: chemical and functional characteristics of the different molecular forms, *J. Endocrin. Invest.*, 12, 279, 1989.

138. **Shimasaki, S. and Ling, N.,** Identification and molecular characterization of insulin-like growth factor binding proteins (IGFBP-1, -2, -3, -4, -5, and -6), *Prog. Growth Factor Res.*, 3, 243, 1991.

139. **Giordano, G., Barreca, A., and Minuto, F.,** Growth factors in the ovary, *J. Endocrin. Invest.*, 15, 689, 1992.

140. **Adashi, E.Y., Resnick, C.E., D'Ercole, A.J., Svoboda, M.E., and van Wyk, J.J.,** Insulin-like growth factors as intraovarian regulators of granulosa cell growth and function, *Endocrine Rev.*, 6, 400, 1985.

141. **Yee, D., Morales, F.R., Hamilton, T.C., and von Hoff, D.D.,** Expression of insulin-like growth factor I, its binding proteins, and its receptor in ovarian cancer, *Cancer Res.*, 51, 5107, 1991.

142. **Bartlett, J.M.S., Rabiasz, G.J., Scott, W.N., Langdon, S.P., Hirst, G.L., Lee, A., Smyth, J.F., and Miller, W.R.,** Growth control of human ovarian carcinoma cells by insulin-like growth factors, *Oncol. Rep.*, 2, 857, 1995.

143. **Hofmann, J., Wegmann, B., Hackenberg, R., Kunzmann, R., Schulz, K., and Havemann, K.,** Production of insulin-like growth factor binding proteins by human ovarian carcinoma cells, *J. Cancer Res. Clin. Oncol.,* 120, 137, 1994.

144. **Krywicki, R.F., Figueroa, J.A., Jackson, J.G., Kozelsky, T.W., Shimasaki, S., von Hoff, D.D., and Yee, D.,** Regulation of insulin-like growth factor binding proteins in ovarian cancer cells by oestrogen, *Eur. J. Cancer,* 29A, 2015, 1993.

145. **Resnicoff, M., Ambrose, D., Coppola, D., and Rubin, R.,** Insulin-like growth factor-1 and its receptor mediate the autocrine proliferation of human ovarian carcinoma cell lines, *Lab Invest*, 69, 756, 1993.

146. **Foekens, J.A., van Putten, W., Portengen, H., Rodenburg, C.J., Reubi, J.C., Henzen-Logmans, S.C., Alexieva-Figusch, J., and Klijn, J.,** Prognostic value of receptors for epidermal growth-factor (EGF-r), insulin-like growth factor-I (IGF-I-r), and somatostatin (SS-r), and of pS2 protein, in patients with breast and ovarian-cancer, *Eur. J. Cancer*, 26, 154, 1990a.

147. **Foekens, J.A., van Putten, W.L.J., Portengen, H., Rodenburg, C.J., Reubi, J.-C., Berns, P.M.J.J., Henzen-Logmans, S.C., van der Burg, M.E.L., Alexieva-Figusch, J., and Klijn, J.G.M.,** Prognostic value of pS2 protein and receptors for epidermal growth factor (EGF-R) insulin-like growth factor-1 (IGF-1-R) and somatostatin (SS-R) in patients with breast and ovarian cancer, *J. Ster. Biochem. Mol. Biol.*, 37, 815, 1990.

148. **Beck, E.P., Russo, P., Gliozza, B., Jaeger, W., Papa, V., Wildt, L., Pezzino, V., and Lang, N.,** Identification of insulin and insulin-like growth factor I (IGF-I) receptors in ovarian cancer tissue, *Gynecol. Oncol.*, 54, 196, 1994.

149. **Weigang, B., Nap, M., Bittl, A., and Jaeger, W.,** Immunohistochemical localization of insulin-like growth factor 1 receptors in benign and malignant tissues of the female genital tract, *Tumor Biol.*, 15, 236, 1994.

150. **Van Dam, P.A., Vergote, I.B., Lowe, D.G., Watson, J.V., Van Damme, P., Van der Auwera, J-C., and Shepherd, J.H.,** Expression of c-erbB-2, c-myc and c-ras oncoproteins, insulin-like growth factor receptor I, and epidermal growth factor receptor in ovarian carcinoma, *J. Clin. Pathol.*, 47, 914, 1994.

151. **Sariban, E., Sitaras, N.M., Antoniades, H.N., Kufe, D.W., and Pantazis, P.,** Expression of platelet-derived growth factor (PDGF)-related transcripts and synthesis of biologically active PDGF-like proteins by human malignant epithelial cell lines, *J. Clin. Invest.*, 82, 1157, 1988.

152. **Henrikson, R., Funa, K., Wilander, E., Backstrom, T., Ridderheim, M., and Oberg, K.,** Expression and prognostic significance of platelet-derived growth factor and its receptors in epithelial ovarian neoplasms, *Cancer Res.*, 53, 4550, 1993.

153. **Versnel, M.A., Haarbrink, M., Langerak, A.W., de Laat, P.A.J.M., Hagemeijer, A., van der Kwast, T.H., van den Berg-Bakker, L. A.M., and Schrier, P.I.,** Human ovarian tumours of epithelial origin express PDGF in vitro and in vivo, *Cancer Genet. Cytogenet.,* 73, 60, 1994.

154. **Di Blasio, A.M., Cremonesi, L., Vigano, P. Ferrari, M., Gospodarowicz, D., Vignali, M., and Jaffe, R.B.,** Basic fibroblast growth factor and its receptor messenger ribonucleic acid are expressed in human ovarian epithelial neoplasms, *Am. J. Obstet. Gynecol.*, 169, 1517, 1993.

155. **Crickard, K., Gross, J.L., Crickard, U., Yoonessi, M., Lele, S., Herblin, W.F., and Eidsvoog, K.,** Basic fibroblast growth factor and receptor expression in human ovarian cancer, *Gynecol. Oncol.*, 55, 277, 1994.

156. **Reynolds, K., Farzaneh, F., Collins, W.P., Campbell, S., Bourne, T.H., Lawton, F., Moghaddam, A., Harris, A.L., and Bicknell, R.,** Association of ovarian malignancy with expression of platelet-derived endothelial cell growth factor, *J. Natl. Cancer Inst.*, 86, 1234, 1994.

157. **Boocock, C.A., Charnock-Jones, D.S., Sharkey, A.M., McLaren, J., Barker, P.J., Wright, K.A., Twentyman, P.R., and Smith, S.K.,** Expression of vascular endothelial growth factor and its receptors flt and KDR in ovarian carcinoma, *J. Natl. Cancer Inst.*, 87, 506, 1995.

158. **Malik, S. and Balkwill, F.,** Epithelial ovarian cancer: a cytokine propelled disease? *Br. J. Cancer*, 64, 617, 1991.

159. **Wu, S., Rodabaugh, K., Martinez-Maza, O., Watson, J.M., Silberstein, D.S., Boyer, C.M., Peters, W.P., Weinberg, J.B., Berek, J.S., and Bast, R.C.,** Stimulation of ovarian tumour cell proliferation with monocyte products including IL-1-alpha, IL-6 and tumor necrosis factor-alpha, *Am. J. Obstet. Gynecol.*, 166, 997, 1992.

160. **Wu, S., Boyer, C.M., Whitaker, R.S., Berchuck, A., Wiener, J.R., Weinberg, J.B., and Bast, R.C.,** Tumor necrosis factor alpha as an autocrine and paracrine growth factor for ovarian cancer: monokine induction of tumor cell proliferation and tumor necrosis factor alpha expression, *Cancer Res.*, 53, 1939, 1993.

161. **Wu, S., Meeker, W.A., Wiener, J.R., Berchuck, A., Bast, R.C., and Boyer, C.,** M., Transfection of ovarian cancer cells with tumor necrosis factor-alpha (TNF-α) antisense mRNA abolishes the proliferative response to interleukin-1 (IL-1) but not TNF-α, *Gynecol. Oncol.*, 53, 59, 1994.

162. **Watson, J., Sensintaffar, J.L., Berek, J.S., and Martinez-Maza, O.,** Constitutive production of interleukin 6 by ovarian cell lines and by primary ovarian tumor cultures, *Cancer Res.*, 50, 6959, 1990.

163. **Berek, J.S., Chung, C., Kaldi, K., Watson, J.M., Knox, R.M., and Martinez-Maz, O.,** Serum IL-6 levels correlate with disease status in epithelial ovarian cancer patients, *Am. J. Obstet. Gynecol.*, 164, 1038, 1991.

164. **Berek, J.S. and Martinez-Maza, O.,** Molecular and biologic factors in the pathogenesis of ovarian cancer, *J. Reprod. Med.*, 39, 241, 1994.

165. **Lidor, Y.J., Xu, F.I., Martinez-Maza, O., Olt, G.J., Marks, J.R., Berchuck, A., Ramakrishnan, S., Berek, J.S., and Bast, R.C.,** Constitutive production of macrophage colony stimulating factor and IL-6 by human ovarian surface epithelial cells, *Exp. Cell. Res.*, 207, 332, 1993.

166. **Watson, J.M., Berek, J.S., and Martinez-Mara, O.,** Growth inhibition of ovarian cancer cells induced by antisense IL-6 oligonucleotides, *Gynecol. Oncol.*, 49, 8, 1993.

167. **Gotlieb, W.H., Abrams, J.S., Watson, J.M., Velu, T.J., Berek, J.S., and Martinez-Maza, O.,** Presence of IL-10 in the ascites of patients with ovarian and other intra-abdominal cancers, *Cytokine*, 4, 385, 1992.

168. **Zlotnik, A. and Moore, K.W.,** Interleukin 10, *Cytokine*, 3, 366, 1991.

169. **Carswell, E.A., Old, L.J., Kassel, R.J., Green, S., Fiore, N., and Williamson, B.,** An endotoxin-induced serum factor that causes necrosis of tumours, *Proc. Natl. Acad. Sci. U.S.A.*, 72, 3666, 1975.

170. **Malik, S.T., Griffin, D.B., Fiers, W., and Balkwill, F.R.,** Paradoxical effects of tumor necrosis factor in experimental ovarian cancer, *Int. J. Cancer*, 44, 918, 1989.

171. **Naylor, M.S., Stamp, G.W., Foulkes, W.D., Eccles, D., and Balkwill, F.R.,** Tumor necrosis factor and its receptors in human ovarian cancer. Potential role in disease progression, *J. Clin. Invest.*, 91, 2194, 1993.

172. **Naylor, M.S., Malik, S.T., Stamp, G.W., Jobling, T., and Balkwill, F.R.,** In situ detection of tumor necrosis factor in human ovarian cancer specimens, *Eur. J. Cancer*, 26, 1027, 1990.

173. **Malik, S.T.A., Naylor, M.S., East, N., Oliff, A., and Balkwill, F.R.,** Cells secreting tumour necrosis factor show enhanced metastasis in nude mice, *Eur. J. Cancer*, 26, 1031, 1990.

174. **Kacinski, B.M., Carter, D., Mittal, K., Yee, L.D., Scata, K.A., Donofrio, L., Chambers, S.K., Wang, K.I., Yang-Feng, T., and Rohrschneider, L.R.,** Ovarian adenocarcinomas express fms-complementary transcripts and fms antigen, often with co-expression of CSF-1, *Am. J. Pathol.*, 137, 135, 1990.

175. **Baiocchi, G., Kavanagh, J.J., Talpaz, M., Wharton, J.T., Gutterman, J.U., and Kurzrock, R.,** Expression of the macrophage colony-stimulating factor and its receptor in gynecologic malignancies, *Cancer*, 67, 990, 1991.

176. **Kacinski, B.M.,** CSF-1 and its receptor in ovarian, endometrial and breast cancer, *Ann. Med.*, 27, 79, 1995.

177. **Filderman, A.E., Bruckner, A., Kacinski, B.M., Deng, N., and Remold, H.G.,** Macrophage colony-stimulating factor (CSF-1) enhances invasiveness in CSF-1 receptor-positive carcinoma cell lines, *Cancer Res.*, 52, 3661, 1992.

178. **Connor, J.P., Squatrito, R.C., Terrell, K.L., Antisdel, B.J., and Buller, R.E.,** In vitro growth effects of colony-stimulating factors in ovarian cancer, *Gynecol. Oncol.*, 52, 347, 1994.

179. **Cimoli, G., Russo, P., Billi, G., Mariani, G.L., Rovini, E., and Venturini, M.,** Human granulocyte-macrophage colony stimulating factor active on human ovarian cancer cells, *Jpn. J. Cancer Res.*, 82, 1196, 1991.

180. **Malik, S.T.A., Knowles, R.G., East, N., Lando, D., Stamp, G., and Balkwill, F.R.,** Antitumor activity of γ-interferon in ascitic and solid tumor models of human ovarian cancer, *Cancer Res.*, 51, 6643, 1991.

181. **Markman, M., Reichman, B., Hakes, T., Curtin, J., Jones, W., Lewis, J.L., Jr., Barakat, R., Rubin, S., Mychalczak, B., and Saigo, P.,** Intraperitoneal chemotherapy in the management of ovarian cancer, *Cancer*, 71, 1565, 1993.

182. **Nicoletto, M.O., Fiorentino, M.V., Vinante, O., Prosperi, A., Tredese, F., Tumolo, S., Cima, G.P., and Monfardini, S.,** Experience with intraperitoneal alpha-2a interferon, *Oncology*, 49, 467, 1992.

183. **Willemse, P.H., deVries, E.G., Mulder, N.H., Aalders, J.G., Bouma, J., and Sleijfer, D.T.,** Intraperitoneal human recombinant interferon alpha-2b in minimal residual ovarian cancer, *Eur. J. Cancer*, 26, 353, 1990.

184. **Iaffaioli, R.V., Frasci, G., Facchini, G., Pagliarulo, C., Pacelli, R., Scala, S., and Espinosa, A.,** Alpha 2b interferon (IFN) by intraperitoneal administration via temporary catheter in ovarian cancer, *J. Clin. Oncol.*, 8, 1036, 1990.

185. **Cappelli, R. and Gotti, G.,** The locoregional treatment of neoplastic ascites with interferon-beta, *Recenti Prog. Med.*, 83, 82, 1992.

186. **Pujade-Lauraine, E., Guastella, J.P., Colombo, N., Francois, E., Fumoleau, P., Monnier, A., Nooy, M.A., Mignot, L., Bugat, R., and Oliviera, C.M.,** Interferon-gamma par voie intraperitoneale. Un complement efficace de la chimiotherapie des cancers de l'ovaire. A propos d'une etude europeenne de 108 patientes, *Bull. Cancer Paris*, 80, 163, 1993.

187. **Colombo, N., Peccatori, F., Paganin, C., Bini, S., Brandely, M., Mangioni, C., Mantovani, A., and Allavena, P.,** Anti-tumor and immunomodulatory activity of intraperitoneal IFN-gamma in ovarian carcinoma patients with minimal residual tumor after chemotherapy, *Int. J. Cancer*, 51, 42, 1992.

188. **Maenpaa, J., Kivinen, S., Raisanen, I., Sipila, P., Vayrynen, M., and Grohn, P.,** Combined intraperitoneal interferon alpha-2b and mitoxantrone in refractory ovarian cancer, *Ann. Chir. Gynaecol.*, Suppl., 208, 25, 1994.

189. **Frasci, G., Tortoriello, A., Facchini, G., Conforti, S., Persico, G., Mastrantonio, P., Cardone, A., and Iaffaioli, R.V.,** Carboplatin and alpha-2b interferon intraperitoneal combination as first-line treatment of minimal residual ovarian cancer. A pilot study, *Eur. J. Cancer*, 30A, 946, 1994.

190. **Nardi, M., Cognetti, F., Pollera, C.F., Giulia, M.D., Lombardi, A., Atlante, G., and Calabresi, F.,** Intraperitoneal recombinant alpha-2-interferon alternating with cisplatin as salvage therapy for minimal residual-disease ovarian cancer: a phase II study, *Eur. J. Gynaecol. Oncol.*, 12, 69, 1991.

191. **Berek, J.S., Welander, C., Schink, J.C., Grossberg, H., Montz, F.J., and Zigelboim, J.,** A phase I-II trial of intraperitoneal cisplatin and alpha-interferon in patients with persistent epithelial ovarian cancer, *Gynecol. Oncol.*, 40, 237, 1991.

192. **Ferrari, E., Maffeo, D.A., Graziano, R., Gallo, M.S., Pignata, S., De Rosa, L., Montella, M., and Pergola, M.,** Intraperitoneal chemotherapy with carboplatin and recombinant interferon alpha in ovarian cancer, *Eur. J. Gynaecol. Oncol.*, 15, 437, 1994.

193. **Frasci, G., Tortoriello, A., Facchini, G., Conforti, S., Cardone, A., Persico, G., Mastrantonio, P., and Iaffaioli, R.V.,** Intraperitoneal (ip)cisplatin-mitoxantrone-interferon-alpha 2b in ovarian cancer patients with minimal residual disease, *Gynecol.-Oncol.*, 50, 60, 1993.

194. **Repetto, L., Chiara, S., Guido, T., Bruzzone, M., Oliva, C., Ragni, N., Conte, P.F., and Rosso, R.,** Intraperitoneal chemotherapy with carboplatin and interferon alpha in the treatment of relapsed ovarian cancer: a pilot study, *Anticancer Res.*, 11, 1641, 1991.

Chapter 8

Oncogenes and Tumor-Suppressor Genes in Ovarian Cancer

C. Michael Steel and Hani Gabra

CONTENTS

1. INTRODUCTION

The ultimate objective of molecular studies in any tumor is to gain an understanding of the fundamental processes of carcinogenesis, but, for the forseeable future, they serve less ambitious, though still important, purposes. In the specific case of ovarian cancer, molecular analysis is proving of value in more accurate classification of tumors, in better definition of prognosis and, very gradually, in unraveling the biology of disease initiation and progression. Despite the limitations of our current knowledge, it is not entirely fanciful to suggest that, in time, the molecular approach will revolutionize the diagnosis and treatment of all cancers.

2. NONEPITHELIAL OVARIAN TUMORS

Though over 90% of ovarian tumors are of epithelial origin, two other tissue elements contribute to neoplasms of this complex organ (in addition, of course, to secondary deposits from other sites). The mesoderm of the gonadal ridge and sex cords gives rise to granulosa cell, Sertoli and Leydig tumors and to thecomas or fibromas while dysgerminomas, teratomas, embryonal carcinomas, and choriocarcinomas all originate in the germ cells.[1] Given the clear differences in histiogenesis, it is inappropriate to group all of these tumor types together when analyzing ovarian cancer at the molecular level. Unfortunately, many published studies have included small numbers of nonepithelial tumors in their series and very few have reported details of the findings in a manner that allows separate assessment of the molecular characteristics of mesodermal or germ cell neoplasms.

Teratomas commonly arise through parthenogenesis and are thus effectively haploid.[2] While this can be useful as a source of "positive control" material in allele imbalance studies,[3,4] it can be misleading if such tumors are included when estimating rates of loss of heterozygosity at specific loci in ovarian cancer. Similar considerations apply to choriocarcinoma which can be triploid (due to dispermy) or haploid.[5] Structural abnormalities of chromosome 12 have been noted on cytogenetic analysis of malignant germ cell tumors of both ovary and testis.[6,7] Fibromas and granulosa cell tumors, even when benign, commonly show a variety of cytogenetic aberrations, particularly trisomy 12 in the former,[8-10] while loss of chromosome X, with or without trisomy 12 and rearrangements of other autosomes have been observed as recurrent features of the latter, though the number of tumors examined is small.[10,11] Preliminary allele imbalance studies of granulosa cell tumors also shows evidence of substantial aberrations at the subchromosomal level.[3,4,12,13]

The cytogenetic findings may point to underlying, but as yet undefined, specific molecular lesions characteristic of these ovarian tumors, some of which could be common to other neoplasms of stromal or germ cell origin and distinct from those that contribute to the initiation or progression of epithelial ovarian cancer.

3. BENIGN AND BORDERLINE (LMP) EPITHELIAL TUMORS

Benign epithelial ovarian tumors have consistently been found to have a normal diploid DNA content and a normal karyotype.[8,14] Nor have they been shown to carry mutations in any oncogene or tumor suppressor gene, except in the case, discussed later, of the morphologically benign epithelium of cysts contiguous with ovarian carcinomas.[15] However, there are a few reports of loss of heterozygosity at certain chromosomal sites (for example, 6q and 17q) in a minority of benign ovarian cystadenomas.[16-19] When more extensive investigations have been undertaken, therefore, these tumors may prove to be characterized by specific molecular lesions. If so, this could shed important light on the vexed question of the role of benign tumors in the evolution of epithelial ovarian cancer.[1,20]

The molecular constitution of borderline or LMP (low malignant potential) ovarian tumors has been a focus of considerable interest in recent years. Standardization of the histological classification of these lesions, particularly for the purpose of multicenter comparisons, has proved difficult.[21] Hence, it is hardly surprising that reported rates of aneuploidy vary from 0 to 52% in different series.[21,22] Nevertheless, most workers agree that aneuploidy is correlated with advancing FIGO stage and with adverse prognosis. In the largest single center study (370 patients) from the Norwegian Radium Hospital,[23] aneuploidy was associated with 50% mortality (over 15 years) even among patients with stage 1A LMP tumors, compared with only 3% for those with similar, but diploid, tumors.

Loss of heterozygosity has been noted in borderline tumors at several loci, though the frequency, at least for some sites, is lower than for fully malignant epithelial ovarian cancers.[3,4,17-19,24-26] Similarly, mutations have been found at the p53 locus in a few instances, certainly less commonly than in malignant tumors.[26] Logic suggests that that these molecular changes will be adverse prognostic signs but the published data are too sparse to confirm or refute that presumption. As discussed later, LOH at specific loci and mutations affecting certain oncogenes and tumor suppressors appear to correlate with histological type, grade and FIGO stage of ovarian cancers and further work is required to establish how LMP tumors fit into the overall picture.

4. FULLY MALIGNANT EPITHELIAL OVARIAN CANCER

4.1 Ploidy

The DNA content of individual cells can be measured by integrated densitometry on Feulgen stained histological sections or, more commonly, by flow cytometry on suspensions of isolated cells or nuclei, stained with a fluorochrome such as ethidium bromide or propidium iodide that binds stoichiometrically to nucleic acids. The latter approach can be applied both to fresh tissue and to archival, formalin fixed, material.[22,27,28] The very extensive literature on analysis of DNA content in ovarian tumors has been well reviewed by van Dam.[22]

It is agreed that DNA content correlates well with modal chromosome number and that aneuploidy (usually seen as an increase in DNA per cell — the DNA index) is more common in advanced (FIGO stage III and IV) than in early ovarian cancers. Even among stage I and II tumors, however, aneuploidy has been recorded in some 20 to 50% of cases. There are correlations between aneuploidy and histological features suggesting poor prognosis (high grade and serous or clear cell type) but in several studies, aneuploidy has emerged as an independent predictor of reduced survival on multivariate analysis.

4.2 S-Phase Fraction

The proportion of cells undergoing DNA synthesis can be measured from the DNA content profile and in some situations (e.g., breast cancer) it is argued that this s-phase fraction provides a better index of tumor aggressiveness than ploidy.[29] This has not been established in the case of ovarian cancer where, in any event, there appears to be close concordance between DNA index and s-phase fraction.[22]

It is, of course, assumed that aneuploidy, visible chromosome aberrations or LOH and other measures of molecular lesions at the subchromosomal level are simply surrogate markers for biologically important mutations affecting specific oncogenes and tumor suppressors. In keeping with this concept, van Dam and colleagues have found that when a number of more precise investigations (levels of expression of epidermal growth factor receptor, insulin-like growth factor 1 receptor, c-erbB-2, c-ras, and c-myc oncogenes) were included in Cox proportional hazard analysis of a large series of epithelial ovarian cancers, DNA ploidy ceased to be a significant independent prognostic variable.[30]

4.3 Cytogenetics

Despite the difficulties of obtaining adequate numbers of analyzable metaphase preparations from solid tumors, there have been many cytogenetic studies of epithelial ovarian cancers[31-44] and some consistency is evident in the reported findings. Among the most frequent aberrations recorded are deletions on chromosome 3p,[33,35] deletions or translocations involving the distal region of 6q[31,32,39,40,44] loss of an entire chromosome 11 or aberrations of its long or short arm[33,34,39,40,43] and loss of an X chromosome.[38,44] In those studies where the information has been sought, the frequency of cytogenetic abnormalities generally increases with tumor stage and grade, being higher for serous than mucinous carcinomas.[31,39,40] This is very much in accord with the findings on DNA index measurement and, as discussed later, on allele imbalance analysis. However, loss of the X chromosome may be an exception, having been reported with relatively high frequency even in early stage cancers.[37,44] Additional chromosomal aberrations occur after chemotherapy.[45] Newer techniques such as fluorescence *in situ* hybridization (FISH), applied both to cytogenetic preparations and interphase cells, are beginning to yield data that may be easier to interpret than conventional banded metaphase spreads and, to date, they seem to corroborate the findings reported above.[46,47]

5. LOSS OF HETEROZYGOSITY (ALLELE IMBALANCE)

The idea that molecular lesions in tumors can be localized to particular regions by comparing paired samples of tumor and constitutional DNA at mapped polymorphic sites is an attractive one and has become more so as the range of available markers has increased.[48] Microsatellite sequences are particularly popular for this kind of study, not only because of their high levels of polymorphism but also because they require negligible amounts of tissue and can be applied to fixed archival specimens.[49] It must, however, be borne in mind that a marked change in the relative intensities of two allelic signals can come about either through loss of one or amplification of the other (or both) and that microsatellite probes are not ideal for distinguishing between these alternatives. A great many allele imbalance studies have been undertaken in epithelial ovarian cancer and several attempts have been made to summarize the results obtained.[25,50,51] Although a consensus does appear to be emerging for the frequencies of allele imbalance at a number of sites, inconsistencies (sometimes striking) between the findings of different groups still exist and may be explained in one or more of several ways. First, small numbers of tumors inevitably generate unreliable figures but, even in large series, the proportions of early and late stage or high and low grade tumors may have an important bearing on the frequency and distribution of molecular lesions. It is possible (though, as yet unproven) that tumors from different geographical areas or ethnic groups may vary systematically in the rates of allele imbalance at specific loci. Second, the ways in which the technique is carried and the data analysed may have an important bearing on the conclusions. Selection of regions of tumor more or less heavily "contaminated" with nonmalignant cells can influence the sensitivity of detection of allele imbalance. The cut-off value for a meaningful change in signal intensity ratio between the pairs of alleles is not standardized and the measurement may be made by eye or by scanning densitometry. The required proportion of tumors showing allele imbalance at any given locus in order to ascribe significance to that locus (i.e. the acceptable level of "background noise") varies widely. Third, there may be wide variety in the choice of markers and density of coverage of any given chromosomal region. Early studies often used a single probe per chromosome arm but evidence is accumulating that allele imbalance may be tightly restricted, and that there can be multiple distinct regions of imbalance on a single chromosome arm. Thus, two probes, though mapping to the same general region, may detect quite different rates of allele imbalance. With these caveats in mind, the chromosomal regions most consistently showing rates of allele imbalance of 30% or greater in epithelial ovarian cancer are listed in Table 1.

Table 1 Chromosomal Regions Showing High Rates of Allele Imbalance (LOH) in Epithelial Ovarian Cancer

Chromosomal Region	Genes Possibly Implicated
3p22-24	
4p	
5q	
6p22-pter	
6p21	? p21 (WAF1, Cip1)
6q26-27	
7p	
8q	
9p	? p16 (CDKN2, MTS1), p15
11p15	p57 Kip2
11p13-q22	? Progesterone receptor
11q23-qter	
13q12.3-14.1	? BRCA 2 (not Rb1)
17p	p53
17q23-25.3	
17 (whole chromosome)	? BRCA 1
18q21-qter	not DCC
19p	
22q	
Xp21.1	

For references to original data see reviews cited (References 25, 50, 51).

As noted earlier, allele imbalance may be a consequence of gene amplification as well as deletion. The technique may therefore localize both oncogenes and tumor suppressor genes. However, it has commonly been associated with the latter and, as discussed later, there is supporting evidence from experiments involving transfection of an intact copy of chromosome 11 that genes carried on it are capable of reversing at least some of the malignant properties of human ovarian carcinoma cell lines (i.e., that they function as tumor suppressors).[52] In most instances, frequency of allele imbalance, like chromosome aberrations and altered DNA index, correlates with other features of aggressive tumor behavior (serous or clear cell type, high histological grade and advanced FIGO stage). This is particularly true for 17p and 11q23-qter, as discussed below, as well as for 13q[53] and 18q.[54] On the other hand, imbalance at 17q and Xp may occur at relatively early stages of tumor evolution.[16,44]

Much of the more detailed information on allele imbalance in epithelial ovarian cancer comes from studies in chromosomes 11 and 17. Whether they carry a higher density of relevant genes or have simply been subjected to more intensive investigations remains to be established. However, the findings to date from these two chromosomes begin to suggest a molecular basis for the clinical and pathological diversity of the disease.

5.1 Chromosome 11

5.1.1 11p LOH

Interest in 11p arose initially from the mapping of the human Harvey ras proto-oncogene.[55] Subsequently, the Wilm's tumor locus was found on the same chromosome arm,[56] and more recently, a tumor suppressor locus implicated in prostate cancer.[57]

Several groups have reported LOH at 11p15 in epithelial ovarian cancer, at rates approaching 50%,[4,25,58-64] though others have found much lower rates of loss.[12,13,65] The site of maximal LOH has been found by some workers to coincide with H-ras, the rate diminishing at more centromeric loci.[61,62] However, SSCP analysis and genomic sequencing of H-ras from ovarian cancers has revealed no mutations.[66] Further evidence for a functional tumor suppressor in this region has come from experiments in which 11p15 subchromosomal DNA fragments were transferred into rhabdomyosarcoma cells, causing growth arrest. The tumor suppressing locus responsible was mapped between the beta globin and insulin genes.[67]

11p13 (the site of the Wilm's tumor gene) generally shows rates of LOH in the range 20 to 30% in ovarian cancer.[4,63,64] SSCP and sequencing have failed to reveal any WT-1 mutations even in ovarian cancers showing LOH at this locus.[68,69] A centromeric site at 11p12 exhibits much higher rates of LOH, and this is particularly interesting in view of KIA1, a new metastasis suppressor involved in prostate cancer isolated by functional cloning. This gene located on chromosome 11p11.2 was shown to suppress metastatic ability of a rat prostate cancer cell, and its expression is reduced in human cell lines derived from metastatic prostate tumors. The gene encodes a 267 aa protein distinct from, but related to, leukocyte surface glycoproteins.[57,70] Several of the published studies note an association between 11p LOH and poor differentiation and/or advanced stage of the ovarian tumors.[4,61,64,65]

5.1.2 11q LOH

It is clear that LOH is far from uniformly distributed over the long arm of chromosome 11. Relatively low rates of imbalance are found for the proximal region (to q14) in ovarian cancer.[13,58,62,71,72] Southern analysis, in ovarian but more particularly in breast cancer, has demonstrated that allele imbalance in the proximal long arm is commonly attributable to amplification. 11q13 contains a region that is often amplified in breast and other cancers (the "11q13 amplicon").[73] This gene-rich segment encodes growth factors such as FGF-3 (Int-2) and at least one cell cycle regulator, cyclin D (PRAD 1, BCL-1).[74] Studies that have concentrated specifically on the subtelomeric region of 11q have recorded much higher rates of LOH (66% LOH at 11q23.3-qter) than are found more proximally.[4,71] On Southern analysis, in both breast and ovarian cancer, imbalance in this region is associated with deletion rather than amplification.

One study noted that LOH at 11q14-q22 was more frequent in early stage than advanced FIGO stage tumors suggesting that some early FIGO stage cancers may have followed a distinct pathway of biological evolution rather than representing mere stages en route to advanced disease.[4] None of eight endometroid cancers in that series had evidence of loss close to the progesterone receptor (11q21), which may be significant since that histological subtype is characterized by higher progesterone receptor content than other ovarian cancers. A follow-up to that study showed a correlation between allele loss close to the progesterone receptor gene and tumor progesterone receptor content.[75] Interestingly, a polymorphism of the progesterone receptor associated with two exon mutations has been shown to confer an increased risk of nonfamilial ovarian cancer.[76]

Foulkes and colleagues[71] noted a nonsignificant trend towards association of 11q24 LOH with advanced stage and high grade ovarian cancers. Another study noted a similar nonsignificant tendency for allelic imbalance at this site to occur preferentially in grade 3 tumors; they also found a significant association of 11q24 LOH with advanced stage and adverse survival using Kaplan-Meier analysis.[4] This has been confirmed in at least one other unpublished series (G. Chenevix-Trench, personal communication). Interestingly, an analogous study in breast cancer also demonstrated adverse survival in patients with metastatic disease where there was 11q24 LOH.[77]

This evidence for a region discriminating advanced FIGO stage, poor prognosis epithelial ovarian cancers from earlier stage, better prognosis tumors, suggests that distal 11q probably carries a late-acting suppressor of tumor progression, perhaps involved in invasion/or metastasis (Figure 1). This region has now been narrowed down to an 8.5 Mb interval between 11q23.3 and q24.3 (Gabra, unpublished). Several candidate genes map to this location, including some known to be implicated in invasion and metastasis. The region appears to be separable from that described recently in breast cancer[78-80] and excludes ATM, potentially a significant determinant of breast cancer risk.[81]

5.2 Chromosome 17

Allele loss on chromosome 17 is extremely common in epithelial ovarian cancer. Several studies have demonstrated that loss of one complete copy of the chromosome occurs frequently and allelotyping shows highly significant statistical associations between losses on 17p and 17q. This has been attributed to non-disjunction, with or without duplication of the remaining homolog. Some of the biological

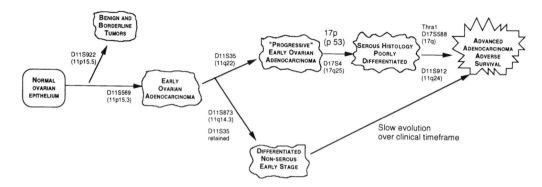

Figure 1 An outline of the possible sequence of molecular events in the evolution of epithelial ovarian cancer based on observations, described in the text, relating specifically to chromosome 11 aberrations.

consequences can be explained in terms of alterations to the p53 gene but there is evidence for other important loci on chromosome 17.

5.2.1 17p and p53

It is difficult to define non-p53 tumor suppressors on 17p because of the overwhelming importance of this gene in tumorigenesis generally and its frequent mutation (up to 50% of cases) in epithelial ovarian cancer.[82]

The p53 gene is composed of 11 exons, the first of which is noncoding and localized 8 to 10 kb upstream of exons 2 through 11 on the short arm of human chromosome 17. There are five conserved domains within the the coding regions of the gene which contain 80% to 90% of described functional p53 mutations; p53 structure is not similar to any other known DNA binding proteins. Two antiparallel beta sheets anchor peptide loops which directly contact DNA and also stabilize the shape of the molecule. P53 monomers assemble into tight dimer pairs, then symmetrical tetramers held together through multiple molecular interactions. The carboxyl end of the p53 protein can also bind to broken DNA ends; p53 has a sequence-specific transactivator domain located in the central "core" of the protein, comprising residues 90–290.

Normal or wild type P53 is involved in several downstream functions: DNA damage detection, apoptosis and growth arrest. p53 protein has the capacity to detect and bind sites of primary DNA damage (in the form of insertion/deletion mismatches or broken DNA ends). This functional ability to detect DNA damage maps to the carboxyl terminus of the protein forming high affinity complexes with the damaged DNA site, and providing a focus for the subsequent assembly of proteins involved in DNA repair. One of these potential partners is a factor implicated in DNA excision repair known as ERCC3. Binding to DNA has also been shown to protect p53 from rapid proteolysis, which is consistent with the observed stabilization and accumulation of p53 protein in DNA-damaged cells.

p53 protein can also mediate cell death by apoptosis. This morphologically distinct form of cell death is an active response to cell stimuli. Cell volume shrinks, chromatin condenses, and the nucleus fragments. Phagocytes ingest and destroy apoptotic cells without provoking inflammation. p53-mediated growth arrest or apoptosis depends partly on the state of cellular activation. Conflicting growth regulatory signals, sustained p53 synthesis due to extensive unrepaired DNA damage or p53 activation after irreversible commitment to replication can drive a cell to apoptosis. Apoptosis can also occur without the presence of p53 via a steroid mediated pathway.

The retinoblastoma protein (pRb) family is implicated in the relationship between p53 mediated growth arrest and apoptosis. pRb is normally required for imposing a p53-mediated G1 arrest. If pRb is functionally inactive, either physiologically or pathologically, the growth-inhibitory activity of wild-type p53 is converted into an apoptotic fate.

As mentioned above, DNA damage stimulates the production of the p53 protein, in turn transactivating genes such as p21 and GADD45 via p53's central core domain. Transcriptional activity is mediated by a domain located at the amino end of the p53 protein. GADD45 is involved with growth arrest in the presence of DNA damage. p21, however, inhibits the activities of cyclin-dependent kinase enzymes which are responsible for pushing the cell through its cycle for division. This allows for cell arrest at the G1/S phase junction and time for DNA repair. On restoration of DNA integrity, proliferation resumes.

Both GADD45 and p21 interact with and inhibit PCNA (proliferating cell nuclear antigen), the function of which is to stimulate DNA synthesis allowing coordination of two separate processes; cell cycling/division and DNA synthesis.

Functional p53 mutations usually affect transactivation domains. More than 90% of the substitution mutations reported so far in malignant tumors are clustered between exons 5 and 8 and most are localized in the evolutionarily conserved regions. Most p53 mutants fall into one of two classes: DNA contact mutants or structural mutants. DNA contact mutants of p53 are unable to bind specific DNA target sequences because of substitution of crucial DNA contact residues, including mutational hot spots, such as arginine-248 and arginine-273. Structural mutants, on the other hand, lose specific DNA binding capacity because of abnormal conformation of the protein. Central core mutations are the main mutations affecting transactivation. It has been suggested that greater than 60% loss of p53 function is needed to result in a tendency towards neoplastic transformation or growth. The presence of transversions (mutation where purines are replaced by pyrimidines or vice versa) represent events related to exogenous carcinogens, whereas transitions (conversion of pyrimidine to pyrimidine, or purine to purine) are more likely to be due to errors of fidelity during DNA replication, which may be a spontaneous process. p53 is frequently mutated in epithelial ovarian cancer and its sequence analysis reveals no transversion "hotspots"; mutation of transition type were predominant (72% of all mutation events detected) and dispersed throughout the gene (predominantly exons 5 to 8).[83] This predominance of transition mutations suggests that the impact of environmental carcinogens to the development of ovarian cancer may be minimal. The role and independence of p53 abnormality as a prognostic factor in ovarian cancer is conflicting, with some investigations showing no association with p53 abnormalities,[84-87] and others showing that p53 status functions as an independent prognostic factor.[88-91] P53 mutation is reported to be significantly associated with serous histology.[17,86] Recent data suggests that those ovarian cancers which have a normal p53 gene by sequence analysis are more likely to respond to chemotherapy than those with p53 mutation.[92] Recently, a p53 polymorphism designated *p53PIN3* has been identified; homozygotes were 9 times more frequent in a cohort of ovarian cancer patients than in a control group. The polymorphism is due to a 16bp insertional tandem repeat in intron 3 of p53.[93] This intriguing finding was not supported, however, in another study.[94]

There may be other clinically relevant cancer genes but the presence of a distinct tumor suppressor locus on the distal portion of 17p is much less clear in ovarian than in breast cancer.

5.2.2 *Non-BRCA1 Molecular Changes on 17q*

It was something of a surprise to find that, in sporadic ovarian cancer, despite a high rate of LOH with microsatellite markers close to BRCA1, there is a very low rate of somatic mutations in the coding sequence.[95,96] Indeed, in the original study of Futreal and colleagues, the few BRCA1 mutations found in ovarian cancers were all of germ-line origin.[97] It may well emerge that down-regulation of BRCA1 expression, rather than structural mutation, is the main route by which that gene becomes involved in ovarian carcinogenesis. It should not be forgotten, however, that other candidate genes still remain in this region. They include NM23-H1 and -H2 which appear to function as metastasis suppressors. In ovarian cancer, as in some other tumors, there is a negative correlation between expression of these genes and the presence of demonstrable metastases.[98] In addition, abutting BRCA1 itself is a gene that encodes the protein sequence of CA125, the glycoprotein marker for ovarian cancer commonly used to monitor response to treatment.[99]

As mentioned earlier, several workers have recorded significant rates of LOH at sites distal to BRCA 1.[3,16,17,24,100,101] Allele imbalance at 17q25 has been reported in benign and borderline tumors in the absence of 17p involvement.[17,25] This 17q telomeric locus has been localized by fine detail mapping to a region of approximately 1 Mb and a YAC contig is currently being constructed (H. Russel — personal communication). Functional evidence from chromosome 17 microcell transfer into a breast cancer cell line expressing normal p53, indicates that tumorigenicity can be suppressed by a gene or genes on this chromosome and, significantly, microsatellite analysis of the retained fragments showed that the BRCA1 region had been lost.[102] Hence, tumor suppressing activity in this instance cannot be ascibed to p53, nor to BRCA1.

5.3 Functional Identification of Tumor-Suppressor Genes in Ovarian Cancer

Two studies have been reported of monochromosome 11 transfer to ovarian cancer. The first, published in abstract form only, claimed that the growth of an ovarian cancer cell line was inhibited but no animal inoculation data were presented. A more recent study[103] reported transfer of chromosomes 3 and 11 into an ovarian carcinoma cell line HEY. Whereas chromosome 3 powerfully suppressed tumorigenicity and altered the phenotype of HEY cells, the effect of chromosome 11 was markedly different: the immortalized phenotype and morphology were unaltered. However, the *in vitro* growth rate and clonogenicity in soft agar were reduced. Tumorigenicity was not consistently controlled by chromosome 11 transfer. This pattern might, in fact, be quite consistent with the action of a putative late-acting tumor suppressor which does not determine the fundamental property of malignancy but has a more subtle influence on the behavior of established tumors, perhaps reflected here in altered growth rate and cloning efficiency. This raises interesting questions about the appropriateness of growth arrest and tumor suppression as the only — or even the major — end points of chromosome transfer experiments designed to look for functionally important tumor suppressor genes, given that the clinical course of cancers may be strongly influenced by genetic events occurring late in the evolution of the disease.

Transfer of chromosome 6 causes senescence in SKOV3, OVCAR3, and several other ovarian cancer cell lines, whereas chromosome 10 and 14 do not.[104] Introduction of the X chromosome also seems to carry a functional senescence gene.[105]

6. ONCOGENES AND SIGNAL TRANSDUCTION

Much of the interest in dominantly acting genes in ovarian tumors has centered on the role of growth factors and their receptors. Epidermal growth factor receptor (EGFr, p170), erbB-2 (HER-2/*neu*, p185) and macrophage colony stimulating factor receptor (FMS, p165) are all transmembrane proteins with tyrosine kinase activity in their cytoplasmic domains and overexpression of each has been shown to correlate with poor prognosis in epithelial ovarian cancer. The effect of ligand binding is to activate the tyrosine kinase domain, which in turn phosphorylates tyrosine residues on other cytoplasmic enzymes such as phospholipase C-γ which cleaves diacylglycerol from membrane phospholipids, generating a second messenger with powerful mitogenic activity.[106,107]

Twenty percent of ovarian cancers exhibit HER-2/*neu* gene amplification, though overexpression at the protein level is demonstrable by immunohistochemistry in a somewhat higher proportion (around 30%).[108,109] Immunohistochemistry is only a semiquantitative technique and its interpretation can be subjective. Nevertheless, it carries the advantage that tumor cells can be assessed on an individual basis without the problems associated with stromal and/or admixed normal epithelial cells encountered in tissue extracts. Developments in automated image analysis may provide a more objective basis for this type of study.[110] Production of anti-p185 (HER-2/neu) antibody has potentially profound therapeutic implications since it inhibits the growth of erbB-2 expressing tumors. This effect has been shown to involve down-regulation of intracellular diacylglycerol production.[111]

In over 80% of epithelial ovarian cancers, cultures established directly from fresh biopsies and examined within 24 hours revealed double minute chromosomes in at least a proportion of the cells.[112] Double minutes are unstable genetic elements carrying amplified portions of the genome that commonly include dominantly activating oncogenes. Their presence at such high frequency in ovarian cancer highlights the importance of oncogenes in this disease. In some instances, the growth factor receptor genes mentioned above are likely to be incorporated in the double minutes but other oncogenes are also implicated in ovarian cancer. Double-minute DNA has been microdissected, PCR amplified, and fluorescently labeled as a FISH probe. The unique amplified sequence hybridized with 8q24, the site of c-myc.[113]

c-Myc amplification and/or overexpression has been detected in over 20% of cases, particularly in serous cancers.[107,114] Where it is amplified or over-expressed, it may present a therapeutic target since, for example, cyclopentone antitumor prostaglandins inhibit c-myc expression, causing a G_1 block and leading to apoptotic death of tumor cells.[115]

Amplification of ras genes does occur[114,116] but more commonly they appear to be activated by point mutations. In some studies, mutation of K-ras was noted more often in borderline than in malignant ovarian tumors[106,107] but other reports suggest that K-ras mutations are usually associated with advanced FIGO stage cancers.[117,118]

A number of growth factors with possible autocrine functions have been investigated in ovarian cancer. The fibroblast growth factor genes *fgf3* and *fgf4* are among these and amplification of *fgf4* in particular appears to correlate with advanced FIGO stage.[119,120] The macrophage colony stimulating factor (CSF-1) is the natural ligand for p165 c-fms and overexpression of either is associated with poor prognosis.[106,121] CSF-1 not only stimulates proliferation of ovarian cancer cell lines but also stimulates invasive behavior *in vitro*, an effect which is probably mediated through urokinase.[122] It is known that the expression of urokinase plasminogen activator correlates strongly with tumor invasiveness *in vivo*.

7. OVARIAN CANCER AND THE CELL CYCLE

Unregulated growth as a result of unchecked progression through the cell cycle is one of the defining features of malignancy. In recent years there have been great advances in our understanding of the subcellular machinery responsible for controlling the cell cycle and its relevance to cancer (Figure 2).[123-125]

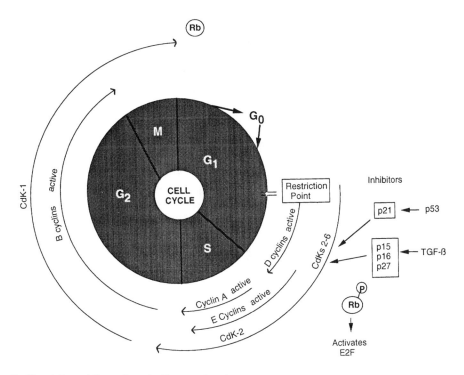

Figure 2 Regulation of the cell cycle. Progression through the phases $G_1 \rightarrow S \rightarrow G_2 \rightarrow M$ appears to be determined by the serial activation of specific cyclins and their accompanying cyclin-dependent kinases (CdKs). The kinases are subject to regulation by inhibitors p15, p16, p21 and p27 which are induced by p53 or TGFβ as shown. One of the critical substrates for cyclin-dependent kinases is the retinoblastoma protein Rb which, when phosphorylated, activates transcription factor E2F.

At the heart of the cell cycle "engine" are phosphorylating enzymes, cyclin-dependent kinases (CDKs) which associate as complexes with activating holoenzymes (cyclins) and are responsible for driving the cell through each of the checkpoints G_1/S, S/G_2 and G_2/M (see Figure 2).[126,127] One of the major substrates through which CDKs mediate their effects at the major G_1/S checkpoint is the retinoblastoma protein Rb. Rb forms a complex with E2F, a promiscuous transcription factor. When Rb is phosphorylated by CDKs, E2F is released to mediate coordinated transcription of genes required for entry into S phase.[125-129]

The CDKs are regulated by specific inhibitors (CDKIs) which suppress cycling by binding either to cyclin/CDK complexes (an activity exemplified by the Cip/Kip group of proteins; p21, p27, p57) or to monomeric CDKs (e.g. the INK4 group of CDKs p15, p16, p18, p19). The latter are more specific since they block the actions of the D-cyclins only. Upstream regulation of the CDKIs is important in signal-mediated control of the cell cycle. For example, p21 is activated by p53 in response to DNA damage while p27 and p15 are induced by TGF-β.[125-131] Recently, evidence has been presented that estradiol can stimulate cyclin/Cdks in ovarian cancer cell lines.[132] Estradiol has been shown to stimulate BRCA1 expression in an ovarian cancer cell line.[133]

The role of these genes in ovarian cancer is now becoming apparent. Deletion (even homozygous deletion) of p16 is found rather frequently in several types of cancer including ovarian cancer — both primary tumors and cell lines. LOH studies show frequent disruption of the p16 region on chromosome

9p but this is rarely accompanied by mutation of the retained p16. It has been postulated that deletion events at this locus are more powerfully tumorigenic (and therefore selected for) because they also involve the adjacent p15 gene.[134,135]

As discussed earlier, amplification at 11q13, though less common than in breast cancer, does occur in ovarian tumors and is usually associated with overexpression of cyclin D1.[136,137] Overexpression of MDM2, which has been described as a mechanism for inactivation of p53, for example in sarcomas,[138] occurs rarely in ovarian cancer[139] and the same is true for inactivating mutations of p21.[140] There is evidence that p21 is associated with better differentiation status and tumor grade, and p21 positive staining was associated strongly with good survival of patients with ovarian cancer when tumors with mutant p53 were excluded.[141]

8. FAMILIAL OVARIAN CANCER

Familial clustering of ovarian cancer has been recorded at least since the 19th century[142] and large-scale epidemiological surveys in the U.S. and Europe have concluded that between 5 and 10% of all cases of epithelial ovarian cancer are attributable to the effect of one or more powerful predisposing genes which are generally inherited as autosomal dominants with limited penetrance.[143-145] In these respects, "ovarian cancer genes" seem to behave in a similar manner to those associated with heritability of other common cancers. There have been rare reports of nonepithelial cancers of the ovary in sisters or mother-daughter pairs but, for the most part, familial ovarian cancer is restricted to the epithelial type, typically with serous histology and adverse prognostic features.[146] One dissenting study recorded an excess of endometrioid cancers within familial clusters.[147] Borderline epithelial ovarian cancers do not appear to feature as part of any familial cancer syndrome.[148,149]

Families with multiple cases of ovarian cancer, without any excess of other malignancies, are recorded but much more commonly ovarian cancer is seen in association with breast, colorectal, endometrial and other epithelial tumors. The combination of breast and ovarian cancer is usually attributed to germline mutation of the BRCA1 gene on chromosome 17 at q21 and indeed that is almost certainly true for the great majority of multi-case breast/ovarian cancer families.[150] Perhaps 20% of them, however, actually carry germline mutations in BRCA2 (chromosome 13q13)[151] and in a very few instances, a similar pattern can be due to germline mutations in p53 (chromosome 17p13).[150,152]

Ovarian cancer is recognized as a feature (albeit a minor one) of the multi-cancer (Lynch type 2) family syndrome[153] in which there is a defect in one of the replication error repair genes, hMSH2, hMLH1, hPMS1 or hPMS2.[154-157] In most instances, colorectal and endometrial cancers figure more prominently among members of these families and the characteristic "RER" phenotype — detected as microsatellite instability on DNA analysis of the tumors from such families — is rare in unselected ovarian cancers, supporting the view that RER gene mutations contribute little to the overall incidence of the disease.

8.1 BRCA1 and Ovarian Cancer

The BRCA1 gene is large, extending over 100 Kb of genomic DNA, with 24 exons (Figure 3). It encodes a protein of 1863 amino acids that is expressed in many tissues during development, including thymus, breast and testis.[158,159] It is strongly expressed in the epithelial cell layer of the adult ovary but expression appears to be reduced in malignant cells.[160] Little information is available yet on the function of BRCA1 protein. It has a ring finger domain consistent with DNA binding and can localise to the nucleus but there is, as yet, no proof that it acts as a transcription factor. Though somatic mutation of BRCA1 is evidently rare, loss of the wild type allele has been found in many studies of ovarian and other tumors occurring in carriers of a germ-line mutation[161,162] and, as noted earlier, LOH on 17q, including the BRCA 1 locus, is a common event in ovarian cancer. BRCA1 therefore has most of the characteristics of a tumor suppressor gene and the aggregate mutation frequency in most Western populations is estimated at between 0.03% and 0.1%.[163] Among Ashkenazi Jews, one particular mutation (185 del AG) is found at a frequency approaching 1%[164] and evidence is accumulating that other specific mutations may characterize different population groups, though present at considerably lower frequencies.

Germ line mutations are found at many different positions within BRCA1. They are demonstrable by a variety of DNA analysis techniques, including single strand conformational polymorphism (SSCP), heteroduplex band shifts, dideoxy nucleotide "fingerprinting" or in vitro translation and protein truncation testing (PTT).[162-167] None of these methods is yet 100% efficient and mutation detection is some way from being a routine diagnostic procedure. The majority of mutations found to date are nonsense or

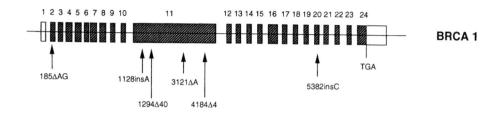

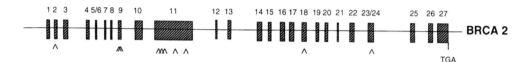

Figure 3 Genomic structure of BRCA1 (24 exons) and BRCA2 (27 exons). Both extend over more than 100 Kb. BRCA2 exons 5/6 and 23/24 are too close together to be separated at this resolution. Some recurrent BRCA1 mutations are indicated, including the AG insertion at position 185 in exon 2, common in Ashkenazi Jews. Sites of mutations recorded to date in BRCA2 are marked by arrow-heads. The coding sequence of BRCA1 comprises 5589 bases and specifies a protein of 1863 amino acids. The corresponding sizes for BRCA2 are > 11385 bases and 3418 amino acids.

frame-shift type, generating a truncated protein but a significant proportion (perhaps 25%) appear to be regulatory changes, not yet characterized at the molecular level, that inhibit transcription of one allele. Overall, there is about a threefold greater risk of breast than ovarian cancer among carriers of BRCA1 mutations but the data suggest that these relative risks are not uniform in different families[163] and there is much interest in the possibility that particular mutations may be associated with a greater or lesser propensity to ovarian malignancy. Findings from a British study[168] indicate that mutations closer to the 5′ end of the gene carry a higher risk of ovarian cancer but this is not apparent in two other surveys (unpublished data from Breast Cancer Linkage Consortium). This is clearly an important issue to resolve since the counseling of female carriers of BRCA1 mutations should, if possible, include information on the absolute risk of ovarian cancer. Recent data suggest that the presence of rare alleles at the H-RAS1 VNTR locus (1Kb downstream of the H-RAS1 proto-oncogene) conferred a 2.1-fold increased risk for ovarian cancer in BRCA-1 carriers compared with BRCA1 carriers harboring common alleles only.[169] Susceptibility to breast cancer was not altered by the presence of the rare alleles in BRCA1 carriers and so the allele appears specifically to modify the ovarian cancer risk in this inherited syndrome.

BRCA1 is expressed in rapidly proliferating cell types undergoing differentiation, and its expression is induced in oophorectomized animals after treatment with 17 β-estradiol and progesterone.[170] An inhibitory role for BRCA1 has now been determined for breast and ovarian cancer cell lines. Transfection of these cell lines with plasmid expressing wild-type BRCA1 inhibits growth, although this is not the case for colon or lung cancer cell lines, demonstrating a tissue specific effect.[171]

An interesting correlate with the above observation that 5′ mutations demonstrate an excess of ovarian cancers is that although transfection of a 3′ BRCA1 mutant (1835 stop) does not inhibit the growth of breast cancer, it still does inhibit the growth of ovarian cancer cells.[171] Furthermore, this 3′ mutation appears to map to the putative granin consensus sequence,[172] and this suggests a mechanism for mutational differences involved in tumors of different tissue origins.

There may be other explanations for non-uniform distribution of breast and ovarian cancer risks in BRCA1 families. One could be the influence of environmental factors, notably the use of exogenous hormones. Prolonged use of the combined oral contraceptive "pill" has been shown to reduce the risk of ovarian cancer[145,173] — probably by reducing the number of ovulatory cycles — and this may be relevant to those families in which cases of breast cancer have appeared recently though past generations have shown an excess of ovarian cancers only. Alternatively, co-inheritance of modifying genes may influence the "BRCA1 phenotype". However, preliminary studies of candidate modifiers of cancer risk, including genes involved in carcinogen metabolism, such as the glutathione S-transferase and cytochrome p450 series have so far yielded no evidence to implicate any of them (unpublished data from EC Concerted Action on breast cancer genetics). Furthermore, an analysis of the distribution of ovarian cancer cases within large breast/ovarian cancer families, failed to demonstrate any clustering that might be consistent with co-inheritance of a risk-modifying factor (DT Bishop and colleagues, data presented at meeting of EC Concerted Action on breast cancer genetics, Lillehammer, March 1995).

The lifetime risk of ovarian cancer in a carrier of a BRCA1 mutation is estimated at between 20 and 80%.[163] The tumors rarely present under age 35 and though, in most studies, the age of onset of the disease has tended to be younger than for sporadic cases, there is some disagreement as to the value of age of onset as a marker for genetic predisposition.[146] Many centers now offer screening (mainly by ultrasound scan) for female members of ovarian or breast/ovarian cancer families but the value of this approach is still unproven.[174] Studies are also in progress to evaluate serum markers such as CA125 and OVX1[175] as predictors of pre-cancer.

Prophylactic oophorectomy is clearly an option for women in this situation, particularly if proven to be mutation carriers. It is, however, uncertain at what age it might be advisable to undergo the procedure. Furthermore, there are several reports of "primary peritoneal carcinoma," histologically indistinguishable from serous ovarian cancer, arising several years after prophylactic oophorectomy in members of cancer families.[176,177] These tumors may result from a "field change" in the peritoneal epithelium, which is embryologically related to the ovarian epithelium, or they may be derived from seedlings of cancer from an unrecognized microscopic primary malignancy already present in the ovary at the time of "prophylactic" surgery. Clinical and pathological data are now being gathered as a matter of some urgency to try to establish what the actual frequency of these peritoneal tumors may be and whether, for example, early oophorectomy may reduce the risk.

8.2 BRCA2 and Ovarian Cancer

It appears that a significant proportion of inherited breast cancer may be attributable to a susceptibility gene located on chromosome 13q12-q13 which has been called BRCA2 .[178] Initial impressions suggest that it does not confer substantial increase in the risk of ovarian cancer, however, 13q LOH in the presence of normally expressed Rb-1 product suggests that it may have some role in ovarian cancer. The gene was cloned in 1995 [179] and consists of an 11–12kb messenger RNA encoded by a 27 — exon gene spanning 70 kb of DNA. It encodes a 3418 aa protein of about 350 kDa. The gene has no obvious strong homologies, and the mouse and human genes share only about 55 to 60% homology. However, it contains eight copies of a highly conserved repeat element dispersed throughout the gene, whose function is as yet unclear. Over 50 mutations have been described in BRCA2, nearly all of which are frameshift or nonsense, and no mutational hotspots have been identified.[179-182] In one of these reports, only 2 of 17 non-BRCA1-linked breast/ovarian cancer families contained a BRCA2 mutation.[181] These findings suggest that perhaps the proportion of inherited ovarian cancer linked to a third (unidentified) susceptibility locus may be greater than previously believed.

8.3 Familial Ovarian Cancers Due to Other Genes

The possibility of a separate ovarian cancer gene, distinct from those mentioned above, responsible for some familial cases has not been formally ruled out, but epidemiological studies had seemed to suggest that it cannot acount for more than a small proportion of cases. Relatively common lower penetrance gene aberrations (notably of ATM, mutations in which cause Ataxia Telangiectasia in the homozygous state)[81] have been invoked to explain a significant proportion of breast and some other cancers.[150,163] However, there is no evidence at present to suggest that they are important in ovarian cancer.

9. CONCLUSION

There is no longer serious doubt that the roots of cancer lie in the genetic material of the cell. The greatest current difficulty is, in fact, the superabundance of molecular lesions in most cancers and our inability to fit these into a coherent framework that matches what is already known of the pathology, epidemiology, and clinical features of any given tumor type. Progress is undoubtedly being made and ovarian cancer provides many illustrations. The relationship between hereditary and acquired mutations, the molecular characteristics of different histological types and the correlation between accumulation of mutations and advancing clinico-pathological stage all appear to make biological sense, even if they do not add up to a complete explanation for all features of the disease.

As understanding grows, we may reasonably expect it to be translated into therapeutic advances. Whether these will come in the form of "gene therapy," "rational drug design," or immunotherapy, all of these or none, only time will tell. In any event, ovarian cancer promises to occupy a prominent place in the development of radical new forms of treatment not only because current management is, in most instances, only a "holding operation" but also because the disease, though disseminated, is commonly localized, for a considerable period, to the peritoneal cavity and is therefore accessible for delivery and assessment of drugs or other agents.

REFERENCES

1. **Langley, F.A. and Fox, H.,** Ovarian tumours: classification, histiogenesis and aetiology, in *Obstetrics and Gynaecologic Pathology*, Fox, H., Ed., Churchill Livingstone, London, 1987.
2. **Kaiser-McCaw, B. and Latt, S.A.,** X-Chromosome replication in parthenogenic benign ovarian teratomas, *Hum. Genet.*, 38, 163, 1977.
3. **Steel, C.M., Eccles, D.M., Gruber, L., Wallace, M., Lessels, A., Mossman, J.M., Gabra, H., Leonard, R.C.F., and Cohen, B.B.,** Allele losses on chromosome 17 in ovarian tumours, in *Ovarian Cancer 3*, Sharp, F., Mason, P., Blackett, T. and Berek, J., Eds., Chapman and Hall, London, 1995, 45.
4. **Gabra, H., Taylor, L., Cohen, B.B., Lessels, A., Eccles, D.M., Leonard, R.C.F., Smyth, J.F., and Steel, C.M.,** Chromosome 11 imbalance and clinicopathological correlates in ovarian tumours, *Br. J. Cancer*, 72, 367, 1995.
5. **Lawler, S.D. and Fisher, R.A.,** Genetic studies in hydatidiform mole with clinical correlations, *Placenta*, 8, 77, 1987.
6. **Murty, V.V., Dmitrovsky, E., Bosl, G.J., and Chaganti, R.S.,** Nonrandom chromosome abnormalities in testicular and ovarian germ cell tumour lines, *Cancer Genet. Cytogenet.*, 50, 67, 1990.
7. **Bosl, G.J., Ilson, D.H., Rodriguez, E., Motzer, R.J., Reuter, V.E., and Chaganti, R.S.,** Clinical relevance of the i (12p) marker chromosome in germ cell tumours, *J. Natl. Cancer Inst.*, 86, 349, 1994.
8. **Pejovic, T., Heim, S., Mandahl, N., Elmfors, B., Floderus, V.M., Furgyik, S., Helm, G., Willen, H., and Mitelman, F.,** Trisomy 12 is a consistent chromosomal aberration in benign ovarian tumours, *Genes Chrom. Cancer*, 2, 48, 1990.
9. **Mrozek, K., Nedoszytko, B., Babinska, M., Mrozek, E., Hrabowska, M., Emerich, J., and Limon, J.,** Trisomy of chromosome 12 in a case of thecoma of the ovary, *Gynecol. Oncol.*, 36, 413, 1989.
10. **Leung, W.-H., Schwartz, P.E., Ng, H.-T., and Yang-Feng, T.L.,** Trisomy 12 in benign fibroma and granulosa cell tumour of the ovary, *Gynecol. Oncol.*, 38, 28, 1990.
11. **Teyssier, J.-R., Adnet, J.-J., Pigeon, F., and Bajolle, F.,** Chromosomal changes in an ovarian granulosa cell tumour: similarity with carcinoma, *Cancer Genet. Cytogenet.*, 14, 147, 1985.
12. **Yang-Feng, T.L., Li, S., Han, H., and Schwartz, P.E.,** Frequent loss of heterozygosity on chromosomes Xp and 13q in human ovarian cancer, *Int. J. Cancer*, 52, 578, 1992.
13. **Sato, T., Saito, H., Morita, R., Koi, S., Lee, J.H., and Nakamura, Y.,** Allelotype of human ovarian cancer, *Cancer Res.*, 51, 5118, 1991.
14. **Iversen, O.E. and Skaarland, E.,** Ploidy assessment of benign and malignant ovarian tumours by flow cytometry: a clinicopathological study, *Cancer*, 60, 82, 1987.
15. **Zheng, J., Benedict, W.F., and Xu, H.-J.,** Genetic disparity between morphologically benign cysts contiguous to ovarian carcinomas and solitary cystadenomas, *J. Natl. Cancer Inst.*, 87, 1146, 1995.
16. **Russel, S.E.H., Hickey, G.I., Lowry, W.S., White, P., and Atkinson, R.J.,** Allele loss from chromosome 17 in ovarian cancer, *Oncogene*, 5, 1581, 1990.
17. **Eccles, D.M., Russel, S.E.H., Haites, N.E., Atkinson, R., Bell, D.W., Gruber, L., Hickey, I., Kelly, K., Kitchener, H., and Leonard, R.,** Early loss of heterozygosity on 17q in ovarian cancer, *Oncogene*, 7, 2069, 1992.
18. **Jacobs, I.J., Smith, S.A., Wiseman, R.W., Futreal, P.A., Harrington, T., Osborne, R.J., Leach, V., Molyneux, A., Berchuck, A., and Ponder, B.A.,** A deletion unit on chromosome 17q in epithelial ovarian tumours distal to the familial breast/ovarian cancer locus, *Cancer Res.*, 53, 1218, 1993.
19. **Orphanos, V., McGown, G., Hey, Y., Thorncroft, M., Santibanez-Koref, M., Russel, S.E., Hickey, I., Atkinson, R.J., and Boyle, J.M.,** Allelic imbalance of chromosome 6q in ovarian tumours, *Br. J. Cancer*, 71, 666, 1995.
20. **Scully, R.E., Bell, D.A., and Abu-Jawdeh, G.M.,** Update on early ovarian cancer and cancers developing in benign ovarian tumours, in *Ovarian Cancers 3*, Sharp, F., Mason, P., Blackett, T. and Berek, J., Eds., Chapman and Hall, London, 1995, 135.
21. **Hoskins, P.J.,** Ovarian tumours of low malignant potential: borderline epithelial ovarian carcinoma, in *Epithelial Cancer of the Ovary*, Lawton, F.G., Neijt, J.P. and Swenerton, K.D., Eds., BMJ, London, 1995, 112.
22. **van Dam, P.A.,** Ploidy in ovarian cancer and prognosis, in *The Biology of Ovarian Cancer*, Leake, R., Gore, M. and Ward, R.H., Eds., RCPG Press, London, 1995, 258.
23. **Kaern, J., Trope, C.G., Abeler, V., and Petterson, E.O.,** Cellular DNA content: the most important prognostic factor in patients with borderline tumours of the ovary. Can it prevent overtreatment? in *Ovarian Cancer 3*, Sharp, F., Mason, P., Blackett, T., and Berek, J., Eds., Chapman and Hall, London, 1995, 181.
24. **Foulkes, W.D., Black, D.M., Stamp, G.W.H., Solomon, E., and Trowsdale, J.,** Very frequent loss of heterozygosity throughout chromosome 17 in sporadic ovarian carcinoma, *Int. J. Cancer*, 54, 220, 1993.

25. **Gallion, H.H., Powell, D.E., Morrow, J.K., Pieretti, M., Case, E., Turker, M.S., De Priest, P.D., Hunter, J.E., and Van Nagell Jr., J.R.,** Molecular genetic changes in human epithelial ovarian malignancies, *Gynecol. Oncol.*, 47, 137, 1992.

26. **Shelling, A.N., Cooke, I.E., and Ganesan, T.S.,** The genetic analysis of ovarian cancer, *Br. J. Cancer*, 72, 521, 1995.

27. **Kaern, J., Wetteland, J., Trope, C.G., Farrants, G.W., Juhng, G.W., Pettersen, E.O., Reith, A., and Danielsen, H.E.,** Comparison between flow cytometry and image cytometry in ploidy distribution assessments in gynecologic cancer, *Cytometry*, 13, 314, 1992.

28. **Vindelov, L.L., Christensen, I.J., Jensen, G., and Nissen, N.I.,** Limits of detection of nuclear DNA abnormalities by flow cytometric DNA analysis. Results obtained by a set of methods for sample storage, staining and internal standardisation, *Cytometry*, 3, 332, 1983.

29. **Hedley, D.W., Clark, G.M., Cornelisse, C.J., Killander, D., Kute, T., and Merkel, D.,** DNA cytometry consensus conference. Consensus review of the clinical utility of DNA cytometry in carcinoma of the breast, *Cytometry*, 14, 482, 1993.

30. **van Dam, P.A., Vergote, I.B., Lowe, D.G., Watson, J.V., van Damme, P., van der Auwera, J.C., and Shepherd, J.H.,** Expression of c-erbB-2, c-myc and c-ras oncoproteins and insulin-like growth factor receptor 1, epidermal growth factor in ovarian carcinoma, *J. Clin. Pathol.*, 1994.

31. **Wake, N., Hreshchyshyn, M.M., Piver, S.M., Matsui, S., and Sandberg, A.A.,** Specific cytogenetic changes in ovarian cancer involving chromosomes 6 and 14, *Cancer Res.*, 40, 4512, 1980.

32. **Trent, J.M. and Salmon, S.E.,** Karyotypic analysis of human ovarian carcinoma cells cloned in short term agar culture, *Cancer Genet. Cytogenet.*, 3, 279, 1981.

33. **Whang-Peng, J., Knutsen, T., Douglass, E.C., Chu, E., Ozols, R.F., Hogan, W.M., and Young, R.C.,** Cytogenetic studies in ovarian cancer, *Cancer Genet. Cytogenet.*, 11, 91, 1984.

34. **Sheer, D., Sheppard, D.M., Gorman, P.A., Ward, B., Whelan, R.D., and Hill, B.T.,** Cytogenetic analysis of four human ovarian carcinoma cell lines, *Cancer Genet. Cytogenet.*, 26, 339, 1987.

35. **Panani, A. and Ferti-Passantanopoulou, A.,** Common marker chromosomes in ovarian cancer, *Cancer Genet. Cytogenet.*, 16, 65, 1985.

36. **Atkin, N.B. and Baker, M.C.,** Abnormal chromosomes, including small metacentrics in 14 ovarian cancers, *Cancer Genet. Cytogenet.*, 26, 355, 1987.

37. **Jenkyn, D.J. and McCartney, A.J.,** A chromosome study of three ovarian tumours, *Cancer Genet. Cytogenet.*, 26, 327, 1987.

38. **Tanaka, K., Boice, C.R., and Testa, J.R.,** Chromosome aberrations in nine patients with ovarian cancer, *Cancer Genet. Cytogenet.*, 43, 1, 1989.

39. **Pejovic, T., Heim, S., Mandahl, N., Baldetorp, B., Elmfors, B., Floderus, V.M., Furgyik, S., Helm, G., Himmelman, A., and Willen, H.,** Consistent occurrence of a 19p+ marker chromosome and loss of 11p material in ovarian seropapillary cystadenocarcinomas, *Genes Chrom. Cancer*, 1, 167, 1989.

40. **Pejovic, T., Heim, S., Mandahl, N., Baldetorp, B., Elmfors, B., Floderus, V.M., Furgyik, S., Helm, G., Himmelman, A., and Willen, H.,** Chromosome aberrations in 35 primary ovarian carcinomas, *Genes Chrom. Cancer*, 4, 58, 1992.

41. **Roberts, C.G. and Tattersall, M.H.N.,** Cytogenetic study of solid ovarian tumours, *Cancer Genet. Cytogenet.*, 48, 243, 1990.

42. **Bello, M.J. and Rey, J.A.,** Chromosome aberrations in metastatic ovarian cancer: relationships with abnormalities in primary tumours, *Int. J. Cancer*, 45, 50, 1990.

43. **Jenkins, R.B., Bartelt, D., and Stalboerger, P.,** Cytogenetic studies of epithelial ovarian carcinoma, Cytogenetic studies of epithelial ovarian carcinoma, *Cancer Genet. Cytogenet.*, 71, 76, 1993.

44. **Thompson, F.H., Emerson, J., Alberts, D., Liu, Y., Guan, X.Y., Burgess, A., Fox, S., Taetle, R., Weinstein, R., and Makar, R.,** Clonal chromosome abnormalities in 54 cases of ovarian carcinoma, *Cancer Genet. Cytogenet.*, 73, 33, 1994.

45. **Islam, M.Q., Kopf, I., Levan, A., Granberg, S., Friberg, L.G., and Levan, G.,** Cytogenetic findings in 111 ovarian cancer patients: therapy-related chromosome aberrations and heterochromatic variants, *Cancer Genet. Cytogenet.*, 65, 35, 1993.

46. **Persons, D.L., Hartmann, L.C., Herath, J.F., Borell, T.J., Cliby, W.A., Keeney, G.L., and Jenkins, R.B.,** Interphase molecular cytogenetic analysis of epithelial ovarian carcinomas, *Am. J. Pathol.*, 142, 733, 1993.

47. **Lastowska, M., Lillington, D.M., Shelling, A.N., Cooke, I., Gibbons, B., Young, B.D., and Ganasan, T.S.,** Fluorescence in situ hybridisation analysis using cosmid probes to define chromosome 6q abnormalities in ovarian carcinoma cell lines, *Cancer Genet. Cytogenet.*, 77, 99, 1994.

48. **Lasko, D., Cavanee, W., and Nordenskjold, M.,** Loss of constitutional heterozygosity in human cancer, *Annu. Rev. Genet.*, 25, 281, 1991.

49. **Weber, J.L. and May, P.E.,** Abundant class of human DNA polymorphisms which can be typed using the polymerase chain reaction, *Am. J. Hum. Genet.*, 44, 388, 1989.

50. **Foulkes, W.D. and Trowsdale, J.,** Isolating tumour suppressor genes relevant to ovarian carcinoma — the role of loss of heterozygosity, in *Ovarian Cancer 3*, Sharp, F., Mason, P., Blackett, T., and Berek, J., Eds., Chapman and Hall, London, 1995, 23.

51. **Steel, C.M.,** Molecular genetics of gynaecological cancer: an overview, in *The Biology of Gynaecological Cancer,* Leake, R., Gore, M., and Ward, R.H., Eds., RCOG Press, London, 1995, 147.

52. **Cao, Q., Cedrone, E., Barrett, C., and Wang, N.,** Suppression of in vitro growth of ovarian carcinoma cells by microcell-mediated chromosome 11 transfer, *Am. J. Hum. Genet.,* 53, 1517, 1993.

53. **Kim, T.M., Benedict, W.F., Xu, H.J., Hu, S.X., Gosewehr, J., Velicescu, M., Yin, E., Zheng, J., D'Ablaing, G., and Dubeau, L.,** Loss of heterozygosity on chromosome 13 is common only in the biologically more aggressive subtypes of ovarian epithelial tumours and is associated with normal retinoblastoma gene expression, *Cancer Res.,* 54, 605, 1994.

54. **Chenevix-Trench, G., Leary, J., Kerr, J., Michel, J., Kefford, R., Hurst, T., Parsons, P.G., Friedlander, M., and Khoo, S.K.,** Frequent loss of heterozygosity on chromosome 18 in ovarian adenocarcinomas which does not always include the DCC locus, *Oncogene,* 7, 1059, 1992.

55. **Chang, E.H., Gonda, M.A., Elis, R.W., Scolnick, E.M., and Lowry, D.R.,** Human genome contains four genes homologous to transforming genes of Harvey and Kirsten murine sarcoma viruses, *Proc. Natl. Acad. Sci. U.S.A.,* 79, 4848, 1982.

56. **Call, C.M., Glaser, T., Ito, C.Y., Buckler, A.J., Pelletier, J., Haber, D.A., Rose, E.A., Kral, A., Yeger, H., and Lewis, W.H.,** Isolation and characterisation of a zinc finger polypeptide gene at the human chromosome 11 Wilm's tumour locus, *Cell,* 60, 509, 1990.

57. **Ichikawa, T., Ichikawa, Y., Dong, J., Hawkins, A.L., Griffin, C.A., Isaacs, W.B., Oshimura, M., Barett, J.C., and Isaacs, J.T.,** Localisation of metastasis suppressor gene(s) for prostatic cancer to the short arm of chromosome 11, *Cancer Res.,* 52, 3486, 1992.

58. **Osborne, R.J. and Leech, V.,** Polymerase chain reaction allelotyping of human ovarian cancer, *Br. J. Cancer,* 69, 429, 1994.

59. **Eccles, D.M., Gruber, L., Stewart, M., Steel, C.M., and Leonard, R.C.F.,** Allele loss on chromosome 11p is associated with poor survival in ovarian cancer, *Dis. Markers,* 10, 95, 1992.

60. **Ehlen, T. and Dubeau, L.,** Loss of heterozygosity on chromosomal segments 3p, 6q and 11p in human ovarian cancer, *Oncogene,* 5, 219, 1990.

61. **Kiechleschwarz, M., Bauknecht, T., Wienker, T., Walz, L., and Pfleiderer, A.,** Loss of constitutional heterozygosity on chromosome 11 in human ovarian cancer — positive correlation with grade of differentiation, *Cancer,* 72, 2423, 1993.

62. **Lee, J.H., Kavanagh, J.J., Wharton, J.T., Wildrick, D.M., and Black, M.,** Allele loss at the c-Ha-ras locus in human ovarian cancer, *Cancer Res.,* 49, 1220, 1989.

63. **Vandamme, B., Lissens, W., Amfo, K., De Sutter, P., Bougain, C., Vamos, E., and De Greve, J.,** Deletion of chromosome 11p13-11p15.5 sequences in human ovarian cancer is a subclonal progression factor, *Cancer Res.,* 52, 6646, 1992.

64. **Viel, A., Giannini, F., Tumiotto, L., Sopracorderole, F., Visentin, M.C., and Boiocchi, M.,** Chromosomal location of 2 putative 11p oncosuppressor genes involved in human ovarian tumours, *Br. J. Cancer,* 66, 1030, 1992.

65. **Zheng, J.P., Robinson, W.R., Ehlen, T., Yu, M.C., and Dubeau, L.,** Distinction of low grade from high grade human ovarian carcinomas on the basis of losses of heterozygosity on chromosome 3, chromosome 6 and chromosome 11 and her-2/neu gene amplification, *Cancer Res.,* 51, 5118, 1991.

66. **Friedman, R.L., Velicescu, M., and Dubeau, L.,** Mutations in an alternatively spliced exon of H-ras are not associated with loss of heterozygosity on chromosome 11p15.5 in ovarian tumourigenesis, *Int. J. Oncol.,* 1, 215, 1993.

67. **Koi, M., Johnson, L.A., Kalikin, L.M., Little, P.F.R., Nakamura, Y., and Feinberg, A.P.,** tumour cell growth arrest caused by subchromosomal transferable fragments from chromosome 11, *Science,* 260, 361, 1993.

68. **Bruening, W., Gros, P., Sato, T., Stanimis, J., Nakamura, Y., Housman, D., and Pelletier, J.,** Analysis of the 11p13 Wilms tumour suppressor gene (wt1) in ovarian tumours, *Cancer Invest.,* 11, 393, 1993.

69. **Viel, A., Giannini, F., Capozzi, E., Canzonieri, V., Scarabelli, C., Gloghini, A., and Boiocchi, M.,** Molecular mechanisms possibly affecting wt1 function in human ovarian tumours, *Int. J. Cancer,* 57, 515, 1994.

70. **Ichikawa, T., Nihei, N., Suzuki, H., Oshimura, M., Emi, M., Nakamura, Y., Hayata, I., Isaacs, J.T., and Shimizaki, J.,** Suppression of metastasis of rat prostatic cancer by introducing human chromosome 8, *Cancer Res.,* 54, 2299, 1994.

71. **Foulkes, W.D., Campbell, I.G., Stamp, G.W.H., and Trowsdale, J.,** Loss of heterozygosity and amplification on chromosome 11q in human ovarian cancer, *Br. J. Cancer,* 67, 268, 1993.

72. **Cliby, W., Ritland, S., Hartmann, L., Dodson, M., Halling, K.C., Keeney, G., Podratz, K.C., and Jenkins, R.B.,** Human epithelial ovarian cancer allelotype, *Cancer Res.,* 53, 2393, 1993.

73. **Karlseder, J., Zeillinger, R., Schneeberger, C., Czerwenka, K., Speiser, P., Kubista, E., Birnbaum, D., Gadray, P., and Theillet, C.,** Patterns of DNA amplification at band q13 of chromosome 11 in human breast cancer, *Genes Chrom. Cancer,* 9, 42, 1994.

74. **Adelaide, J., Mattei, M.G., Marics, I., Raybaud, F., Planche, J., Lapeyriere, O., and Birbaum, D.,** Chromosomal localisation of the hst oncogene and its co-amplification with the Int-2 oncogene in human melanoma, *Oncogene,* 2, 413, 1988.

75. **Gabra, H., Langdon, S.P., Watson, J.E.V., Hawkins, R.A., Cohen, B.B., Taylor, L., Mackay, J., Steel, C.M., Leonard, R.C.F., and Smyth, J.F.,** Loss of heterozygosity at 11q22 correlates with low progesterone receptor content in epithelial ovarian cancer, *Clin. Cancer Res.,* 1, 945, 1995.

76. **Kieback, D.G., Tong, X.-W., Konig, R., Korner, W., Blankenburg, K., Agoulnik, I., Runnebaum, I.B., and Atkinson, E.N.,** A complex of mutations in the human progesterone receptor gene is associated with increased risk of non-familial breast and ovarian cancer but not of uterine cancer, *Proc. Am. Assoc. Cancer Res.*, 37, 250, 1996.

77. **Winqvist, R., Hampton, G.M., Mannermaa, G., Blanco, G., Alavaikko, M., and Kiviniemi, H.,** Loss of heterozygosity for chromosome 11 in primary human breast tumours is associated with poor survival after metastasis, *Cancer Res.*, 55, 2660, 1995.

78. **Negrini, M., Rasio, D., Hampton, G.M., Sabbioni, S., Rattan, S., Carter, S.L., Rosenberg, A.L., Schwartz, G.F., Shiloh, Y., and Cavanee, W.K.,** Definition and refinement of chromosome 11 regions of loss of heterozygosity in breast cancer: identification of a new region at 11q23.3, *Cancer Res.*, 55, 3003, 1995.

79. **Carter, S.L., Negrini, M., Baffa, R., Gillum, D.R., Rosenberg, A.L., Schwartz, G.F., and Croce, C.M.,** Loss of heterozygosity at 11q22-23 in breast cancer, *Cancer Res.*, 54, 6270, 1994.

80. **Hampton, G.M., Mannermaa, A., Winqvist, R., Alavaikko, M., Blanco, G., Tashinen, P.J., Kiviniemi, H., Newsham, I., Cavanee, W.K., and Evans, G.A.,** Loss of heterozygosity in sporadic human breast carcinoma: a common region between 11q22 and 11q23.3, *Cancer Res.*, 54, 4586, 1994.

81. **Savitsky, K., Bar-Shira, A., Gilad, S., Rotman, G., Ziv, Y., Vanagaite, L., Tagle, D.A., Smith, S., Vziel, T., and Sfez, S.,** A single ataxia telangiectasia gene with a product similar to PI-3 kinase, *Science*, 268, 1749, 1995.

82. **Eccles, D.M., Brett, L., Lessels, A., Gruber, L., Lane, D., Steel, C.M., and Leonard, R.C.,** Loss of function in the p53 tumour suppressor gene in advanced ovarian carcinoma, *Br. J. Cancer*, 65, 40, 1992.

83. **Kohler, M.F., Kerns, B.J.M., Humphrey, P.A., Marks, J.R., Bast, R.C., and Berchuck, A.,** Mutation and overexpression of p53 in early-stage epithelial ovarian-cancer, *Obstet. Gynecol.*, 81, 643, 1993.

84. **Frank, T.S., Bartos, R.E., Haefner, H.K., Roberts, J.A., Wilson, M.D., and Hubbell, G.P.,** Loss of heterozygosity and overexpression of the p53 gene in ovarian-carcinoma, *Mod. Pathol.*, 7, 3, 1994.

85. **Kohler, M.F., Marks, J.R., Wiseman, R.W., Jacobs, I.J., Davidoff, A.M., Clarke-Pearson, D.L., Soper, J.T., Bast R., Jr., and Berchuck, A.,** Spectrum of mutation and frequency of allelic deletion of the p53 gene in ovarian cancer, *J. Natl. Cancer Inst.*, 85, 1513, 1993.

86. **Niwa, K., Itoh, M., Murase, T., Morishita, S., Itoh, N., Mori, H., and Tamaya, T.,** Alteration of p53 gene in ovarian-carcinoma — clinicopathological correlation and prognostic significance, *Br. J. Cancer*, 70, 1191, 1994.

87. **Sheridan, E., Silcocks, P., Smith, J., Hancock, B.W., and Goyns, M.H.,** P53 mutation in a series of epithelial ovarian cancers from the UK, and its prognostic significance, *Eur. J. Cancer*, 11, 1701, 1994.

88. **Henriksen, R., Strang, P., Wilander, E., Backstrom, T., Tribukait, B., and Oberg, K.,** P53 expression in epithelial ovarian neoplasms — relationship to clinical and pathological parameters, ki-67 expression and flow-cytometry, *Gynecol. Oncol.*, 53, 301, 1994.

89. **Klemi, P.J., Takahashi, S., Joensuu, H., Kiilholma, P., Narimatsu, E., and Mori, M.,** Immunohistochemical detection of p53 protein in borderline and malignant serous ovarian-tumours, *Int. J. Gynecol. Pathol.*, 13, 228, 1994.

90. **Levesque, M.A., Katsaros, D., Yu, H., Zola, P., Sismondi, P., Giardina, G., and Diamandis, E.P.,** Mutant p53 protein overexpression is associated with poor outcome in patients with well or moderately differentiated ovarian-carcinoma, *Cancer*, 75, 1327, 1995.

91. **Vanderzee, A.G.J., Hollema, H., Suurmeijer, A.J.H., Krans, M., Sluiter, W.J., Willemse, P.H.B., Aalders, J.G., and Devries, E.G.E.,** Value of p-glycoprotein, glutathione-s-transferase-pi, c-erbb-2, and p53 as prognostic factors in ovarian carcinomas, *J. Clin. Oncol.*, 13, 70, 1995.

92. **Al-Azraqi, A., Chapman, C., Challen, C., Aswad, S., Sinha, D., Calvert, A.H., and Lunec, J.,** P53 mutations in primary human ovarian cancer as a determinant of resistance to carboplatin, *Proc. Am. Assoc. Cancer Res.*, 36, 228, 1995.

93. **Wang, S., Konig, R., Kohler, T., Stickeler, E., Tong, X.W., Kieback, D.G., Kreienberg, R., and Runnebaum, I.B.,** p53 polymorphism P53PIN3 in ovarian and breast cancer patients, *Proc. Am. Assoc. Cancer Res.*, 37, 262, 1996.

94. **Steel, C.M., Cohen, B.B., Lessels, A., Williams, A., and Gabra, H.,** Gene Alterations in Ovarian Cancer 4, Sharp, F., Blackett, T., Leake, R., and Beck, J., Eds., Chapman & Hall, London, 1996, 61.

95. **Takahashi, H., Behbakht, K., McGovern, P.E., Chiu, H.C., Couch, F.J., Weber, B.L., Friedman, L.S., King, M.C., Furusato, M., and Livolsi, V.A.,** Mutation analysis of BRCA1 gene in ovarian cancers, *Cancer Res.*, 55, 2998, 1995.

96. **Merajver, S.D., Pham, T.M., Carduff, R.F., Chen, M., Poy, E.L., Cooney, K.A., Weber, B.L., Collins, F.S., Johnston, C., and Frank, T.S.,** Somatic mutations in the BRCA1 gene in sporadic ovarian tumours, *Nat. Genet.*, 9, 439, 1995.

97. **Futreal, P.A., Liu, Q., Shattuck-Eidens, D., Cochran, C., Hershman, K., Tavtigian, S., Bennett, L.M., Haugen-Strano, A., Swensen, J., and Miki, Y.,** BRCA1 mutations in primary breast and ovarian carcinomas, *Science*, 266, 120, 1994.

98. **Mandai, M., Konishi, I., Komatsu, T., Mori, T., Arao, S., Nomura, H., Kanda, Y., Hiai, H., and Fukumoto, M.,** Mutation of the nm23 gene, loss of heterozygosity at the nm23 locus and K-ras mutation in ovarian carcinoma: correlation with tumour progression and nm23 gene expression, *Br. J. Cancer*, 72, 691, 1995.

99. **Brown, M.A., Nicolai, H., Xu, C.-H., Griffiths, B.L., Jones, K.A., Solomon, E., Hoskink, L., Trowsdale, J., Black, D.M., and McFarlane, R.,** Regulation of BRCA1, *Nature*, 372, 733, 1994.

100. **Eccles, D.M., Cranston, G., Steel, C.M., Nakamura, Y., and Leonard, R.C.F.,** Allele losses on chromosome 17 in human epithelial ovarian cancer, *Oncogene*, 5, 1599, 1990.

101. **Yangfeng, T.L., Han, H., Chen, K.C., Li, S.B., Claus, E.B., Carcngiu, M.L., Chambers, S.K., Chambers, J.T., and Schwartz, P.E.,** Allelic loss in ovarian cancer, *Int. J. Cancer*, 54, 546, 1993.
102. **Theile, M., Hartmann, S., Scherthan, H., Arnold, W., Deppert, W., Frege, R., Glaab, F., Haensch, W., and Scherneck, S.,** Suppression of tumourigenicity of breast cancer cells by transfer of human chromosome 17 does not require transferred BRCA1 and p53 genes, *Oncogene*, 10, 439, 1995.
103. **Rimessi, P., Gualandi, F., Morelli, C., Trabanelli, C., Wu, Q., Possati, L., Montesi, M., Barret, J.C., and Barbanti-Brodano, G.,** Transfer of human chromosome 3 to an ovarian carcinoma cell line identifies 3 regions on 3p involved in ovarian cancer, *Oncogene*, 9, 3467, 1994.
104. Sandhu et al.
105. **Horikawa, I., Choi, C., Cho, S., Koi, M., and Barrett, J.C.,** Positional and functional analyses of ovarian cancer suppressor genes on chromosome X, *Proc. Am. Assoc. Cancer Res.*, 37, 549, 1996.
106. **Bast, R.C., Boyer, C.M., Jacobs, I., Xu, F.J., Wu, S., Wiener, J., Kohler, M., and Berchuck, A.,** Cell growth regulation in epithelial ovarian cancer, *Cancer*, 71, 1597, 1993.
107. **Berchuck, A., Elbendary, A., Havrilesky, L., Rodriguez, G.C., and Bast, R.C.,** Pathogenesis of ovarian cancers, *J. Soc. Gyn. I*, 1, 181, 1994.
108. **Slamon, D.J., Godolphin, W., Jones, L.A., Holt, J.A., Wong, S.G., Keith, D.E., Levin, W.J., Stuart, S.G., Udove, J., and Ullrich, A.,** Studies of the Her-2/neu proto-oncogene in human breast and ovarian cancer, *Science*, 244, 707, 1989.
109. **Berchuck, A., Kamel, A., Whitaker, R., Kerns, B., Olt, G., Kinney, R., Soper, J.T., Dodge, R., Clarke-Pearson, D.L., and Marks, P.,** Overexpression of Her-2/neu is associated with poor survival in advanced epithelial ovarian cancer, *Cancer Res.*, 50, 4087, 1990.
110. **Aikens, R.S., Agard, D.A., and Sedat, J.W.,** Solid state imagers for microscopy. *Meth. Cell. Biol.*, 29A, 291, 1989.
111. **Deshane, J., Loechel, F., Conry, R.M., Siegal, G.P., King, C.R., and Curiel, D.T.,** Intracellular single chain antibody directed against erbB2 down-regulates cell surface erbB2 and exhibits a selective anti-proliferative effect in erbB2 over-expressing cancer cell lines, *Gene Ther.*, 1, 332, 1994.
112. **McGill, J.R., Beitzel, B.F., Nielsen, J.L., Walsh, J.T., Drabek, S.M., Meador, R.J., and Von Hoff, D.D.,** Double minutes are frequently found in ovarian carcinomas, *Cancer Genet. Cytogenet.*, 71, 125, 1993.
113. **McGill, J., Mattern, V., Beitzel, B., Leach, R., Morrow, M., Hodge, J., Johnson-Pais, T., von Hoff, D., and Eckhardt, S.,** Microdissected double minute DNA facilitates discovery of genomic alterations in human ovarian cancer, *Proc. Am. Assoc. Cancer Res.*, 37, 126, 1996.
114. **Yokota, J., Tsunetsugu-Yokota, Y., Battifora, H., Lefevre, C., and Cline, M.J.,** Alterations of myc, myb and ras-Ha proto-oncogenes in cancers are frequent and show clinical correlation, *Science*, 231, 261, 1986.
115. **Kikuchi, Y., Kita, T., Hirata, J., and Fukushima, M.,** Preclinical studies of antitumour prostaglandins by using human ovarian cancer cells, *Cancer Metastasis Rev.*, 13, 309, 1994.
116. **Filmus, J.E. and Buick, R.N.,** Stability of c-K-ras amplification during progression in a patient with adenocarcinoma of the ovary, *Cancer Res.*, 45, 4468, 1985.
117. **Chien, C.H. and Chow, S.N.,** Point mutation of the ras oncogene in human ovarian cancer, *DNA Cell. Biol.*, 12, 623, 1993.
118. **Park, J.S., Kim, H.K., Han, S.K., Lee, J.M., Namkoong, S.E., and Kim, S.J.,** Detection of c-K-ras point mutation in ovarian cancer, *Int. J. Gynecol. Cancer*, 5, 107, 1995.
119. **Rosen, A., Sevelda, P., Klein, M., Dobianer, K., Hruza, C., Czerwenka, K., Hanak, H., Vavra, N., Salzer, H., and Leodolters, S.,** First experience with FGF-3 (Int-2) amplification in women with epithelial ovarian cancer, *Br. J. Cancer*, 67, 1122, 1993.
120. **Jaakkola, S., Salmikangas, P., Nylund, S., Partanen, J., Amstronge, E., Pyrhonen, S., Lehtovista, P., and Nevanlinna, H.,** Amplification of FGFR4 gene in human breast and gynecological cancers, *Int. J. Cancer*, 54, 378, 1993.
121. **Baiocchi, G., Kavanagh, J.J., Talpaz, M., Wharton, J.T., Gutterman, J.U., and Kurzrock, R.,** Expression of the macrophage colony stimulating factor and its receptor in gynecologic malignancies, *Cancer*, 67, 990, 1991.
122. **Chambers, S.K., Wang, Y., Gertz, R.E., and Kacinski, B.M.,** Macrophage colony stimulating factor mediates invasion of ovarian cancer cells through urokinase, *Cancer Res.*, 55, 1578, 1995.
123. **Marx, J.,** How cells cycle towards cancer, *Science*, 263, 319, 1994.
124. **Clurman, B.E. and Roberts, J.M.,** Cell cycle and cancer, *J. Natl. Cancer Inst.*, 87, 149, 1995.
125. **Kamb, A.,** Cell cycle regulators and cancer, *Trends Genet,* 11, 136, 1995.
126. **Morgan, D.O.,** Principles of CDK regulation, *Nature*, 374, 131, 1995.
127. **Nigg, E.A.,** Cyclin-dependent protein kinases: key regulators of the eukaryotic cell cycle, *Bioessays*, 17, 471, 1995.
128. **Cobrinik, D., Dowdy, S.F., Hinds, P.W., Mittnacht, S., and Weinberg, R.A.,** The retinoblastoma protein and the regulation of cell cycling, *TIBS*, 17, 312, 1992.
129. **Weinberg, R.A.,** The retinoblastoma protein and cell cycle control, *Cell*, 81, 323, 1995.
130. **Karp, J.E. and Broder, S.,** Molecular foundations of cancer: new targets for intervention, *Nat. Med.*, 1, 309, 1995.
131. **Toyoshima, H. and Hunter, T.,** p27, a novel inhibitor of G1 cyclin-CDK protein kinase activity, is related to p21, *Cell*, 78, 67, 1994.
132. **Ahamed, S., Wang, T.H., Foster, J.S., Bukovsky, A., and Wimalasena, J.,** Estradiol activates cyclin/Cdks in ovarian cancer cells, *Proc. Am. Assoc. Cancer Res.*, 37, 218, 1996.

133. **Romagnola, D., Annab, L.A., Risinger, J.I., Terry, L.A., Lyon, T.T., and Barrett, J.C.,** Regulation of expression of BRCA-1 by estrogen in breast MCF-7 and ovarian BG-1 cancer cells, *Proc. Am. Assoc. Cancer Res.*, 37, 516, 1996.

134. **Nobori, T., Miura, K., Wu, D.J., Lois, A., Takabayashi, K., and Carson, D.A.,** Deletions of the cyclin-dependent kinase-4 inhibitor gene in multiple human cancers, *Nature*, 368, 753, 1994.

135. **Schultz, D.C., Vanderveer, L., Buetow, K.H., Boente, M.P., Ozols, R.F., Hamilton, T.C., and Godwin, A.K.,** Characterisation of chromosome 9 in human ovarian neoplasia identifies frequent genetic imbalance on 9q and rare alterations involving 9p, including cdkn2, *Cancer Res.*, 55, 2150, 1995.

136. **Zukerberg, L.R., Yang, W.I., and Gadd, M.,** Cyclin D1 (prad 1) protein expression in breast cancer — approximately 1/3 of infiltrating mammary carcinomas show overexpression of the cyclin D1 oncogene, *Mol. Pathol.*, 8, 560, 1995.

137. **Motokura, T. and Arnold, A.,** Cyclins and oncogenesis, *Biochim. Biophys. Acta*, 1155, 63, 1993.

138. **Leach, F.S., Tokino, T., Meltzer, P., Burrell, M., Oliner, J.D., Smith, S., Hill, D.E., Sidransky, D., Kinzler, K.W., and Vogelstein, B.,** p53 mutation and MDM2 amplification in human soft tissue sarcomas, *Cancer Res.*, 53, 2231, 1993.

139. **Foulkes, W.D., Stamp, G.W.H., Afzal, S., Lalani, N., McFarlane, C.P., Trowsdale, J., and Campbell, I.G.,** MDM2 overexpression is rare in ovarian carcinoma irrespective of p53 mutation status, *Br. J. Cancer*, 72, 883, 1995.

140. **Shiohara, M., El-Deiry, W.S., Wada, M., Nakamaki, T., Takeuchi, S., Yang, R., Chen, D.L., Vogelstein, B., and Koeffler, H.P.,** Absence of WAF_1 mutations in a variety of human malignancies, *Blood*, 84, 3781, 1994.

141. **Al-Azraqi, A., Challen, A., McKenna, D., George, M., Ghazal-Aswad, S., Angus, B., Sinha, D., Calvert, A.H., and Lunec, J.,** p21WAF1 expression and p53 status as determinants of response to carboplatin in primary human ovarian cancer, *Proc. Am. Assoc. Cancer Res.*, 37, 195, 1996.

142. **Olshausen, G.,** Die krankheiten den ovarien, Stuttgart, F Enke, 1877.

143. **Houlston, R.S., Collins, A., Slack, J., Campbell, S., Collins, W.P., Whitehead, M.I., and Morton, N.E.,** Genetic epidemiology of ovarian cancer: segregation analysis, *Ann. Hum. Genet.*, 55, 291, 1991.

144. **Easton, D.F., Ford, D., Matthews, F.E., and Peto, J.,** The genetic epidemiology of ovarian cancer, in *Ovarian Cancer 3*, Sharp, F., Mason, P., Blackett, T. and Berek, J., Eds., Chapman and Hall Medical, London, 1995, 3.

145. **Narod, S.A.,** Genetics of breast and ovarian cancer, *Br. Med. Bull.*, 50, 656, 1994.

146. **Steel, C.M.,** Cancer of the breast and female reproductive tract, in *Emery and Rimoin's Principles and Practice of Medical Genetics*, 3rd Edition, Churchill Livingstone, New York, in press, 1996.

147. **Schildkraut, J.M. and Thompson, W.D.,** Relationship of epithelial ovarian cancer to other malignancies within families, *Genet. Epidemiol.*, 5, 355, 1988.

148. **Bewtra, C., Watson, P., Conway, T., Read-Hippee, C., and Lynch, H.T.,** Hereditary ovarian cancer: a clinico-pathological study, *Int. J. Gynecol. Pathol.*, 11, 180, 1992.

149. **Piver, M.S., Baker, T.R., Jishi, M.F., Sandecki, A.M., Tsukada, Y., Natarajan, N., Mettlin, C.J., and Blake, C.A.,** Familial ovarian cancer. A report of 658 families from the Gilda Radner Familial Ovarian Cancer Registry 1981-1991, *Cancer*, 71, 582, 1993.

150. **Ford, D. and Easton, D.F.,** The genetics of breast and ovarian cancer, *Br. J. Cancer*, 72, 805, 1995.

151. **Wooster, R., Neuhausen, S.L., Mangion, J., Quirk, Y., Ford, D., Collins, N., Nguyen, K., Seal, S., Tran, T., and Averill, D.,** Localisation of a breast cancer gene, BRCA2, to chromosome 13q12-13, *Science*, 265, 2088, 1994.

152. **Prosser, J., Porter, D., Coles, C., Condie, A., Thompson, A.M., Chetty, U., Steel, C.M., and Evans, H.J.,** Constitutional p53 mutation in a non-Li Fraumeni family, *Br. J. Cancer*, 65, 527, 1992.

153. **Watson, P. and Lynch, H.T.,** Extracolonic cancer in hereditary nonpolyposis colorectal cancer, *Cancer*, 71, 677, 1993.

154. **Leach, F.S., Nicolaides, N.C., Papadopoulos, N., Liu, B., Jen, J., Parsons, R., Peltomaki, P., Sistonen, P., Aaltonen, L.A., and Nystrom-Lahti, M.,** Mutations of a mutS homolog in hereditary nonpolyposis colorectal cancer, *Cell*, 75, 1215, 1993.

155. **Bronner, C.E., Baker, S.M., Morrison, P.T., Warren, G., Smith, L.G., Lescoe, M.K., Kane, M., Earabino, C., Lipford, J., Lindblom, A., Tannergard, P., Bollag, R.J., Godwin, A.R., Ward, D.G., Nordenskjold, M., Fishel, R., Kolodner, R., and Liskay, R.M.,** Mutation in the DNA mismatch repair gene homologue hMLH1 is associated with hereditary non-polyposis colon cancer, *Nature*, 368, 258, 1994.

156. **Nicolaides, N.C., Papadopolous, N., Liu, B., Wei, Y.-F., Carter, K.C., Ruben, S.M., Rosen, C.A., Haseltine, W.A., Fleischmann, R.D., Fraser, C.M., Adams, M.D., Venter, J.C., Dunlop, M.G., Hamilton, S.R., Petersen, G.M., de la Chapelle, A., Kinzler, K.W., and Vogelstein, B.,** Mutation of two fPMSFf homologues in hereditary non-polyposis colon cancer, *Nature*, 371, 75, 1994.

157. **Papadopolous, N., Nicolaides, N.C., Wei, Y.-F., Ruben, S.M., Carter, K.C., Rosen, C.A., Haseltine, W.A., Fleischmann, R.D., Fraser, C.M., Adams, M.D., Venter, J.C., Hamilton, S.R., Petersen, G.M., Watson, P., Lynch, H.T., Peltomaki, P., Mecklin, J.-P., de la Chapelle, A., Kinzler, K.W., and Vogelstein, B.,** Mutation of a mutL homolog in hereditary colon cancer, *Science*, 263, 1625, 1994.

158. **Miki, Y., Swensen, J., Shattuck-Eidens, D., Futreal, P.A., Harshman, K., Tavtigian, S., Liu, Q., Cochran, C., Bennett, L.M., and Ding, W.,** A strong candidate for the breast and ovarian cancer susceptibility gene BRCA1, *Science*, 266, 66, 1994.

159. **Simard, J., Tonin, P., Durocher, F., Morgan, K., Rommens, J., Gingras, S., Samson, C., Leblanc, J.F., Belanger, C., and Dion, F.,** Common origins of BRCA1 mutations in Canadian breast and ovarian families, *Nat. Genet.*, 8, 392, 1994.

160. **Thompson, M.E., Jensen, R.A., Obermiller, P.S., Page, D.L., and Holt, J.T.,** Decreased expression of BRCA1 accelerates growth and is often present during breast cancer progression, *Nat. Genet.*, 9, 444, 1995.

161. **Cornelis, R.S., Neuhausen, S.L., and Johannson, O.,** High allele loss rates at 17q12-21 in breast and ovarian tumours from BRCA1-linked families, *Genes Chrom. Cancer*, 13, 203, 1995.

162. **Takahashi, H., Behbakht, K., McGovern, P.E., Chiu, H.C., Couch, F.J., Weber, B.L., Friedman, L.S., King, M.C., Furusato, M., and Livolsi, V.A.,** Mutation analysis of the BRCA1 gene in ovarian cancers, *Cancer Res.*, 55, 2998, 1995.

163. **Easton, D., Ford, D., and Peto, J.,** Inherited susceptibility to breast cancer, *Cancer Surv.*, 18, 95, 1993.

164. **Struewing, J.P., Abeliovich, D., Peretz, T., Avishai, N., Kaback, M.M., Collins, F.S., and Brody, L.C.,** The carrier frequency of the BRCA1 185delAG mutation is approximately 1 percent in Ashkenazi Jewish individuals, *Nat. Genet.*, 11, 198, 1995.

165. **Castilla, L.H., Couch, F.J., and Erdos, M.R.,** Mutations in the BRCA1 gene in families with early onset breast and ovarian cancer, *Nat. Genet.*, 8, 387, 1994.

166. **Shattuck-Eidens, D., McClure, M., and Simard, J.,** A collaborative survey of 80 mutations in the BRCA1 breast and ovarian cancer susceptibility gene, *JAMA*, 273, 535, 1995.

167. **Hogervorst, F.B.L., Cornelis, R.S., Bout, M., van Vliet, M., Oosterwijk, J.C., Olmer, R., Bakker, B., Klijn, J.G.M., Vasen, H.F.A., Meijers-Hejiboes, H., Menko, F.H., Cornelisse, C.J., den Dunner, J.T., Devilee, P., and van Ommen, G.J.B.,** Rapid detection of BRCA1 mutations by the protein truncation test, *Nat. Genet.*, 10, 208, 1995.

168. **Gayther, S., Warren, W., Mazoyer, S., Russel, P.A., Harrington, P.A., Chiano, M., Seal, S., Hamoudi, R., van Rensburg, E.J., Dunning, A.M., Love, R., Evans, G., Easton, D., Clayton, D., Stratton, M.R., and Ponder, B.A.J.,** Germline mutations in the BRCA1 gene in breast and ovarian cancer families provides evidence for a genotype-phenotype correlation, *Nat. Genet.*, 11, 428, 1995.

169. **Phelan, C.M., Rebbeck, T.R., Weber, B.L., Devilee, P., Ruttledge, M.H., Lynch, H.T., Lenoir, G.M., Stratton, M.R., Easton, D.F., Ponder, B.A., Cannon-Albright, L., Larsson, C., Goldgar, D.E., and Narod, S.A.,** Ovarian cancer risk in BRCA1 carriers is modified by the HRAS1 variable number of tandem repeat (VNTR) locus, *Nat. Genet.*, 12, 309, 1996.

170. **Marquis, S.T., Rajan, J.V., Wynshaw-Boris, A., Xu, J., Yin, G.Y., Abel, K.J., Weber, B.L., and Chodosh, L.A.,** The developmental pattern of Brca1 expression implies a role in differentiation of the breast and other tissues, *Nat. Genet.*, 11, 17, 1995.

171. **Holt, J.T., Thompson, M.E., Szabo, C., Robinson-Benion, C., Arteaga, C.L., King, M.C., and Jensen, R.A.,** Growth retardation and tumour inhibition by BRCA1, *Nat. Genet.*, 12, 298, 1996.

172. **Jensen, R.A., Thompson, M.E., Jetton, T.L., Szabo, C.I., van der Meer, R., Helou, B., Tronick, S.R., Page, D.L., King, M.C., and Holt, J.T.,** BRCA1 is secreted and exhibits properties of a granin, *Nat. Genet.,* 12, 223, 1996.

173. **Parazzini, F., Franchesi, S., La Vecchia, C., and Fasoli, M.,** The epidemiology of ovarian cancer, *Gynecol. Oncol.*, 43, 9, 1991.

174. **Mackey, S.E. and Creasman, W.T.,** Screening: potential benefits and adverse consequences, in *Epithelial Cancer of the Ovary*, Lawton, F.G., Neijt, J.P., and Swenerton, K.D., Eds., BMJ Publishing Group, London, 1995, 36.

175. **Bast, R.C., Xu, F., Woolas, R.P., Yu, Y., Conaway, M., O'Briant, K., Daly, L., Oram, D.H., Berchuck, A., Clarke-Pearson, D.L., Soper, J.T., Rodriguez, G., and Jacobs, I.J.,** Complementary and co-ordinate markers for detection of epithelial ovarian cancers, in *Ovarian Cancer 3*, Sharp, F., Mason, P., Blackett, T., and Berek, J., Eds., Chapman and Hall, London, 1995, 189.

176. **Tobacman, J.K., Tucker, M.A., Kase, R., Greene, M.H., Costa, J., and Fraumeni, J.F.,** Intra-abdominal carcinomatosis after prophylactic oophorectomy in ovarian cancer prone families, *Lancet,* 2, 795, 1982.

177. **Piver, M.S., Jishi, M.F., Tsukada, Y., and Nava, G.,** Primary peritoneal carcinoma after prophylactic oophorectomy in women with a family history of ovarian cancer, *Cancer*, 71, 2751, 1993.

178. **Wooster, R., Neuhausen, S.L., Mangion, J., Quirk, Y., Ford, D., Collins, N., Nguyen, K., Seal, S., Tran, T., and Averill, D.,** Localization of a breast cancer susceptibility gene, BRCA2, to chromosome 13q12-13, *Science*, 265, 2088, 1994.

179. **Wooster, R., Bignell, G., Lancaster, J., Swift, S., Seal, S., Mangion, J., Collins, N., Gregory, S., Gumbs, C., and Micklem, G.,** Identification of the breast cancer susceptibility gene BRCA2, *Nature*, 378, 789, 1995.

180. **Phelan, C.M., Lancaster, J.M., Tonin, P., Gumbs, C., Cochran, C., Carter, R., Ghadiran, P., Perret, C., and Moslehi, R.,** Muation analysis of the BRCA2 gene in 49 site-specific breast cancer families, *Nat. Genet.*, 13, 120, 1996.

181. **Couch, F., Farid, L.M., Deshano, M.L., Tavitigan, S.V., Calzone, K., and Campeau, L.,** BRCA2 germline mutations in male breast cancer cases and breast cancer families, *Nat. Genet.*, 13, 123, 1996.

182. **Neuhausen, S., Gilewski, T., Norton, L., Tran, T., McGuire, P., Swenson, J., Hampel, H., Borgen, P., and Brown, K.,** Recurrent BRCA2 6174 delT mutations in Ashkenazi Jewish women affected by breast cancer, *Nat. Genet.*, 13, 126, 1996.

Chapter 9

Ovarian Cancer: Biological Basis of Therapy

Jacqueline S. Beesley, Katherine V. Ferry, Stephen W. Johnson, Andrew K. Godwin, and Thomas C. Hamilton

CONTENTS

1. INTRODUCTION

Ovarian cancer is largely asymptomatic and difficult to detect at an early stage. Most patients with ovarian cancer therefore present with advanced stage disease which has spread throughout the peritoneal cavity. Treatment of this advanced stage disease consists of cytoreductive surgery followed by combination chemotherapy.[1] Since the serendipitous discovery of cisplatin and its application to the treatment of cancer,[2] many clinical trials have established platinum-based combination chemotherapy as superior to combinations that do not include platinum.[3] In fact, use of aggressive regimens including cisplatin or its analog carboplatin results in complete response rates of 60 to 80% in ovarian cancer patients. Unfortunately, 30 to 40% of these complete responders will subsequently relapse with tumors that are resistant not only to platinum compounds but also to a wide range of other chemotherapeutic agents including alkylating agents and natural products.[1,4] The possible exception appears to be in responses to the natural product taxol, which promotes and stabilizes the formation of microtubules leading to mitotic arrest and cell death.[3,5,6] These responses and recurrence data suggest that the manifestation of resistance to chemotherapy is often an integral part of the biology of ovarian cancer. Hence an understanding of mechanisms of resistance may provide better strategies for therapy.

Here, current knowledge with regard to anti-neoplastic drug resistance in ovarian cancer will be presented. The recent arrival of taxol in the arena of ovarian cancer treatment makes discussion of resistance to this drug premature. Limited information with regard to taxol resistance in other cancer cell models is however available, see for example work by Horwitz, Saijo, and Willingham.[7-9] As cisplatin and carboplatin are historically the most active drugs for the treatment of ovarian cancer, focus will be placed on how resistance to these important agents develops. This information also provides several insights into why resistance to platinum compounds is consequently accompanied by broad cross-resistance to other drugs. Furthermore, we will describe how available information on resistance mechanisms, although incomplete, is allowing the development of strategies to deal with the problem in clinical management.

In aqueous solution, cisplatin and carboplatin form an identical reactive species due to the displacement of their leaving groups (chloride for cisplatin and the cyclobutanedicarboxylato ring for carboplatin) (Figure 1).[10] This reactive species is a potent electrophile which reacts with sulfhydryl groups on proteins and nucleophilic groups on nucleic acids. The cytotoxicity of platinum drugs is generally accepted to occur through the formation of DNA adducts, which include DNA-protein crosslinks, DNA monoadducts and both intrastrand bidentate adducts and interstrand DNA crosslinks.[11,12] There continues to be controversy as to which of the DNA lesions is most critical to cytotoxicity, however it is apparent that any cellular process that alters the frequency, types or effectiveness of these lesions may have an impact on the sensitivity of a cell to platinum compounds. In general terms, there are several cellular processes which may intervene to limit platinum-induced lethality. First, intracellular concentrations of the drug adequate to produce death may not be achieved if uptake into the cell is decreased and/or efflux is

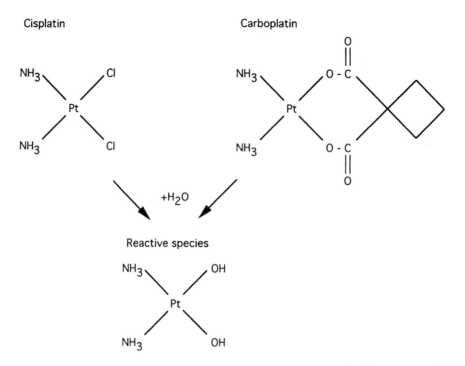

Figure 1 Structures of cisplatin and carboplatin and the identical reactive species these compounds form in aqueous solution.

increased. Second, the reactive species may be more readily inactivated if metabolic pathways capable of the detoxification of electrophiles become upregulated. Third, platinum-induced DNA damage may be more efficiently repaired or may become less effective as a trigger for cell death. Available information on the relevance of each of these theoretical physiological changes will be discussed below.

The investigation of platinum resistance mechanisms is a complex process. Although valuable information can be obtained by the analysis of biopsy material, much work has initially been directed toward the development of resistance models. In their simplest form, they include cell lines from ovarian cancer patients relatively sensitive to platinum compounds which are exposed to cisplatin *in vitro* to either select or induce resistant variants. Another approach has been to establish cell lines from patients before and after the development of resistance. In some cases these can be from the same patient.[13] Additionally, *in vivo* models of acquired resistance to platinum compounds have been developed using human tumor xenografts grown in athymic nude mice[14,15] or carcinogen-treated rat ovarian epithelial cells.[16] All of these systems have strengths and weaknesses. The final determinants of whether any of them are relevant will be to establish not only if a putative mechanism of drug resistance occurs in clinical samples but also whether it is possible to manipulate it and thus decrease resistance.

2. DRUG ACCUMULATION

Several workers have reported decreased platinum accumulation in cisplatin-resistant cell lines.[17-19] This could be due to decreased uptake and/or increased efflux of the drug. The relative significance of passive diffusion and carrier-mediated transport has not been established for platinum accumulation. Cell association of platinum occurs at a linear rate even at high drug concentrations but not against a concentration gradient suggesting carrier-independent transport occurs. However, there is evidence that some uptake may occur by an energy-dependent Na^+, K^+-ATPase.[20] Both decreases and increases in as yet poorly characterized membrane proteins have been described in platinum resistant cells.[21,22] Whether these proteins have any role in accumulation has not yet been determined. It should be noted however, that the 170 kDa transmembrane glycoprotein product of the MDR1 gene, P-glycoprotein, which is a drug efflux pump associated with resistance to natural products e.g., vinca alkaloids, anthracyclines, epipodophyllotoxins, and taxanes, does not correlate with either *in vitro* resistance or *in vivo* response to cisplatin.[23] Based on the substantial evidence that decreases in cell associated platinum commonly occur in resistant cells, new platinum analogs such as the Pt (IV)-mixed amine dicarboxylates have been developed. These drugs have been shown to have increased cytotoxicity in ovarian cancer cell lines due to their more lipophilic nature and increased accumulation rates.[24-26] Interestingly, in ovarian carcinoma cell line panels, these compounds appear selectively cytotoxic to cells that are resistant to cisplatin.[24] JM216, an effective analog in this class of compounds, is currently undergoing phase II clinical evaluation.[24,27,28]

3. CELLULAR INACTIVATION

The intracellular thiols metallothionein and glutathione have been proposed to play a role in the detoxification of platinum-based drugs and therefore their upregulation has been considered a likely possibility in mediation of resistance to these compounds.

Metallothionein is a small sulfhydryl protein which is involved in the detoxification of heavy metals such as zinc, copper and possibly platinum.[29] Its role in platinum resistance has not been clearly established. Although its levels are higher in tumor compared with normal cells, metallothionein gene expression does not appear to correlate with the sensitivity of ovarian cancer cells to cisplatin.[30,31] In contrast, the resistance of small cell lung carcinoma cell lines to cisplatin does appear to relate to levels of this protein[32] as does cisplatin resistance in squamous carcinoma cells,[33] suggesting that metallothionein may be involved in resistance to cisplatin but that its role has some tissue specificity.

Glutathione (GSH) is the most abundant non-protein thiol in the cell and can often be present in tumor cells at levels as high as 1 to 10 mM. It reacts with a variety of electrophiles usually making them less reactive and more water soluble.[34,35] GSH-mediated drug inactivation has been implicated in the acquisition of resistance to a number of drugs including cisplatin. In fact, GSH has been shown to quench the formation of cisplatin-DNA adducts in a cell-free system[36] and several groups have reported that cells resistant to cisplatin have small to large increases in their GSH levels compared to their sensitive counterparts.[37-39] These increased GSH levels are also associated with decreased sensitivity to alkylating agents, such as melphalan, and natural products, such as adriamycin.[40] These circumstantial data are suggestive that an increase in GSH is an important mechanism by which cells become resistant to a broad range of drugs. Additionally, there are data which show that GSH levels are greater in tumor biopsies from patients not responding to cisplatin than in biopsies obtained prior to treatment.[41]

Much has recently been learned about the regulation of GSH levels within the cell, in particular the enzymes involved and how these may be altered in drug-resistant cells (Figure 2).[42] In normal cellular systems GSH synthesis, from its constituent amino acids glutamate, cysteine and glycine, is achieved by the action of γ-glutamylcysteine synthetase (γGCS) followed by GSH synthetase. GSH synthetase is the product of a single gene located on chromosome 20 at q11.2.[43] γGCS, which catalyzes the rate-limiting step, consists of 2 subunits which are the products of different genes, on chromosomes 6 at p12[44] and 1 (personal communication, M.L. Freeman), respectively. The heavy subunit has catalytic activity and its expression in normal cellular systems is inhibited by GSH in a negative feedback loop. The light subunit has been suggested to have a regulatory function.[45] GSH is broken down in a pathway which includes action by γ-glutamyl transpeptidase (γGT), a cell surface enzyme which cleaves extracellular GSH allowing salvage of cysteine moieties across the membrane for intracellular GSH synthesis. Expression of γGT is very low in the normal ovary. Interestingly, it is detectable in a high percentage of ovarian tumors. In fact, assessment of γGT has been suggested as a possible prognostic indicator.[46]

Not surprisingly, high GSH levels in resistant cells are associated with increased expression of the heavy subunit of γGCS. Interestingly, this suggests that the normal feedback loop is altered or absent from resistant cells. In both melphalan-resistant human prostate cancer cells and cisplatin-resistant ovarian cancer cells this increase in the heavy subunit of γGCS has been shown to be due to an increase in its rate of transcription.[39,47-51] As yet, specific proof of how this increase in transcriptional activity is accomplished has not been presented. However, the promoter of the heavy subunit of GCS has recently been cloned. This region contains several motifs consistent with transcription factor binding elements and limited data suggest that enhanced binding to an AP-1 element within the promoter may be important in upregulation of the gene in resistant cells.[50,52] It should be noted that at the moment there is a lack of evidence that increases in GSH and γGCS expression directly cause resistance and are not just a consequence of drug-induced stress. The circumstantial evidence however is persuasive. For example, bacteria with a low capacity for GSH synthesis that are radiation sensitive are made much less so by increasing their γGCS and GSH synthetase levels.[53]

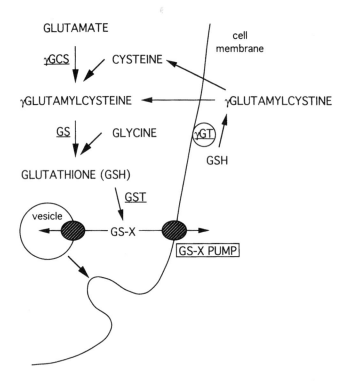

Figure 2 Pathways of glutathione (GSH) metabolism which may be altered in drug-resistant cells. γGCS, c-glutamylcysteine; GS, GSH synthetase; GST, glutathione S-transferase; GS-X, GSH-drug conjugate; γGT, c-glutamyl transpeptidase.

Cisplatin-resistant ovarian cancer cells expressing high levels of GSH have also been shown to have enhanced γGT messenger RNA expression and activity.[37,39] This suggests that this enzyme, as well as γGCS, may be involved in maintaining high GSH levels and consequently may effect resistance. It has been observed that increases in γGT and GSH can be as a consequence of activation of proto-oncogenes[54] and that resistance to ionizing radiation can result.[55]

The kinetics of the formation of GSH-platinum conjugates has been shown to be slow.[11] Hence, the question arises as to whether the efficiency of this process can be increased. The glutathione S-transferase (GST) family of enzymes are capable of facilitating the conjugation of some drugs to GSH. At present, five classes of GST isozymes have been described which have the distinctive ability of being able to recognize diverse chemical structures. GST levels are often raised in tumors, especially GSTπ, although other isozymes may be important in particular tissues.[56] While some chemotherapeutic drugs, such as the alkylating agent melphalan, have been clearly identified as GST substrates,[57] there is considerable question as to whether cisplatin is a substrate.[56] Furthermore, a consistent relationship between GST expression and platinum resistance in ovarian cancer has not been found. For example, a correlation between expression of GSTπ, the predominant isozyme present in the ovary, and lack of response to treatment has been observed in some clinical studies[58] but not others.[59] Similarly, no relationship was observed between sensitivity of ovarian cancer cell lines to platinum drugs and GST activity,[38] while in one convincing study Chinese hamster ovary (CHO) cells transfected with GSTπ, were shown to be threefold more resistant to cisplatin than the parental cells.[60] Nevertheless, recent work on the impact of elevated GSTs on carcinogen-DNA adduct formation clearly shows a functional role for these enzymes. In these studies, cells transfected with GSTπ, were found to have a lower adduct formation than control cells thus demonstrating at a molecular level the ability of GSTs to impact on the formation of potentially cytotoxic DNA damage.[61]

It should be noted that the conjugation of a drug to GSH can produce a molecule of greater toxicity than the original compound. Some data suggest that this is the case for platinum compounds.[62] Hence, the question arises as to how this toxic material is eliminated from the cell. The GS-X pump has been proposed to serve this purpose. Experimental data suggest that this pump is an ATP-dependent transport protein involved in the elimination from cells of organic anions, cysteinyl leukotrienes, glutathione disulphide and glutathione S-conjugates.[63] As such, it has a critical role in cellular detoxification since it prevents the accumulation of glutathione S-conjugates. Clearly, the GS-X pump has the potential to affect sensitivity to cisplatin as it has been demonstrated that cisplatin reacts with GSH both *in vitro* and in leukemia cells and that the resulting complex may be actively removed via the GS-X pump.[62] Furthermore, cisplatin-resistant leukemia cells functionally overexpress the GS-X pump.[64] Interestingly, the activity of the GS-X pump in these leukemia cells, whether cisplatin-sensitive or not, is down-regulated during differentiation and decreased proliferation. This suggests a potential link between drug detoxification and the cell cycle. Until recently little was known about the molecular structure of the GS-X pump; however, overexpression of the gene for the multidrug resistance-associated protein (MRP), which is a member of the ATP-binding cassette transmembrane transporter superfamily,[65] has been shown to result in an increase in ATP-dependent glutathione S-conjugate transport strongly suggesting that MRP is a member of the GS-X pump family.[66] As yet, no correlation between MRP levels and resistance to cisplatin has been described.[40] It should be noted that MRP is often abundantly expressed in ovarian cancer cells. Hence, it is interesting to speculate that the pump may not be rate limiting in ovarian cancer but rather that GSH amounts could be critical.

GSH may not only inactivate platinum in the cytoplasm but it may also influence platinum sensitivity at the level of the nucleus directly and indirectly. In fact, GSH levels are higher in the nucleus than the cytoplasm.[67] This relative nuclear abundance of GSH could favor its potential to quench the remaining active site of platinum in platinum-DNA monoadducts. These may be less cytotoxic and more readily repaired. Furthermore, GSH may have a role in repairing DNA damage induced by the electrophile. It may facilitate repair by acting as a cofactor and stabilizing DNA polymerase and also promoting the formation of deoxyribonucleotides. In support of this function it has been shown that reducing intracellular GSH levels can decrease repair activity.[68] The role of GSH as an antioxidant may also be important. For example, interactions between the high-mobility group (HMG) family of proteins and DNA, which are thought to help maintain chromatin structure,[69] may be altered by the redox balance of the cell.[70]

In conclusion, it is apparent that platinum-resistant cells are altered in a number of ways and that not all of the described changes in GSH-mediated detoxification and homeostasis are necessarily causally involved in resistance but may only be as a consequence of this phenotype. However, use of buthionine sulfoxime (BSO) has provided some of the strongest evidence of a role for GSH in platinum resistance. BSO, which irreversibly inhibits γGCS activity and thus depletes GSH levels,[71] can potentiate the effect of cisplatin in ovarian cancer cells resistant to cisplatin, melphalan, and adriamycin.[32,72-74] *In vivo* it can also reduce GSH levels by 90% in tumor cells as well as normal tissues and in combination with melphalan increase survival in a nude mouse model of ovarian cancer.[75] Such data formed the basis for phase I clinical trials of BSO and melphalan which have demonstrated that this combination of drugs can be safely used. However the data on depletion of GSH levels in these studies is contradictory. BSO given intravenously at 12-hour intervals at a dose of up to 13 g/m^2 consistently reduced GSH levels in both normal and tumor tissues by about 90% in one study[76] while in another, similar dosing schedules achieved no more than a 40% inhibition.[77] Subsequent studies will obviously be required to resolve this discrepancy.

4. DNA REPAIR

The available evidence indicates that the formation of lesions in DNA is the major means by which cisplatin and its analogs kill cells. If mechanisms which can prevent or limit the amount of DNA damage are inactive or insufficient to subvert platinum-induced lethality, a cell must repair or tolerate this DNA damage in order to survive. DNA repair and/or damage tolerance might therefore appropriately be considered a survival/resistance mechanism of last resort to the cancer cell. Such a mechanism could also explain the existence of cross-resistance to other chemotherapeutic agents and irradiation which exert their cytotoxic effects by damaging DNA.

The majority of adducts formed between cisplatin and DNA *in vitro* are intrastrand. This involves the coordination of the N7 atoms of two adjacent purines, either Pt[GG] or Pt[AG] (Figure 3).[10-12] The percentage of these adducts formed in biological systems (i.e., cultured cells and *in vivo*), however, may be lower. The major lesion Pt[GG] typically acounts for 60 to 70% of the platinum-DNA adducts formed *in vitro*, whereas in DNA isolated from treated cells only about 35% of lesions are of this type. It is hypothesized that this low value is a reflection of a slower rate of formation of intrastrand bidentate

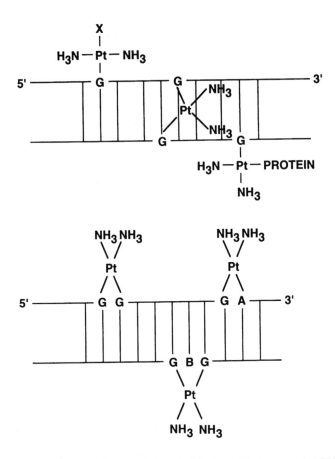

Figure 3 Schematic diagram of cisplatin binding to double-stranded DNA.

adducts which originate from platinum-DNA monoadducts. These monoadducts account for approximately 39% of the total platinum-DNA lesions in cultured cells.[78] Interstrand Pt[GG] crosslinks have been estimated to constitute approximately 1 to 5% of the total genomic platinum-DNA lesions.[11,79]

Removal of platinum-DNA intrastrand lesions and mono-adducts is believed to occur by the process of nucleotide excision repair. Interstrand crosslink repair is also known to involve the excision process, however it is believed that a recombination event must occur in order for these lesions to be fully removed.[80] Nucleotide excision repair in bacteria is well understood and requires six proteins (UvrA, UvrB, UvrC, UvrD, Pol1, and Lig).[81] Although rapidly increasing, our knowledge of nucleotide excision repair in eurkyotic cells is still not complete. Briefly, during eukaryotic nucleotide excision repair (Figure 4), as presented in more detail below, a multiprotein system locates a lesion in DNA and enzymatically cleaves sites in the damaged strand flanking the adduct. The damaged, excised portion of DNA is then displaced along with the incision proteins, a new strand is synthesized by DNA polymerases and repair is completed by a DNA ligase.[82-84]

Identification and elucidation of the function of the many proteins that play a role in eukaryotic nucleotide excision repair has been facilitated by the study of mammalian cells which are defective in various aspects of DNA repair. Nucleotide excision repair defective mammalian cell lines have been established from xeroderma pigmentosum (XP) patients. These individuals are hypersensitive to sunlight and predisposed to cancer due to various defects in DNA repair. Seven XP complementation groups named XP-A through XP-G have been identified and most of the genes responsible for the individual defects have been cloned. Repair-deficient, UV-sensitive mutant cells have also been isolated from rodents and have been assigned to 11 complementation groups. Human genes that correct the repair defect in these rodent mutant cells are denoted ERCC (excision repair cross-complementing) 1 to 11. Additionally, cells isolated from patients with the hereditary disease Cockayne's syndrome (CS), which is characterized by sunlight sensitivity, neurological dysfunction and severe developmental abnormalities, comprise two repair-deficient complementation groups denoted CS-A and CS-B.[83] Some of the XP genes as well as

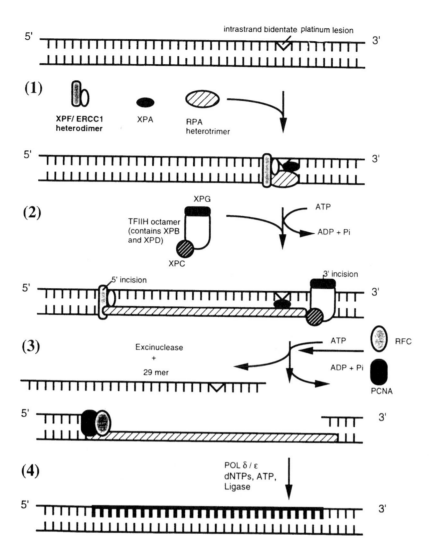

Figure 4 Schematic diagram of human nucleotide excision repair. The sequence of events is as follows (see text for details) : (1) Recognition of platinum lesion and binding of XPA, RPA and XPF/ERCC1 to the lesion site; (2) Unwinding of the DNA helix by TFIIH and formation of incisions by XPF/ERCC1 and XPG; (3) Binding of PCNA and RFC and displacement of the excinuclease proteins and the excised oligomer; (4) Gap-filling DNA repair synthesis by a DNA polymerase and ligation. (Adapted from Sancar, A., *Science*, 266, 1955, 1994.)

the cloned CSB gene have been found to be identical to ERCC genes (Table 1). Of interest, the extract from cells of one complementation group can correct the deficiency of another, allowing repair to occur as demonstrated in a cell-free assay system. This analysis involves incubating damaged, closed-circular DNA with cell extracts in a reaction mixture containing radiolabelled deoxynucleoside triphosphates. This assay has assisted in the purification of the defective proteins and the study of their function. [84-87]

Knowledge of the genes responsible for the repair defects described above has allowed a hypothetical view of the normal process of nucleotide excision repair in mammalian cells to be constructed (Figure 4). This information suggests the range of changes possible that could yield DNA repair-mediated cisplatin resistance and also provides ideas as to possible cytotoxic targets. Although the order of their association is not clear, XPA, the ERCC1-XPF heterodimer, the heterotrimeric replication protein RPA and the transcription factor TFIIH are all recruited to the lesion site. It is thought that the initial recognition of the DNA lesion involves XPA and possibly XPE since these proteins bind damaged DNA. It should be noted that, although poorly understood, the damage recognition process is believed to be linked to a signaling pathway which can lead to growth arrest and, in some cases, apoptosis.[88,89] TFIIH, which

Table 1 Properties of Proteins Involved in Nucleotide Excision Repair. ERCC Rodent Genes which Are Homologous to Human Genes Are Shown in Parentheses

Protein	Properties	Ref.
XPA	Affinity for damaged DNA, single-stranded DNA binding protein	111, 112
XPB (ERCC3)	DNA helicase, component of TFIIH transcription factor	113, 114
XPC	Affinity for single-stranded DNA	115
XPD (ERCC2)	DNA helicase (probable), component of TFIIH transcription factor	113, 116
XPE	Affinity for damaged DNA	117
XPF (ERCC4)	Component of single-stranded DNA endonuclease	87, 118, 119
ERCC1	Component of single-stranded DNA endonuclease	87, 119
XPG (ERCC5)	Single-stranded DNA endonuclease, 3′incision maker	120, 121
CSB (ERCC6)	DNA helicase (probable), functions primarily in repair of active genes, nonessential function	122
PCNA	Component of DNA polymerase holoenzyme	90
RPA	Single-stranded DNA binding protein, needed to form stable incisions	123

consists of at least eight polypeptides including the DNA helicases XPB and XPD, is believed to facilitate the repair process by unwinding DNA. This allows the endonucleases XPF and XPG to make 5′ and 3′ incisions, respectively, which encompass the lesion. XPG is loosely associated with TFIIH as is the single stranded DNA binding protein XPC. Following the incision events, a subset of the proteins is displaced along with the excised oligomer. This displacement event is facilitated by proliferating cell nuclear antigen (PCNA) and possibly replication factor C (RFC). These proteins are also required for the final stage of nucleotide excision repair during which a DNA polymerase fills in the gap using the unmodified strand as a template and DNA ligase completes the process (Table 1).[82] Since they are PCNA-dependent, δ and ε are therefore considered the most likely DNA polymerases to function in this process.[86,90] Interestingly, the p53-regulated DNA-damaged induced protein Gadd45 has recently been shown to bind PCNA and stimulate nucleotide excision repair in a cell-free system,[91] suggesting that it may also be a member of the family of eukaryotic nucleotide excision repair proteins.

In addition to studying cells defective in repair another approach to examine this process has been to identify proteins which bind to various types of DNA lesion as measured by a damaged DNA affinity assay. Among the proteins identified by this method are HMG1 and HMG2 which bind to cisplatin-damaged DNA.[69] These proteins are members of a family of highly conserved, low molecular weight eukaryotic proteins believed to play a role in maintaining chromatin structure.[92] Recently, binding of the HMG proteins has been shown to inhibit *in vitro* repair.[93] It is hypothesized that shielding of adducts by the HMG proteins in tumor cells could modulate the repair of cisplatin-induced lesions or possibly alter a signaling mechanism leading to programmed cell death and, therefore, affect responsiveness to chemotherapy. It is unclear how this postulate relates to normal cellular response vs. development of resistance. It is important to note that nucleotide excision repair can occur preferentially in actively transcribed genes relative to the overall genome.[80] This intragenomic heterogeneity in the DNA repair process has been demonstrated with various DNA damaging agents (e.g., UV-induced pyrimidine dimers, nitrogen mustard, methylnitrosourea, and psoralens).[94] Gene products which have been implicated in this process thus far include TFIIH, XPB, XPD and CSB.[95] With regard to platinum-induced DNA damage, cisplatin intrastrand adducts are preferentially repaired in active genes versus inactive genes and the overall genome in hamster cells[96] and in active genes versus the overall genome in human ovarian cancer cells.[97] Conversely, the removal of cisplatin interstrand crosslinks from active, inactive and non-coding genomic regions occurs at a similar rate in hamster and human ovarian cell lines.[94,96,98]

There is a substantial literature associating increased DNA repair with decreased sensitivity to platinum compounds. Resistance to cisplatin in human ovarian cancer cell lines has been found to be associated with increased DNA repair by a variety of methods. Evidence for increased repair of platinum damage in resistant ovarian cancer cells was first demonstrated by the measurement of unscheduled DNA synthesis[99,100] and subsequently by reactivation of cisplatin-damaged plasmid DNA[101,102] and atomic absorption spectrophotometry.[19,98,103] Although knowledge is limited as to which components are critical in producing this enhanced repair, altered expression of some DNA repair genes in association with resistance has been reported. For example, repair-deficient CHO cells transfected with ERCC1 lose their hypersensitivity to cisplatin.[104] Furthermore, messenger RNA levels for ERCC1 and XPA have been shown to be higher in malignant tissue from ovarian cancer patients resistant to platinum-based therapy compared to those responsive to treatment.[105]

The finding that DNA repair is likely to be an important mechanism of platinum resistance has resulted in the investigation of various strategies to inhibit the process. Aphidicolin, an inhibitor of the DNA polymerases α, δ, and ε, has been shown to inhibit DNA repair and to potentiate platinum cytotoxicity in cisplatin-resistant human ovarian cell lines.[99,100] Furthermore, aphidicolin glycinate has been shown to enhance the efficacy of cisplatin against both murine tumor models and a human ovarian cancer xenograft from a cisplatin refractory patient.[106] The cytidine nucleoside analog gemcitabine inhibits the activity of ribonucleotide reductase and inhibits DNA synthesis in its triphosphate form by incorporation into DNA. As such gemcitabine has been shown to inhibit DNA repair of cisplatin-damaged DNA in an *in vitro* repair assay and to synergize with cisplatin in causing reduced clonogenic survival in cisplatin-resistant colon tumor cells. The information on interactions of gemcitabine with cisplatin has served as a basis for the initiation of clinical trials with this drug combination.[107]

One possible mechanism of resistance that has received relatively little attention involves enhanced tolerance of DNA damage. It has been shown that the level of platinum-DNA adducts required to kill 50% of ovarian cancer cells can be significantly higher in cisplatin-resistant cells compared to their sensitive counterparts.[98] This tolerance of DNA damage may result from an increased ability to replicate DNA past the site of a lesion. In fact, replicative bypass has been shown to be enhanced in platinum-resistant ovarian cancer cell lines.[108] A limitation to the concept of replicative bypass as a putative resistance mechanism occurs when the DNA synthesis machinery encounters an interstrand crosslink. This lesion must, at a minimum, be partially repaired in order for DNA synthesis to continue. This supports the conclusions of some investigators who believe that interstrand crosslinks are the most cytotoxic cisplatin lesions and suggests the importance of their removal as part of DNA repair-mediated resistance. In fact, cisplatin-resistant cell lines have been shown to remove cisplatin interstrand crosslinks more efficiently than their cisplatin-sensitive counterparts.[97,98]

Damage tolerance in resistant cells may also be a consequence of the disruption of a pathway linking DNA damage recognition to induction of programmed cell death or apoptosis. In CHO cells, cisplatin has been shown to induce growth arrest in the G_2 phase of the cell cycle.[89] It is thought that this block in the cycle may allow a cell more time to assess its level of DNA damage and repair platinum-induced lesions before proceeding to mitosis. If the amount of DNA damage at this point is too great, apoptosis may be triggered. Since resistant cells can tolerate more damage to their DNA, it is conceivable that signals leading to apoptosis have been changed in these cells. A possible candidate for alteration of this pathway in drug resistance is Bcl-2. This oncoprotein has been shown to block apoptosis in a number of cell types. In fact, the expression of Bcl-2 can protect neuroblastoma and leukemia cell lines from apoptosis induced by cisplatin.[109,110] Hence, it is apparent that various forms of damage tolerance, as well as DNA repair, may contribute to decreased sensitivity to cisplatin. As our understanding of DNA repair and apoptosis increases this may ultimately lead to a greater insight into how these processes may be perturbed to clinical advantage in drug-resistant cells.

5. CONCLUSION

In conclusion, we have presented evidence supporting the existence of several mechanisms by which sensitivity to platinum compounds may be decreased in ovarian cancer. This has included where possible a description of how this information is being used in an attempt to alter clinical outcome. The changes so far detected which may confer resistance include: decreased drug accumulation, upregulation of detoxification pathways and increased repair of DNA damage and/or its tolerance. Clearly development of resistance to platinum is a complex process. It may involve some or all of these mechanisms, possibly in combination, and the extent to which each contributes in clinical samples has not as yet been established. Strategies to establish the role of specific genes in clinical resistance are therefore warranted. Furthermore, it is necessary to consider whether fundamental changes in transcriptional regulation in resistant cells are at the root of the wide variety of genes whose expression is altered.[50] If so, the transcription factors responsible may prove to be the best therapeutic targets.

REFERENCES

1. **Ozols, R.,** Ovarian cancer, Part II: Treatment, *Current Problems in Cancer,* 16, 63, 1992.
2. **Rosenberg, B., van Camp, L., and Krigas, T.,** Inhibition of cell division in *Escherichia coli* by electrolysis products from a platinum electrode, *Nature,* 205, 698, 1965.
3. **Ozols, R. and Young, R.,** Chemotherapy of ovarian cancer, *Semin. Oncol.,* 18, 222, 1991.

4. **Ozols, R. and Young, R.,** Chemotherapy of ovarian cancer, *Semin. Oncol.*, 11, 251, 1984.
5. **Kohn, E., Sarosy, G., Bicher, A., Link, C., Christian, M., Steinberg, S., Rothenberg, M., Orvis Adamo, D., Davis, P., Ognibene, F., Cunnion, R., and Reed, E.,** Dose-intense taxol: high response rate in patients with platinum-resistant recurrent ovarian cancer, *J. Natl. Cancer Inst.*, 86, 18, 1994.
6. **Runowicz, C., Wiernik, P., Einzig, A., Goldberg, G., and Horwitz, S.,** Taxol in ovarian cancer, *Cancer,* 71, 1591, 1993.
7. **Ohta, S., Nishio, K., Kubota, N., Ohmori, T., Funayama, Y., Ohira, T., Nakajima, H., Adachi, M., and Saijo, N.,** Characterization of a taxol-resistant human small cell lung cancer cell line, *Jpn. J. Cancer Res.*, 85, 290, 1994.
8. **Haber, M., Burkhart, C., Regl, D., Madafiglio, J., Norris, M., and Horwitz, S.,** Taxol resistance in murine J774.2 cells is associated with altered expression of specific b-tubulin isotypes, *Proc. Am. Assoc. Cancer Res.*, 36, 318, 1995.
9. **Bhalla, K., Huang, Y., Tang, C., Self, S., Ray, S., Mahoney, M., Ponnathpur, V., Tourkina, E., Ibrado, A., Bullock, G., and Willingham, M.,** Characterization of a human myeloid cell line highly resistant to taxol, *Leukemia*, 8, 465, 1994.
10. **Lippard, S.,** Chemistry and molecular biology of platinum anticancer drugs, *Pure Appl. Chem.*, 59, 731, 1987.
11. **Eastman, A.,** The formation, isolation and characterization of DNA adducts produced by anticancer platinum complexes, *Pharmacol. Ther.*, 34, 155, 1987.
12. **Reedijk, J.,** The mechanism of action of platinum anti-tumor drugs, *Pure Appl. Chem.*, 59, 181, 1987.
13. **Langdon, S., Lawrie, S., Hay, F., Hawkes, M., McDonald, A., Hayward, I., Schol, D., Hilgers, J., Leonard, R., and Smyth, J.,** Characterization and properties of nine human ovarian adenocarcinoma cell lines, *Cancer Res.*, 48, 6166, 1988.
14. **Hamilton, T., Young, R., Louie, K., Behrens, B., McKoy, W., Grotzinger, K., and Ozols, R.,** Characterization of a xenograft model of human ovarian cancer which produces ascites and intraabdominal carcinomatosis, *Cancer Res.*, 44, 5286, 1984.
15. **Jones, M., Siracky, J., Kelland, L., and Harrap, K.,** Acquisition of platinum drug resistance and platinum cross resistance patterns in a panel of human ovarian carcinoma xenografts, *Br. J. Cancer*, 67, 24, 1992.
16. **Nakata, T., Suzuki, K., Fujii, J., Ishikawa, M., Tatsumi, H., Sugiyama, T., Nishida, T., Shimizu, T., Yakushiji, M., and Taniguchi, N.,** High expression of manganese superoxide dismutase in 7,12-dimethylbenz[a]anthracene-induced ovarian cancer and increased serum levels in tumor-bearing rats, *Carcinogenesis*, 13, 1941, 1992.
17. **Johnson, S., Perez, R., Godwin, A., Yeung, A., Handel, L., Ozols, R., and Hamilton, T.,** Role of platinum-DNA adduct formation and removal in cisplatin resistance in human ovarian cancer cell lines, *Biochem. Pharmacol.*, 47, 689, 1994.
18. **Andrews, P., Velury, S., Mann, S., and Howell, S.,** cis-Diamminedichloroplatinum (II) accumulation in sensitive and resistant human ovarian carcinoma cells, *Cancer Res.*, 48, 68, 1988.
19. **Parker, R., Eastman, A., Bostick-Bruton, F., and Reed, E.,** Acquired cisplatin resistance in human ovarian cancer cells is associated with enhanced repair of cisplatin-DNA lesions and reduced drug accumulation, *J. Clin. Invest.*, 87, 772, 1991.
20. **Andrews, P., Mann, S., Huynh, H., and Albright, K.,** Role of the Na+,K+-adenosine triphosphate in the accumulation of cis-diamminedichloroplatinum (II) in human ovarian carcinoma cells, *Cancer Res.*, 51, 3677, 1991.
21. **Bernal, S., Speak, J., Boeheim, K., Dreyfuss, A., Teicher, B., Rosowsky, A., Tsao, S.-W., and Wong, Y.-C.,** Reduced membrane protein associated with resistance if human squamous carcinoma cells to methotrexate and cis-platinum, *Mol. Cell. Biochem.*, 95, 61, 1990.
22. **Kawai, K., Kamatani, N., Georges, E., and Ling, V.,** Identification of a membrane glycoprotein overexpressed in murine lymphoma sublines resistant to cis-diamminedichloroplatinum (II), *J. Biol. Chem.*, 265, 13137, 1990.
23. **Veneroni, S., Zaffaroni, N., Daidone, M., Benini, B., Villa, R., and Silvestrini, R.,** Expression of P-glycoprotein and in vitro or in vivo resistance to doxorubicin and cisplatin in breast and ovarian cancers, *Eur. J. Cancer*, 30A, 1002, 1994.
24. **Harrap, K.,** Initiatives with platinum- and quinazoline-based antitumor molecules — Fourteenth Bruce F. Cain Memorial Award Lecture, *Cancer Res.*, 55, 2761, 1995.
25. **Loh, S., Mistry, P., Kelland, L., Abel, G., and Harrap, K.,** Reduced drug accumulation as a major mechanism of acquired resistance to cisplatin in a human ovarian cancer cell line: circumvention studies using novel platinum (II) and (IV) ammine/amine complexes, *Br. J. Cancer*, 66, 1109, 1992.
26. **Mistry, P., Kelland, L., Loh, S., Abel, G., Murrer, B., and Harrap, K.,** Comparison of cellular accumulation and cytotoxicity of cisplatin with that of tetraplatin and amminedibutyratodichloro(cyclohexylamine)platinum (IV) (JM221) in human ovarian carcinoma cell lines, *Cancer Res.*, 52, 6188, 1992.
27. **Harrap, K.,** Initiatives with platinum- and quinazoline-based antitumour molecules, *Proc. Am. Assoc. Cancer Res.*, 36, 648, 1995.
28. **Schilder, R., LaCreta, F., Perez, R., Johnson, S., Brennan, J., Rogatko, A., Nash, S., McAleer, C., Hamilton, T., Roby, D., Young, R., Ozols, R., and O'Dwyer, P.,** Phase I and pharmacokinetic study of ormaplatin (tetraplatin, NSC 363812) administered on a day 1 and day 8 schedule, *Cancer Res.*, 54, 709, 1994.
29. **Hamer, D.,** Metallothionein, *Annu. Rev. Biochem.*, 55, 913, 1986.
30. **Murphy, D., McGown, A., Crowther, D., Mander, A., and Fox, B.,** Metallothionein levels in ovarian tumours before and after chemotherapy, *Br. J. Cancer*, 63, 711, 1991.

31. **Schilder, R., Hall, L., Monks, A., Handel, L., Fornace, A., Ozols, R., Fojo, A., and Hamilton, T.,** Metallothionein gene expression and resistance to cisplatin in human ovarian cancer, *Int. J. Cancer*, 45, 416, 1990.

32. **Kasahara, K., Fujiwara, Y., Nishio, K., Ohmori, T., Sugimoto, Y., Komiya, K., Matsuda, T., and Saijo, N.,** Metallothionein content correlates with the sensitivity of human small cell lung cancer cell lines to cisplatin, *Cancer Res.*, 51, 3237, 1991.

33. **Kelley, S., Basu, A., Teicher, B., Hacker, M., Hamer, D., and Lazo, J.,** Overexpression of metallothionein confers resistance to anticancer drugs, *Science*, 241, 1813, 1988.

34. **Meister, A. and Anderson, M.,** Glutathione, *Annu. Rev. Biochem.*, 52, 711, 1983.

35. **Meister, A.,** Metabolism and function of glutathione, in *Glutathione: Chemical, Biochemical and Medical Aspects — Part A*, Dolphin, D., Poulson, R. and Avramovic, O., Eds., Wiley and Sons, 367, 1989.

36. **Eastman, A.,** Crosslinking of glutathione to DNA by cancer chemotherapeutic platinum coordination complexes, *Chem. Biol. Interact.*, 61, 241, 1987.

37. **Lewis, A., Hayes, J., and Wolf, C.,** Glutathione and glutathione-dependent enzymes in ovarian adenocarcinoma cell lines derived from a patient before and after the onset of drug resistance: intrinsic differences and cell cycle effects, *Carcinogenesis*, 9, 1283, 1988.

38. **Mistry, P., Kelland, L., Abel, G., Sidhar, S., and Harrap, K.,** The relationships between glutathione, glutathione-S-transferase and cytotoxicity of platinum drugs and melphalan in eight human ovarian carcinoma cell lines, *Br. J. Cancer*, 64, 215, 1991.

39. **Godwin, A., Meister, A., O'Dwyer, P., Huang, C., Hamilton, T., and Anderson, M.,** High resistance to cisplatin in human ovarian cancer cell lines is associated with marked increase of glutathione synthesis, *Proc. Natl. Acad. Sci. U.S.A.*, 89, 3070, 1992.

40. **Hamaguchi, K., Godwin, A., Yakushiji, M., O'Dwyer, P., Ozols, R., and Hamilton, T.,** Cross-resistance to diverse drugs is associated with primary cisplatin resistance in ovarian cancer cell lines, *Cancer Res.*, 53, 5225, 1993.

41. **Britten, R., Green, J., and Warrenius, H.,** Cellular glutathione (GSH) and glutathione S-transferase (GST) activity in human ovarian tumor biopsies following exposure to alkylating agents, *Int. J. Rad. Oncol. Biol. Phys.*, 24, 527, 1992.

42. **Meister, A.,** Glutathione, ascorbate and cellular protection, *Cancer Res.*, 54, 1969s, 1994.

43. **Board, P., Gali, R., Vaska, V., and Webb, G.,** Molecular genetic characterization of human glutathione synthetase, in *Glutathione S-Transferase: Structure, Function and Clinical Implications*, Vermulen, N., Mulder, G., Niewenhuyse, H., Peters, W., and van Bladeren, P., Eds., Taylor and Francis, New York, 1996, 153.

44. **Sierra-Rivera, E., Summar, M., Krishnamani, M., Phillips, J., and Freeman, M.,** Chromosomal assignment of the human gene that encodes the catalytic subunit of γ-glutamyl synthetase to chromosome 6, *Proc. Am. Assoc. Cancer Res.*, 36, 313, 1995.

45. **Huang, C.-S., Anderson, M., and Meister, A.,** Amino acid sequence and function of the light subunit of rat kidney γ-glutamylcysteine synthetase, *J. Biol. Chem.*, 268, 20578, 1993.

46. **Hanigen, M., Frierson, H., Brown, J., Lovell, M., and Taylor, P.,** Human ovarian tumors express -glutamyl transpeptidase, *Cancer Res.*, 54, 286, 1994.

47. **Mulcahy, R., Untawale, S., and Gipp, J.,** Transcriptional upregulation of γ-glutamylcysteine synthetase gene expression in melphalan-resistant human prostate carcinoma cells, *Mol. Pharmacol.*, 46, 909, 1994.

48. **Bailey, H., Gipp, J., Ripple, M., Wilding, G., and Mulcahy, R.,** Increase in γ-glutamylcysteine synthetase activity and steady-state messenger RNA levels in melphalan-resistant DU-145 human prostate carcinoma cells expressing elevated glutathione levels, *Cancer Res.*, 52, 5115, 1992.

49. **Untawale, S., Gipp, J., Bailey, H., and Mulcahy, R.,** Transcriptional upregulation of the γ-glutamylcysteine synthetase gene in melphalan-resistant prostate carcinoma cells expressing elevated glutathione, *Proc. Am. Assoc. Cancer Res.*, 35, 377, 1994.

50. **Yao, K.-S., Godwin, A., Johnson, S., Ozols, R., O'Dwyer, P., and Hamilton, T.,** Evidence for altered regulation of γ-glutamylcysteine synthetase gene expression among cisplatin-sensitive and -resistant human ovarian cancer cell lines: changes in transcription factor expression and function, *Cancer Res.*, 56, 1731, 1996.

51. **Hamilton, T., Yao, K.-S., Beesley, J., Godwin, A., O'Dwyer, P., and Ozols, R.,** The relationship and regulation of glutathione in cancer cells resistant to chemotherapy, in *Glutathione S-Transferase: Structure, Function and Clinical Implications*, Vermulen, N., Mulder, G., Niewenhuyse, H., Peters, W., and van Bladeren, P., Eds., Taylor and Francis, New York, 1996, 173.

52. **Wu, A.-L. and Moye-Rowley, W.,** GSH1, which encodes γ-glutamylcysteine synthetase, is a target gene for yAP-1 transcriptional regulation, *Mol. Cell. Biol.*, 14, 5832, 1994.

53. **Moore, W., Anderson, M., Meister, A., Murata, K., and Kimura, A.,** Increased capacity for glutathione synthesis enhances resistance to radiation in *Escherichia coli*: a possible model for mammalian cell protection, *Proc. Natl. Acad. Sci. U.S.A*, 86, 1461, 1989.

54. **Sinha, S., Hockin, L., and Neal, G.,** Transformation of a rat liver cell line: neoplastic phenotype and regulation of gamma glutamyl transpeptidase in tumour tissue, *Cancer Lett.*, 35, 215, 1987.

55. **Vincenzini, M., Marraccini, P., Iantomasi, T., Favilli, F., Pacini, S., and Ruggiero, M.,** Altered metabolism of glutathione in cells transformed by oncogenes which cause resistance to ionizing radiations, *FEBS Letts.*, 320, 219, 1993.

56. **Tew, K.,** Glutathione-associated enzymes in anticancer drug resistance, *Cancer Res.*, 54, 4313, 1994.

57. **Hall, A., Matheson, E., Hickson, I., Foster, S., and Hogarth, L.,** Purification of an a class glutathione S-transferase from melphalan-resistant Chinese hamster ovary cells and demonstration of its ability to catalyze melphalan-glutathione adduct formation, *Cancer Res.*, 54, 3369, 1994.

58. **Hamada, S.-I., Kamada, M., Furumoto, H., Hirao, T., and Aono, T.,** Expression of glutathione S-transferase-π in human ovarian cancer as an indicator of resistance to chemotherapy, *Gynecol Oncol.*, 52, 313, 1994.

59. **van der Zee, A., van Ommen, B., Meijer, C., Hollema, H., van Bladeren, P. and de Vries, E.,** Glutathione S-transferase activity and isozyme composition in benign ovarian tumours, untreated malignant ovarian tumours and malignant ovarian tumours after platinum/cyclophosphamide chemotherapy, *Br. J. Cancer*, 66, 930, 1992.

60. **Miyazaki, M., Kohno, K., Saburi, Y., Matsuo, K., Ono, M., Kuwano, M., Tsuchida, S., Sato, K., Sakai, M., and Muramatsu, M.,** Drug resistance to cis-diamminedichloroplatinum (II) in Chinese hamster ovary cell lines transfected with glutathione S-transferase pi gene, *Biochem. Biophys. Res. Commun.*, 166, 1358, 1990.

61. **Fields, W., Li, Y., and Townsend, A.,** Protection by transfected glutathione S-transferase isozymes against carcinogen-induced alkylation of cellular macromolecules in human MCF-7 cells, *Carcinogenesis*, 15, 1155, 1994.

62. **Ishikawa, T. and Ali-Osman, F.,** Glutathione-associated cis-diamminedichloroplatinum (II) metabolism and ATP-dependent efflux from leukemia cells: molecular characterization of glutathione-platinum complex and its biological significance, *J. Biol. Chem.*, 268, 20116, 1993.

63. **Ishakawa, T.,** The ATP-dependent glutathione S-conjugate export pump, *Trends Biochem. Sci.,* 17, 463, 1992.

64. **Ishikawa, T., Wright, C., and Ishizuka, H.,** GS-X pump is functionally overexpressed in cis-diamminedichloro-platinum (II)-resistant human leukemia HL-60 cells and down-regulated by cell differentiation, *J. Biol. Chem.*, 269, 29085, 1994.

65. **Cole, S., Bhardwaj, G., Gerlach, J., Mackie, J., Grant, C., Almquist, K., Stewart, A., Kurtz, E., Duncan, A., and Deeley, R.,** Overexpression of a transport gene in a multidrug-resistant human lung cancer cell line, *Science*, 258, 1650, 1992.

66. **Muller, M., Meijer, C., Xaman, G., Birst, P., Scheper, R., Mulder, N., de Vries, E., and Jansen, P.,** Overexpression of the gene encoding the multidrug resistance-associated protein results in increased ATP-dependent glutathione S-conjugate transport, Proc. Natl. Acad. Sci. U.S.A., 91, 13033, 1994.

67. **Bellomo, G., Vairetti, M., Stivala, L., Mirabelli, F., Richelmi, P., and Orrenius, S.,** Demonstration of nuclear compartmentalization of glutathione in hepatocytes, *Proc. Natl. Acad. Sci. U.S.A.*, 89, 4412, 1992.

68. **Lai, G.-M., Ozols, R., and Hamilton, T.,** Role of glutathione on DNA repair in cisplatin resistant human ovarian cancer cell lines, *J. Natl. Cancer Inst.*, 81, 535, 1989.

69. **Billings, P., Davis, R., Engelsberg, B., Skov, K., and Hughes, E.,** Characterization of high mobility group protein binding to cisplatin-damaged DNA, *Biochem. Biophys. Res. Commun.*, 188, 1286, 1992.

70. **Kohlstaedt, L. and Cole, R.,** Effect of pH on interactions between DNA and high mobility group protein HMG1, *Biochemistry*, 33, 12702, 1994.

71. **Griffith, O. and Meister, A.,** Potent and specific inhibition of glutathione synthesis by buthionine sulfoxomine (S-n-butyl homocysteine sulfoxime), *J. Biol. Chem.*, 254, 7558, 1979.

72. **Hamilton, T., Winker, M., Louie, K., Batist, G., Behrens, B., Tsuruo, T., Grotzinger, K., McKoy, W., Young, R., and Ozols, R.,** Augmentation of adriamycin, melphalan and cisplatin cytotoxicity in drug-resistant and -sensitive human ovarian cancer cell lines by buthionine sulfoximine mediated glutathione depletion, *Biochem. Pharmacol.*, 34, 2583, 1985.

73. **Andrews, P., Murphy, M., and Howell, S.,** Characterization of cisplatin resistant human ovarian carcinoma cells, *Eur. J. Cancer Clin. Oncol.*, 25, 619, 1989.

74. **Batist, G., Behrens, B., Makuch, R., Hamilton, T., Katki, A., Louie, K., Myers, C., and Ozols, R.,** Serial determination of glutathione levels and glutathione related enzyme activities in human tumor cells in vitro, *Biochem. Pharmacol.,* 35, 2257, 1986.

75. **Ozols, R., Louie, K., Plowman, J., Behrens, B., Fine, R., Dykes, D., and Hamilton, T.,** Enhanced melphalan cytotoxicity in human ovarian cancer in vitro and in tumor bearing nude mice by buthionine sulfoximine depletion of glutathione, *Biochem. Pharmacol.,* 36, 147, 1987.

76. **O'Dwyer, P., Hamilton, T., Young, R., LaCreta, F., Carp, N., Tew, K., Padavic, K., Comis, R., and Ozols, R.,** Depletion of glutathione in normal and malignant human cells *in vivo* by buthionine sulfoximine: clinical and biochemical results, *J. Natl. Cancer Inst.*, 84, 264, 1987.

77. **Bailey, H., Mulcahy, T., Tutsch, K., Arzoomanian, R., Alberti, D., Tombes, M., Wilding, G., Pomplun, M., and Spriggs, D.,** Phase I clinical trial of intravenous L-buthionine sulfoximine and melphalan: an attempt at modulation of glutathione, *J. Clin. Oncol.*, 12, 194, 1994.

78. **Plooy, A., Fichtinger-Scepman, A., Schutte, H., van Dijk, M., and Lohman, P.,** The quantitative detection of various Pt-DNA-adducts in Chinese hamster ovary cells treated with cisplatin: application of immunochemical techniques, *Carcinogenesis*, 6, 561, 1985.

79. **Larminat, F., Zhen, W., and Bohr, V.,** Gene-specific DNA repair of interstrand cross-links induced by chemotherapeutic agents can be preferential, *J. Biol. Chem.*, 268, 2649, 1993.

80. **Bohr, V., Smith, C., Okumoto, D., and Hanawalt, P.,** DNA repair in an active gene: removal of pyrimidine dimers from the DHFR gene of CHO cells is much more efficient than in the genome overall, *Cell*, 40, 359, 1985.

81. **Sancar, A. and Tang, M.,** Nucleotide excision repair, *Photochem. Photobiol.*, 57, 905, 1993.

82. **Sancar, A.,** Mechanisms of DNA excision repair, *Science*, 266, 1954, 1994.

83. **Hoeijmakers, J.,** Nucleotide excision repair II: from yeast to mammals, *Trends Genet.,* 9, 211, 1993.

84. **Wood, R.,** Studying nucleotide excision repair of mammalian DNA in a cell-free system, *Ann. N.Y. Acad. Sci.,* 726, 274, 1994.

85. **Wood, R., Robins, P., and Lindahl, T.,** Complementation of the xeroderma pigmntosum DNA repair defect in cell-free extracts, *Cell,* 53, 97, 1988.

86. **Aboussekhra, A., Biggerstaff, M., Shivji, M., Vilpo, J., Moncollin, V., Podust, V., Protic, M., Hubscher, U., Egly, J.-M., and Wood, R.,** Mammalian DNA nucleotide excision repair reconstituted with purified protein components, *Cell,* 80, 859, 1995.

87. **Biggerstaff, M., Szymkowski, D., and Wood, R.,** Co-correction of the ERCC1, ERCC4 and xeroderma pigmentosum group F DNA repair defect *in vitro, EMBO J.,* 12, 3685, 1993.

88. **Evans, D. and Dive, C.,** Effects of cisplatin on the induction of apoptosis in proliferating hepatoma cells and nonproliferating immature thymocytes, *Cancer Res.,* 53, 2133, 1993.

89. **Eastman, A.,** Activation of programmed cell death by anticancer drugs: cisplatin as a model system, *Cancer Cells,* 2, 275, 1990.

90. **Shivji, M., Kenny, M., and Wood, R.,** Proliferating cell nuclear antigen is required for DNA excision repair, *Cell,* 69, 367, 1992.

91. **Smith, M., Chen, I.-T., Zhan, Q., Bae, I., Chen, C.-Y., Gilmer, T., Kastan, M., O'Connor, P., and Fornace, A.,** Interaction of the p53-regulated protein Gadd45 with proliferating cell nuclear antigen, *Science,* 266, 1376, 1994.

92. **Bustin, M., Lehn, D., and Lanfsman, D.,** Structural features of the HMG chromosomal proteins and their genes, *Biochem. Biophys. Acta.,* 1049, 231, 1990.

93. **Huang, J.-C., Zamble, D., Reardon, J., Lippard, S., and Sancar, A.,** HMG-domain proteins specifically inhibit the repair of the major DNA adduct of the anticancer drug cisplatin by human excision nuclease, *Proc. Natl. Acad. Sci. U.S.A.,* 91, 10394, 1994.

94. **Bohr, V.,** Gene specific DNA repair, *Carcinogenesis,* 12, 1983, 1991.

95. **Hanawalt, P., Donahue, B., and Sweder, K.,** Collision or collusion? While some proteins have distinct responsibilities in both transcription and DNA repair, additional proteins are needed to couple these essential DNA transactions in expressed genes, *Curr. Biol.,* 4, 518, 1994.

96. **Jones, J., Zhen, W., Reed, E., Parker, R., Sancar, A., and Bohr, V.,** Gene-specific formation and repair of cisplatin intrastrand adducts and interstrand cross-links in Chinese hamster ovary cells, *J. Biol. Chem.,* 266, 7101, 1991.

97. **Zhen, W., Link, C., O'Connor, P., Reed, E., Parker, P., Howell, S., and Bohr, V.,** Increased gene-specific repair of cisplatin interstrand cross-links in cisplatin-resistant human ovarian cancer cell lines, *Mol. Cell. Biol.,* 12, 3689, 1992.

98. **Johnson, S., Swiggard, P., Handel, L., Brennan, J., Godwin, A., Ozols, R., and Hamilton, T.,** Relationship between platinum-DNA adduct formation and removal and cisplatin-sensitive and -resistant human ovarian cancer cells, *Cancer Res.,* 54, 5911, 1994.

99. **Masuda, H., Ozols, R., Lai, G.-M., Fojo, A., Rothenberg, M., and Hamilton, T.,** Increased DNA repair as a mechanism of acquired resistance to cisdiamminedichloroplatinum (II) in human ovarian cancer cell lines, *Cancer Res.,* 48, 5713, 1988.

100. **Lai, G.-M., Ozols, R., Smyth, J., Young, R., and Hamilton, T.,** Enhanced DNA repair and resistance to cisplatin in human ovarian cancer, *Biochem. Pharmacol.,* 37, 4597, 1988.

101. **Chao, C., Lee, Y., and Lin-Chao, S.,** Phenotypic reversion of cisplatin resistance in human cells accompanies reduced host cell reactivation of damaged plasmid, *Biochem. Biophys. Res. Commun.,* 170, 851, 1990.

102. **Sheibani, N., Jennerwein, M., and Eastman, A.,** DNA repair in cells sensitive and resistant to cis-diamminedichloroplatinum(II): host cell reactivation of damaged plasma DNA, *Biochemistry,* 28, 3120, 1989.

103. **Masuda, H., Tanaka, T., Matsuda, H., and Kusuba, I.,** Increased removal of DNA-bound platinum in a human ovarian cancer cell line resistant to cis-diamminedichloroplatinum (II), *Cancer Res.,* 50, 1863, 1990.

104. **Lee, K., Parker, R., Bohr, V., Cornelison, T., and Reed, E.,** Cisplatin sensitivity/resistance in UV repair-deficient Chinese hamster ovary cells of complementation groups 1 and 3, *Carcinogenesis,* 14, 2177, 1993.

105. **Dabholkar, M., Vionnet, J., Bostick-Bruton, F., Yu, J., and Reed, E.,** Messenger RNA levels of XPAC and ERCC1 in ovarian cancer tissue correlate with response to platinum-based chemotherapy, *J. Clin. Invest.,* 94, 703, 1994.

106. **O'Dwyer, P., Moyer, J., Suffnes, M., Harrison, S., Cysyk, R., Hamilton, T., and Plowman, J.,** Antitumor activity and biochemical effects of aphidicolin glycinate (NSC 303812) alone and in combination with cisplatin in vivo, *Cancer Res.,* 54, 724, 1994.

107. **Yang, L., Li, L., Liu, X., Keating, M., and Plunkett, W.,** Gemcitabine suppresses the repair of cisplatin adducts in plasmid DNA by extracts of cisplatin-resistant human colon carcinoma cells, *Proc. Am. Assoc. Cancer Res.,* 36, 357, 1995.

108. **Mamenta, E., Poma, E., Kaufmann, W., Delmastro, D., Grady, H., and Chaney, S.,** Enhanced replicative bypass of platinum-DNA adducts in cisplatin-resistant human ovarian carcinoma cell lines, *Cancer Res.,* 54, 3500, 1994.

109. **Dole, M., Nunez, G., Merchant, A., Maybaum, J., Rode, C., Bloch, C., and Castle, V.,** Bcl-2 inhibits chemotherapy-induced apoptosis in neuroblastoma, *Cancer Res.,* 54, 3253, 1994.

110. **Miyashita, T. and Reed, J.,** Bcl-2 oncoprotein blocks chemotherapy-induced apoptosis in a human leukemia cell line, *Blood,* 81, 151, 1993.

111. **Robins, P., Jones, C., Biggerstaff, M., Lindahl, T., and Wood, R.,** Complementation of DNA repair in xeroderma pigmentosum group A cell extracts by a protein with affinity for damaged DNA, *EMBO J.*, 10, 3913, 1991.

112. **Eker, A., Vermeulen, W., Miura, N., Tanaka, K., Jaspers, N., Hoeijmakers, J., and Bootsma, D.,** Xeroderma pigmentosum group A correcting protein from calf thymus, *Mutation Res.*, 274, 211, 1992.

113. **Schaeffer, L., Moncollin, V., Roy, R., Staub, A., Mezzina, M., Sarasin, A., Weeda, G., Hoeijmakers, J., and Egly, J.,** The ERCC/DNA repair protein is associated with the class II BTF2/TFIIH transcription factor, *EMBO J.*, 13, 2388, 1994.

114. **Weeda, G., van Ham, R., Masurel, R., Westerveld, A., Odijk, H., de Wit, J., Bootsma, D., van der Eb, A. and Hoeijmakers, J.,** Molecular cloning and biological characterization of the human excision repair gene ERCC-3, *Mol. Cell. Biol.*, 10, 2570, 1990.

115. **Legerski, R. and Peterson, C.,** Expression cloning of a human DNA repair gene involved in xeroderma pigmentosum group C, *Nature*, 359, 70, 1992.

116. **Weber, C., Salazar, E., Stewart, S., and Thompson, L.,** ERCC2: cDNA cloning and molecular characterization of a human nucleotide excision repair gene with high homology to yeast RAD3, *EMBO J.*, 9, 1437, 1990.

117. **Hwang, B. and Chu, G.,** Purification and characterization of a human protein that binds to damaged DNA, *Biochemistry*, 32, 1657, 1993.

118. **Bardwell, A., Bardwell, L., Tomkinson, A., and Friedberg, E.,** Specific cleavage of model recombination and repair intermediates by the yeast Rad1-Rad10 DNA endonuclease, *Science*, 265, 2082, 1994.

119. **van Vuuren, A., Appeldoorn, E., Odijk, H., Yasui, A., Jaspers, N., Bootsma, D., and Hoeijmakers, J.,** Evidence for a repair enzyme complex involving ERCC1 and complementing activities of ERCC4, ERCC11 and xeroderma pigmentosum group F, *EMBO J.*, 12, 3693, 1993.

120. **O'Donovan, A. and Wood, R.,** Identical defects in DNA repair in xeroderma pigmentosum group G and rodent ERCC group 5, *Nature*, 363, 185, 1993.

121. **O'Donovan, A., Davies, A., Moggs, J., West, S., and Wood, R.,** XPG endonuclease makes the 3′ incision in human DNA nucleotide excision repair, *Nature*, 371, 432, 1994.

122. **Troelstra, C., Heseen, W., Bootsma, D., and Hoeijmakers, J.,** Structure and expression of the excision repair gene ERCC6 involved in the human disorder Cockayne's syndrome B, *Nucleic Acid Res.*, 21, 419, 1993.

123. **Coverly, D., Kenny, M., Lane, D., and Wood, R.,** A role for the human single-stranded DNA binding protein HSSB/RPA in an early stage of nucleotide excision repair, *Nucleic Acid Res.*, 20, 3873, 1992.

Chapter 10

Biologic Strategies in the Therapy of Gynecologic Cancers

J. Michael Mathis, Carolyn Y. Muller, Vivian E. von Gruenigen, and David Scott Miller

CONTENTS

1. INTRODUCTION

The conventional strategy in the treatment of female malignancies has relied on the use of surgical extirpation, radiation, chemotherapy, or a combination of these modalities, which have been most successful in the primary treatment of early malignancies. Unfortunately, despite advances in chemotherapy, radiation delivery, and surgical techniques, survival rates for primary advanced cancers of the genital tract and salvage rates of recurrent disease remain poor. For example, in the treatment of advanced epithelial ovarian carcinoma, improved operative techniques and novel combination chemotherapy strategies have resulted in higher initial complete response rates. However, the overall survival rate has improved little over the past 25 years.

New treatment strategies to alter the biology of human cancers is a recent area of investigation that shows a great potential for improving survival. Exploitation of recent advances in our emerging understanding of the molecular genetics involved in this disease is the basis of such strategies. One approach includes the use of molecular modifiers of the host immune system both for diagnosis and therapy. A second approach involves direct cellular alteration at the gene level involving transcription and translation that is tumor specific. Additional uses of molecular technology involve gene marking of tumor cells, and molecular alterations of both normal and cancer cells to augment traditional conventional chemotherapy.

Two clinical applications of molecular technology to alter the biology of human cancers have now emerged: somatic cell therapy and gene therapy. Somatic cell therapy is the administration of living cells that have been modified in their biologic characteristics. This method involves the removal of tumor cells from the patient and *ex vivo* modifying or stimulating a biologic change in the cells then reintroducing them as therapy. Gene therapy involves the deliberate insertion or alteration of genetic material into the cell for the purpose of diagnosis, prevention, or treatment of disease. This can be done in combination with somatic cell therapy *ex vivo* or by direct injection *in vivo*. This chapter is dedicated to discussing the strategies of biologic therapeutics as they apply to gynecologic cancers, and future possibilities for improving cancer survival.

1.1 Molecular Mechanisms

The cellular alterations involved in the initiation and progression of carcinogenesis are extremely complex. Cancer arises from the accumulation of mutations in multiple combinations of genes.[1] These changes involve dysregulation of cellular proto-oncogenes,[2] such as k-*ras*, c-*fos*, c-*myc*, and HER2/*neu*. Other alterations involve inactivation of tumor suppressor genes or gene function.[3] Germline mutations in tumor suppressor genes give rise to hereditary cancer predisposition syndromes,[4] such as Li-Fraumeni Syndrome (p53 gene mutations), Wilms' Tumor (WT-1 gene mutations), Hereditary Nonpolyposis Colon Cancer (hMSH2, hMLH1 and other gene mutations), and Familial Breast-Ovarian Cancer (BRCA-1 gene mutations). Alternatively, molecular changes may allow tumor cells to avoid host immune responses.[5] For example, tumors may have decreased expression of major histocompatibility complex (MHC) class 1 molecules which limits their recognition by cytotoxic T lymphocytes (CTLs).[6] Finally, some tumors evolve, in part, from the expression of foreign viral gene products, such as the human papilloma virus E6 and E7 protein in cervical cancer.[7]

Carcinogenesis involves a random accumulation of errors in genes that normally regulate cellular growth and differentiation.[8] Individual tumors are heterogeneous and clonal populations can arise within the patient that exhibit different patterns of cellular abnormalities. This makes a universal biologic approach to cancer therapy quite challenging. However, in spite of cancer's polygenic nature, there is strong evidence that correction of only one of the genetic defects in cancer cells *in vitro* can lead to the inhibition of cell growth and reversal of tumorigenicity.[9-12] By identifying common alterations such as the loss of p53 tumor suppressor gene function, strategies can be designed that may be effective in a majority of cancers. However, combination gene strategies for each tumor type may ultimately be necessary to enhance patient survival in clinical trials. Regardless of these obstacles, advances in animal and early clinical trials are promising and suggest that biologic therapeutics will be a part of clinical practice in the future of gynecologic oncology.

1.2 Principles of Gene Therapy

The principles of gene therapy are based on the ability to deliver DNA or related nucleic acid derivatives into a cell as anticancer agents. The DNA can be delivered to tumor cells or normal cells by infectious or non-infectious vectors. This manipulation may alter a variety of normal or abnormal cellular processes, or even introduce new biochemical pathways which did not exist previously. These changes can reverse the tumorigenic phenotype, or cause the cell to undergo apoptosis and die. Alternatively, these changes may cause the cell to be more susceptible to death by conventional treatment, produce a protein that will stimulate host immunity, or simply mark tumor cells to follow the progress of conventional cancer treatments.

The design of the biologic cancer therapy includes the choice of the vector, the route of delivery, the therapeutic gene or nucleic acid sequence to be expressed, the target tissue, and the desired target cell response.[13] The choice of gene or nucleic acids to be transferred or regulated is based on the altered cellular mechanism that is targeted. Present technology constraints cannot allow the transfer of an entire gene, but artificial genes can be constructed from cloned cDNA sequences. This approach is feasible when the therapeutic goal is to replace a wild-type (normal) gene into cells deficient of a critical gene product. This strategy may also be useful for the overexpression of normal gene products, such as cytokines, that can affect a significant host biologic response.

Alternative strategies to gene expression utilize much smaller sequences of DNA or RNA that are transferred into the nucleus or cytoplasm of the cell. These alternatives include oligonucleotide gene therapy,[14] antisense gene therapy,[15] and ribozyme gene therapy.[16] The oligonucleotide approach utilizes

DNA sequences designed to interfere in the natural machinery of gene expression. Small custom synthesized DNA oligonucleotides can be designed to target a specific gene sequence within a major groove in the DNA double helix. This allows for formation of a triple helix, which then prevents transcription of the target gene. Several technological hurdles must still be overcome, however, in order to produce oligodeoxynucleotides that efficiently enter the cell and reach the nucleus without becoming degraded by normal cellular DNAses. An alternative approach to antisense therapy uses RNA antisense sequences designed to hybridize to an mRNA transcript. The double stranded RNA-RNA complex can inhibit mRNA transport, block translation, and stimulate ribonucleases to degrade the complex. The RNA antisense sequences can also block the complementary DNA template strand to prohibit further gene transcription. Ribozyme gene therapy is a third approach that utilizes ribo-oligomers designed to directly target an mRNA, forming a hammerhead secondary structure that catalyzes RNA cleavage and degradation. These three strategies are useful in reversing overexpression of oncoproteins.

The most commonly used DNA delivery systems in clinical trials are the viral vectors, which exploit the inherent ability of these agents to infect host cells with foreign DNA.[17] Genetically engineered replication deficient adenoviruses and retroviruses have been used successfully to transfer genes into mammalian cells *in vitro* and *in vivo*. Each viral system has distinct advantages and disadvantages. Retroviruses are positive strand RNA viruses which must be reverse transcribed and integrated into the host genome for gene expression to occur. Retroviral vector integration into the host genome results in stable DNA propagation and expression. However, retroviruses efficiently infect only dividing cells. In addition, because retroviral DNA integration into the host genome is random, this event is potentially oncogenic if an important host genomic region is altered. Adenoviruses are linear, double stranded DNA viruses. Adenoviral vectors infect both dividing and nondividing cells with great efficiency and have nearly 100% transfection efficiency, but the expression of the transferred gene is limited as the gene remains episomal.

Nonviral (physical) delivery systems useful in gene therapy include liposome mediated gene transfer, direct DNA injection, and DNA bombardment.[18] The technology of liposome mediated gene transfer uses the properties of electrical charge in the DNA, cell membranes and cationic lipids. Liposomes (monocationic lipids) are nonimmunogenic, are of commercial purity, and are approved for human use. Liposomes spontaneously bind nucleic acids (DNA or RNA) and can fuse to cell membranes, delivering the nucleic acids into the cell via pinocytosis or endocytosis. The ratio of lipid to nucleic acid used, however, is critical, since significant cellular toxicity can be seen with high doses of liposomes. Additional nonviral delivery techniques in development utilize direct DNA injection or bombardment of cells with gold particles coated with DNA. These direct physical approaches are not yet practical, however, in a clinical therapeutic setting. In theory the nonviral delivery systems have many advantages over viral based delivery systems. For example, gene transfer of larger nucleic acid fragments is possible, there is no capacity to form infectious agents, and there is a lower potential to evoke host inflammatory or immune responses. Poor efficiency of gene transfer, however, is the major disadvantage of current nonviral delivery systems.

Understanding the biological nature of targeted cancer cells is the final component of designing a gene therapy strategy. Care must be taken to determine the best route of delivery to affect peak gene transfer efficiency and limited host toxicity. Retrovirus and adenovirus vectors have been administered systemically,[22,23] intratrachealy,[24,25] intraperitoneally,[9] intramuscularly,[25] intratumorally,[27] in the portal vein,[28] and stereotactically to the brain in animal models,[29] and there are no reports of toxic or clinical infectious response to the viral treatments. The safety and efficacy of virus mediated gene therapy has been proven in multiple animal studies and in early clinical trials,[19-21] although problems have been encountered for each system. These include difficulty to penetrate bulk disease and, more significantly, inability to bypass the host immune system against the viral antigens. Host immunity may potentially limit the use of multiple exposures to the vectors in human trials. Both early and late inflammatory responses have been reported in non-immunocompromised animal studies.[20] Some active preclinical and clinical gene therapy trials for the treatment of gynecologic malignancies are summarized in Table 1. It is anticipated that these studies will further define the roadblocks that must be circumvented in order to attain effective biologic therapies for the gynecologic oncology patient.

2. MUTATION COMPENSATION

Tumor suppressor genes encode cellular proteins that negatively regulate cell growth. Loss of function of these genes by mutational events have been shown to play a role in many human cancers. Thus, it is

Table 1 Current Site-Specific Clinical Protocols for the Biotherapy of Female Cancers

Site	Therapy	Protocol	Method of Transfer	Institution	Ref.
Breast	Immunotherapy	IL-2 cDNA	Lipofection	Duke University	82
Breast	Drug resistance	MDR cDNA	Retrovirus	UT MD Anderson	82
			Retrovirus	NIH	
Breast	Gene marker	Neomycin resistance cDNA	Retrovirus	NIH	82
			Retrovirus	Fred Hutchinson Cancer Research Center	
			Retrovirus	University of Southern California	
Breast	Gene therapy	Antisense c-*fos* cDNA Antisense c-*myc* cDNA	Retrovirus	Vanderbilt University	82
Ovary	Immunotherapy	T-cell receptor cDNA	Retrovirus	NIH	82
Ovary	Immunotherapy	IL-2 cDNA	Lipofection	Duke University	82
Ovary	Gene therapy	anti-*erb*B-2 Single-chain Antibody gene	Adenovirus	University of Alabama	82
Ovary	Drug resistance	MDR cDNA	Retrovirus	UT MD Anderson	82
Ovary	Drug sensitivity	HSV-TK cDNA	Retrovirus	University of Rochester School of Medicine	82
			Retrovirus	Human Gene Ther. Res. Ctr., Des Moines, IA	
Ovary	Gene marker	Neo resistance cDNA	Retrovirus	UT MD Anderson	82

conceivable that reintroduction of a functional tumor suppressor gene into a tumor cell can alter its phenotype or cause cell cycle arrest.[9-12] The most characterized tumor suppressor gene is p53, a 53 kDa phosphoprotein that functions in transcription regulation, cell cycle control, cell arrest, and apoptosis. Other tumor suppressor gene mutations or protein inactivation characterized to be involved in gynecologic malignancies include the retinoblastoma gene RB1,[30-33] and BRCA1.[34-36] Loss of heterozygosity studies in gynecologic malignancies suggest that additional tumor suppressor genes have yet to be isolated on other chromosomes.[37-40] Somatic cell hybrid studies showed that whole chromosome transfer carrying a functional tumor suppressor gene results in a reversal of cell growth and tumorigenicity.[41] Transfection studies of wild-type tumor suppressor genes, such as p53, into cancer cell lines showed that replacement of a single defective gene can revert the tumorigenic phenotype, demonstrating the feasibility of *in vivo* gene therapy.[9] The following are examples of two tumor suppressor genes that are currently being evaluated for clinical application in gene therapy.

2.1 Augmentation of Tumor-Suppressor Genes
2.1.1 p53
The p53 tumor suppressor gene is the most commonly altered gene yet characterized in solid tumors. Mutations in the p53 gene have been described in breast,[42] endometrial,[43] ovarian,[44,45] and cervical cancers.[32] The mechanism of wild-type p53 growth inhibition in tumor cells is complex and may involve cell cycle control via induction of target genes, or may involve induction of apoptosis or programmed cell death.[46] Recently, apoptosis has been the proposed mechanism by which p53 inhibits tumorigenicity *in vivo*.[47] Inhibition of tumorigenesis by apoptosis was observed by Yang et al.[48] in an *ex vivo* treatment of human prostate cancer cells using adenovirus-based p53 gene therapy.

The adenovirus vector enters the cell by endocytosis through interaction with specific cell surface receptors, moves into a cytoplasmic endosome, breaks out, and delivers its viral DNA genome into the nucleus. Here the DNA functions in an epichromosomal fashion to direct the expression of its recombinant gene product. (Figure 1A). Introduction of a wild-type p53 gene using an adenovirus construct (Ad-CMV-p53) was shown *in vitro* to specifically arrest growth of an ovarian cancer cell line.[9] In order to confirm the expression of wild-type p53 in Ad-CMV-p53 infected cells, the level of p53 expression was evaluated by immunoelectrophoretic (Western blot) analysis using human H358 lung tumor cells, which contain a homozygous deletion of the p53 gene (Figure 1B). The level of p53 was undetectable in cells infected with a control adenovirus construct containing the *E. Coli* β-galactosidase gene (Ad-

CMV-βgal). However, infection of cells with Ad-CMV-p53 resulted in the transient appearance of a protein band at 53 kDa corresponding to p53.

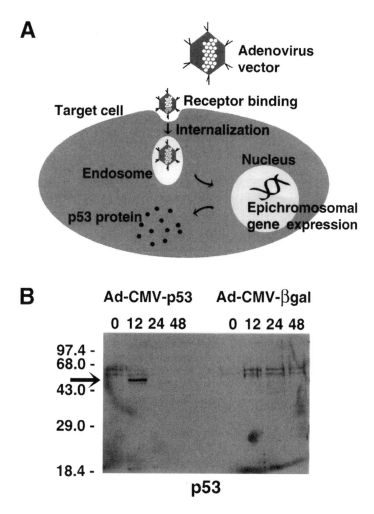

Figure 1 Transfer of the wild-type p53 gene into cells using an adenovirus vector. (A) Schematic representation of adenovirus-mediated gene transfer. (B) Content of p53 protein in H358 Cells after Adenovirus Infection. The amount of p53 protein in H358 cells maintained for up to 48 hours after infection with Ad-CMV-βgal or Ad-CMV-p53 was determined by immunoelectrophoretic (Western blot) analysis. Lanes 1 to 4: 30 μg of total cellular protein from H358 cells at 0, 12, 24, and 48 hours after infection with Ad-CMV-p53. Lanes 5 to 8: 30 μg of total cellular protein from H358 cells at 0, 12, 24, and 48 hours after infection with Ad-CMV-βgal. The nitrocellulose containing lanes 1 to 5 was treated with mouse anti-human p53.

Adenovirus vectors containing the wild-type p53 gene have been studied *in vitro* and/or *in vivo* in lung cancer,[24,46] head and neck cancer,[49,50] cervix cancer,[51] and ovarian cancer.[24] Clayman et al.,[49] utilizing a microscopic disease model of head and neck cancers, found a significant inhibition of tumorigenesis *in vivo* between Ad-CMV-p53 treated animals and control animals. However, inhibition of tumorigenesis in the cell lines used was independent of the mutation status of the endogenous p53 gene. An explanation for these findings is the possibility that an inflammatory response to adenovirus injection caused a non-specific inhibition of tumor cells implantation and growth. Zhang et al.[52] described both an early and late inflammatory response to adenovirus infection in the lung after transtracheal injection in BALB/c mice. It is also possible that overexpression of a foreign gene (either human p53 or *E. coli* β-galactosidase) by the adenovirus vector may have induced a weak host immune response that inhibited residual tumor formation. There are still many unanswered questions that must be addressed before adenoviral-based gene therapy becomes a realistic clinical therapeutic modality.

2.1.2 RB

The RB1 gene, also a nuclear phosphoprotein, initially was thought to function as a tissue-specific transcription factor and function as a tumor suppressor gene in retinoblastoma. However, current evidence indicates that RB1 plays a much broader role in regulating cellular DNA synthesis, and cell growth in many different cell types.[53,54] Inactivation of the RB1 gene plays a role in tumor development in a minority of ovarian cancer patients.[54] More common is the inactivation and degradation of the RB1 gene product by a direct protein-protein interaction with the HPV oncoprotein E7, in a large subset of cervical cancers.[7] Experimental animal models are under investigation, in order to establish whether overexpression of RB1 in normal cells will protect them from tumor development.[55] Overexpression of wild-type RB1 in tumor cells using recombinant viral vectors is also under investigation.[56]

2.2 Ablation of Oncogenes

Proto-oncogenes code for the protein products required for cellular signaling. These oncoproteins function as growth factors, protein kinases, transcription factors and GTP binding proteins, all of which together tightly regulate cell cycle control.[2] Overexpression, amplification, point mutations or chromosomal translocation are mechanisms in which proto-oncogene control is lost and cellular proliferation, transformation, immortalization or metastases can occur.[1-2] Proto-oncogenes known to play a significant role in solid tumors include the *ras* family of proteins (H-*ras*, N-*ras*, and K-*ras*),[57] HER2/*neu* (erbB-2),[58] and c-*myc*.[59] There are a number of proto-oncogenes now identified whose significance in gynecologic malignancies has yet to be determined.[60] Strategies targeting the uncontrolled expression of these genes are complex and are designed to exploit this knowledge of the cellular machinery. Engineered oligonucleotides,[14] antisense ribonucleotides,[15] and highly specific RNAses (ribozymes)[16] are a few techniques found to inhibit overproduction of oncoproteins *in vitro* and *in vivo*. To date, however, few are in early clinical trials.

2.2.1 HER2/neu

HER2/*neu* is a trans-membrane tyrosine kinase which when overexpressed induces malignant transformation.[61] Overexpression has been found in 25 to 30% of breast and ovarian cancers and up to 50% in endometrial cancers.[61] Several studies have shown that overexpression of HER2/*neu* is an independent predictor of poor prognosis and survival.[63,64] Therefore attempts at prevention or correction of overexpression of this oncoprotein may improve survival in a subset of patients where conventional therapy has failed miserably. The adenovirus gene product, E1A, has been shown to block the HER2/*neu* promoter and repress transcription of the oncoprotein and subsequently repress the transforming activity.[65] The E1A gene has been successfully introduced into SKOV-3 ovarian cancer cells utilizing two different strategies. Cationic liposomes were designed to transfect the E1A gene into SKOV-3 ovarian cancer cells *in vitro* and *in vivo*.[65] In nude mice treated with E1A-liposomes intraperitoneally, 75% survived 6 months as compared to none of the control animals.[66] Adenovirus mediated E1A transfection produced similar results *in vitro* and *in vivo*.[67]

2.2.2 K-ras

The *ras* family of oncogenes (N-*ras*, H-*ras*, and K-*ras*) produce membrane bound GTP binding phosphoproteins of approximately 21 kDa.[68] The wild-type *ras* gene products play an important role in normal signaling transduction pathway. However, *ras* mutations play a significant role in cellular metastatic potential.[69] *Ras* overexpression has been identified in up to 71% of post-menopausal endometrial cancers and a portion of these are likely due to point mutations in the K-*ras* gene.[70] These point mutations frequently occur in codon 12 or 61.[70] The presence of *ras* mutations signify a poor prognosis and correlate with metastatic lymph node spread.[71] Therefore, ablation of mutant K-*ras* gene expression may be an important target for gene therapy.

Antisense RNA was targeted against a K-*ras* mRNA mutated at codon 61 in human lung cancer cell lines H460a.[72] Only the antisense RNA expressed clones demonstrated marked decrease in both the mutant mRNA transcripts and of the protein as compared with controls. Soft agar growth as well as tumorigenicity in nude mice were also significantly suppressed. Antisense oligonucleotide therapy has also been directed at N-*ras* and H-*ras* in human hepatoma, bladder, and breast carcinoma cell lines.[73] Utilizing antisense ribozyme RNA gene therapy targeting an endogenous mutant *ras* transcript at codon 12, tumorigenicity was inhibited in the EJ human bladder carcinoma cell line.[74] Since the spectrum of *ras* mutations in gynecologic cancers is similar to that of other tumors studied, this strategy may be directly applicable to treatment of gynecologic malignancies.

3. MOLECULAR CHEMOPOTENTIATION

3.1 Drug Resistance/Sensitivity Genes

Chemotherapeutic agents have a nonselective distribution *in vivo* which results in damage to normal, viable cells. Side effects of chemotherapeutic agents in patients include bone marrow depression, alopecia, peripheral neuropathy, and gastrointestinal mucositis.[75] One alternative approach to enhance conventional chemotherapy is the use of gene therapy to target the cytotoxic effects of chemotherapeutic drugs toward tumor cells with suicide genes.[76] Another approach to augment conventional chemotherapy is to modify normal hematopoietic precursor cells using gene therapy to increase resistance to toxic chemotherapeutic agents.[77]

3.1.1 HSV-TK

The strategy of suicide genes is to modify tumor cells so they become sensitive to agents that are otherwise nontoxic to normal host cells. Suicide genes encode enzymes that convert a nontoxic prodrug into a highly toxic agent. Cancer cells can be transduced to express a suicide gene which causes a metabolic death in the presence of the appropriate prodrug. Three metabolic suicide genes have been described: the herpes simplex virus thymidine kinase gene (HSV-TK), the *E. coli* cytosine deaminase gene (CD), and the varicella zoster thymidine kinase gene (VZ-TK). HSV-TK phosphorylates gancyclovir (a nucleoside analogue of 2′-deoxyguanosine).[78] The phosphorylated product both competitively inhibits DNA polymerase and is directly incorporated irreversibly into DNA. The prodrug gancyclovir requires the viral kinase for phosphorylation and will kill only cells which express the enzyme. In the CD gene system, the final chemotherapeutic product generated by the expression of *E. Coli* CD is 5-FU.[79] Varicella virus also expresses a unique viral thymidine kinase, whose chemotherapeutic product generated by the suicide gene is ara-M.[80] Both the CD and TK systems have an additional advantage of a bystander effect in tumor killing.[81] *In vitro* and *in vivo* studies have shown that tumor cells which do not express the suicide gene but which are in proximity to transfected cells are also killed by addition of the prodrug. Exploitation of the bystander effect could be used to enhance the efficacy of gene therapy. Clinical trials have begun employing the HSV-TK strategy in ovarian cancer.[82]

3.1.2 MDR

The second novel approach in altering the phenotype of cells involves the multiple drug resistance (MDR) gene.[83] This gene therapy involves modifying early hematopoietic progenitor cells with the MDR cDNA and *in vitro* selection of genetically modified hematopoietic cells which are resistant to chemotherapy. The genetically modified cells are then transferred to the patient's bone marrow before treatment with toxic agents, where the modified progenitor cells would be protected and thus be able to repopulate after high dose chemotherapy. Clinical trials are underway in a variety of tumors including ovarian cancer.[84]

Another strategy in modifying tumor cells is the use of selectable markers to augment chemotherapy.[85] Gene marking studies were the first approved clinical protocols introducing exogenous genetic material into human cells.[86] The gene marking technique involves the transfer of a foreign, nonhuman gene into a human host cell. The gene or gene product can be easily detected and serves as a marker. For example, the bacterial neomycin phosphoryltranferase (*neo*) gene is a frequently used marker gene. It encodes an enzyme which inactivates neomycin, and can be used to select cells in culture. There are many gene marking clinical protocols underway, including a pilot study involving the use of a retroviral vector to study the trafficking patterns of ovarian tumor infiltrating lymphocytes (TILs).

3.2 Monoclonal Antibodies

A number of research efforts over the past decade have focused on the identification of tumor-associated or tumor-specific antigens. The neoplastic state, resulting from the expression of mutated or unregulated oncogenes or tumor-suppressor genes, leads to complex changes in cancer cells with a concomitant expression of certain cellular proteins that are repressed or expressed at much lower levels in normal cells. These proteins include tumor antigens that are recognized as foreign by the host resulting in specific immune responses to tumor cells. To date, a number of tumor-associated antigens have been identified: those that are unique to a given tumor, those that are common to all tumors of a histopathologic type, those that are associated with a given organ or tissue, those that may represent embryonic or transplantation-like antigens, as well as those that are ubiquitous in normal allogenic and xenogeneic cells. These tumor antigens can be classified into four major categories. The first category of antigens mediate resistance to tumor growth, and were observed by transplanting chemically or virally induced tumors between inbred mice preimmunized with the same tumor.[87] Tumor transplantation experiments also helped to identify a second category of tumor-specific antigens that are presented on the cell surface in association with major histocompatability (MHC) class I molecules.[88] Antigens of virally induced tumors constitute a third category of tumor-associated antigens.[89] A fourth category of antigens, the oncofetal antigens, such as carcinoembryonic antigen (CEA) and alpha-fetoprotein (AFP), are expressed during fetal development, but not in normal adult tissues, and are reexpressed in transformed cells.[90]

3.2.1 Immunodiagnosis

One of the most important clinical applications involving the production of monoclonal antibodies prepared from mice immunized with tumor cells or tumor fragments has been the identification of tumor-associated antigens.[91] One such monoclonal antibody, OC-125,[92] recognizes a high molecular weight glycoprotein antigen, CA-125, which is elevated in most ovarian adenocarcinomas. Serum CA-125 levels are elevated in 80% to 85% of patients with invasive serous ovarian cancer at diagnosis, and serial CA-125 levels correlate with the clinical course and response to treatment in greater than 90% of patients whose tumors express the antigen.[93] The usefulness of CA-125 in screening for preclinical cancer is limited, however, due to problems with assay sensitivity and elevated levels in a variety of benign conditions, particularly endometriosis.

3.2.2 Immunodetection

Monoclonal antibodies conjugated to radionuclides have now been used to assess the extent of disease in patients with solid tumors.[94] Most of these studies in the gynecologic cancers have focused on patients with ovarian cancer utilizing OC-125 antibody directed against CA-125[95] or utilizing B72.3 antibody directed against TAG-72,[96] a high molecular weight glycoprotein expressed on most adenocarcinomas. Much of the work in immunodetection has been done using antibodies conjugated to ^{123}I or ^{131}I, ^{99}Tc, or [111]In. At its present state of development, however, immunodetection should be viewed as a complement, rather than a replacement, to conventional imaging techniques such as CT scan or ultrasound and remains investigational in gynecologic malignancies.

3.2.3 Immunotherapy

Binding of antibody to tumor cells does not necessarily affect their growth. For immunologically mediated killing of tumor cells, recruitment of complement components or effectors for antibody-dependent cell-mediated cytotoxicity (ADCC) are required. The requirement for recruitment of sufficient complement components, as well as of adequate numbers of functionally competent effector cells *in vivo*, has limited the use of unmodified monoclonal antibodies in immunotherapy.[97] In the absence of appropriate effector cells, immunoconjugates could provide more potent antitumor activity. Monoclonal antibodies have been linked to radionuclides, immunotoxins, and cytotoxic drugs in an effort to increase the antitumor activity.[98]

4. THERAPEUTIC IMMUNOPOTENTIATION

4.1 Tumor Vaccines

4.1.1 Autologous Tumor Vaccines

The concept of autologous tumor vaccines has been around for decades and has not achieved significant clinical success. There has been a resurgence of interest, however, with genetically modified tumor vaccines, which may prove more effective in some circumstances.[99] Cytokine gene-modified autologous and allogeneic tumor cells are the predominant approaches being studied in phase I trials in patients with different histologic tumor types.[100] Most of these studies are based on the successful priming of specific antitumor immune responses employing cytokines in preclinical trials.

4.1.2 DNA Vaccines

Direct injection of plasmid DNA encoding a variety of genes has been shown to result in protein expression of these genes.[101] In addition, inoculation with plasmids can generate both T cell and antibody responses to the neo-antigens encoded by the plasmids.[102] Thus, plasmid DNA administered *in vivo* may provide a novel means of active specific immunization. Live attenuated viruses constitute effective vaccines, in part, because a proportion of the viral proteins expressed *de novo* in infected cells are degraded in the cytosol and transported into the endoplasmic reticulum where the peptide degradation products associate with MHC class I molecules before display on the cell surface. Here they interact with CD8+ precursor cells, activating them, and leading to the production of clones of CD8+ cytolytic T cells. Peptide products derived by cytosol degradation of fragments of tumor-associated antigens encoded in plasmid DNA, expressed *de novo*, might gain access to the presentation pathway, mimicking the presentation of viral proteins in infected cells. Presentation as neo-antigens or surrogate antigens in this context may be a means of breaking immunologic tolerance, and may lead to the generation of tumor-specific immune responses. Peptides derived from the viral genes can induce cytotoxic T-cell responses *in vitro*.[103] Attractive targets for T-cell mediated immunity stimulated by DNA vaccination in gynecologic malignancies are the E6, E7, and L1 genes from oncogenic human papillomaviruses HPV16 and HPV18. Clinical trials are already in progress utilizing ribozymes directed against HPV E6 and E7 mRNA in patients with extensive condyloma.[104]

4.2 Cytokine Genes

Interest in the applicability of cytokines in the treatment of gynecologic cancers is related to the multiple mechanisms of anti-tumor effects they exert *in vivo* including: (1) direct destruction of tumor cells; (2) destruction of tumor blood vessels; (3) mobilization of host inflammatory cells which invade and destroy tumor tissue; and (4) activation of host anti-tumor immunity.[105] Cytokines also appear to have the ability to directly inhibit tumor cell growth *in vitro*.[106] In addition, cellular anti-tumor effector cells, such as T-lymphocytes, NK cells, neutrophils, and macrophages, can be activated by cytokines. Therefore, tumor cells which are not sensitive to the direct effects of cytokines may be sensitive to the host response evoked by cytokines. The effectiveness of such a host-induced immune response depends on the assumption that tumor cells express tumor-specific antigens, which are not recognized due to an inadequate cytokine milieu or T-cell help. High levels of local cytokine production from cytokine gene transfer can attract effector cells to the tumor site to detect and kill tumor cells. Thus, tumor cells expressing these cytokines might be important in adjuvant tumor therapy. A substantial number of clinical protocols using cytokines genes, including IL-2, IL-4, IL-12, γ-IFN, and GM-CSF, have been initiated.[107] Two examples are described below.

4.2.1 IL-2

IL-2 is known to be a potent immune mediator with a variety of *in vitro* and *in vivo* effects that have been demonstrated in preclinical trials.[108] Its anti-cancer effect is likely based on its ability to induce a local inflammatory response leading to activation of both helper and cytotoxic T cells. In one *ex vivo* study, irradiated autologous tumor cells were modified to express IL-2 using a retrovirus vector, and were subsequently implanted subcutaneously in the donor.[109] In some individuals, this treatment led to the recruitment of NK cells, cytotoxic T cells, and eosinophils.

4.2.2 IL-4

In addition to IL-2, IL-4 is currently being tested in clinical gene therapy trials for its effects on tumor growth. IL-4 has broad immunoregulatory properties affecting both B and T cells as well as hematapoietic cells.[110] Its anti-cancer effect is likely based on its ability to recruit and activate CD4+ antigen presenting T cells, as well as its ability to induce infiltration and activation of macrophages and eosinophils. In one trial, IL-4 transduced irradiated autologous fibroblasts were mixed with irradiated autologous tumor cells, and the cells were implanted subcutaneously.[111] In some recipients, this evoked infiltration with tumor-specific CD4+ T cells at the immunization site.

5. CONCLUDING COMMENTS

Molecular advances in both diagnostics and therapeutics will undoubtedly change the practice of medicine as it is known today. This chapter introduces only some of the principles and strategies under active investigation. It is unlikely that gene therapy will replace conventional treatment strategies for human cancers, rather it will likely become a very important adjuvant line of therapy. As more advances occur in the study of dedifferentiation, angiogenesis, invasion and metastasis, drug resistance, and the cellular pathways that define normal versus abnormal growth and development, new targets will be identified. However, it would be naive to think that correction of any one gene defect will act universally or that any attempt at therapeutic gene correction will effect only the gene targeted and not introduce unwanted effects in the tightly orchestrated cellular mechanics. The gene therapeutics discussed in this chapter are directed at correction of somatic errors and therefore will not alter germline mutations. Germline mutations, which account for approximately 5% of all gynecologic cancers, would require a permanent gene replacement in the genome of the patient, a possibility with techniques such as homologous recombination. Alteration of genetic material that would be propagated in generations to come, would result in significant social and ethical implications to be resolved.

Many practical questions arise from the use of viral vectors or immune regulators in the introduction of non-human DNA sequences into the genome. Standardization of clinical grade reagents with impeccable quality control must govern industry as commercialization emerges. Thus, biomedical oversight and regulation must also be addressed with these technological advancements. Presently, the governmental regulatory bodies in the United States have effectively, albeit painstakingly, evolved to address the issues raised from the advancement of gene therapy strategies.[112] The Recombinant DNA Advisory Committee (RAC) of the National Institute of Health publicly evaluates new technology, particularly with regard to social and ethical issues. In 1985 the Human Gene Therapy Subcommittee (HGTS) was created by RAC to guide procedures and safety protocols for the work place in somatic cell and gene therapy investigations. The HGTS officiated the "Points to Consider" document as a guide for investigators who would ultimately approach the HGTS and RAC for approval of somatic cell of gene therapy clinical trials.[113] Any investigator involved with an organization funded in any amount by the NIH must have approval from RAC. Industry was urged by public pressure to seek approval from the RAC as well, however the first private company to bypass this regulatory committee was reported in 1992.[114] In addition, all protocols must be approved under an Investigational New Drug Application by the FDA.

The goal of the FDA is to assure safety of the biologic product, which involves regulation of the biologic source, production methods of the biologic product, and quality of the bulk final product. The system of regulations appears to be effective despite initial difficulties encountered in the first gene therapy clinical trial.[115] Importantly, public opinion remains favorable toward gene therapy. A new thrust for commercialization of biologic products is now underway, and was marked most recently by the assignment of a patent to the NIH for the most common type of gene therapy.[116] Biotechnology companies are emerging with active and highly visible roles in genetic research, as seen in the cloning and rapid commercialization of mutation screening of the BRCA1[117] and BRCA2 genes.[118]

The Human Genome Project is predicted to be completed in the year 2004, at that time all estimated 100,000 human genes will be sequenced. Further details into cell signaling and regulation will likely be described and novel pathways for therapeutic intervention will likely be characterized. Until now, oncogenes and tumor suppressor genes have been used primarily as molecular markers to predict prognosis and survival[119] or drug resistance.[120] As biologic therapies based on these genes reach clinical application, we must carefully proceed under stringent regulatory and ethical guidelines.

REFERENCES

1. **Weinberg, R.A.,** The molecular basis of oncogenes and tumor suppressor genes. *Ann. N.Y. Acad. Sci.,* 758, 331, 1995.
2. **Baserga, R.,** Oncogenes and the strategy of growth factors. *Cell,* 79, 927, 1994.
3. **Pejovic, T.,** Genetic changes in ovarian cancer. *Ann. Med.,* 27, 73, 1995.
4. **Lynch, H.T., Fusaro, R.M., and Lynch, J.,** Hereditary cancer in adults. *Cancer Detection Prev.,* 19, 219, 1995.
5. **Doherty, P.C., Tripp, R.A., and Sixbey, J.W.,** Evasion of host immune responses by tumours and viruses. *Ciba Foundation Symp.,* 187, 245, 1994.
6. **Cohen, E.P. and Kim, T.S.,** Neoplastic cells that express low levels of MHC class I determinants escape host immunity, *Semi. Cancer Biol.,* 5, 419, 1994.
7. **Tommasino, M. and Crawford, L.,** Human papillomavirus E6 and E7: proteins which deregulate the cell cycle. *Bioessays,* 17, 509, 1995.
8. **Colman, W.B. and Tsongalis, G.J.,** Multiple mechanisms account for genomic instability and molecular mutation neoplastic transformation. *Clin. Chem.,* 41, 644, 1995.
9. **Santoso, J.T., Tang, D.C., Lane, S.B., Hung, J., Reed, D.J., Muller, C.Y., Carbone, D.P., Lucci, J.A., Miller, D.S., and Mathis, J.M.,** Adenovirus-based p53 gene therapy in ovarian cancer. *Gynecol. Oncol.,* 59, 171, 1995.
10. **Liu, T.J., el-Nagger, A.K., McDonnell, T.J., Steck, K.D., Wang, M., Taylor, D.L., and Clayman, G.L.,** Apoptosis induction mediated by wild-type p53 adenoviral gene transfer in squamous cell carcinoma of the head and neck. *Cancer Res.,* 55, 3117, 1995.
11. **Janicek, M.F., Sevin, B.U., Nguyen, H.N., and Averette, H.E.,** Combination anti-gene therapy targeting c-myc and p53 in ovarian cancer cell lines. *Gynecol. Oncol.,* 59, 87, 1995.
12. **Rosenfeld, M.R., Meneses, P., Dalmau, J., Drobnjak, M., Cordon-Cardo, C., and Kaplitt, M.G.,** Gene transfer of wild-type p53 results in restoration of tumor-suppressor function in a medulloblastoma cell line. *Neurology,* 45, 1533, 1995.
13. **Herrman, F.,** Cancer gene therapy: principles, problems, and perspectives. *J. Mol. Med.,* 73, 157, 1995.

14. **Scanlon, K.J., Ohta, Y., Ishida, H., Kijima, H., et al.,** Oligonucleotide-mediated modulation of mammalian gene expression. *FASEB J.,* 9, 1288, 1995.

15. **Mercola, D. and Cohen, J.S.,** Antisense approaches to cancer gene therapy. *Cancer Gene Ther.,* 2, 47, 1995.

16. **Poeschla, E. and Wong-Staal, F.,** Antiviral and anticancer ribozymes. *Curr. Opin. Oncol.,* 6, 601, 1994.

17. **Jolly, D.,** Viral vector systems for gene therapy. *Cancer Gene Ther.,* 1, 51, 1994.

18. **Schofield, J.P. and Caskey, C.T.,** Non-viral approaches to gene therapy. *Br. Med. Bull.,* 51, 56-71, 1995.

19. **Zhang, W.W., Alemany, R., Wang, J., Koch, P., Ordonex, N.G., and Roth, J.A.,** Safety evaluation of AD5CMV-p53 *in vitro* and *in vivo. Hum. Gene Ther.,* 6, 1555, 1995.

20. **Boris-Lawrie, K. and Temin, H.M.,** The retroviral vector. Replication cycle and safety considerations for retrovirus-mediated gene therapy. *Ann. N.Y. Acad. Sci.,* 716, 59, 1994.

21. **Ostrove, J.M.,** Safety testing programs for gene therapy viral vectors. *Cancer Gene Ther.,* 1, 125, 1994.

22. **Herz, J. and Gerard, R.D.,** Adenovirus-mediated transfer of low-density lipoprotein receptor gene acutely accelerates cholesterol clearance in normal mice. *Proc. Natl. Acad. Sci. U.S.A.,* 90, 2812, 1993.

23. **Lemarchand, P., Jones, M., Yamada, I., and Crystal, R.G.,** In vivo gene transfer and expression in normal uninjured blood vessels using replication-deficient recombinant adenovirus vectors. *Circulation Res.,* 72, 1132, 1993.

24. **Zhang, W.W., Fang, X., Mazur, W., French, B.A., Georges, R.N., and Roth, J.A.,** High-efficiency gene transfer and high-level expression of wild-type p53 in human lung cancer cells mediated by recombinant adenovirus. *Cancer Gene Ther.,* 1, 5, 1994.

25. **Zhang, W.W., Alemany, R., Wang, J., Koch, P., Ordonex, N.G., and Roth, J.A.,** Safety evaluation of Ad5CMV-p53 *in vitro* and *in vivo. Hum. Gene Ther.,* 6, 1555, 1995.

26. **Ragot, T., Vincent, N., Chafey, P., Vigne, E., et al.,** Efficient adenovirus-mediated transfer of a human minidystophin gene to skeletal muscle of *mdx* mice. *Nature,* 361, 647, 1993.

27. **Liu, T.J., Zhang, W.W., Taylor, D.L., Roth, J.A., Goepfert, H., and Clayman, G.L.,** Growth suppression of human head and neck cancer cells by the introduction of a wild-type p53 gene via a recombinant adenovirus. *Cancer Res.,* 54, 662, 1994.

28. **Jaffe, H.A., Danel, C., Longenecker, G., Metzger, M., et al.,** Adenovirus-mediated gene transfer and expression in normal rat liver. *Nat. Genet.,* 1, 372, 1993.

29. **LeGal La Salle, G., Robert, J.J., Bernard, S., Ridoux, V., et al.,** Defective and non-defective adenovirus vectors for expressing foreign genes *in vitro* and *in vivo. Gene,* 101, 195, 1991.

30. **Kurzrock, R., Ku, S., and Talpaz, M.,** Abnormalities in the PRAD1 (Cyclin D1/bcl-1) oncogene are frequent in cervical and vulvar squamous cell lines. *Cancer,* 75, 584, 1995.

31. **Paquette, R.L., Lee, Y.Y., Wilczynski, S.P., Karmakar, A., Kizaki, M., Miller, C.W., and Koeffler, H.P.,** Mutations of p53 and human papillomavirus infection in cervical carcinoma. *Cancer,* 72, 1272, 1993.

32. **Berns, E.M., de Klein, A., van Putten, W.L., van Staveren, I.L., Bootsma, A., Klijn, J.G., and Foekens, J.A.,** Association between RB-1 gene alterations and factors of favorable prognosis in human breast cancer, without effect on survival. *Int. J. Cancer,* 64, 140, 1995.

33. **Scheffner, M., Munger, K., Byrne, J.C., and Howley, P.M.,** The state of p53 and retinoblastoma genes in human cervical carcinoma cell lines. *Proc. Natl. Acad. Sci. U.S.A.,* 88, 5523, 1991.

34. **Schildkraut, J.M., Collins, N.K., Dent, G.A., Tucker, J.A., Barrett, J.C., Berchuck, A., and Boyd, J.,** Loss of heterozygosity on chromosome 17q11-21 in cancers of women who have both breast and ovarian cancer. *Am. J. Obstet. Gynecol.,* 172, 908, 1995.

35. **Takahashi, H., Behbakht, K., McGovern, P.E., Chiu, H.C., Couch, F.J., Weber, B.L., Friedman, L.S., King, M.C., Furusato, M., LiVolsi, V.A., et al.,** Mutation analysis of the BRCA1 gene in ovarian cancers. *Cancer Res.,* 55, 2998, 1995.

36. **Ford, D., Easton, D.F., and Peto, J.,** Estimates of the gene frequency of BRCA1 and its contribution to breast and ovarian cancer incidence. *Am. J. Hum. Genet.,* 57, 1457, 1995.

37. **Weitzel, J.N., Patel, J., Smith, D.M., Goodman, A., Safaii, H., and Ball, H.G.,** Molecular genetic changes associated with ovarian cancer. *Gynecol. Oncol.,* 5, 245, 1994.

38. **Jesudasan, R.A., Rahman, R.A., Chandrashekharappa, S., Evans, G.A., and Srivatsan, E.S.,** Deletion and translocation of chromosome 11q13 sequences in cervical carcinoma cell lines. *Am. J. Hum. Genet.,* 56, 705, 1995.

39. **Mitra, A.B., Murty, V.V., Li, R.G., Pratap, M., Luthra, U.K., and Chaganti, R.S.,** Allelotype analysis of cervical carcinoma. *Cancer Res.,* 54, 4481, 1994.

40. **Peiffer, S.L., Herzog, T.J., Tribune, D.J., Mutch, D.G., Gersell, D.J., and Goodfellow, P.J.,** Allelic loss of sequences from the long arm of chromosome 10 and replication errors in endometrial cancers. *Cancer Res.,* 55, 1922, 1995.

41. **Anderson, M.J. and Stanbridge, E.J.,** Tumor suppressor genes studied by cell hybridization and chromosome transfer. *FASEB J.,* 7, 826, 1993.

42. **Cox, L.A., Chen, G., and Lee, E.Y.,** Tumor suppressor genes and their roles in breast cancer. *Breast Cancer Res. Treat.,* 32, 19, 1994.

43. **Kohler, M.F., Berchuck, A., Davidoff, A.M., Humphrey, P.A., Dodge, R.K., Iglehart, J.D., Soper, J.T., Clarke-Pearson, D.L., Bast, R.C., Jr., and Marks, J.R.,** Overexpression and mutation of p53 in endometrial carcinoma. *Cancer Res.,* 52, 1622, 1992.

44. **Marks, J.R., Davidoff, A.M., Kerns, B.J., Humphrey, P.A., Pence, J.C., Dodge, R.K., Clarke-Pearson, D.L., Iglehart, J.D., Bast, R.C., Jr., and Berchuck, A.,** Overexpression and mutation of p53 in epithelial ovarian cancer. *Cancer Res.,* 51, 2979, 1991.

45. **Berchuck, A., Kohler, M.F., Marks, J.R., Wiseman, R., Boyd, J., and Bast, R.C., Jr.,** The p53 tumor suppressor gene is frequently altered in gynecologic cancers. *Am. J. Obstet. Gynecol.,* 170, 246, 1994.

46. **Fujiwara, T., Grimm, E.A., Mukhopadhyay, T., Zhang, W.W., Owen-Schaub, L.B., and Roth, J.A.,** Induction of chemosensitivity in human lung cancer cells *in vivo* by adenovirus-mediated transfer of the wild-type p53 gene. *Cancer Res.,* 54, 2287, 1994.

47. **Fujiwara, T., Grimm, E.A., Mukhopadhyay, T.E.A., De, W.C., Owen-Shaub, L.B., and Roth, J.A.,** A retroviral wild-type p53 expression vector penetrates human lung cancer spheroids and inhibits growth by inducing apoptosis. *Cancer Res.,* 53, 4129, 1993.

48. **Yang, C., Cirielli, C., Capogrossi, M.C., and Passaniti, A.,** Adenovirus-mediated wild-type p53 expression induces apoptosis and suppresses tumorigenesis of prostatic tumor cells. *Cancer Res.,* 55, 4210, 1995.

49. **Clayman, G.L., El-Naggar, A.K., Roth, J.A., Zhang, W.W., Goepfert, H., Taylor, D.L., and Liu, T.J.,** *In vivo* molecular therapy with p53 adenovirus for microscopic residual head and neck squamous carcinoma. *Cancer Res.,* 55, 1, 1995.

50. **Liu, T.J., Zhang, W.W., Taylor, D.L., Roth, J.A., Goepfert, H., and Clayman, G.L.,** Growth suppression of human head and neck cancer cells by the introduction of a wild-type p53 gene via a recombinant adenovirus. *Cancer Res.,* 54, 3662, 1994.

51. **Hamanda, K., Zhang, W.W., Alemany, R., Roth, J.A., Wolf, J., and Mitchell, M.F.,** Growth inhibition of human cervical cancer cells by the recombinant adenovirus-mediated transfer of a wild-type p53 gene. *Gynecol. Oncol.,* 56, 130, 1995.

52. **Zhang, W.W., Fang, X., Mazur, W., French, B.A., Georges, R.N., and Roth, J.A.,** High-efficiency gene transfer and high-level expression of wild-type p53 in human lung cancer cells mediated by recombinant adenovirus. *Cancer Gene Ther.,* 1, 5, 1994.

53. **Skuse, G.R. and Ludlow, J.W.,** Tumour suppressor genes in disease and therapy. *Lancet,* 345, 902, 1995.

54. **Whyte, P.,** The retinoblastoma protein and its relative. *Sem. Cancer Biol.,* 6, 83, 1995.

55. **Bignon, Y.J. and Rio, P.,** The retinoblastoma gene: will therapeutic use of its tumor suppressive properties be possible? *Bull. Cancer,* 80, 704, 1993.

56. **Deissroth, A.B., Hanania, E.G., Claxton, D., Andreeff, M., Champlin, R., Kavanagh, J., Hortobagyi, G., Holmes, F., Reading, C., et al.,** Genetic therapy of human neoplastic disease, *J. Hemather.,* 2, 373, 1993.

57. **Burgring, B.M. and Bos, J.L.,** Regulation of Ras-mediated signalling: more than one way to skin a cat. *Trends Biochem. Sci.,* 20, 18, 1995.

58. **Hynes, N.E. and Stern, D.F.,** The biology of erbB-2/neu/HER-2 and its role in cancer. *Biochim. Biophys. Acta,* 1198, 165, 1994.

59. **Garte, S.J.,** The c-myc oncogene in tumor progresion. *Crit. Rev. Oncog.,* 4, 435, 1993.

60. **Pitot, H.C.,** The molecular biology of carcinogenesis. *Cancer,* 72, 962, 1993.

61. **Dougall, W.C., Qian, X., Peterson, N.C., Miller, M.J., Samanta, A., and Greene, M.I.,** The neu-oncogene: signal transduction pathways, transfotion mechanisms and evolving therapies. *Oncogene,* 9, 2109, 1994.

62. **Hung, M.C., Matin, A., Zhang, Y., Xing, X., Sorgi, F., Huang, L., and Yu, Y.,** HER-2/neu-targeting gene therapy—a review. *Gene,* 159, 65, 1995.

63. **Ravdin, P.M. and Chamness, G.C.,** The c-erbB-2 proto-oncogene as a prognostic and predictive marker in breast cancer: a paradigm for the development of other macromolecular markers—a review. *Gene,* 159, 19, 1995.

64. **Brandt-Rauf, P.W., Pincus, M.R., and Carney, W.P.,** The c-erbB-2 protein in oncogenesis: molecular structure to molecular epidemiolgy *Crit. Rev. Oncogenesis,* 5, 313, 1994.

65. **Yu, D., Wolf, J.K., Scanlon, M., Price, J.E., and Hung, M.C.,** Enhanced c-erbB-2/neu expression in human ovarian cancer cells correlates with more severe malignancy that can be suppressed by E1A. *Cancer Res.,* 53, 891, 1993.

66. **Matin, A. and Hung, M.C.,** Negative regulation of the neu promoter by the SV40 large T antigen. *Cell Growth Differen.,* 4, 1051, 1993.

67. **Zhang, Y., Yu, D., and Hung, H.,** HER2/neu-targeting cancer therapy via adenovirus-mediated E1A delivery in an animal model. *Oncogene,* 10, 194, 1995.

68. **Pronk, G.J. and Bos, J.L.,** The role of p21 ras in receptor tyrosine kinase signalling. *Biochim. Biophys. Acta,* 1198, 131, 1994.

69. **Modie, S.A., Wolfman, A., and The, T.,** 3Rs of life: Ras, Raf and growth regulation. *Trends Genet.,* 10, 44, 1994.

70. **Caduff, R.F., Johnston, C.M., and Frank, T.S.,** Mutations of the Ki-ras oncogene in carcinoma of the endometrium. *Am. J. Pathol.,* 146, 18, 1995.

71. **Fujimoto, I., Shimizu, Y., Hirai, Y., Chen, J.T., Teshima, H., Hasumi, K., Masubuchi, K., and Takahashi, M.,** Studies on ras oncogene activation in endometrial carcinoma. *Gynecol. Oncol.,* 48, 196, 1993.

72. **Zhang, Y., Mukhopadhyay, T., Donehower, L.A., Georges, R.N., and Roth, J.A.,** Retroviral vector-mediated transduction of K-ras antisense RNA into human lung cancer cells inhibits expression of the malignant phenotype. *Hum. Gene Ther.,* 4, 451, 1993.

73. **Schwab, G., Duroux, I., Chavany, C., Helene, C., and Saison-Behmoaras, E.,** An approach for new anticancer drugs: oncogene-targeted antisense DNA. *Ann. Oncol.,* 5, 55, 1994.

74. **Zhang, J.Y. and Schultz, R.M.,** Fibroblasts transformed by different ras oncogenes show disimilar patterns of protease gene expression and regulation. *Cancer Res.,* 52, 6682, 1992.

75. **Moolten, F.L.,** Drug sensitivity ("suicide") genes for selective cancer chemotherapy. *Cancer Gene Ther.,* 1, 279, 1994.

76. **Deonarain, M.P., Spooner, R.A., and Epenetos, A.A.,** Genetic delivery of enzymes for cancer therapy. *Gene Ther.,* 2, 235, 1995.

77. **Brenner, M.K., Cunningham, J.M., Sorrentino, B.P., and Helop, H.E.,** Gene transfer into human hemopoietic progenitor cells. *Br. Med. Bull.,* 51, 167, 1995.

78. **Mullen, C.A.,** Metabolic suicide genes in gene therapy. *Pharmacol. Therapeut.,* 63, 199, 1994.

79. **Mullen, C.A. and Blaese, R.M.,** Gene therapy of cancer. *Cancer Chemoth. Biol. Response Modifiers,* 1, 176, 1994.

80. **Biron, K.K., de Miranda, P., Burnette, T.C., and Krenitsky, T.A.,** Selective anabolism of 6-methoxy purine arabinoside in varicella-zoster virus-infected cells. *Antimicrob. Agents Chemoth.,* 35, 2116, 1991.

81. **Wu, J.K., Can, W.G., Meylaerts, S.A., Qi, P., Vrionis, F., and Cherington, V.,** Bystander tumoricidal effect in the treatment of experimental brain tumors. *Neurosurgery,* 35, 1094, 1994.

82. Clinical Protocols. *Cancer Gene Ther.,* 2, 225, 1995.

83. **Whartenby, K.A., Abboud, C.N., Marrogi, A.J., Ramesh, R., and Freeman, S.M.,** The biology of cancer gene therapy. *Lab. Investig.,* 72, 131, 1995.

84. **Goldstein, L.J.,** Clinical reversal of drug resistance. *Curr. Probl. Cancer,* 19, 65, 1995.

85. **Licht, T., Pastan, I., Gottsman, M., and Herrmann, F.,** P-glycoprotein-mediated multidrug resistance in normal and neoplastic hematopoietic cells. *Ann. Hematol.,* 69, 159, 1994.

86. **Cai, Q., Rubin, J.T., and Lotze, M.T.,** Genetically marking human cells — results of the first clinical transfer study. *Cancer Gene Ther.,* 2, 125, 1995.

87. **DeLeo, A.B.,** 1992. Tumor rejection inducing antigens of mouse sarcomas: Biochemical and immunological characterization with monoclonal antibodies and CTL lines. In: *Tumor Immunology: Basic Mechanisms and Prospects for Therapy,* Goldfarb, R.H., Heisler, J., Whiteside, T.L., Eds., W.B. Sanders Company, Philadelphia, pp. 312-334.

88. **Urban, J.L. and Schreiber, H.,** Tumor Antigens. *Annu. Rev. Immunol.,* 10, 617, 1992.

89. **Klein, G.,** Immunovirology of transforming viruses. *Curr. Opin. Immunol.,* 3, 665, 1991.

90. **Bookman, M.A. and Bast, R.C., Jr.,** The immmunobiology and immunotherapy of ovarian cancer. *Semin. Oncol.,* 18, 270, 1991.

91. **Urban, J.L. and Schreiber, H.,** Tumour antigens, *Annu. Rev. Immunol.,* 10, 617, 1992.

92. **Bast, R.C. and Knapp, R.C.,** in Griffiths, C.T. and Fuller, A.F., Eds., *Gynecologic Oncology,* Martinus Nijhoff, Boston, MA, 1983, 187-72.

93. **Niloff, J.M.,** The role of the CA125 assay in the management of ovarian cancer *Oncology,* 2, 67, 1988.

94. **Delaloye, A.B. and Delaloye, B.,** Radiolabelled monoclonal antibodies in tumour imaging and therapy: out of fashion?. *Eur. J. Nuclear Med.,* 22, 571, 1995.

95. **Farghaly, S.A.,** Tumor markers in gynecologic cancer. *Gynecol. Obstet. Invest.,* 34, 65, 1992.

96. **Kuiper, M., Peakman, M., and Farzaneh, F.,** Ovarian tumor antigens as potential targets for immune gene therapy. *Gene Ther.,* 2, 7, 1995.

97. **Kuzel, T.M. and Rosen, S.T.,** Antibodies in the treatment of human cancer. *Curr. Opin. Oncol.,* 6, 622, 1994.

98. **Bast, R.C., Jr., Xu, F., Yu, Y., Crews, J., Argon, Y., Maier. L., Lidor, Y., Berchuck, A., and Boyer, C.M.,** Additive and synergistic interactions of monoclonal antibodies and immunotoxins reactive with breast and ovarian cancer. *Immunol. Series,* 61, 23, 1994.

99. **Rosenthal, F.M., Zier, K.S., and Gansbacher, B.,** Human tumor vaccines and genetic engineering of tumors with cytokine and histocompatibility genes to enhance immunogenicity. *Curr. Opin. Oncol.,* 6, 611, 1994.

100. **Herrmann, F.,** Cancer gene therapy: principles, problems, and perspectives, *J. Mol. Med.,* 73, 157, 1995.

101. **Nabel, E.G., Yang, Z., Muller, D., Chang, A.E., Gao, X., Huang, L., Cho, K.J., and Nabel, G.J.,** Safety and toxicity of catheter gene delivery to the pulmonary vasculature in a patient with metastatic melanoma. *Hum. Gene Ther.,* 5, 1089, 1994.

102. **Whalen, R.G. and Davis, H.L.,** DNA-mediated immunization and the energetic immune response to hepatitis B surface antigen. *Clin. Immunol. Immunopathol.,* 75, 1, 1995.

103. **Hines, J.F., Ghim, S., Schlegel, R., and Jenson, A.B.,** Prospects for vaccine against human papillomavirus. *Obstet. Gynecol.,* 86, 860, 1995.

104. **Shillitoe, E.J., Kamath, P., and Chen, Z.,** Papillomaviruses as targets for cancer gene therapy. *Cancer Gene Ther.,* 1, 193, 1994.

105. **Pardoll, D.M.,** Paracrine cytokine adjuvants in cancer immunotherapy. *Annu. Rev. Immunol.,* 13, 399, 1995.

106. **Forni, G., Giovarell, M., Cavallo, F., Consalvo, M., Allione, A., Modesti, A., Musiani, P., and Colombo, M.P.,** Cytokine-induced tumor immunogenicity: from exogenous cytokines to gene therapy. *J. Immunother.,* 14, 253, 1993.

107. **Miller, A.R., McBide, W.H., Hunt, K., and Economou, J.S.,** Cytokine mediated gene therapy for cancer. *Ann. Surg. Oncol.,* 1, 436, 1994.

108. **Foa, R., Guarini, A., Cignetti, A., Cronin, K., Rosenthal, F., and Gansbacher, B.,** Cytokine gene therapy: a new strategy for the management of cancer patients. *Nat. Immun.,* 13, 65, 1994.

109. **Crystal, R.G.,** Transfer of genes to humans: early lessons and obstacles to success. *Science,* 270, 404, 1995.

110. **Pippin, B.A., Rosentein, M., Jacob, W.F., Chiang, Y., and Lotze, M.T.,** Local IL-4 delivery enhances immune reactivity to murine tumors: gene therapy in combination with IL-2. *Cancer Gene Ther.,* 1, 35, 1994.

111. **Lotze, M.T., Rubin, J.T., Carty, S., Edington, H., Ferson, P., Landreneau, R., Pippin, B., Posner, M., Rosenfelder, D., and Watso, C. et al.,** Gene therapy of cancer: a pilot study of IL-4-gene-modified fibroblasts admixed with autologous tumor to elicit an immune response. *Hum. Gene Ther.,* 5, 41, 1994.

112. **Kessler, D.A., Siegel, J.P., Noguchi, P.D., Zoon, K.C., Feiden, K.L., and Woodcock, J.,** Regulation of somatic-cell therapy and gene therapy by the food and drug administration. *N. Engl. J. Med.,* 329, 1169, 1993.

113. **McGarrity, G.J. and Anderson, W.F.,** Human gene therapy protocols: RAC review. *Science,* 268, 1261, 1995.

114. **Anderson, C.,** U.S. company's gene therapy trial is first to bypass RAC for approval. *Nature,* 357, 615, 1992.

115. **Carmen, I.H.,** Debates, divisions, and decisions: recombinant DNA advisory committee (RAC) authorization of the first human gene transfer experiments. *Am. J. Hum. Genet.,* 50, 24, 1992.

116. **Nowak, R.,** Gene therapy. Patent stirs a controversy. *Science,* 267, 1899, 1995.

117. **Goldberg, K.B. and Goldberg, P.,** The first BRCA1 test hits the market; are oncologists, patients ready? *Cancer Lett.,* 22, 1, 1996.

118. **Goldberg, K.B. and Goldberg, P.,** Myriad files for patent on full sequence for BRCA2 *Cancer Econom.* January 1996.

119. **Mansour, E.G., Ravdin, P.M., and Dressler, L.,** Prognostic factors in early breast carcinoma. *Cancer,* 74, 381, 1994

120. **Roninson, I.B.,** From amplification to function: the case of the MDR1 gene. *Mutation Res.,* 276, 151, 1992.

Part IV.
Endometrial Cancer

Chapter 11

Endometrial Cancer: An Introduction

Roger J.B. King

CONTENTS

1. INTRODUCTION

Endometrial cancer is the most common female genital cancer in Western societies and is considered to be one of the less aggresive tumors in that it has about one fifth the incidence of breast cancer but causes only one tenth the mortality of the latter in the U.S. In large part, this reflects early diagnosis because of the cancer's visible effect on vaginal bleeding. Endometrial cancer can arise from several cell types but the glandular epithelium is the major progenitor. It is uncommon below the age of 40 with a peak incidence by about 70 years; mortality climbs steadily from about age 50.

Mortality has been dropping steadily since 1950 in most countries, whereas incidence has changed little in the U.K. and is still increasing in the U.S. Such a pattern is seen at all ages although it is most evident in lower age-groups (Figure 1); however, care must be taken in considering such data because of variable reporting methods for mortality in different countries. Cancer of the uterine body is mainly endometrial with about one fifth being of other types such as sarcomas and carcinosarcomas. If unspecified, the numbers can include the cervix which, because of its different etiology, complicates interpretation.[1] The reason for the decline with time is unclear with oral contraceptives (OC) use being one candidate. From the epidemiological data indicating a protective effect of OC use, one can model a 60% decrease in mortality but this positive link is counteracted by the fact that falling mortality preceded wide-spread OC use. Other possible contributing factors include hormone replacement therapy, increased hysterectomy, diet, and treatment effects.[1]

Endometrial cancer is most common in Western countries and low in Asia but this pattern is changing as the low incidence in Japan has been steadily increasing from 1970 to the present time. This is associated with dietary changes but, fortunately, it is not accompanied by an increase in mortality.[2]

2. PATHOLOGY

Adenocarcinomas are the predominant type of cancer, accounting for 80 to 90% of uterine tumors, with sarcomas and mixed carcinosarcomas making up the other 10 to 20%. Only adenocarcinomas will be discussed here and the reader is referred to Reference 3 for details of the other forms.

184

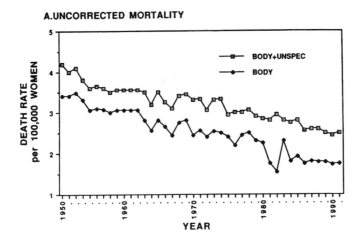

A. UNCORRECTED MORTALITY

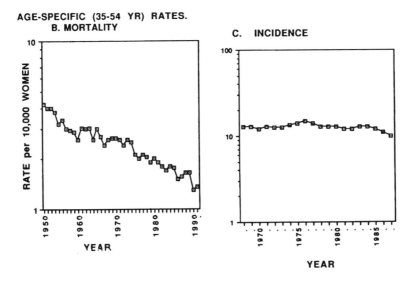

AGE-SPECIFIC (35-54 YR) RATES.
B. MORTALITY

C. INCIDENCE

Figure 1 Incidence and mortality of endometrial cancer in England and Wales. Body — uterine corpus; unspec — unspecified site within the reproductive tract. (Adapted from Reference 1.)

2.1 Adenocarcinomas

About a quarter of adenocarcinomas exhibit some degree of squamous differentiation with terms such as adenocanthoma and adenosquamous carcinoma being applied according to the degree of benign and malignant components respectively present. The significance of these changes is unclear, and they are little used in deciding treatment options. Other variants such as papillary, secretory, and ciliated tumors will not be dealt with as separate variants.

Adenocarcinomas are graded 1 to 3, according to their differentiation status with low grade indicating a well-differentiated tumor. The FIGO system incorporates nuclear pleiomorphism, number of mitoses and degree of retention of glandular structures and has prognostic use, in that 5-year survival rates were higher for women with Grade 1 than Grade 3 tumors.[4] In 1988 a revised system was introduced, incorporating architectural features and vascular invasion which is more precise and also has prognostic value (Table 1).

Table 1 Five-Year Survival Rates for Patients with Various Prognostic Factors

Prognostic Factor	% Surviving	Ref.
Stage*		
1 A	94	4
1 B	95	
1 C	75	
II	83	
III	43	
Grade*		
1	89	3
2	73	
3	61	
Myometrial invasion*		
None	80	28
Superficial	85	
Deep	50	
Ploidy		
Diploid	85	32
Non-diploid	60	
Proliferation index		
Low	90	32
High	65	
Estrogen/progestin receptors*		
ER-negative	54	24
ER-positive	82	
PR-negative	46	
PR-positive	76	
HER-2 overexpression		
None	95	33
High	56	
HSP27		
Low	92	34
High	60	

* Values averaged from several trials.

2.2 Hyperplasias

The terminology suggested by the International Society for Gynecological Oncology, which divides hyperplasias into simple and complex categories with subdivisions within each type, will be used here.[3]

Simple hyperplasias include cystic hyperplasia and have an elevated proportion of glands to stroma but the stroma is still abundant while the glandular epithelium has no abnormal features and minimal mitotic activity. The complex variant of simple hyperplasia has a higher density of glands whose luminal surface becomes irregular; nuclear morphology remains normal. The majority of simple hyperplasias remain static or regress which is compatible with the very low number that progress to adenocarcinoma (1% simple; 3% complex simple) if left untreated.[3]

Complex or atypical hyperplasias are characterized by abnormal luminal and nuclear morphology and a decreased proportion of stroma. The simple variant has an enlarged nucleus containing a prominent nucleolus and irregular chromatin but stroma is evident between glands. The complex subcategory presents more extreme features with nuclear pleiomorphism and luminal protrusions that can form intraglandular epithelial bridges; stroma can be virtually absent with a 'back-to-back' appearance of the glands. Cytological atypia is the best indicator of the likelihood of progression to invasive cancer.[3] It is sometimes difficult to distinguish atypical hyperplasias from grade 1 tumors although presence or absence of stromal invasion is considered the best discriminant. On this criterion, carcinoma in-situ is categorized as a complex, atypical hyperplasia. Progression to invasive cancer occurs in about 8 and 29% of atypias with simple and complex features, respectively.[4,5]

3. PATHOGENESIS

The development of invasive carcinoma from normal endometrium occurs as a continuum of changes (Figure 2) with the caveats that stages defined by histological criteria can be bypassed and that only atypical hyperplasia be considered preneoplastic. Estrogens are the main driving force for carcinogenesis with progestins reversing or preventing some of those changes (see Section 4). As tumors become more dedifferentiated (increased grade), they are less likely to respond to high-dose progestins and more likely to metastasize (see Section 5).

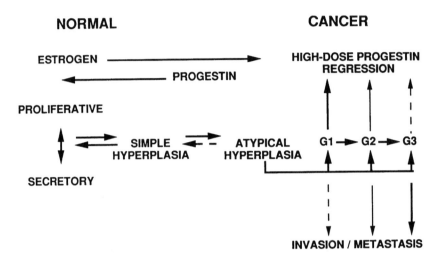

Figure 2 Hormone involvement in the pathogenesis of endometrial cancer. G1 to G3 — histological grade. The different types of arrow indicate increased (bold) or decreased (dashed) likelihood of an event.

3.1 Morphological and Clinical Criteria

The pathway described in Figure 2 is compatible with the increasing risk of carcinomatous progression associated with increasing severity of hyperplasia (Section 2.2). In part, this is a tautological argument given that the criteria used to define hyperplasias are based on increasing degrees of morphological abnormality so some justification must be provided to warrant the linking arrows in Figure 2. Clear-cut evidence is lacking but a consensus of opinion based on four types of data indicate the validity of the model with the caveats mentioned above. The four data sets are: association of different hyperplasias in individual samples, reversibility of simple and some complex hyperplasias, clinical results from exogenous hormone use, and evidence from cell and molecular biological research.

Foci of atypical hyperplasia are often seen in association with both simple hyperplasias and invasive carcinoma, while hyperestrogenic states result in elevated proportions of both hyperplasias and neoplasias.[6] Simple hyperplasias are reversible either spontaneously or with progestins and the latter are effective with some complex hyperplasias.[5,7] However, some irreversible changes have occurred by the atypical stage as seen in the changed ratios of estrogen (ER) and progestin (PR) receptor[8] and altered subcellular organelles.[9] The word reversible is used advisedly as it is unclear as to whether the cells are in fact reversed, as opposed to being killed.

Although invasive cancers become progressively more dedifferentiated with time and low-stage tumors are more likely to be Grade 1 than advanced cancers, the multiple arrows between atypical hyperplasia and different grades of tumor should be noted. Clinical results with exogenous hormone use (see Section 4) and high-dose progestin treatments (see Section 5) are compatible with the model.

Such data are most easily explained by the progression model but do not rule out the possibility of field changes within which separate, divergent pathways occur. This may be the case both with some simple hyperplasias and in the generation of high-grade tumors from complex hyperplasias.

3.2 Biochemical and Molecular Criteria

Laboratory studies on *in-vitro* carcinogenesis, oncogene and repressor actions have established two fundamentally important criteria about carcinogenesis; it is a multistep process that involves progressive accumulation of errors in regulatory pathways, and second, that a single type of cancer can be generated by different mechanisms. The elegant work on colon carcinogenesis[10] illustrates the first point, while analysis of differential p53 mutations in both lung and liver cancers point to variable causative agents.[11] The limited amount of relevant work with endometrium is compatible with both these points (Section 3.3).

3.3 Relevance of Experimental Studies to Human Endometrial Carcinogenesis

Everything we know about physiological estrogen and progestin effects on endometrium is compatible with clinical experience with these compounds and with the model shown in Figures 2 and 3, but it is pertinent to ask how estrogens actually generate cancers? Being mitogenic, the appearance of simple hyperplasias is understandable, but why should additional changes occur? Experimental work with rodents generated the concept of initiation and promotion with estrogens classically being defined as promotional agents.[12] Their proliferative effects and, with the important exception of diethylstilboestrol-induced uterine tumors in embryonic mice,[13] inability to damage DNA[14] are compatible with promotion, but this is a tautological argument, in that the same data are used to justify the promotion model.

LUMEN

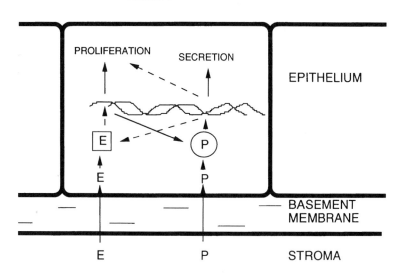

Figure 3 Receptor-mediated effects on endometrial epithelium Estrogen (E) and progestin (P) diffuse from stromal blood vessels through the basement membrane into the glandular epithelial cells. Within the cells, the hormones bind to their respective receptors (□, estrogen receptor or ○, progestin receptor) and activate gene transcription. This can result in stimulation (→) or inhibition (⋅→) of the entities shown.

The initiation/promotion model should not be applied uncritically to human endometrium as no initiating agents have yet been identified and it has been hypothesized that proliferation alone is the engine that drives endometrial carcinogenesis.[15] The proliferation-alone idea may be strengthened by inclusion of two features emerging from studies with other cancers: generation of genetic instability and the importance of programmed cell death or apoptosis. The former idea, based on analysis of mutation rates, suggests that increased proliferation assists in the generation of errors such that genetic changes are accelerated.[16-17] Patients with Lynch's syndrome have an increased risk of endometrial cancer in addition to the more frequent colon cancer. This familial condition is accompanied by mutations in DNA repair proteins, MLH1, MSH2, PMS1, and PMS2, loss of which results in increased error accumulation that can be detected by DNA microsatellite instability; endometrial carcinomas exhibit such instability.[16-17] Thus, mitogenic estrogens could provoke genetic instability without invoking additional properties or unknown agents.

It is now clear that cell proliferation and apoptotic death are closely linked in normal cells and that this link is disrupted in cancer cells; changes that sever the link could accelerate carcinogenesis. It has long been known that estrogens induce apoptosis in endometrial cells,[18] so a disruptive change in that link would have important consequences. A pivotal gene linking proliferation and apoptosis is the repressor gene, p53 which is inactivated in about 20% of endometrial carcinomas.[4] This is an underestimate, being mainly based on imunohistochemical studies that only detect approximately half of all mutations but tumors such as sarcomas have normal p53 that is inactivated by combination with other proteins. It is noteworthy that the gene for one such protein, MDM2, is amplified by the estrogen, diethylstilbestrol, as part of the latter's carcinogenic effect on embryonic mouse uterus.[13] These speculations should be taken for what they are, but overall, they do indicate that estrogen-induced proliferation may indeed be sufficient to explain its carcinogenic potential in the uterus. However, a desire for simplicity should not overide what may be a more complex situation!

Finally, it should be remembered that normal exposure to female sex hormones followed by their withdrawal results in monthly shedding of the functionalis part of the endometrium. Abnormal cells in this part of the endometrium would be lost at this time so menstruation may be protective. As menstruation is a consequence of progesterone production and withdrawal,[19] all the known endometrial risk factors could be accommodated within an inadequate menstrual loss model. Menstruation is unique to primates, so laboratory models would miss what might be an important, albeit incomplete, regulatory mechanism.

4. HORMONE SENSITIVITY

A clear-cut and relatively simple hormonal involvement in endometrial carcinogenesis has been established from epidemiological, laboratory, and clinical data: estrogens promote carcinogenesis and progestins counteract that effect (Figure 2). Separate intracellular receptors mediate their actions. Estrogens sensitize the cell to progestin action by increasing PR levels while progestins downregulate estrogen effects by decreasing ER levels (Figure 3), by increasing the catabolism of estradiol by induction of estradiol dehydrogenase and by inducing differentiation (secretion).[20]

4.1 Epidemiology

The main risk factors for endometrial cancer are obesity, estrogen use, not using hormonal contraceptives, clinical conditions causing hyperestrogenism, anovulation, and nulliparity, all of which can be accommodated in the estrogen-bad, progestin-good hypothesis.[21] The two most clear-cut sets of evidence come from exogenous sex-steroid use, oral contraceptives (OCs)[22] and hormone replacement therapies[6] involving either estrogen alone (ERT) or plus a progestin (HRT) (Figure 4). Most OCs are a combination of estrogen and progestin and can generate a 60% decreased risk if used for more than one year. Once established, the protection can last for up to 15 years after stopping.[22] ERT rapidly increases cancer risk (Figure 4.B), an effect that is negated by progestins.

Key and Pike[21] have effectively modeled the OC effect on the basis that estrogens are mitogenic and progestins block proliferation over the period of use with normal epithelial activity resuming at cessation of OC use (Figure 5). Their model, in which the altered incidence rate at the normal menopause is accounted for by decreased ovarian steroid production, also applies to ERT which increases endometrial cancer risk (Figures 4 and 5) and is counteracted by inclusion of a progestin for 10 to 12 days each month.[6] As well as providing proof of estrogen and progestin functions in endometrial carcinogenesis, these data indicate the relative rapidity with which these cancers can be generated as risk increases within two years use and then continues to rise with duration of use (Figure 4B). It should be noted that modern OC's contain progestin levels that are near-maximally effective in blocking endometrial proliferation and yet do not provide 100% protection (Figure 4) so factors other than hormone exposure cannot be discounted.

Obesity increases risk and, in some parts of the world, may account for up to a third of endometrial cancers. This effect has been linked with the ability of fat to synthesize estrogens but additional dietary influences may be important. Migration studies have indicated that Japanese women moving to the U.S. retain a low risk provided they maintain a Japanese lifestyle but that risk rises to U.S. levels if they westernize their diet. There is probably a multifactorial explanation for these changes involving fat, fruit, and vegetables.

The increased risk associated with anovulation is best explained by the associated hypoprogestogenic environment whilst nulliparity may be linked to anovulation, although an additional protective effect of pregnancy may occur.

4.2 Clinical Data

The high proportion of both simple and atypical hyperplasias in women on ERT can be eliminated with 12 days per month progestin, shorter times being less effective.[6] This fits with the observation that simple and some atypical hyperplasias can be normalized with progestins.[6-7] Most invasive adenocarcinoma contain estrogen (ER) and progesterone (PR) receptors, the presence of which correlates with responses

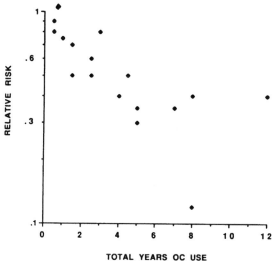

A. COMBINED ESTROGEN/PROGESTIN CONTRACEPTIVE PILL

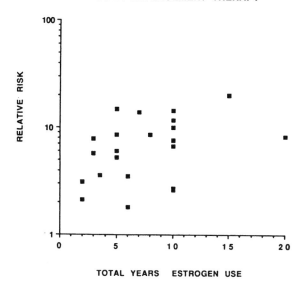

B. ESTROGEN REPLACEMENT THERAPY

Figure 4 Effect of exogenous steroids on endometrial cancer risk. A. Combined estrogen plus progestin oral contraceptive pill. (Data from Reference 22.) B. Estrogen replacement therapy. (Data from Reference 6.) Where that paper refers to greater than a set number of years use, it has been plotted as that duration of use.

to progestins. This hormone sensitivity of the tumors is less marked in poorly differentiated tumors (Figure 2).[24]

Tamoxifen, a partial estrogen antagonist, increases endometrial cancer because of its agonist activity.[25] Interestingly, tamoxifen and ERT generate a higher proportion of well-differentiated tumors than appear in the general population and it has been suggested that these exogenous estrogen-induced cancers, being less malignant, be considered as separate entities.

4.3 Laboratory Data

Both estrogens and progestins influence proliferation via their own receptors which are nuclear transcription factors.[27] Many genes have been identified whose transcription is modulated by these hormones

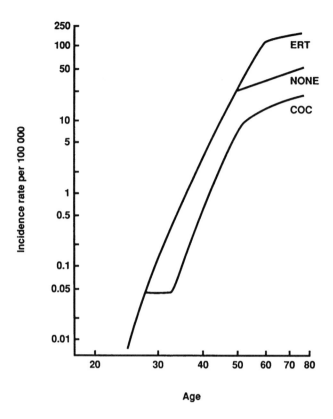

Figure 5 Predicted effect of 5-years combined oral contraceptive (COC) or estrogen replacement therapy (ERT) on endometrial cancer. (Data from Reference 21.)

but current information on which ones are involved in the proliferative response are sparse; growth factors and their receptors as well as oncogenes are certainly involved. These aspects are discussed elsewhere in this volume.

5. TREATMENT

Endometrial cancers are very vascular and superficial bleeding is an early event that leads to diagnosis by curettage; because of this early detection, about 75% of tumors are diagnosed as being Stage 1 with the attendant good 5-year survival rate.[28] Hysterectomy is the first line of treatment but careful staging is crucial to subsequent management. The establishment by FIGO of staging and grading criteria with prognostic value (Table 1) has benefited treatment decisions.[3] The main classification features are histological grade, depth of myometrial invasion, lymph node involvement, and presence of tumor outside the uterus, each of which provide independant information about likelihood of survival (Table 1). Degree of vascular invasion is not one of the FIGO criteria but, given its prognostic value, it has been advocated as an additional marker.[4]

A number of biochemical markers of survival have been identified (Table 1) which can be interpreted in terms of DNA abnormalities (ploidy) or proliferation. Proliferation index, the sum of S, G_2 and M phase cells, is the most direct measure of proliferation but each of the other markers have been linked to that parameter.

Stage 1 disease with good histopathological features like low grade has an excellent prognosis and adjuvant treatment is not required whereas extensive invasion plus high-grade tumor suggests a more aggresive situation that might benefit from adjuvant chemo- or hormone therapy additional to radiotherapy.[28] Unfortunately, current evidence with adjuvant progestin or chemotherapy do not look promising[4,29] even though estrogen and progestin receptors do help identify hormone responsive tumors.[24]

With advanced disease, both hormone and chemotherapy generate responses albeit of short duration. A review of early trials with progestins such as medroxyprogesterone acetate, megestrol, or 17 hydroxy progesterone caproate indicated an overall response rate of 34% with an 18 to 33 month survival, but recent data look less optimistic with a less than 20% response rate.[30] Intramuscular injection of MPA has no advantage over the oral route and there are no benefits in using massive doses. Steroid receptor content indicates increased probability of response with the presence of both ER and PR being better than only one receptor type.[24] The partial estrogen antagonist, tamoxifen, also generates a response in about a quarter of cases but this masks a range varying from 0 to 53% in different trials.[30] The benefical effect of tamoxifen is tempered by the fact that it also has some agonist activity and will actually generate endometrial cancers. However, women with metastatic endometrial cancer have limited life-span so this is not a major problem. With single-agent progestin or antiestrogen therapies, best responses are seen with steroid receptor-positive, low grade tumors.[30]

Gonadotrophin-releasing hormone analogs are currently being tested with results similar to those seen with other hormone therapies.[30] Their mode of action remains to be established in postmenopausal women, but a direct effect of the polypeptides on endometrial cells remains a possibilty.

Chemotherapy is effective in about one third of patients with advanced disease but the duration of response rarely extends beyond 12 months. Agents such as cisplatin, carboplatin, or doxorubicin are the most effective and combinations of cisplatin and doxorubicin may extend survival marginally beyond that seen with single agents.[31] Combinations of chemotherapy and hormones have no apparent advantages over chemotherapy.[31]

ACKNOWLEDGMENTS

This work was supported by Charitable Trusts administered by Marshall's solicitors, Godalming, U.K.

REFERENCES

1. **Mant, J.W.F. and Vessey, M.P.,** Ovarian and endometrial cancers, *Cancer Surv.*, 19, 287, 1994.
2. **Henderson, M.M.,** International differences in diet and cancer incidence, *J. Natl. Cancer Inst.*, 12, 59, 1992.
3. **Gordon, M.D. and Ireland, K.,** Pathology of hyperplasia and carcinoma of the endometrium, *Sem. Oncol.*, 21, 64, 1994.
4. **Homesley, H.D. and Zaino, R.,** Endometrial cancer: prognostic factors, *Sem. Oncol.*, 21, 71, 1994.
5. **Richardson, G.S. and MacLaughlin, D.T.,** Hormonal biology of endometrial cancer, *UICC Techn. Rep. Ser.*, 42, 49, 1978.
6. **Grady, D., Rubin, S.M., Petitti, D.B., Fox, C.S., Black, D., Ettinger, B., Ernster, V.L., and Cummings, S.R.,** Hormone therapy to prevent disease and prolong life in postmenopausal women, *Ann. Int. Med.*, 117, 1016, 1992.
7. **Richardson, G.S. and MacLaughlin, D.T.,** Hormonal biology of endometrial cancer, *UICC Tech. Rep. Ser.*, 42, 155, 1978.
8. **King, R.J.B.,** Effects of female sex hormones on human endometrium in relation to neoplasia, in *Mechanisms of Steroid Action*, Lewis, G.P. and Ginsburg, M., Eds., Biological Council, 19, 1981, chap. 4.
9. **Ferenczy, A.,** Endometrial morphologic response to endogenous and exogenous hormonal stimuli in health and disease: normal endometrium during the menstrual cycle; endometrial hyperplasia and carcinoma, *Prog. Cancer Res. Ther.*, 14, 489, 1980.
10. **Fearon, E.R. and Jones, P.A.,** Progressing towards a molecular description of colorectal cancer development, *FASEB J.*, 6, 2783, 1992.
11. **Greenblatt, M.S., Bennett, W.P., Hollstein, M., and Harris, C.C.,** Mutations in the p53 tumor suppressor gene: clues to cancer etiology and molecular pathogenesis, *Cancer Res.*, 54, 4855, 1994.
12. **King, R.J.B.,** Biology of female sex hormone action in relation to contraceptive agents and neoplasia, *Contraception*, 43, 527, 1991.
13. **Risinger, J.I., Terry, L.A., and Boyd, J.,** Use of representational difference analysis for the identification of mdm2 oncogene amplification in diethylstilbestrol-induced murine uterine adenocarcinomas, *Mol. Carcinogenesis*, 11, 13, 1994.
14. World Health Organization/International Agency for Research on Cancer. Genetic and related effects: an updating of selected IARC monographs from vols 1-42. IARC monographs on the evaluation of carcinogenic risks to humans, supplements 6: 250, 293, 369, 426, 437, 1987.
15. **Preston-Martin, S., Pike, M.C., Ross, R.K., Jones, P.A., and Henderson, B.E.,** Increased cell division as a cause of human cancer, *Cancer Res.*, 50, 7415, 1990.
16. **Loeb, L.A.,** Microsatellite instability: marker of a mutator phenotype in cancer, *Cancer Res.*, 54, 5059, 1994.
17. **Eshleman, J.R. and Markowitz, S.D.,** Microsatellite instability in inherited and sporadic neoplasms, *Curr. Opinion Oncol.*, 7, 83, 1995.
18. **Martin, L., Finn, C.A., and Trinder, G.,** Hypertrophy and hyperplasia in the mouse uterus after oestrogen treatment: an autoradiographic study. *J. Endocrinol.*, 56, 133, 1973.

19. **King, R.J.B. and d'Arcangues, C.,** Steroid hormone effects on uterine blood vessels, in *Steroid Hormones and Uterine Bleeding*, Alexander, N.J. and d'Arcangues, C., Eds, AAAS Press, 1992, 15.

20. **King, R.J.B. and Whitehead, M.I.,** Estrogen and progestin effects on epithelium and stroma from pre- and postmenopausal endometria: application to clinical studies of the climacteric syndrome, in *Steroids in Endometrial Cancer, Progress Cancer Res. Ther.*, 25, 105, 1983.

21. **Key, T.J.A. and Pike, M.C.,** The dose-effect relationship between 'unopposed' oestrogens and endometrial mitotic rate: its central role in explaining and predicting endometrial cancer risk, *Br. J. Cancer*, 57, 205, 1988.

22. **Schlesselman, J.J.,** Oral contraceptives and neoplasia of the uterine corpus, *Contraception,* 43, 557, 1991.

23. **La Vecchia, C.,** Cancers associated with high-fat diets, *J. Natl. Cancer Inst.,* 12, 79, 1992.

24. **Chambers, J.T., MacLusky, N., Eisenfield, A., Kohorn, E.I., Lawrence, R., and Schwartz, P.E.,** Estrogen and progestin receptor levels as prognosticators for survival in endometrial cancer, *Gynecol. Oncol.*, 31, 65, 1988.

25. **van Leeuwen, F.E., Benraadt, J., Coebergh, J.W.W., Kiermeney, L.A.L.M., Gimbrere, C.H.F., Otter, R., Schouten, L.J., Damhuis, R.A.M., Bontenbal, M., Diepenhorst, F.W., van den Belt-Dusebout, A. W. and van Tinteren, H.,** Risk of endometrial cancer after tamoxifen treatment of breast cancer, *Lancet*, 343, 448, 1994.

26. **Seoud, M.A.-F., Johnson, J., and Weed, J.C.,** Gynecologic tumors in tamoxifen-treated women with breast cancer, *Obstet. Gynecol.*, 82, 165, 1993.

27. **King, R.J.B.,** Effects of steroid hormones and related compounds on gene transcription, *Clin. Endocrinel.*, 36, 1, 1992.

28. **Hoskins, W.J., Perez, C.A., and Young, R.C.,** Gynecologic Tumors, in *Cancer, Principles and Practice of Oncology*, DeVita, V.T., Hellman, S. and Rosenberg, S.A., Eds, Lippincott, Philadelphia, 1993, 1152.

29. **Burke, T.W. and Wolfson, A.H.,** Limited endometrial carcinoma: adjuvant therapy, *Sem. Oncol.*, 21, 84, 1994.

30. **Lentz, S.S.,** Advanced and recurrent endometrial carcinoma: hormonal therapy, *Sem. Oncol.*, 21, 100, 1994.

31. **Muss, H.B.,** Chemotherapy of metastatic endometrial cancer, *Sem. Oncol.*, 21, 107, 1994.

32. **Podratz, K.C., Wilson, T.O., Gaffey, T.A., Cha, S.S., and Katzmann, J.A.,** Deoxyribonucleic acid analysis facilitates the pretreatment identification of high risk endometrial cancer patients, *Am. J. Obstet. Gynecol.*, 168, 1206, 1993.

33. **Hertzel, D.J., Wilson, T.O., Keeney, G.L., Roche, P.C., Cha, S.S., and Podratz, K.C.,** Her-2/neu expression: a major prognostic factor in endometrial cancer, *Gynecol. Oncol.*, 47, 179, 1992.

34. **Raju, K.S., King, R.J.B., Kaern, J., Sumner, D., Abeler, V.M., Mandalaya, S., and Trope, C.,** Influence of HSP27 and steroid receptor status on provera sensitivity, DNA-ploidy and survival of females with endometrial cancer. *Int. J. Gynecol. Cancer*, 5, 94, 1995.

Chapter 12

Endometrial Carcinoma: Steroid Hormones, Growth Factors, and Cytokines

Pondichery G. Satyaswaroop and Siamak Tabibzadeh

CONTENTS

1. INTRODUCTION

Epithelial cells of the human endometrium are highly vulnerable to neoplastic transformation. Endometrial carcinoma is the most common gynecologic malignancy in the Western world, and in the U.S., an annual occurrence of 31,000 new cases and 5,700 deaths were reported in 1993.[1] Predominantly a postmenopausal disease, its incidence appears to have increased during the past 50 years, presumably due to increased life expectancy in women and improved detection methods. Surgery and radiation result in 80 to 90% cure rates in early stages. Most of the deaths occur in patients with advanced, recurrent, or metastatic disease.

Hormonal or chemotherapy is the treatment of choice for recurrent or metastatic endometrial carcinoma. Progestins have been commonly used in the treatment of the metastatic disease with about 25% response rates[2,3] while 30% response rates were observed with cytotoxic chemotherapy.[4] Unfortunately, there are no accurate tests to predict the response of a given endometrial tumor to hormonal or chemotherapy. Hence, the major focus of research studies in this area is the determination of the sensitivity of endometrial carcinomas to steroids, growth factors, and cytotoxic agents. Development of *in vitro* tests of drug sensitivity and preclinical *in vivo* models for studying therapeutic strategies would obviously result in the design of rational and effective treatment approaches for this malignancy.

In this review, we will address the actions of steroid hormones on endometrial carcinomas, with special emphasis on progestin effects, mediation of progesterone receptors (PR), levels of heterogeneity in these tumors and the inherent difficulties in developing predictive tests for patient selection with various treatment modalities. The effects of growth factors and cytokines in normal and neoplastic endometrium are also discussed.

2. SEX STEROIDS AND NORMAL ENDOMETRIUM

Human endometrium is remarkably responsive to the ovarian sex steroids, 17β-estradiol (E_2) and progesterone (P). These steroids regulate the menstrual cycle-related cyclic changes of proliferation, differentiation, tissue removal, and regeneration in the endometrium of the reproductive female. E_2 stimulates the proliferation of the endometrium as evidenced by the dramatic increase in the thickness of this tissue during the proliferative phase of the menstrual cycle. During the secretory phase, under the influence of increasing levels of P, the proliferative activity of epithelial cells is inhibited and the tissue undergoes secretory differentiation. The morphological, biochemical, and physiological effects of E_2, P and their synthetic analogs and the mediation of steroid receptors in these processes have been widely investigated.[5] Biochemically, effects such as the synthesis and accumulation of glycogen, and the induction of steroid metabolizing enzymes, E_2 dehydrogenase (E_2DH) and 20 α-dihydroprogesterone dehydrogenase, have been extensively investigated.[6-8] Morphologically, the accumulation of glycogen can be readily visualized as subnuclear vacuolization in histologic sections stained with PAS.[9] These effects are potentially useful in determining the responses of endometrial carcinomas to sex steroids.

3. STEROID RECEPTORS AND HORMONE ACTION

Effects of E_2 and P on the uterus are mediated through the intracellular receptors which are steroid-specific, saturable, high affinity binding proteins.[10,11] The interaction of a steroid with its specific receptor protein is the key event that initiates defined responses within the target tissue. The receptor molecule within the responsive cell sequesters and retains the steroid hormone. Several studies have established a positive correlation between the tissue-receptor concentration and the magnitude of steroid response. Cells which lack specific steroid receptors fail to respond to the particular steroid. Purification and generation of antibodies to steroid receptors paved the way for cloning and sequencing of the complementary DNA for various steroid receptors. Comparison of the nucleotide sequences of different steroid receptors showed that they shared a similar modular domain structure among themselves and with thyroid and retinoid receptors. It became evident that these receptors belong to a superfamily of ligand-activated transcription factors.[12,13] The receptor protein functions as a transducer, transmitting the steroid signal to specific genes. The responsive genes contain precise nucleotide sequences within their regulatory regions, termed the hormone responsive elements. Once the steroid binds to its cognate receptor within the cell, the receptor molecules dimerize, enabling the homodimer to interact with its hormone response element located in responsive genes, influencing its transcription.

The observation that specific steroid receptor molecules must be present in target cells to elicit hormonal response forms the basis for the use of receptor measurements as a tool in understanding various hormonal disease processes, especially those within endocrine tissues. Determination of E_2 and P receptor status has been shown to be of clinical value in the management of patients with breast cancer.[14] In general, tumors which are receptor-positive are more likely to respond, while those devoid of receptors are less likely to respond to hormonal therapy.[15,16] Extension of these concepts to endometrial carcinoma indicated that determination of ER and PR levels in these tumors could be useful in the selection of patients for hormonal therapy.

There are several studies on the determination of ER and PR concentrations in endometrial carcinomas.[17-20] Essentially, all endometrial carcinomas are positive for ER, while PR levels vary with the histologic grade of tumor. The well-differentiated adenocarcinomas invariably contain higher PR concentrations compared to poorly differentiated tumors which exhibit low or no progestin binding activity. The correlation, however, is far from perfect since highly differentiated tumors with low PR and poorly differentiated tumors with high receptor levels have been reported. These apparently conflicting findings can be explained by the commonly observed tissue and tumor heterogeneity described later. Limited numbers of correlative studies on the relationship between tumor PR levels and response to progestin treatment in a few patients showed that while higher response rates were consistently seen in patients with PR-positive tumors, variable responses were witnessed in both PR-positive and negative endometrial carcinomas.[21,22]

Absence of a predictable correlation between steroid receptor levels and response of endometrial cancers to progestin therapy led to the notion that functional tests may be better indicators of progestin response. Several tests which monitor the end-product of P action, demonstrating the presence of a functional PR were examined. As noted earlier, various actions of P and synthetic progestins, such as glycogen accumulation and induction of steroid dehydrogenases in normal endometrium were identified, and were investigated for their utility in establishing the responsiveness of endometrial carcinomas.[23,24] Endometrial carcinoma explants were cultured *in vitro* in nutrient medium in the presence or absence of different concentrations of progestin for 2 to 3 days and the progestin effects determined by morphologic or biochemical assays. It is noteworthy that these studies on endometrial cancer explants are essentially extensions of responses to normal endometrium. Therefore, presence of normal endometrium within the "cancer specimen" might lead to an erroneous interpretation of tumor response.

4. STABILITY OF PR IN NORMAL AND NEOPLASTIC ENDOMETRIUM

Steroid effects on endometrial carcinomas were widely examined in established cell lines and tumor explants. Unlike the studies in breast cancer, essentially all the established endometrial carcinoma cells are devoid of steroid receptors or fail to respond to added steroids in culture, limiting their use in studies of steroid effects. Endometrial carcinoma cells with ER and PR which respond to steroids were reported only recently.[25,26]

Induction of E_2DH by the synthetic progestin, medroxyprogesterone acetate (MPA) was compared in normal and neoplastic endometrium *in vitro*.[27] Explants of normal endometrium consistently exhibited a twofold increased enzymic activity when exposed to MPA compared to untreated controls (Table 1). On the other hand, explants of endometrial carcinoma tissues failed to respond to added progestin. The endometrial carcinoma tissue contained high PR levels at the start of *in vitro* culture, suggesting that lack of response to added progestin was not due to the absence of PR. Examination of the stability of PR in normal and neoplastic tissues under identical conditions indicated that while the receptor levels were maintained for up to 48 hours of culture in the nonmalignant tissue minces, the PR concentrations in tumor tissue explants dropped precipitously within 2 hours of culture (Figure 1). Thus, the failure of carcinoma specimens to respond to added progestin under culture conditions was not surprising. Notwithstanding the receptor stability, the *in vitro* studies are further subject to the problem of tissue and tumor heterogeneity described below.

Table 1 Induction of Estradiol Dehydrogenase by Progestin in Normal and Neoplastic Endometrium *In Vitro*

Endometrium	Estrone Formed (nmol/mg protein/h)	
	Control	MPA
Proliferative (n = 8)	2.4 ± 0.9*	5.3 ± 2.1
Carcinoma (n = 9)	2.7 ± 1.8	2.9 ± 2.1

*Mean ± S.D.

Note: Explants of proliferative endometrium and endometrial carcinoma were cultured in nutrient medium for 2 days at 37°C in a humidified air atmosphere at 5% CO_2 in the absence (control) and presence of 500 ng medroxyprogesterone acetate (MPA) per milliliter. The enzyme activities were assayed in 800 g supernatants of homogenates of these explants.

From Satyaswaroop, P.G. and Mortel, R., *Cancer Res.*, 42, 1323, 1982. With permission.

5. TUMOR HETEROGENEITY, PR DISTRIBUTION, AND PROGESTIN RESPONSIVENESS

The complexity of tissue and tumor heterogeneity in human endometrial carcinomas is one of the major limitations in the development of a reliable test of progestin-responsiveness. Steroid receptor assays are routinely carried out on tissue specimen surgically removed from endometrial cancer patients. Invariably, the gross examination of surgical specimen is unreliable in identifying the presence of normal tissue within the cancer tissue submitted for biochemical receptor analysis. Nonmalignant and malignant tissues are usually found within the same uterus.[28] Normal and hyperplastic endometria contained in the "tumor specimen" can be anticipated to differentially contribute and interfere in receptor estimations based on ligand-binding assays, performed on tissue homogenates. In addition to the presence of normal tissue within the cancer specimens, heterogeneity of different tumor cell populations of varying histologic grades are also observed within a single tumor. Well-differentiated endometrial carcinoma is frequently identified adjacent to poorly differentiated or squamous adenocarcinoma (Figure 2). In such situations, it is practically impossible to determine if the biologic characterisic of the tumor represents one or the other of the various tumor cell populations. As indicated earlier, well differentiated tumors generally exhibited higher PR levels compared to anaplastic tumors. The ligand-binding assays of tumor tissues containing heterogenous cell populations of normal, hyperplastic and neoplastic cells of different degrees of differentiation will invariably lead to erroneous interpretation of receptor data. Various tests of progestin sensitivity are also subject to this complexity of tissue heterogeneity. Localization of receptors within various cell populations *in situ* can be expected to overcome this limitation.

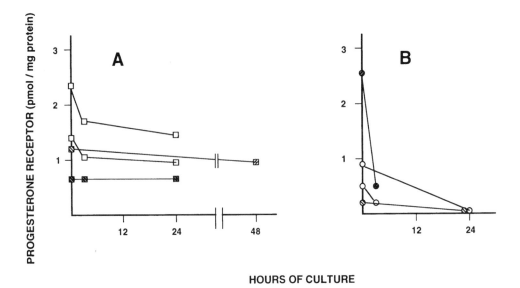

Figure 1 Stability of cytosol PR in (A) proliferative endometrium and (B) endometrial carcinoma under culture conditions. Symbols indicate separate tumor samples. (From Satyaswaroop, P.G. and Mortel, R., *Cancer Res.*, 42, 1322, 1982. With permission.)

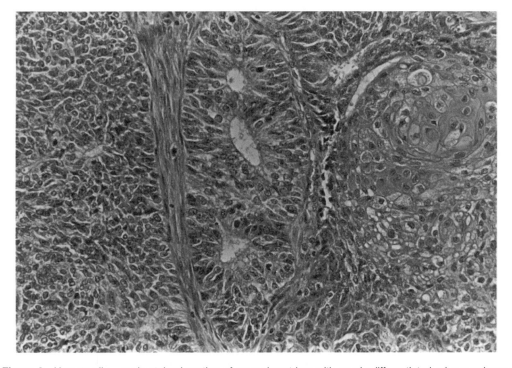

Figure 2 Hematoxylin - eosin stained section of an endometrium with poorly differentiated adenocarcinoma on the left and well-differentiated adenocarcinoma on the right. Tumor heterogeneity is reflected in the intermingling of well-differentiated gland-forming portions of a tumor with areas of solid growth or squamous differentiation. Original magnification 100X. (From Mortel, R., Zaino, R.J., and Satyaswaroop, P.G. *Cancer*, 53, 113, 1984. With permission.)

6. HETEROGENEITY OF PR DISTRIBUTION IN ENDOMETRIAL CARCINOMAS

Monoclonal antibodies generated against human PR have been helpful in the immunolocalization of PR on histological sections of endometrial carcinomas.[29,30] Immunolocalization on tumor sections has permitted the identification of heterogeneous distribution of PR in endometrial carcinomas. A comparative study of the ligand-binding assay with immunolocalization of PR in the same tumor specimens indicated a good correlation between the two procedures, with a few exceptions.[30] The exceptions were clearly related to tumor and tissue heterogeneity. In general, the tumor cells exhibiting high intensity immunostaining for PR had high PR values in the biochemical assays. The discordancies between the two assays in some cases were due to positive PR by biochemical assay which were reflected not in the nuclear staining of tumor cells, but the adjacent benign stromal or myometrial cells within tissue sections (Table 2).

Table 2 Comparison of Immunohistochemistry for PR with Biochemical Determination of PR

Tumor No.	Histologic Grade	Immunohistochemistry for PR							PR Content[b] (fmol/mg pr)	Corr[c]
		% Tumor Stained at Each Intensity[a]					Intensity of Stain			
		3+	2+	1+	Tr	Neg	Myo	Str		
1018	I	30	20	30		20	2+	Tr	2610	+
1025	I	90	10				2+	1+	1654	+
1005	I		20		40	40	1+	Abs	467	+
941	I		30	20		50	1+	Tr	1537	+
1063	I	5	20	15	30	30	2+	1=	916	+
1031	I					100	2+	Abs	1203	−
925	II			90	5	5	2+	1+	433	+
921	II					100	Tr	Neg	c	+
951	II					100	Abs	Abs	c	+
971	II		20	30		50	Abs	Tr	878	+
994	II					100	1+	Abs	c	+
1041	II		80			20	Abs	Tr	364	+
950	II					100	1+	Abs	502	−
891	III					100	Abs	Abs	c	+
893	III					100	Neg	Neg	c	+
903	III					100	1+	Neg	c	+
907	III		20	50	30		1+	Neg	c	+
910	III	100					Abs	Abs	2097	+
995	III		10			90	Abs	Abs	963	+
912	III					100	Abs	Abs	c	+
1028	III					100	Abs	Abs	c	+
1046	III		20			80	Abs	2+	263	+
970	III			2		98	Abs	Abs	c	−
1029	III					100	Tr	Abs	122	−

[a] – Intensity of reaction graded as: Neg, negative; Tr, trace; 1+, weak; 2+, moderate; 3+, strong.

[b] c, <50 fmol/mg cytosol protein.

[c] +, Immunohistochemical staining of tumor PR and positive biochemical assay for PR, or absence of staining of tumor PR and a negative biochemical assay for PR.

–, Lack of correlation.

From Zaino, R.J., Clarke, C.L., Mortel, R., and Satyaswaroop, P.G., *Cancer Res.*, 48, 1889, 1988. With permission.

The above studies further indicated that PR localization on tumor sections yielded information not evident from the quantitative biochemical assays of PR. The ability to determine the distribution, proportion, and relative intensity of immunostaining within individual tumors of heterogenous cells is extremely valuable. The PR immunolocalization studies further suggested that "failure" of some PR-positive endometrial carcinoma patients to respond to progestin therapy might be due to (1) PR-negative tumors being falsely designated as PR-positive resulting from contamination by PR-rich non-malignant myometrium or stroma, and (2) failure of PR-negative subpopulations within a tumor containing PR-positive cells to respond.

7. STEROID/ANTISTEROID EFFECTS ON ENDOMETRIAL CARCINOMA IN EXPERIMENTAL MODELS

The development of an *in vivo* experimental model was helpful in systemically examining progestin effects and the role of PR in eliciting responses from human endometrial carcinomas to progestin therapy.[31] The experimental model comprised of subcutaneous growth of human endometrial carcinomas of various histologic grade and steroid receptor status under controlled steroid hormonal milieu and their continued maintenance by serial transplantation.[32] Investigations in this experimental system provided insights into the biologic behavior of individual subtypes of tumors. Studies using this model system indicated that E_2 enhanced the growth of steroid receptor-positive endometrial tumors while it had no influence on the receptor-negative tumors. The non-steroidal agent, tamoxifen, which is widely used in the treatment of breast carcinomas for its antitumor activity, paradoxically enhanced the growth of endometrial carcinomas.[33] Combined treatment of receptor-positive endometrial carcinomas with tamoxifen and progestin dramatically inhibited the growth of these tumors.[34] However, after several weeks of growth suppression, the tumors began to regrow during continued progestin therapy, reminiscent of the clinically observed "resistance" to progestin therapy. Detailed investigations in this model system further indicated that the apparent resistance was due to PR downregulation resulting from sustained progestin exposure.[35] Further studies showed that progestin withdrawal led to resynthesis of PR, suggesting that an intermittent progestin administration may better control the growth of steroid responsive endometrial carcinomas.[36] A clinical study based on these findings is underway as part of a Gynecologic Oncology Group study.

8. TAMOXIFEN AND ENDOMETRIAL CARCINOMA

As mentioned earlier, tamoxifen is widely used to treat women with breast cancer and its antiestrogenic effect on the growth of breast cancers is universally accepted. Therefore, the observation that tamoxifen increased human endometrial carcinoma growth in nude mice was intriguing. The fact that tamoxifen was without any effect on steroid receptor-negative endometrial tumors in this experimental model suggested the mediation of ER. Based on these observations, it was inferred that tamoxifen may potentially increase the rate of growth of an already existing receptor-positive endometrial tumor during treatment for breast cancer. It was further suggested that breast cancer patients undergoing tamoxifen therapy must be routinely monitored for endometrial abnormalities.[33] Later studies with other receptor-positive endometrial carcinomas and experiments where breast and endometrial tumors were grown in the opposite flanks of the same nude mice have confirmed the paradoxic growth effects of tamoxifen on these two target tissues.[37-39] Recently, several studies have reported on the increased incidence of endometrial polyps, cysts, hyperplasias, and endometrial carcinomas in postmenopausal women on tamoxifen therapy.[40-44] A randomized trial of tamoxifen as an adjunct to surgery for breast cancer in 1846 postmenopausal patients reported an increased relative risk of 6.4 of detecting new primary endometrial carcinomas compared to a placebo group.[44] A greater than twofold increased risk of endometrial cancer was reported from the Netherlands Cancer Registry and the National Cancer Institute (U.S.) — sponsored NSABP study.[45,46] Recently, three major clinical prevention trials with tamoxifen in women, who presently do not have breast cancer but who may be highly susceptible, were initiated in the U.S., Italy and England. That women, free of any malignancy, are on tamoxifen breast cancer prevention trial while this drug has the potential to induce several endometrial abnormalities is troubling. In light of the paradoxical growth effect of tamoxifen in breast and endometrial tissues, understanding the mechanism underlying the differential responses of the two target tissues to tamoxifen assumes greater importance. Studies where breast cancer tissue was transplanted in the opposite flank of athymic mice carrying endometrial carcinoma did not indicate any differential metabolism of tamoxifen.[34] Comparison of the early immediate gene expressions in endometrial tumors exposed to E_2 or tamoxifen again indicated similar patterns on proto-oncogene expression *in vivo*.[47] Understanding the basis for the differential growth effects of tamoxifen in breast and endometrial tissues is essential for developing strategies for protecting women from the undesired deleterious endometrial abnormalities.

9. GROWTH FACTORS AND ENDOMETRIAL CARCINOMA

While the role of ovarian steroids as key regulators in normal and neoplastic endometrium has been well established, the likelihood that proliferation and differentiation of this tissues may be partly mediated by local production of peptide growth factors has been widely discussed. This expectation was fostered by the findings in human breast cancers and rodent uteri. Uterine expression of various growth factors, their receptors and responses to added growth factors were examined in rodents. The expression of EGF, EGF receptors, EGF binding and TGF-α and their induction by E_2 were widely reported. Similar interactions between steroids and growth factors were anticipated in human endometrium during the menstrual cycle and in hormone-responsive endometrial cancers. Expression of various growth factors, their receptors and the effects of growth factors have been examined in normal cycling endometrium, endometrial carcinoma tissues and in established cell lines and are summarized in Table 3. Although the presence of various growth factors, their receptors, binding proteins and modulation by steroid hormones in various cell populations of the endometrium have been identified, their role and importance in the regulation of endometrial carcinoma cell proliferation is far from certain. Some of these factors, for example, EGF receptor, c-erbB2, CSF-1 and its receptor were also examined for their role in pathogenesis and prognostic importance in endometrial carcinomas. Although these studies are of potential clinical significance, more definitive studies are needed to establish the participation of growth factors in endometrial carcinoma growth.

Table 3 Growth Factors and Their Receptors in Normal and Neoplastic Endometrium

Growth Factor/Growth Factor Receptor	Endometrium	Ref.
EGF expression	Normal	56
	Carcinoma	56
EGF receptor expression	Normal	48,49,51,52,55
	Carcinoma	48,53,55,61
	Ca cell line	49
EGF effect	Ca cell line	54
IGF-1 expression	Normal	50
	Ca cell line	60
IGF-1 receptor expression	Normal	62
	Ca cell line	60
TGF-α expression	Ca cell line	54,57-59
TGF-β1 expression	Ca cell line	74
TGF-β1 effect	Ca cell line	58,74,75
TNF-α expression	Normal	63,66
TNF-α receptor	Normal	63,66
PDECGF expression	Normal	64
LH/hCG receptor expression	Normal	65
	Carcinoma	65
CSF-1/c-fms	Normal	68
	Carcinoma	67,69

10. CYTOKINES AND ENDOMETRIAL CARCINOMA

Cytokines were originally identified as immunomodulatory and inflammatory proteins, based on their specific actions. Later studies demonstrated their pleiotropic role in paracrine, autocrine, and endocrine functions, modulating growth and differentiation. The list of cytokines is continually expanding and includes interleukins, colony stimulating factors, tumor necrosis factor, interferons, chemotactic, and angiogenic factors and there is considerable overlap in the listing of various growth factors and cytokines. In this section, on the role and interaction of cytokines in normal and malignant human endometrium, we will discuss the functions of some of the interleukins, interferons and colony stimulating factors.

Human endometrium was recently shown to be an active site of cytokine production and action.[70] Table 4 lists some of the functions of cytokines relevant to the endometrium. Information on the role of cytokines in human endometrium and interaction of cytokines with gonadal steroids were derived from *in vitro* studies performed in endometrial tissue explants, primary cell cultures, and established endometrial carcinoma cell lines. Reverse transcription-polymerase chain reaction (RT-PCR) data indicated the expression of mRNA of various cytokines and immunohistochemical staining identified the presence of IL-1, IL-6, IL-1 receptor antagonist (IRAP), and TNF-α proteins in the human endometrium.[71] Several studies suggest that IFN-γ exhibits a paracrine role in human endometrium.[72]

Table 4 Functions of Cytokines Relevant to Human Endometrium

Function	Cytokine
Modulation of proliferation	IFN-γ, IL-1, IL-6, TNF-α, EGF, TGF-α, TGF-β, CSF-1
Induction of HLA-DR and ICAM-1	IFN-γ
Alteration of morphology	IFN-γ, IL-1, TNF-α, TGF-β
Gland formation	TGF-β
Induction of adhesion molecules	IFN-γ, IL-1, TNF-α, TGF-β
Chemotaxis and induction of lymphoid infiltration	IFN-γ, IL-1, TNF-α, TGF-β
Edema	IL-1, TNF-α
Deposition of extracellular matrix	TGF-β
Induction of PGE_2	IL-1, TNF-α
Activation of T cells	IL-1, IL-6
Activation of polymorphonuclear leukocytes	TNF-α
Injury to uterine vessels	TNF-α
Angiogenesis	TGF-β
Elevation of body temperature	IL-1, IL-6, TNF-α

From Tabibzadeh, S., *Endocrine Rev.*, 2, 272, 1991. With permission.

The T cells of the lymphoid aggregates present within the endometrium are the source of IFN-γ while receptors for IFN-γ are present in the epithelium throughout the menstrual cycle. IFN-γ induces HLA-DR and inhibits epithelial cell proliferation.[72] Endometrial stromal cells, free of epithelial, endothelial and lymphoid cells secrete IL-6 and IL-1, TNF-α and IFN-γ enhances stromal cell secretion of IL-6, while E_2 inhibited IL-6 production by these cells.[71] Several studies have recently shown the expression of colony stimulating factor-1 in endometrial glands of the normal endometrium through the menstrual cycle and the overexpression of CSF-1 and its receptor, c-fms, in endometrial carcinomas.[67-69] A significant correlation between the overexpression of CSF-1 and c-fms and poor prognosis has been identified by a recent study.[69]

The pleiotropic nature of cytokines and their profound local effects have limited investigations on experimental *in vivo* systems and there are few reports on the effects of cytokines on endometrial carcinoma growth.

11. SUMMARY

Endometrial carcinoma is the most common gynecologic malignancy in the Western world. This is primarily a postmenopausal disease and high cure rates are obtained by surgery alone. Fatalities in endometrial carcinoma are primarily due to recurrence and metastatic disease. Laboratory investigations and studies in experimental model system with human tumor have significantly enhanced our understanding of the steroid hormone modulation of growth of hormone responsive endometrial carcinomas. However, knowledge on the mediation and the role of peptide growth factors and cytokines in regulating endometrial carcinoma growth is relatively limited. Further investigations on the effects of steroids, peptide growth factors, and cytokines in human endometrium is anticipated to provide information and aid in the design of effective treatment strategies for this disease.

REFERENCES

1. Cancer Facts and Figures — 1993, American Cancer Society, Inc., 1993.
2. **Kelley, R.M. and Baker, W.H.,** Progestational agents in the treatment of carcinoma of the endometrium, *N. Engl. J. Med.*, 264, 216, 1960.
3. **Satyaswaroop, P.G., Podczaski, E.S., and Mortel, R.,** Hormonal interactions in gynecological malignancies, Principles and practice of gynecologic oncology, in *Principles and Practice of Gynecologic Oncology*. Hoskins, W. J., Perez, C.A. and Young, R.C., Eds., J.B. Lippincott Co., Philadelphia, chap. 9, 1992.
4. **Deppe, E.,** Chemotherapy of endometrial carcinoma, in *Chemotherapy of Gynecologic Cancer*, Alan Liss, New York, 1984.
5. **Richardson, G.S. and MacLaughlin, D.T.,** Hormonal biology of endometrial cancer. UICC Technical report series, 42, 1978. International Union against Cancer, Geneva.
6. **Hughes, E.C., Demers, L.M., Csermely, T., and Jones, D.B.,** Organ culture of human endometrium, *Am. J. Obstet. Gynecol.*, 105, 707, 1969.

7. **Shapiro, S.S., Dyer, R.D., and Colas, A.E.,** Progesterone-induced glycogen accumulation in human endometrium, *Am. J. Obstet. Gynecol.,* 136, 419, 1980.

8. **Tseng, L. and Gurpide, E.,** Induction of human endometrial estradiol dehydrogenase by progestins, *Endocrinology,* 97, 825, 1975.

9. **Kohorn, E.I. and Tchao, R.,** Conversion of proliferative endometrium to secretory endometrium by progesterone in organ culture, *J. Endocrinol.,* 45, 401, 1969.

10. **Jensen, E.U. and DeSombre, E.R.,** Mechanism of action of steroid hormone action, *Annu. Rev. Biochem.,* 41, 203, 1972.

11. **Gorski, J. and Ganong, F.,** Current models of steroid action: a critique, *Annu. Rev. Biochem.,* 38, 425, 1976.

12. **Evans, R.,** The steroid and thyroid hormone receptor superfamily, *Science,* 240, 889, 1988.

13. **Green, S. and Chambon, P.,** Nuclear receptors enhance our understanding of transcriptional regulation, *Trends Genet.,* 4, 309, 1988.

14. **McGuire, W.L., Carbone, P.O., and Vollmer, E.P.,** *Estrogen Receptors in Human Breast Cancer,* Raven Press, New York, 1975.

15. **Allegra, J.C., Lippman, M.E., and Thompson, E.B.,** Relationship between the progesterone, androgen and glucocorticoid receptor and response rate to endocrine therapy in metastatic breast cancer, *Cancer Treat. Rep.,* 62, 1281, 1978.

16. **McGuire, W.L.,** *Hormones, Receptors and Breast Cancer,* Raven Press, New York, 1978.

17. **Martin, P.M., Rolland, P.H., Gammere, M., Serment, H., and Toga, M.,** Oestradiol and progesterone receptors in normal and neoplastic endometrium: correlations between receptor, histopathological examinations and clinical responses under progestin therapy, *Int. J. Cancer,* 24, 324, 1979.

18. **Benraad, T.J., Frieberg, L.G., Koenders, A.J.M., and Kullander, S.,** Do oestrogen and progesterone receptors in metastasising cancers predict the response to gestagen therapy? *Acta Obstet. Gynecol. Scand.,* 59, 155, 1980.

19. **Creaseman, W.T., McCarty, K.S., Sr., Barton, T.K., and McCarty, K.S., Jr.,** Clinical correlates of estrogen and progesterone binding proteins in human endometrial adenocarcinoma, *Obstet. Gynecol.,* 55, 363, 1980.

20. **Mortel, R., Zaino, R.J., and Satyaswaroop, P.G.,** Heterogeneity and progesterone receptor distribution in endometrial adenocarcinoma, *Cancer,* 53, 113, 1984.

21. **Kauppila, A., Kujansuu, E., and Vihko, R.,** Cytosol estrogen and progesterone receptors in endometrial carcinoma of patients treated with surgery, radiotherapy and progestin: clinical correlates., *Cancer,* 50, 2157, 1982.

22. **Martin, P.M.,** Endometrial cancer: correlations between oestrogen and progestin receptor status, histopathological findings, and clinical responses during progestin therapy, *Excerpta Med. Int. Can. Ser.,* 611, 333, 1982.

23. **Tseng, L., Gusberg, S.B., and Gurpide, E.,** Estradiol receptor and 17 β-dehydrogenase in normal and abnormal human endometrium, *Ann. N.Y. Acad. Sci.,* 286, 190, 1977.

24. **Holinka, C.F., Deligdisch, L., and Gurpide, E.,** Histological evaluation of in vitro responses of endometrial adenocarcinoma to progestins and their relation to progesterone receptor levels, *Cancer Res.,* 44, 293, 1994.

25. **Nishida, M., Kasahara, K., Kaneko, M., and Wasaki, H.,** Establishment of a new human endometrial adenocarcinoma cell line, Ishikawa cells, containing estrogen and progesterone receptor, *Acta Obstet. Gynecol. Jpn.,* 37, 1103, 1985.

26. **Satyaswaroop, P.G., Sivarajah, A., Zaino, R.J., and Mortel, R.,** Hormonal control of growth of human endometrial carcinoma in the nude mouse model, in *Progress in Cancer Research Therapy,* Bresciani, F., King, R.J.B., and Lippman, M.E., Eds., Raven Press, New York, 430, 1988.

27. **Satyaswaroop, P.G. and Mortel, R.,** Failure of progestins to induce estradiol dehydrogenase activity in endometrial carcinoma in vitro, *Cancer Res.,* 42, 1323, 1982.

28. **Mortel, R., Zaino, R.J., and Satyaswaroop, P.G.,** Heterogeneity and progesterone receptor distribution in endometrial adenocarcinoma, *Cancer,* 53, 113, 1984.

29. **Clarke, C.L., Zaino, R.J., Feil, P.D., Miller, J.V., Steek, M.E., and Ohlsson-Wilhelm, B.M., and Satyaswaroop, P. G.,** Monoclonal antibodies to human progesterone receptor: characterization by biochemical and immunohistochemical techniques, *Endocrinology,* 121, 1123, 1987.

30. **Zaino, R.J., Clarke, C.L., Mortel, R., and Satyaswaroop, P.G.,** Heterogeneity of progesterone receptor distribution in human endometrial adenocarcinoma, *Cancer Res.,* 48, 1889, 1988.

31. **Satyaswaroop, P.G., Zaino, R.J., and Mortel, R.,** Human endometrial adenocarcinoma transplanted into nude mice: growth regulation by estradiol, *Science,* 219, 58, 1983.

32. **Satyaswaroop, P.G., Zaino, R.J., and Mortel, R.,** Steroid receptors and human endometrial carcinoma: studies in a nude mouse model, *Cancer Metastasis Rev.,* 6, 223, 1987.

33. **Satyaswaroop, P.G., Zaino, R.J., and Mortel, R.,** Estrogen-like effects of tamoxifen on human endometrial carcinoma transplanted into nude mice, *Cancer Res.,* 44, 4006, 1984.

34. **Zaino, R.J., Mortel, R., and Satyaswaroop, P.G.,** Hormonal therapy of human endometrial carcinoma in a nude mouse model, *Cancer Res.,* 45, 539, 1985.

35. **Satyaswaroop, P.G., Clarke, C.L., Zaino, R.J., and Mortel, R.,** Apparent resistance of human endometrial carcinoma during combination treatment with tamoxifen and progestin may result from desensitization following downregulation of tumor progesterone receptor, *Cancer Lett.,* 62, 107, 1992.

36. **Mortel, R., Zaino, R.J., and Satyaswaroop, P.G.,** Designing a schedule of progestin administration in the control of endometrial carcinoma growth in the nude mouse model, Am. J. Obstet. Gynecol., 162, 928, 1990.

37. **Clarke, C.L. and Satyaswaroop, P.G.,** Photoaffinity labeling of progesterone receptor from human endometrial carcinoma, *Cancer Res.*, 45, 5417, 1985.
38. **Gottardis, M.M., Robinson, S.P., Satyaswaroop, P.G., and Jordan, V.C.,** Contrasting actions of tamoxifen on endometrial and breast tumor growth in the athymic mouse, *Cancer Res.*, 48, 812, 1988.
39. **Gottardis, M.M., Ricchio, M., Satyaswaroop, P.G., and Jordan, V.C.,** Effect of steroidal and non-steroidal antiestrogens on the growth of a tamoxifen-stimulated human endometrial carcinoma (EnCa-101) in athymic mice, *Cancer Res.*, 50, 3189, 1990.
40. **Nuovo, M.A., Nuovo, G.J., McCaffrey, R.M., Levine, R.U., Barron, B., and Winkler, B.,** Endometrial polyps in postmenopausal patients receiving tamoxifen, *Int. J. Gynecol. Pathol.*, 8, 125, 1989.
41. **Neven, P., DeMuylder, X., and Van Belle, Y.,** Hysteroscopic follow-up during tamoxifen treatment, *Eur. J. Obstet. Gynecol. Rep. Biol.*, 35, 235, 2990.
42. **Lahti, E., Guillermo, B., and Kauppila, A.,** Endometrial changes in postmenopausal breast cancer patients receiving tamoxifen, *Obstet. Gynecol.*, 81, 660, 1993.
43. **Killackey, M.A., Hakes, T.B., and Pierce, V.K.,** Endometrial carcinoma in breast cancer patients receiving antiestrogens, *Cancer Treat. Rep.*, 69, 237, 1985.
44. **Fornander, T., Rutqvist, L.E., Cedermark, B., Glas, U., Mattsson, A., Silfversward, C., Skoog, L., Somell, A., Theve, T., Wilking, N., Askergren, J., and Hjalmar, M.-L.,** Adjuvant tamoxifen in early breast cancer: occurrence of new primary cancers, *Lancet*, 1, 117, 1989.
45. **van Leewen, F.E., Benraadt, J., Coebergh, J.W.W., Kiemeney, L.A.L., Gibrere, C.H.F., Otter, R., Schouten, L.J., Damhuis, R.A.M., Bontebel, M., Diepenhorst, F.W., Belt-Dusebout, A.M.V., and van Tinteren, H.,** Risk of endometrial cancer after tamoxifen treatment of breast cancer, *Lancet*, 343, 448, 1994.
46. **Fisher, B., Constantino, J.P., Redmond, C.K., Fisher, E.R., Wickerham, D.L., Cronin, W.M., and** other NSABP contributors, Endometrial cancer in tamoxifen-treated breast cancer patients: findings from the National Surgical Adjuvant Breast and Bowel Project (NSABP)B-14, *J. Natl. Cancer Inst.*, 86, 527, 1994.
47. **Sakakibara, K., Kan, N., and Satyaswaroop, P.G.,** Both estradiol and tamoxifen induce c-fos messenger ribonucleic acid expression in human endometrial carcinoma grown in nude mice, *Am. J. Obstet. Gynecol.*, 166, 206, 1992.
48. **Berchuck, A., Soisson, A.P., Olt, G., Soper, J.T., Clarke-Pearson, D.I., Bast, R.C., and McCarty, K.S.,** Epidermal growth factor expression in normal and malignant endometrium, *Am. J. Obstet. Gynecol.*, 161, 1247, 1992.
49. **Reynolds, R.K., Talavera, F., Roberts, J.A., Hopkins, M.P., and Menon, M.J.K.,** Regulation of epidermal growth factor and insulin-like growth factor-1 receptors by estradiol and progesterone in normal and neoplastic endometrial cell cultures, *Gynec. Oncol.*, 38, 396, 1990.
50. **Sheets, E.E., Tsibris, J.C.M., Cook, N.I., Virgin, S.D., DeMay, R.M., and Spellacy, W.N.,** In vitro binding of insulin and epidermal growth factor to human endometrium and endocervix, *Am. J. Obstet. Gynecol.*, 153, 60, 1985.
51. **Chegini, N., Rao, C.V., Wakim, N., and Sanfillipo, J.,** Binding of ^{125}I-epidermal growth factor in human uterus, *Cell. Tissue Res.*, 246, 543, 1986.
52. **Hofmann, G.E., Rao, C.V., Barrows, G.H., Schultz, G.S., and Sanfillipo, J.S.,** Binding sites for epidermal growth factor in human uterine tissues and leiomyomas, *J. Clin. Endocrin. Metab.*, 58, 880, 1984.
53. **Srkalovic, G., Wittliff, J.L., and Schally, A.V.,** Detection and partial characterization of receptors for [D-Trp^6]-luteinizing hormone releasing hormone and epidermal growth factor in human endometrial carcinoma, *Cancer Res.*, 50, 1841, 1990.
54. **Korc, M., Haussler, C.A., and Trookman, N.S.,** Divergent effects of epidermal growth factor and transforming growth factors on a human endometrial carcinoma cell line, *Cancer Res.*, 47, 4909, 1987.
55. **Reynolds, R.K., Talavera, F., Roberts, J.A., Hopkins, M.P., and Menon, K.M.J.,** Characterization of epidermal growth factor receptor in normal and neoplastic human endometrium, *Cancer*, 66, 1967, 1990.
56. **Rigby, R.M., Li, A.X., Bomalaski, J., Stechman, F.B., Look, K.Y., and Sutton, G.P.,** Immunohistochemical study of Her2/neu, epidermal growth factor and steroid receptor expression in normal and malignant endometrium, *Obstet. Gynecol.*, 79, 95, 1992.
57. **Anzai, Y., Gong, Y., Holinka, C.F., Murphy, I.J., Murphy, L.C., Kuramoto, H., and Gurpide, E.,** Effects of transforming growth factors and regulation of their mRNA levels in two human adenocarcinoma cell lines, *J. Ster. Biochem. Mol. Biol.*, 42, 449, 1992.
58. **Murphy, L.J., Gong, Y., and Murphy, L.C.,** Regulation of transforming growth factor gene expression in human endometrial adenocarcinoma cells, *J. Ster. Biochem. Mol. Biol.*, 41, 309, 1992.
59. **Gong, Y., Murphy, L.C., and Murphy, L.J.,** Hormonal regulation of transforming growth factor gene expression in human endometrial adenocarcinoma xenograft, *J. Ster. Biochem. Mol. Biol.*, 50, 13, 1994.
60. **Pekononen, F., Nyman, T., and Rutanen, E.M.,** Human endometrial adenocarcinoma cell lines HEC iB and KLE secrete insulin-like growth factor binding protein-1 and contain IGF-1 receptors, *Mol. Cell. Endocrinol.*, 75, 81, 1991.
61. **Wang, D., Konishi, I., Koshiyma, M., Mandai, M., Nanbu, Y., Ishikawa, Y., Mori, T., and Fujii, S.,** Expression of c-erbB-2 protein and epidermal growth factor receptor in endometrial carcinomas, *Cancer*, 72, 2628, 1993.
62. **Vollenhoven, B.J., Herington, A.C., and Healy, D.L.,** Messenger ribonucleic acid expression of the insulin-like growth factors and their binding proteins in uterine fibroids and myometrium, *J. Clin. Endocrin. Metab.*, 76, 1106, 1993.
63. **Tabibzadeh, S.,** Ubiquitous expression of TNF-/cachectin immunoreactivity in human endometrium, *Am. J. Reprod. Immunol.*, 26, 1, 1991.

64. **Osuga, Y., Toyoshima, H., Mitsuhashi, N., and Taketani, Y.,** The presence of platelet-derived endothelial cell growth factor in human endometrium and its characteristic expression during the menstrual cycle and early gestation, *Mol. Hum. Reprod.*, 1, 989, 1995.

65. **Lin, J., Lei, S., Lojun, S., Rao, C.V., Satyaswaroop, P.G., and Day, T.G.,** Increased expression of luteinizing hormone/human chorionic gonadotropin receptor gene in human endometrial carcinomas, *J. Clin. Endocrin. Metab.*, 79, 1483, 1994.

66. **Tabibzadeh, S., Zupi, E., Babaknia, A., Liu, R., Marconi, D., and Romanini, C.,** Site and menstrual cycle-dependent expression of proteins of the tumor necrosis factor receptor family, and BCL-2 oncoprotein and phase-specific production of TNF- in human endometrium, *Hum. Reprod.*, 10, 277, 1995.

67. **Kacinski, B.M., Chambers, S.K., Stanley, E.R., Carter, D., Tseng, P., Scata, K.A., Chang, D., Pirro, M.H., Nguyen, J.T., Ariza, A., Rohrschneider, A.R., and Rothwell, V.M.,** The cytokine CSF-1 expressed by endometrial carcinomas in vivo and in vitro, may also be a circulating tumor marker of neoplastic disease activity in endometrial carcinoma patients, *Int. J. Radiat. Oncol. Biol. Phys.*, 19, 619, 1990.

68. **Pampfer, S., Tabibzadeh, S., Chuan, F.-C., and Pollard, J.W.,** Expression of colony stimulating (CSF-1) messenger RNA in human endometrial glands during the menstrual cycle: molecular cloning of a novel transcript that predicts cell surface form of CSF-1, *Mol. Endocrinol.*, 5, 1931, 1991.

69. **Smith, H.O., Anderson, P.S., Kuo, D.Y.S., Goldberg, G.L., DeVictoria, C.L., Runowicz, C.D., Stanley, E.R., and Pollard, J.W.,** The role of colony stimulating factor 1 and its receptor in the etiopathogenesis of endometrial adenocarcinoma, *Clin. Cancer Res.*, 1, 313, 1995.

70. **Tabibzadeh, S.,** Human endometrium: an active site of cytokine production and action, *Endocrine Rev.*, 2, 272, 1991.

71. **Tabibzadeh, S. and Sun, X.Z.,** Cytokine expression in human endometrium through the menstrual cycle, *Hum. Reprod.*, 7, 1214, 1992.

72. **Tabibzadeh, S., Satyaswaroop, P.G., and Rao, P.N.,** Antiproliferative effect of interferon gamma in human endometrial epithelial cells in vitro: potential local modulatory role in endometrium, *J. Clin. Endocrin. Metab.*, 67, 131, 1988.

73. **Tabibzadeh, S., Kaffka, K.L., Kilian, P.L., Satyaswaroop, P.G., and EnCa, E.C.,** 101 AE and ECC-1 cell lines: suitable models for studying cytokine actions in endometrium, *in Vitro Cell. Dev. Biol.*, 26, 1173, 1991.

74. **Murphy, L.J., Gong, Y., Murphy, L.C., and Bhavnani, B.,** Growth factors in normal and malignant uterine tissue, *Ann. N.Y. Acad. Sci.*, 622, 145, 1991.

75. **Boyd, J.A. and Kaufman, D.G.,** Expression of transforming growth factor β 1 by human endometrial carcinoma cell lines: inverse correlation with effects of growth rate and morphology, *Cancer Res.*, 50, 3394, 1990.

Chapter 13

Alterations of Oncogenes and Tumor-Suppressor Genes in Endometrial Cancer

Andrew Berchuck, Anthony C. Evans, and Jeff Boyd

CONTENTS

1. INTRODUCTION

Chronic stimulation of the endometrium by estrogens, without the differentiating effects of progestins, is the primary etiologic factor associated with the development of hyperplasia and subsequent progression to adenocarcinoma. Both endogenous (anovulation) and exogenous (hormonal therapy) sources of unnopposed estrogen increase the risk of endometrial adenocarcinoma. Although this cause and effect relationship is widely accepted, the underlying molecular mechanisms remain unclear. In addition, some endometrial cancers appear to arise in the absence of a hormonally induced premalignant hyperplastic phase.

Recently, several lines of evidence have suggested that most human cancers arise due to sequential damage to genes that encode proteins involved in regulation of cellular proliferation and differentiation.[1] Two classes of genes have been implicated in this process — oncogenes and tumor suppressor genes.[2] Oncogenes encode proteins that participate in growth stimulatory pathways in normal cells. Conversely, tumor-suppressor gene products normally inhibit unrestrained proliferation.[3,4] Although all the causes of damage to growth regulatory genes are not known with certainty, most damage is thought to represent spontaneous errors that elude cellular DNA repair mechanisms. Regardless of the etiology of genetic damage, endometrial cancers and others that arise due to accumulation of mutations generally occur in older individuals, presumably because several "hits" to critical genes are required to elicit malignant transformation.

The first evidence that genetic alterations occur during the process of endometrial carcinogenesis came from studies of total cellular DNA content (ploidy) and cytogenetic analyses. Several groups have shown that approximately 20% of endometrial adenocarcinomas have an increased DNA content (aneuploidy) relative to normal cells.[5-7] Aneuploidy is associated with advanced stage, adverse histologic features and poor survival. In addition, cytogenetic studies have described gross chromosomal alterations in endometrial cancers including changes in the number of copies of specific chromosomes.[8] Although 80% of endometrial cancers have a normal DNA content, sub-microscopic alterations in oncogenes and tumor-suppressor genes have been identified in both aneuploid and diploid endometrial cancers.

2. ONCOGENES

Oncogenes encode proteins that ordinarily participate in growth stimulatory pathways in normal cells.[1] It has been demonstrated that amplification, translocation or mutation of these genes facilitates malignant transformation by increasing the abilty of cells to proliferate in an unrestrained fashion. Several classes of oncogene products are involved in transmitting growth stimulatory signals from the periphery of the cell towards the nucleus. The role of each of these classes of molecules in growth regulation and endometrial carcinogenesis is reviewed below (See Table 1).

Table 1 Growth Regulatory Genes Reported To Be Altered in Endometrial Adenocarcinomas

Gene	Class	Activation	Approximate Frequency
EGF receptor	Tyrosine kinase	Altered expression	Unclear
HER-2/*neu*	Tyrosine kinase	Amplification/overexpression	10–20%
K-*ras*	G protein	Mutation	10–30%
c-*myc*	Transcription factor	Amplification/overexpression	30%
p53	Tumor suppressor	Mutation/overexpression	20%
MSH2	DNA repair	Mutation	?
MLH1	DNA repair	Mutation	?

2.1 Peptide Growth Factors and Their Receptors

Peptide growth factors in the extracellular space stimulate the cascade of molecular events that lead to proliferation by binding to cell membrane receptors. Unlike hormones, which often are secreted into the blood stream to act in distant target organs, peptide growth factors usually act in the local environment where they have been secreted. Although increased production of stimulatory growth factors may play a role in enhancing proliferation associated with malignant transformation, growth factors are involved also in development, stromal-epithelial communication, tissue regeneration and wound healing.

Cell membrane receptors that bind peptide growth factors are composed of an extracellular ligand binding domain, a membrane spanning region and a cytoplasmic tyrosine kinase domain.[9] Binding of a growth factor to the extracellular domain results in dimerization and conformational shifts in the receptor and activation of the inner tyrosine kinase. This kinase phosphorylates tyrosine residues on both the growth factor receptor and targets in the cell interior, which leads to generation of secondary signals that propagate the mitogenic stimulus towards the nucleus. Although over a dozen receptor tyrosine kinases have been identified, thus far most studies in endometrial cancer have focused on the epidermal growth factor (EGF) receptor, HER-2/*neu* (erbB-2) and c-*fms*.

2.1.1 Epidermal Growth Factor Receptor

EGF was named because it was found to be involved in stimulating growth of epidermal appendages in developing mice.[10] EGF and its receptors were among the first growth factor/receptor tyrosine kinases to be characterized at a molecular level. EGF receptor is present in glandular and stromal cells of the endometrium in both the proliferative and secretory phase in cycling women.[11,12] Expression is also maintained in atrophic endometrium after the menopause. Although some squamous cancers overexpress EGF receptor due to amplification of the EGF receptor gene,[13] amplification has not been noted in endometrial adenocarcinomas. It appears that loss of EGF receptor may occur during endometrial carcinogenesis. In this regard, using a radioreceptor assay, Reynolds et al. showed that grade 1 to 2 cancers had, on average, a 34% decrease and grade 3 cancers, a 90% decrease, in EGF receptor expression relative to normal endometrium.[12] Several other authors also have examined the relationship between EGF receptor expression and clinicopathologic features and survival.[11,14-19] These studies are difficult to compare because various radioreceptor and immunohistochemical techniques were employed that differ with respect to their sensitivity. The incidence of detectable EGF receptor varies between 25 and 75% in these reports, but there always appears to be a range of expression from low to high. In most studies, EGF receptor expression was not associated with clinical features or survival.

In contrast, two groups found that EGF receptor expression correlated with poor prognostic features and worse survival. In an Italian study, a ligand binding technique revealed that 13/60 cases (22%) had high level expression, which was associated with poor grade and decreased disease-free survival.[17] In the second study, from the University of Oklahoma, immunostaining for EGF receptor was performed in paraffin blocks of 69 cancers.[18] Immunostaining was observed in 49% of cases and was an independent variable predictive of the presence of metastatic disease. In addition, EGF receptor expression correlated with poor grade, deep invasion, older age, and poor survival. Further studies are needed to determine the significance of changes in EGF receptor expression during endometrial carcinogenesis. Although growth factors that bind to and stimulate the EGF receptor (EGF and transforming growth factor-α) are present in normal and malignant endometrium, studies published to date have not addressed the issue of whether there are alterations in co-expression of receptor and ligand.

2.1.2 HER-2/neu

The HER-2/*neu* (erbB-2) gene encodes a receptor tyrosine kinase that is similar in structure to the EGF receptor and a ligand that binds to HER-2/*neu* (heregulin) has been discovered.[20] Amplification and overexpression of HER-2/*neu* has been noted in approximately 20-30% of breast and ovarian cancers[21,22] and in many studies overexpression has been associated with poor survival. In addition, several studies have suggested that this oncogene product is overexpressed in 10 to 20 % of endometrial cancers.[7,19,23-30] In some, but not all, initial studies overexpression was associated with poor outcome.

The group at the Mayo Clinic has performed a study of HER-2/*neu* expression in paraffin blocks from 247 endometrial cancers.[24] Expression was scored as high in 15% of cases, mild in 58% and absent in 27%; 5-year progression-free survival was 56%, 83%, and 95% in these groups respectively. Among stage I cases, 26 (13%) had high expression and 5-year progression-free survival was 62% compared to 97% in cases with lesser expression. The incidence of overexpression was higher in advanced stage cases (11/44; 25%). In addition, multivariate analysis revealed that high expression was an independent variable associated with poor survival.

We also performed a study to determine whether HER-2/neu is an independent variable associated with poor progression-free survival.[7] Similar to the group at the Mayo Clinic, we found high expression in 12% of 100 cases. Overexpression was more common in stage III/IV cases (8/34; 24%) relative to stage I/II cases (4/66; 6%) and was associated with poor progression-free survival in univariate analysis (Figure 1). In the multivariate analysis, however, HER-2/*neu* was found to be an independent variable only if DNA ploidy was excluded from the statistical model.

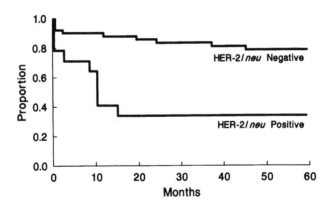

Figure 1   Relationship between HER-2/*neu* expression and disease-free survival in endometrial cancer. (HER-2/*neu* positive n = 12, HER-2/*neu* negative n = 88).

2.1.3 fms

The *fms* oncogene, which was first identified as the transforming gene of a feline retrovirus, has also been shown to encode a receptor tyrosine kinase that serves as a receptor for macrophage-colony stimulating factor (M-CSF). Kacinski et al. examined expression of *fms* in 21 endometrial cancers using *in situ* hybridization.[31] Expression of *fms* complementary mRNA was found to correlate with advanced stage, poor grade and deep myometrial invasion. The association of fms expression with adverse prognostic factors was confirmed by Leiserowitz et al. at the Mayo Clinic.[32] Subsequently, it was shown that fms and its ligand (M-CSF) usually were co-expressed in endometrial cancers and it was proposed that this receptor-ligand pair might mediate an autocrine stimulatory pathway.[33] In support of this hypothesis, M-CSF serum levels are increased in patients with endometrial cancer. In addition, M-CSF increases the invasiveness of cancer cell lines that express significant levels of *fms*, but has no effect on cell lines with low levels of the receptor.[34] It remains unclear, however, whether increased production of M-CSF or other peptide growth factors play a role in eliciting malignant transformation or alternatively are the result of other causitive alterations.

2.2 G Proteins

The *ras* family of G proteins (N-*ras*, H-*ras*, K-*ras*) are thought to play a critical role in regulation of cellular proliferation. They are located on the inner aspect of the cell membrane and have intrinsic GTPase activity that catalyzes the exchange of GTP for GDP. In their active GTP bound form, *ras* proteins activate cytoplasmic serine/threonine kinases, which then relay the mitogenic signal towards the nucleus. Conversely, hydrolysis of GTP to GDP, which is stimulated by GTPase activating proteins (GAPs), leads to inactivation of G proteins.[9] Until recently, the factors that stimulate G protein activity have been a mystery. It has now become clear, however, that activation of *ras* is a final common pathway following binding of growth factors to receptor tyrosine kinases. In many types of cancers, *ras* genes often are found to have undergone point mutations in codons 12, 13, or 61, which results in a constitutively activated molecule.

Ras proteins are expressed in normal endometrium and levels may be upregulated in some endometrial cancers.[35,36] Boyd et al. examined codons 12, 13 and 61 of the K-*ras*, H-*ras* and N-*ras* genes in 11 immortalized endometrial cancer cell lines.[37] Mutations in codon 12 of K-*ras* were seen in four cell lines whereas three had mutations in codon 61 of H-*ras*. Similarly, they also examined codons 12, 13, and 61 of the three *ras* genes in 10 primary endometrial cancers.[38] A mutation in codon 12 of K-*ras* was found in one case whereas the other *ras* genes were not mutated. Subsequent studies of primary endometrial adenocarcinomas have confirmed that codon 12 of K-*ras* is the most frequent site of mutations. Enomoto et al. reported from Japan that 15/52 (29%) endometrial cancers had mutations in codon 12 of K-*ras*.[39-41] Two other studies from Japan reported that codon 12 mutations were present in 10/45 (22%) and 5/49 (10%) cases.[42,43] In two studies of American endometrial cancers, 3/30 (10%) and 7/60 (12%) cases had mutations in codon 12.[38,44] In the latter study, 65% of patients were Hispanic; mutations were seen in 7/39 (18%) Hispanic cases and 2/9 (11%) Caucasian cases. Overall, in the above noted studies, 30 mutations were described in 146 Japanese cases (21%) compared to 10 mutations in 90 American cases (11%). Likewise, mutations in codon 13 of K-*ras* were seen in 4/101 (4%) Japanese cases and in 2/90 (2%) American cases.

This apparent difference in frequency of codon 12 K-*ras* mutations between Japanese and American endometrial cancers was confirmed by Sasaki et al. in a study in which cases from both countries were examined.[45] Mutations were seen in cancers from 5/36 (14%) white Americans, 0/5 black Americans and 10/43 (23%) Japanese. Striking differences exist in the incidence of endometrial cancer between America (approximately 20 cases per 100,000) and Japan (approximately 5 cases per 100,000). In contrast, mortality from the disease is roughly equivalent in the two countries. Epidemiologic studies have suggested that the higher mortality per case in Japan occurs because few Japanese women are obese and develop estrogen-dependent well differentiated favorable lesions. In view of this, it is possible that K-*ras* mutation might be associated with poorly differentiated lesions with unfavorable prognosis. In this regard, one Japanese study found that survival was 93% in 43 cases with normal K-*ras* compared to 50% in 6 cases with mutations.[43] In contrast, Enomoto did not find a correlation with histologic grade[41] and the American study that included a large Hispanic population did not find a correlation between K-*ras* mutation and survival.[44] Furthermore, Sasaki et al. found that none of the 22 patients in their study who died of recurrent disease had K-*ras* mutations.[45] In summary, although it appears that the incidence of codon 12 mutations in endometrial cancers is significantly higher in Japan than in America, there is no consistent association of mutation with specific pathologic features or prognosis.

Finally, K-*ras* mutations have been identified in some endometrial hyperplasias.[40,44,45] The frequency of mutations in hyperplasias is similar to that seen in endometrial cancers, which suggests that K-*ras* mutation may be a relatively early event in the development of some endometrial cancers. Mutations are found more frequently as the severity of the hyperplasia increases — from 10% in simple to 14% in adenomatous to 22% in atypical adenomatous hyperplasias.[45] Since only a minority of endometrial hyperplasias contain K-*ras* mutations, other genes also must play a role in their development.

2.3 Nuclear Transcription Factors

If proliferation is to occur in response to signals generated at the periphery of the cell, these events must lead to changes in gene expression and DNA synthesis. In this regard, a family of genes whose products bind to DNA and regulate gene transcription has been described. Transcription of several of these genes increases dramatically within minutes of treatment of normal cells with peptide growth factors. Once induced, the products of these genes bind to specific DNA regulatory elements and result in transcription of genes involved in DNA synthesis and cell division. When inappropriately overexpressed, however, these transcription factors can act as oncogenes.

Among the nuclear transcription factors involved in stimulating proliferation, amplification of members of the *myc* family has most often been implicated in the development of human cancers. It has been shown that c-*myc* is expressed in normal endometrium[35] and endometriosis,[46] with higher expression in the proliferative phase relative to the secretory phase. Monk et al. found that c-myc was amplified in 11% of 37 frozen endometrioid cancers and amplification correlated with poor grade.[25] Similarly, another small study suggested that c-myc may be amplified in a fraction of endometrial cancers.[47]

3. TUMOR-SUPPRESSOR GENES

Tumor-suppressor genes encode proteins that normally inhibit proliferation.[3,48] Because inactivation of both copies is required to eliminate the inhibitory effect of a tumor suppressor gene, these genes are referred to as recessive cancer causing genes. Loss of tumor-suppressor function may occur via several mechanisms including complete deletion of a gene, mutations or partial deletions that cripple the gene product, lack of transcription of a gene, or inactivation of the corresponding protein. Although initially tumor-suppressor genes were thought to be involved primarily in the development of rare hereditary cancers in humans, it has been shown that these genes also may be inactivated in common types of sporadic cancers.

3.1 p53

Loss of p53 tumor-suppressor gene function is the most frequent genetic event described thus far in human cancers.[49-51] Normally, p53 protein inhibits proliferation by binding to transcriptional regulatory elements in DNA. Binding of p53 to DNA results in expression of several growth inhibitory genes including WAF1, which blocks the action of cyclin dependent kinases that are required for cell cycle progression.[52] Beyond simply inhibiting proliferation, normal p53 is thought to play an active role in preventing cancer. In this regard, p53 functions as a surveillance mechanism in which cells that have undergone genetic damage are arrested in the G_1 phase of the cell cycle to allow for DNA repair.[53] If DNA repair is inadequate, p53 can trigger programmed cell death, also known as apoptosis. If the p53 gene has been inactivated, apoptosis may not occur appropriately, allowing cells that have undergone significant DNA damage to survive.

Many cancers have point mutations in one copy of the p53 gene, which result in an inactive protein product that cannot bind to DNA.[49-51] As is the case for other tumor-suppressor genes, mutation of one copy of the p53 gene often is accompanied by deletion of the other copy, leaving the cancer cell with only mutant p53 protein. On the other hand, if the cancer cell retains one normal copy of the p53 gene, mutant p53 can complex with normal p53 protein and prevent it from interacting with DNA. Because inactivation of both p53 alleles is not required to abrogate p53 function, p53 mutations are said to have a dominant negative effect. Finally, while normal cells have low levels of p53 protein because it is rapidly degraded, mutant p53 proteins are resistant to degradation and overaccumulate in the nucleus. This relative overexpression of mutant p53 protein can be detected immunohistochemically.

Mutation of the p53 gene, with resultant overexpression of the p53 protein, has been found in a wide range of cancers including carcinomas, sarcomas and hematologic malignancies — and represents the most common molecular genetic event described thus far in human cancers.[50,51,54] The p53 mutations in most types of cancer are diverse, but occur in areas of the gene that encode functionally important parts of the p53 molecule that have been highly conserved throughout evolution.[54] With the exception of the rare Li-Fraumeni syndrome, in which affected individuals develop multiple primary cancers at a young age, and a small percentage of childhood sarcomas, it appears that p53 mutations in human cancers represent acquired rather than inherited defects.

We found that mutant p53 protein was overexpressed in 20% of 107 frozen primary endometrial adenocarcinomas, including 9% of stage I/II and 41% of stage III/IV cancers.[55] In a univariate analysis, p53 overexpression in endometrial adenocarcinomas was associated with several known prognostic factors including poor grade, advanced stage, and the absence of progesterone receptor.[55] In addition, survival of patients whose cancers overexpressed p53 was worse than that of patients whose cancers did not overexpress p53 (Figure 2). In a multivariate analysis, however, p53 overexpression was not an

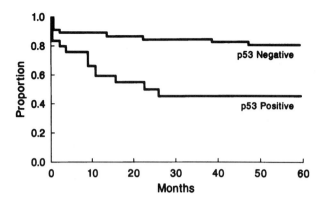

Figure 2 Relationship between p53 overexpression and disease-free survival (p53 positive n = 21, p53 negative n = 79).

independent variable associated with recurrence, whereas FIGO stage, histologic grade, myometrial invasion and DNA ploidy were.[7]

Other studies have confirmed the association between p53 overexpression and poor prognostic features.[19,28,56-58] A correlation with poor survival also has been demonstrated in an immunohistochemical study of paraffin blocks from Japanese endometrial cancers.[59] Overall, 21% of 221 cases were found to overexpress p53. Among patients with stage I/II disease, recurrence developed in 50% of 22 patients whose cancers overexpressed p53 compared to only 15% of 156 patients whose cancers did not overexpress p53. In addition, multivariate analysis suggested that p53 overexpression was an independent predictor of poor survival in early stage cases.

Endometrial cancers that overexpress p53 protein have been shown to harbour point mutations in conserved regions of the gene that result in amino acid substitutions in the protein (Figure 3).[41,55,60-62] In our study, in four of five endometrial cancers with mutations, only the mutant allele was transcribed, indicating that the wild-type allele likely was deleted. Allele loss in concert with mutation of the p53 gene in endometrial cancers also has been observed by Okamoto et al.[62] Overall, 17 mutations in exons 5-8 of the p53 gene have been described in endometrial adenocarcinomas (Table 2). Similar to many other cancers, point mutations in endometrial cancers occur throughout conserved regions in exons 5-9 of the p53 gene. Interestingly, 4/17 (24%) reported mutations are in codon 248 and 2/17 (12%) in codon 282, which are among the most frequently affected codons in other types of cancers as well.[50,51,54] It has been postulated that mutations in these codons are particularly effective in negating the normal tumor-suppressor function of the p53 gene product.

Although little is known regarding molecular alterations in uterine sarcomas, we found that overexpression of p53 frequently occurs in mixed mesodermal sarcomas of the uterus (17/23; 74%).[63] In addition, using DNA sequencing techniques, point mutations in the p53 gene were identified in ten cases (9 transitions, 1 transversion). Mutation and overexpression of p53 were also seen in 16/34 leiomyosarcomas.[64] It appears that p53 alterations in uterine cancers are most frequent in mixed mesodermal sarcomas and advanced stage adenocarcinomas, both of which have a poor prognosis. Similarly, p53 overexpression has been associated with poor outcome in a number of other cancers including ovarian[65] and breast cancer.[66]

Since malignant transformation is believed to be a multistep process, several groups have sought to determine when in the sequence of events mutation of the p53 gene occurs. Thus far, these studies suggest that the timing of p53 mutation varies between different tumor types. In colon cancer, which demonstrates a well-defined progression from adenoma to carcinoma, mutation of the p53 gene is found in 60% of cancers, but is far less commonly seen in adenomas.[67] As a result, Vogelstein has postulated that p53 mutation is a relatively late event in colon carcinogenesis.[68] In contrast, it has been shown that mutation of the p53 gene occurs prior to the development of invasive cancer in some circumstances. For example, mutations of the p53 gene have been noted in squamous carcinoma in-situ of the lung and esophagus.[69,70]

To determine whether alteration of p53 is an early event in endometrial carcinogenesis, we investigated whether mutation of this gene is a feature of endometrial hyperplasias, which are thought to represent precursors of some endometrial carcinomas.[71] Using the technique of single stranded conformation

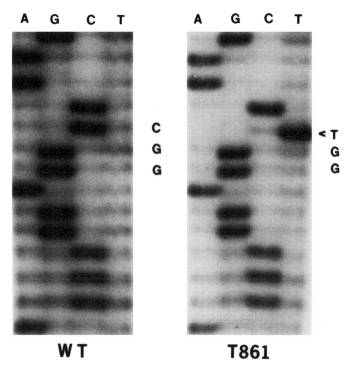

Figure 3   Point mutation of codon 248 of the p53 gene (C → T) in endometrial cancer (WT = normal wild-type sequence, T861 = endometrial cancer).

Table 2   P53 Mutations in Endometrial Adenocarcinomas

Case	Codon	Exon	Mutation	Type Mutation
Okamoto et al. (1991)[62]				
1	282	8	CGG → TGG	Transition
Kohler et al. (1992)[55]				
2	199	6	GGA → GTA	Transversion
3	242	7	TGC → TTC	Transversion
4	248	7	CGG → TGG	Transition
5	282	8	CGG → TGG	Transition
6	241	7	TCC → TTC	Transition
Risinger et al. (1992)[60]				
7	176-8	5	Insertion of C	Insertion
8	248	7	CGG → TGG	Transition
9	248	7	CGG → TGG	Transition
Naito et al. (1992)[87]				
10	239	7	AAC → GAC	Transition
	254		ATC → AGC	Transversion
Enomoto et al. (1993)[41]				
11	204	6	GAG → TAG	Transversion
12	233	7	CAC → CACC	Insertion
13	149	5	TCC → CC	Deletion
14	248	7	CGG → CAG	Transition
15	193	6	CAT → GAT	Transversion
16	181	5	CGC → TGC	Transition
17	175	5	CGC → TGC	Transition

polymorphism analysis we did not identify p53 mutations in any of 117 endometrial hyperplasias including 41 atypical adenomatous cases.[71] In addition, consistent with this finding, p53 immunostaining was not seen in any of 44 endometrial hyperplasias. In another study, Enomoto found p53 mutation in only 1/13 atypical hyperplasias.[41] The rarity of p53 mutations in endometrial hyperplasias and the correlation of p53 alterations with extrauterine disease in invasive endometrial cancers suggests that mutation of the p53 gene is a relatively late event in endometrial carcinogenesis. Alternatively , it is possible that acquisition of a p53 mutation leads to development of a virulent cancer that does not pass through a phase of hyperplasia and is associated with rapid spread of disease.

3.2 Other Tumor-Suppressor Genes

We have sought to determine whether other known tumor-suppressor genes are altered in endometrial cancers. For the most part, these studies have been disappointing. Alterations in the Retinoblastoma (Rb), Wilm's tumor (WT1), von Hippel Lindau (VHL), and Adenomatous Polyposis Coli (APC) tumor-suppressor genes have not been identified (unpublished data). One strategy for locating genetic loci that might harbor tumor-suppressor genes of importance is to perform a broad survey to identify specific chromosomal loci that frequently are deleted in a particular type of cancer. This sort of analysis is done by examining polymorphic areas of DNA that often vary in normal cells, allowing the two copies of a gene to be distinguished. If an individual is heterozygous for a given gene, loss of one copy of the gene is manifest as loss of heterozygosity (LOH). Traditionally, LOH analysis has been performed by exploiting restriction fragment length polymorphisms within genes. The informativity rate using this technique usually is significantly less than 50%, however. More recently, LOH analysis has been performed using microsatellite markers, which are simply repetitive sequences that are interspersed throughout the genome mostly in the "junk DNA" between genes. Although their functional significance is unknown, they are particularly useful for analysis of LOH because the number of repeats in a given microsatellite is highly variable, and thus they yield a high rate of informativity.

We examined 70 CA repeat microsatellite markers encompassing all chromosome arms in matched normal/endometrial cancer DNA from 60 patients who received treatment at Duke University.[72] The frequency of LOH on most chromosomal arms was less than 15% (Figure 4). The highest frequencies of LOH were seen on 3p (18%), 8p (21%), 9p (21%), 14q (19%), 16q (21%), and 18q (33%). This frequency of allele loss was much lower than that seen in ovarian,[73] colon[74] and other cancers. Similarly, Okamoto et al. found a relatively low frequency of allele loss in endometrial cancer.[62] The proportion of informative cases was much lower in this prior study, which used Southern analysis of restriction length fragment polymorphisms, however. In other studies that employed microsatellite markers, the highest frequencies of allele loss were seen on 3p and 10q.[75,76]

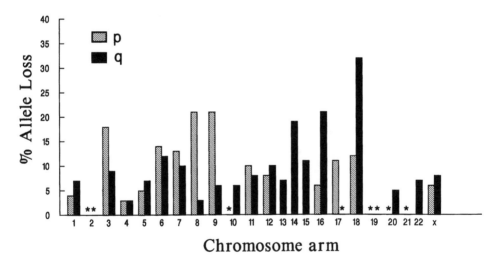

Figure 4 Allelotype of endometrial cancer. Percentages of allele loss represent the number of cancers that had loss of one or more alleles on a given arm divided by the total number of cancers in which loci were examined on that arm (chromosomes 13, 14, 15, 21, and 22 do not have a p arm, asterisk = no allele loss seen.).

The low frequency of allele loss in endometrial cancers may reflect the fact that most of these cancers are confined to the uterus at diagnosis. In this regard, it has been shown in other types of cancers, including ovarian cancer,[73] that advanced cancers usually have a higher frequency of allele loss relative to early stage cancers. It also has been shown that deletions in advanced cancers often involve loss of entire chromosomal arms, rather than discrete loci.[77] Thus, allele losses in advanced cancers may reflect generalized genetic instability rather than inactivation of specific tumor-suppressor genes. In contrast, when allele loss occurs in endometrial cancer, other loci on the same chromosomal arm usually are retained[72] suggesting that critical genes are selectively affected.

The highest frequency of allele loss in our study (33%) was on chromosome 18q at the locus where the DCC (Deleted in Colon Carcinoma) gene resides.[78] The DCC gene originally was identified by virtue of its localization to a region of chromosome 18q that is subject to allelic deletion in a significant fraction of human colorectal carcinomas. The subsequent observation that introduction of a normal human chromosome 18 into a human colorectal carcinoma cell line results in upregulation of DCC expression and suppression of tumorigenicity further supports the hypothesis that DCC functions as a tumor-suppressor. Cloning of the DCC gene revealed homology to NCAMs (neural cell adhesion molecules) and the DCC gene product also is thought to be involved in cell-cell adhesion.[78] Because of the high frequency of allele loss at DCC in endometrial cancers, we screened 60 cases for DCC mutations using SSCP. Analysis of 20 of 29 known DCC exons has not revealed mutations in any of the endometrial carcinomas with DCC allele loss (unpublished data).

An alteration in the DCC gene has been described in colon cancers in which there is an insertion in the intronic region downstream from exon 8 that may disrupt transcription of the gene leading to an aberrant protein product. This insertion mutation also has been noted in 5/11 immortalized endometrial cancer cell lines,[79] but in only 2/60 primary cancers (unpublished data). Although few mutations in DCC have been seen in endometrial cancers, it is possible that haploinsufficiency of the DCC gene product is in itself an adequate driving force for allelic loss during neoplastic progression. In fact, small changes in expression of related NCAMs have been shown to dramatically affect cell adhesion.[80]

4. DNA REPAIR GENES

Our group and others have noted that some endometrial cancer DNA samples contain microsatellite alleles that do not correspond to either allele from the matched normal DNA.[81,82] In our study, among 36 sporadic endometrial cancers in which numerous markers were examined, 17%, were found to have widespread evidence of new microsatellite alleles throughout the genome.[81] Endometrial cancers that exibited microsatellite instability were diploid and had a favorable prognosis. Similarly, Duggan et al. found microsatellite instability in 9/45 (20%) cases.[82] In their study, mutations in the K-*ras* oncogene were more common in cases with instability (56%) compared to cases in which instability was not seen (14%).

Microsatellite instability initially was noted in colon cancers of patients with hereditary nonpolyposis colon cancer (HNPCC), also known as Lynch syndrome type II.[83,84] Endometrial cancer is the second most common malignancy observed in these families, but ovarian, breast and gastrointestinal tract malignancies also occur. We found that 3/4 endometrial cancers from members of HNPCC families had microsatellite instability.[81] Subsequently, it was shown that 60% of affected individuals in HNPCC families have germline mutations in the MSH2 gene on chromosome 2p, which is involved in DNA mismatch repair.[59] An additional 30% of these families have been shown to carry mutations in another DNA repair gene on chromosome 3p (MLH1),[59] and the remainder are thought to be due to mutations in either PMS1 or PMS2. In bacteria and yeast, mutations in these DNA repair enzymes also lead to microsatellite instability, confirming the cause and effect relationship between these events.

It is thought that, in addition to causing microsatellite instability, the inability to repair DNA may lead to an increased rate of genetic damage throughout the genome. This then may increase the likelihood of tumorigenesis due to alterations in oncogenes and tumor-suppressor genes. In this regard it has been shown that colorectal cancers with microsatellite instability have inactivating mutations in the type II transforming growth factor-β receptor gene, but similar mutations have not been seen in endometrial cancers.[85] Thus, DNA repair genes represent a new family of cancer causing genes, distinct from those previously described. Because microsatellite instability has been noted in some sporadic endometrial cancers,[81] several groups have attempted to identify acquired mutations in DNA repair genes in these cancers. In most cases, however, mutations in these genes have not been found in endometrial cancers.[86]

REFERENCES

1. **Bishop, J.M.,** Molecular themes in oncogenesis, *Cell*, 64, 235, 1991.
2. **Cooper, J.M.,** Oncogenes. Boston, Jones and Bartlett, Boston, 1990.

3. **Levine, A.J.,** The tumor-suppressor genes, *Annu. Rev. Biochem.*, 62, 623, 1993.

4. **Weinberg, R.A.,** Tumor-suppressor genes, *Science*, 254, 1138, 1992.

5. **Britton, L.C., Wilson, T.O., Gaffey, T.A., Cha, S.S., Wieand, H.S., and Podratz, K.C.,** DNA ploidy in endometrial carcinoma: Major objective prognostic factor, *Mayo Clin. Proc.*, 65, 643, 1990.

6. **Podratz, K.C., Wilson, T.O., Gaffey, T.A., Cha, S.S., and Katzmann, J.A.,** Deoxyribonucleic acid analysis facilitates the pretreatment identification of high-risk endometrial cancer patients, *Am. J. Obstet. Gynecol.*, 168, 1206, 1993.

7. **Lukes, A.S., Kohler, M.F., Pieper, C.F., Kerns, B.J., Bentley, R., Rodriguez, G.C., Soper, J.T., Clarke-Pearson, D.L., Bast, R.C., and Berchuck, A.,** Multivariable analysis of DNA ploidy, p53, and HER-2/*neu* as prognostic factors in endometrial cancer, *Cancer*, 73, 2380, 1994.

8. **Shah, N.K., Currie, J.L., Rosenshein, N., Campbell, J., Long, P., Abbas, F., and Griffin, C.A.,** Cytogenetic and FISH analysis of endometrial carcinoma, *Cancer Genet. Cytogenet.*, 73, 142, 1994.

9. **Cantley, L.C., Auger, K.R., Carpenter, C., Duckworth, B., Graziani, A., Kapeller, R., and Soltoff, S.,** Oncogenes and signal transduction, *Cell*, 64, 281, 1991.

10. **Carpenter, G. and Cohen, S.,** Epidermal Growth Factor, *J. Biol. Chem.*, 265:14, 7709, 1990.

11. **Berchuck, A., Soisson, A.P., Olt, G.J., Soper, J.T., Clarke-Pearson, D.L., Bast, R.C., and McCarty, K.S.,** Epidermal growth factor receptor expression in normal and malignant endometrium, *Am. J. Obstet. Gynecol.*, 161, 1247, 1989.

12. **Reynolds, K., Talavera, F.R., Hopkins, M., and Menon, K.,** Characterization of epidermal growth factor receptor in normal and neoplastic human endometrium, *Cancer*, 66, 1967, 1990.

13. **Ozanne, B., Richards, C.S., Hendler, F., Burns, D., and Gusterson, B.,** Over-expression of the EGF receptor is a hallmark of squamous cell carcinomas, *J. Pathol.*, 149, 9, 1986.

14. **Lelle, R.J., Talavera, F., Gretz, H., Roberts, J.A., and Menon, K.,** Epidermal growth factor receptor expression in three different human endometrial cancer cell lines, *Cancer*, 72, 519, 1993.

15. **Birmelin, G., Zimmer, V., Sauerbrei, W., Pfleiderer, A., and Bauknecht, T.,** Relationship between epidermal growth factor receptor (EGF-R) and various prognostic factors in human endometrial carcinoma, *Int. J. Gynecol. Cancer*, 2, 66, 1995.

16. **Nyholm, H.C.J., Nielsen, A., and Ottesen, B.,** Expression of epidermal growth factor receptors in human endometrial carcinoma, *Int. J. Gynecol. Pathol.*, 12, 241, 1993.

17. **Scambria, G., Panici, P.B., Ferrandina, G., Battaglia, F., Distefano, M., D'Andrea, G., De Vincenzo, R., Maneschi, F., Ranelletti, F. O., and Mancuso, S.,** Significance of epidermal growth factor receptor expression in primary human endometrial cancer, *Int. J. Cancer*, 56, 26, 1994.

18. **Khalifa, M.A., Abdoh, A., Mannel, R., Haraway, S., Walker, J., and Min, K.,** Prognostic utility of epidermal growth factor receptor overexpression in endometrial adenocarcinoma, *Cancer*, 73, 370, 1995.

19. **Khalifa, M.A., Mannel, R., Haraway, S., Walker, J., and Min, K.,** Expression of EGFR, HER-2/*neu*, p53, and PCNA in endometrioid, serous papillary, and clear cell endometrial adenocarcinomas, *Gynecol. Oncol.*, 53, 84, 1994.

20. **Lupu, R., Colomer, R., Zugmaier, G., Sarup, J., Shepard, M., Slamon, D.J., and Lippman, M.E.,** Direct interaction of a ligand for the *erb*B2 oncogene product with the EGF receptor and p185*erb*B2, *Science*, 249, 1552, 1990.

21. **Slamon, D.J., Godolphin, W., Jones, L.A., Holt, J.A., Wong, S.G., Keith, D.E., Levin, L.J., Stuart, S.G., Udove, J., Ullrich, A., and Press, M.F.,** Studies of HER-2/*neu* proto-oncogene in human breast and ovarian cancer, *Science*, 244, 707, 1989.

22. **Berchuck, A., Kamel, A., Whitaker, R., Kerns, B., Olt, G., Kinney, R., Soper, J.T., Dodge, R., Clarke-Pearson, P., Marks, S., McKenzie, S., Yin, S., and Bast, R.C.,** Overexpression of HER-2/*neu* is associated with poor survival in advanced epithelial ovarian cancer, *Cancer Res.*, 50, 4087, 1990.

23. **Berchuck, A., Rodriguez, G., Kinney, R.B., Soper, J.T., Dodge, R.K., Clarke-Pearson, D.L., and Bast, R.C.,** Overexpression of HER-2/*neu* in endometrial cancer is associated with advanced stage disease, *Am. J. Obstet. Gynecol.*, 164, 15, 1991.

24. **Hetzel, D.J., Wilson, T.O., Keeney, G.L., Roche, P.C., Cha, S.S., and Podratz, K.C.,** HER-2/*neu* expression: A major prognostic factor in endometrial cancer, *Gynecol. Oncol.*, 47, 179, 1992.

25. **Monk, B.J., Chapman, J.A., Johnson, G., Brightman, B., Wilczynski, S., Schell, M.J., and Fan, H.,** Correlation of c-*myc* and HER-2/*neu* amplification and expression with histopathologic variables in uterine corpus cancer, *Am. J. Obstet. Gynecol.*, 171, 1193, 1994.

26. **Bigsby, R., Aixin, L., Bomalaski, J., Stehman, F., Look, K., and Sutton, G.,** Immunohistochemical study of HER-2/*neu*, epidermal growth factor receptor, and steroid receptor expression in normal and malignant endometrium, *Obstet. Gynecol.*, 79, 95, 1992.

27. **Wang, D., Konishi, I., Koshiyama, M., Mandai, M., Nanbu, Y., Ishikawa, Y., Mori, T., and Fujii, S.,** Expression of c-*erb*B-2 protein and epidermal growth factor receptor in endometrial carcinomas, *Cancer*, 72, 2628, 1995.

28. **Pisani, A.L., Barbuto, D.A., Chen, D., Ramos, L., LaGasse, L.D., and Karlan, B.Y.,** HER-2/neu, p53, and DNA analyses as prognosticators for survival in endometrial carcinoma, *Obstet. Gynecol.*, 85, 729, 1995.

29. **Saffari, B., Jones, L.A., El-Naggar, A., Felix, J.C., George, J., and Press, M.F.,** Amplification and overexpression of HER-2/neu (c-*erb*B2) in endometrial cancers: correlation with overall survival, *Cancer Res.*, 55, 5693, 1995.

30. **Nazeer, T., Ballouk, F., Malfetano, J.H., Figge, H., and Ambros, R.A.,** Multivariate survival analysis of clinico-pathologic features in surgical stage I endometroid carcinoma including analysis of HER-2/*neu* expression, *Am. J. Obstet. Gynecol.*, 173, 1829, 1995.

31. **Kacinski, B.M., Carter, D., Kohorn, E.I., Mittal, K., Bloodgood, R.S., Donahue, J., Donofrio, L., Edwards, R., and Schwartz, P.E.,** High level expression of *fms* proto-oncogene mRNA is observed in clinically aggressive endometrial adenocarcinomas, *Int. J. Radiat. Oncol. Biol. Phys.*, 15, 823, 1988.

32. **Leiserowitz, G., Harris, S., Subramaniam, M., Keeney, G., Podratz, K., and Spelsberg, T.,** The proto-oncogene c-*fms* is overexpressed in endometrial cancer, *Gynecol. Oncol.*, 49, 190, 1993.

33. **Kacinski, B.M., Chambers, S., Stanley, E., Carter, D., Tseng, P., Scata, K., Chang, D.Y., Pirro, M.H., Nguyen, J., Ariza, A., Rohrschneider, L., and Rothwell, V.,** The cytokine CSF-1 (M-CSF), expressed by endometrial carcinomas *in vivo* and *in vitro*, may also be a circulating tumor marker of neoplastic disease activity in endometrial carcinoma patients, *Int. J. Radiat. Oncol. Biol. Phys.*, 19, 619, 1990.

34. **Filderman, A., Bruckner, A., Kacinski, B.M., and Remold, H.,** Macrophage colony-stimulating factor (CSF-1) enhances invasiveness in CSF-1 receptor-positive carcinoma cell lines, *Cancer Res.*, 52, 3661, 1992.

35. **Odom, L.D., Barrett, J., Pantazis, C.G., Stoddard, L., and McDonough, P.,** Immunocytochemical study of *ras* and *myc* proto-oncogene polypeptide expression in the human menstrual cycle, *Am. J. Obstet. Gynecol.*, 161, 1663, 1989.

36. **Scambia, G., Benedetti-Panici, P., Ferrandina, G., Battaglia, F., Giovannini, G., Piffanelli, A., and Mancuso, S.,** Expression of *ras* p21 oncoprotein in normal and neoplastic human endometrium, *Gynecol. Oncol.*, 50, 339, 1993.

37. **Boyd, J.A. and Risinger, J.I.,** Analysis of oncogene alterations in human endometrial carcinoma: Prevalence of *ras* mutations, *Mol. Carcinog.*, 4, 189, 1991.

38. **Ignar-Trowbridge, D., Risinger, J.I., Dent, G.A., Kohler, M.F., Berchuck, A., McLachlan, J.A., and Boyd, J.,** Mutations of the Ki-*ras* oncogene in endometrial carcinoma, *Am. J. Obstet. Gynecol.*, 167, 227, 1992.

39. **Enomoto, T., Inoue, M., Perantoni, A.O., Terakawa, N., Tanizawa, O., and Rice, J.M.,** K-*ras* activation in neoplasms of the human female reproductive tract, *Cancer Res.*, 50, 6139, 1990.

40. **Enomoto, T., Inoue, M., Perantoni, A.O., Buzard, G.S., Miki, H., Tanizawa, O., and Rice, J.M.,** K-*ras* activation in premalignant and malignant epithelial lesions of the human uterus, *Cancer Res.*, 51, 5308, 1991.

41. **Enomoto, T., Fujita, M., Inoue, M., Rice, J.M., Nakajima, R., Tanazawa, O., and Nomura, T.,** Alterations of the p53 tumor-suppressor gene and its association with activation of the c-K-*ras*-2 protooncogene in premalignant and malignant lesions of the human uterine endometrium, *Cancer Res.*, 53, 1883, 1993.

42. **Fujimoto, T., Shimizu, Y., Hirai, Y., Chen, J., Teshima, H., Hasumi, K., Masubuchi, K., and Takahashi, M.,** Studies on *ras* oncogene activation in endometrial carcinoma, *Gynecol. Oncol.*, 48, 196, 1993.

43. **Mizuuchi, H., Nasim, S., Kudo, R., Silverberg, S.G., Greenhouse, S., and Garrett, C.T.,** Clinical implications of K-*ras* mutations in malignant epithelial tumors of the endometrium, *Cancer Res.*, 52, 2777, 1992.

44. **Duggan, B., Felix, J., Muderspach, L., Tsao, J., and Shibata, D.,** Early mutational activation of the c-Ki-*ras* oncogene in endometrial carcinoma, *Cancer Res.*, 54, 1604, 1994.

45. **Sasaki, H., Nishii, H., Tada, A., Furusato, M., Terashima, Y., Siegal, G.P., Parker, S.L., Kohler, M.F., Berchuck, A., and Boyd, J.,** Mutation of the Ki-*ras* protooncogene in human endometrial hyperplasia and carcinoma, *Cancer Res.*, 53, 1906, 1993.

46. **Schenken, R.S., Johnson, J.V., and Riehl, R.M.,** c-*myc* protooncogene polypeptide expression in endometriosis, *Am. J. Obstet. Gynecol.*, 164, 1031, 1991.

47. **Borst, M.P., Baker, V.V., Dixon, D., Hatch, K.D., Shingleton, H.M., and Miller, D.M.,** Oncogene alterations in endometrial carcinoma, *Gynecol. Oncol.*, 38, 346, 1990.

48. **Knudson, A.G.,** Anticoncogenes and human cancer, *Proc. Natl. Acad. Sci. U.S.A.*, 90, 10914, 1993.

49. **Levine, A.J., Momand, J., and Finlay, C.A.,** The p53 tumour suppressor gene, *Nature*, 351, 453, 1991.

50. **Greenblatt, M.S., Bennett, W.P., Hollstein, M., and Harris, C.C.,** Mutations in the p53 tumor-suppressor gene: Clues to cancer etiology and molecular pathogenesis, *Cancer Res.*, 54, 4855, 1994.

51. **Berchuck, A., Kohler, M.F., Marks, J.R., Wiseman, R., Boyd, J., and Bast, R.C.,** The p53 tumor-suppressor gene frequently is altered in gynecologic cancers, *Am. J. Obstet. Gynecol.*, 170, 246, 1994.

52. **El-Deiry, W.S., Tokino, T., Velculescu, V.E., Levy, D.B., Parsons, R., Trent, J.M., Lin, D., Mercer, W.E., Kinzler, K.W., and Vogelstein, B.,** WAF1, a potential mediator of p53 tumor suppression, *Cell*, 75, 817, 1993.

53. **Kastan, M.B., Onyekwere, O., Sidransky, D., Vogelstein, B., and Craig, R.W.,** Participation of the p53 protein in the cellular response to DNA damage, *Cancer Res.*, 51, 6304, 1991.

54. **Hollstein, M., Sidransky, D., Vogelstein, B., and Harris, C.C.,** p53 mutations in human cancers, *Science*, 253, 49, 1991.

55. **Kohler, M.F., Berchuck, A., Davidoff, A.M., Humphrey, P.A., Dodge, R.K., Iglehart, J.D., Soper, J.T., Clarke-Pearson, D.L., Bast, R.C., and Marks, J.R.,** Overexpression and mutation of p53 in endometrial carcinoma, *Cancer Res.*, 52, 1622, 1992.

56. **Inoue, M., Okayama, A., Fujita, M., Enomoto, T., Sakata, M., Tanizawa, O., and Ueshima, H.,** Clinicopatho-logical characteristics of p53 overexpression in endometrial cancers, *Int. J. Cancer*, 58, 14, 1994.

57. **Hachisuga, T., Fukuda, K., Uchiyama, M., Matsuo, N., Iwasaka, T., and Sugimore, H.,** Immunohistochemical study of p53 expression in endometrial carcinomas: correlation with markers of proliferating cells and clinicopatho-logic features, *Int. J. Gynecol. Cancer*, 3, 363, 1993.

58. **Ito, K., Watanabe, K., Nasim, S., Sasano, H., Sato, S., Yajima, A., Silverberg, S., and Garrett, C.T.,** Prognostic significance of p53 overexpression in endometrial cancer, *Cancer Res.*, 54, 4667, 1994.

59. **Service, R.F.,** Research news: Stalking the start of colon cancer, *Science*, 263, 1559, 1994.

60. **Risinger, J.I., Dent, G.A., Ignar-Trowbridge, D., McLachlan, J.A., Tsao, M.S., Senterman, M., and Boyd, J.A.,** Mutations of the p53 gene in human endometrial carcinoma, *Mol. Carcinog.*, 5, 250, 1992.

61. **Yaginuma, Y. and Westphal, H.,** Analysis of the p53 gene in human uterine carcinoma cell lines, *Cancer Res.*, 51, 6506, 1991.

62. **Okamoto, A., Sameshima, Y., Yamada, Y., Teshima, S., Terashima, Y., Terada, M., and Yokota, J.,** Allelic loss on chromosome 17p and p53 mutations in human endometrial carcinoma of the uterus, *Cancer Res.*, 51, 5632, 1991.

63. **Liu, F.S., Kohler, M.F., Marks, J.R., Bast, R.C., Boyd, J., and Berchuck, A.,** Mutation and overexpression of the p53 tumor-suppressor gene frequently occurs in uterine and ovarian sarcomas, *Obstet. Gynecol.*, 83, 118, 1994.

64. **Neimann, T.H., Raab, S.S., Lenel, J.C., Rodgers, J.R., and Robinson, R.A.,** p53 protein overexpression in smooth muscle tumors of the uterus, *Hum. Pathol.*, 26, 375, 1995.

65. **Hartmann, L., Podratz, K., Keeney, G., Kamel, N., Edmonson, J., Grill, J., Fu, J., Katzmann, J., and Roche, P.,** Prognostic significance of p53 immunostaining in epithelial ovarian cancer, *J. Clin. Oncol.*, 12, 64, 1994.

66. **Thor, A.D., Moore, D.H., Edgerton, S.M., Kawaski, E.S., Reihsaus, E., Lynch, H.T., Marcus, J.N., Schwartz, L., Chen, L.C., Mayall, B.H., and Smith, H.S.,** Accumulation of the p53 tumor-suppressor gene protein: An independent marker of prognosis in breast cancers, *J. Natl. Cancer Inst.*, 84, 845, 1992.

67. **Baker, S.J., Preisinger, A.C., Jessup, J.M., Paraskeva, C., Markowitz, S., Willson, J.K., Hamilton, S., and Vogelstein, B.,** p53 gene mutations occur in combination with 17p allelic deletions as late events in colorectal tumorigenesis, *Cancer Res.*, 50, 7717, 1990.

68. **Vogelstein, B., Fearon, E.R., Hamilton, S.R., Kern, S.E., Preisinger, A.C., Leppert, M., Nakamura, Y., White, R., Smits, A.M., and Bos, J.L.,** Genetic alterations during colorectal-tumor development, *N. Engl. J. Med.*, 319, 525, 1988.

69. **Sozzi, G., Miozzo, M., Donghi, R., Pilotti, S., Cariani, C.T., Pastorino, U., Della Porta, G., and Pierotti, M.A.,** Deletions of 17p and p53 mutations in preneoplastic lesions of the lung, *Cancer Res.*, 52, 6079, 1992.

70. **Bennett, W.P., Hollstein, M.C., Metcalf, R.A., Welsh, J.A., He, A., Zhu, S., Kusters, I., Resau, J.H., Trump, B.F., Lane, D.P., and Harris, C.C.,** p53 mutation and protein accumulation during multistage human esophageal carcinogenesis, *Cancer Res.*, 52, 6092, 1992.

71. **Kohler, M.F., Nishii, H., Humphrey, P.A., Sasaki, H., Boyd, J.A., Marks, J.R., Bast, R.C., Clarke-Pearson, D.L., and Berchuck, A.,** Mutation of the p53 tumor-suppressor gene is not a feature of endometrial hyperplasias, *Am. J. Obstet. Gynecol.*, 169, 690, 1993.

72. **Fujino, T., Risinger, J.I., Collins, N.K., Liu, F.S., Nishii, H., Takahashi, H., Westphal, E.M., Barrett, C.J., Sasaki, H., Kohler, M.F., Berchuck, A., and Boyd, J.A.,** Allelotype of endometrial carcinoma, *Cancer Res.*, 54, 4294, 1994.

73. **Dodson, M.K., Hartmann, L.C., Cliby, W.A., DeLaceey, K.A., Keeney, G.L., Ritland, S.R., Su, J.Q., Podratz, K.C., and Jenkins, R.B.,** Comparison of loss of heterozygosity patterns in invasive low-grade and high-grade epithelial ovarian carcinomas, *Cancer Res.*, 53, 4456, 1993.

74. **Vogelstein, B., Fearon, E.R., Kern, S.E., Hamilton, S.R., Preisinger, A.C., Nakamura, Y., and White, R.,** Allelotype of colorectal carcinomas, *Science*, 244, 207, 1989.

75. **Jones, M.H., Koi, S., Fujimoto, I., Hasumi, K., Kato, K., and Nakamura, Y.,** Allelotype of uterine cancer by analysis of RFLP and microsatellite polymorphisms: frequent loss of heterozygosity on chromosome arms 3p, 9q, 10q, and 17p, *Genes Chromosome Cancer*, 9, 119, 1994.

76. **Pfeiffer, S.L., Herzog, T.J., Tribune, D.J., Mutch, D.G., Gersell, D.J., and Goodfellow, P.J.,** Allelic loss of sequences from the long arm of chromosome 10 and replication errors in endometrial cancers, *Cancer Res.*, 55, 1922, 1995.

77. **Jacobs, I.J., Smith, S., Wiseman, R., Futreal, A., Berchuck, A., Ponder, B.J., and Bast, R.C.,** A deletion unit on chromosome 17q defined by loss of heterozygosity in epithelial ovarian cancer and a benign serous cystadenoma, *Cancer Res.*, 53, 1218, 1993.

78. **Fearon, E., Cho, K., Nigro, J., Kern, S., Simons, J., Ruppert, J., Hamilton, S., Preisinger, A., and Thomas, G.,** Identification of a chromosome 18q gene that is altered in colorectal cancers, *Science*, 247, 49, 1990.

79. **Cho, K.R., Oliner, J.D., Simons, J.W., Hedrick, L., Fearon, E.R., Preisinger, A.C., Hedge, P., Silverman, G.A., and Vogelstein, B.,** The DCC gene: structural analysis and mutations in colorectal carcinomas, *Genomics*, 19, 525, 1994.

80. **Lawlor, K.G. and Narayanan, R.,** Persistent expression of the tumor-suppressor gene DCC is essential for neuronal differentiation, *Cell. Growth Diff.*, 3, 609, 1992.

81. **Risinger, J.I., Berchuck, A., Kohler, M.F., Watson, P., Lynch, H.T., and Boyd, J.,** Genetic instability of microsatellites in endometrial carcinoma, *Cancer Res.*, 53, 5100, 1993.

82. **Duggan, B.D., Felix, J.C., Muderspach, L.I., Tourgeman, D., Zheng, J., and Shibata, D.,** Microsatellite instability in sporadic endometrial carcinoma, *J. Natl. Cancer Inst.*, 86, 1216, 1994.

83. **Peltomaki, P., Aaltonen, L.A., Sistonen, P., Pylkkanen, L., Mecklin, J., Jarvinen, H., Green, J.S., Jass, J.R., Weber, J.L., Leach, F.S., Petersen, G.M., Hamilton, S.R., de la Chapelle, A., and Vogelstein, B.,** Genetic mapping of a locus predisposing to human colorectal cancer, *Science*, 260, 810, 1993.

84. **Thibodeau, S.N., Bren, G., and Schaid, D.,** Microsatellite instability in cancer of the proximal colon, *Science*, 260, 816, 1993.

85. **Katabuchi, H., Van Rees, B., Lambers, A.R., Ronnett, B.M., Blazes, M.S., Leach, F.S., Cho, K.R., and Hedrick, L.,** Mutations in DNA mismatch repair genes are not responsible for microsatellite instability in most sporadic endometrial carcinomas, *Cancer Res.*, 55, 5556, 1995.

86. **Myeroff, L.L., Parsons, R., Kim, S.J., Hedrick, L., Cho, K.R., Orth, K., Mathis, M., Kinzler, K.W., Lutterbaugh, J., Park, K., Bang, Y.J., Lee, H.Y., Park, J.G., Lynch, H.T., Roberts, A.B., Vogelstein, B., and Markowitz, S.D.,** A transforming growth factor ß receptor type II gene mutation common in colon and gastric but rare in endometrial cancers with microsatellite instability, *Cancer Res.*, 55, 5545, 1995.

87. **Naito, M., Satake, M., Sakai, E., Hirano, Y., Tsuchida, N., Kanzaki, H., Ito, Y., and Mori, T.,** Detection of p53 gene mutations in human ovarian and endometrial cancers by polymerase chain reaction-single strand conformation polymorphism analysis, *Jpn. J. Cancer Res.*, 83, 1030, 1992.

Part V.
Cervical Cancer

Chapter 14

Cervical Cancer and Its Precursors — An Introduction

Thomas C. Wright, Jr. and Tjoung Won Park

CONTENTS

0-8493-9443-0/97/$0.00+$.50

1. INTRODUCTION

Cervical cancer is one of the leading causes of cancer-related death in women worldwide and is the leading cause of cancer-related death in women in developing countries. In much of the world, cervical cancer is diagnosed in women still in their 30s and has a major impact on the stability and social adhesion of families. It is also expected that the prevalence of invasive cervical cancer will increase over the next several decades as the number of women infected with human immunodeficiency virus (HIV) increases, and as the female population of developing countries begins to age.[1]

2. INCIDENCE OF CERVICAL CANCER

It is estimated that each year there are approximately 465,000 new cases of cervical cancer in the world.[2] The regions of the world with the highest risk include sub-Saharian Africa, South and Central America, and Southeast Asia. An age-standardized incidence of 83.2 has been reported in Recife, Brazil and 48.2 in Cali, Colombia. High rates are also reported from India where in the state of Madras there is an age-standardized incidence of 46.1 per 100,000 and in Bangalore an age-standardized incidence of 40.2.[3] Extremely low rates are reported from portions of the Middle East. Non-Jews living in Israel have an age-standardized incidence of 3.0 and Jews living in Israel have an incidence of 4.0 which is similar to that of Kuwaitians living in Kuwait with an age-standardized incidence of 3.9.[4] The incidence and mortality rates from invasive cervical cancer have been increasing in young women over the last decade in Canada, Great Britain, New Zealand, and the United States.[5-7] In the United States this increase has been most apparent for invasive cervical cancers in white women aged 35 to 54 years old.[7] In 1992, there were approximately 13,500 new cases of cervical cancer diagnosed in the United States and approximately 4000 deaths from invasive cervical cancer.[8] The average age-adjusted incidence of cervical cancer in the United States is somewhat in between high-incidence areas in the developing world and low-incidence areas in the Middle East. For the United States as a whole, the age-adjusted incidence in 1990 was 8.8 per 100,000 women with a corresponding age-adjusted mortality rate of 3.0.

Since the introduction of widespread cytologic screening programs for cervical cancer and its precursors, there has been a substantial decrease in the incidence of invasive cervical cancer. In the United States, the incidence per 100,000 has declined from 32.6 in the 1940s to 8.3 in the 1980s.[9,10] A similar decline has not been observed for black women until much more recently, and there remains a higher age-adjusted incidence for blacks as compared to whites in the United States.[11] However most of this difference is currently attributed to a higher age-adjusted incidence among older black women. A higher incidence is also observed in the United States in Hispanics and native Americans. To what extent these differences reflect differential access to cytologic screening programs or different exposures to risk factors such as human papillomavirus (HPV) remains unclear.

Several studies have demonstrated a strong inverse association between socio-economic indicators, such as income and educational status, and cervical cancer incidence rates. These relationships have been observed among both whites and blacks and adjustment for them reduces the excess risk for cervical cancer observed among black women in the United States.[12] Some of the differences observed between whites, blacks and native Americans may also be due to genetic differences between the populations. Recent studies have demonstrated an association between specific HLA haplotypes and cervical cancer.[13-15]

Despite the geographic diversity in cervical cancer rates, the age-specific incidence curves are similar in most populations.[16] Incidence increases from the age of 20 to a plateau at about 40 to 50 years of age. The level of this plateau varies, however, depending on the regional age-standardized incidence. It is approximately 10 per 100,000 in women living in Israel and Kuwait, and 100 to 200 per 100,000 in women living in Colombia and parts of Brazil.

3. ETIOLOGY OF CERVICAL CANCER AND ITS PRECURSORS

Risk factors for cervical cancer are shown in Table 1. These include socio-demographic factors such as older age, residence in certain parts of the world, and lower socio-economic status. They also include indicators of sexual activity such as number of sexual partners and age at first sexual intercourse, as well as a history of sexually transmitted diseases, especially human papillomavirus infections, genital warts, and genital herpes.[17-19] Additionally, a number of behavioral factors including long-term cigarette smoking, oral contraceptive use, and diets low in carotene and vitamin C have been risk factors in some

Table 1 Risk Factors for Cervical Cancer

Older age
Race (i.e., Black, Hispanic, American Indian)
Low socioeconomic status
Country of origin (i.e., Latin America, Asia or Africa)
Multiple pregnancies
Young age of first vaginal intercourse
Human papillomavirus infection
Multiple sexual partners
Exposure to a "high-risk" male partner
History of sexually transmitted diseases (i.e., herpes, genitalis and genital warts)
History of cigarette smoking
History of oral contraceptive use
Diet low in selected vitamins (i.e., vitamin C and carotene)
Extended length of time since last Pap smear

studies.[17-19] Although the risk factors for cervical cancer and its precursor lesions are similar, the strength of association between these risk factors and invasive cervical cancer is generally stronger than the strength of association between the risk factors and cervical cancer precursors.

3.1 Sexual Activity

Since the studies of Rigoni-Stern in 1842, it has been recognized that there is an association between sexual behavior and genital cancer. Women who develop cervical cancer more frequently report having had multiple sexual partners than do control women, and risk appears to increase directly with the number of sexual partners. Women who have had 10 or more partners are at approximately three times greater risk for developing invasive cervical cancer than are women with one or no sexual partners.[19] It has been estimated by Slattery et al. that the population attributable risk, or the proportion of new cases of invasive cancer that are attributable to having 10 or more sexual partners, is approximately 36%.[20] In addition to number of partners being a risk factor, characteristics of the partner(s) also appear to have a role.

Cervical cancer and penile cancer frequently demonstrate geographic clustering. Using data from the Puerto Rican Cancer Registry, Martinez identified 889 cases of invasive cervical cancer in the wives of men diagnosed as having penile cancer between 1950 and 1968. For comparison, no cases of invasive cervical cancer were detected among the wives of an equal number of age-matched control men with cancers of other sites.[21] Based on the age-adjusted incidence rates for invasive cervical cancer in the same population, only a single case of invasive cervical cancer would have been predicted among the wives of the group of men with cervical cancer. Similar results have been reported from Japan where there is a significant correlation between the death rate for uterine cancer and penile cancer within the 48 prefectures.[22]

The notion that a "male factor" may be important has also been supported by studies of the wives of men previously married to women with cervical cancer. These studies have reported elevated rates of both cervical cancer and its precursors among the second wives of men whose first wives died of cervical cancer.[23] Further support for the existence of a "male factor" has come from studies comparing the sexual and behavioral characteristics of husbands of women with cervical disease with the husbands of women without cervical disease.[24,25] These studies have found that the husbands of women with cervical disease have had significantly more sexual partners than did the husbands of control patients and the male partners were more likely to have a history of genital warts, gonorrhea, or genital herpes. In addition, the type of sexual activity that the male engaged in was important. Excessive risk was associated with engaging in sexual activity with prostitutes and a lower risk was associated with the frequent use of condoms.[25] The role of circumcision status is controversial. Early studies found that women who developed invasive cervical cancer were substantially more likely to have uncircumcised spouses than were control women. However, the results of more recent studies are conflicting. Some studies show no significant effect of male circumcision status, whereas others report that the wives of circumcised men are at significantly lower risk for developing invasive cervical cancer.[24,25]

Age at first sexual intercourse has been an independent risk factor for both invasive cervical disease and cervical cancer precursors in most studies. Women who have had their first sexual intercourse before the age of 16 have twice the risk of developing cervical cancer than do women who had their first sexual intercourse after the age of 20 years.[19] Young age at first sexual intercourse appears to be a stronger risk factor for invasive cancer than it is for cervical cancer precursors.[19,26] The biological explanation for early age at sexual intercourse being a risk factor is that the cervix during early life is covered by an immature metaplastic squamous epithelium, which is thought to be particularly susceptible to the transforming effects of a sexually transmitted agent.[27]

3.2 Sexually Transmitted Agents

The concept that cervical cancer and its precursor lesions are caused by a sexually transmitted agent(s) has led to the investigation of a number of different pathogens. These include *chlamydia trachomatis*, *neisseria gonorrhea*, *gardnerella vaginalis*, cytomegalovirus, herpes simplex virus 2 (HSV-2), and, more recently, human papillomaviruses (HPV). Most recent studies have focused on HSV-2 and HPV. The data supporting a role for other sexually transmitted pathogens in the development of invasive cervical cancer and its precursor lesions is much less convincing.

3.2.1 Herpes Simplex Virus Type 2 (HSV-2)

A large number of studies have suggested an etiologic role for HSV-2 in cervical cancer. Herpes virus antigens have been detected using immunofluorescence techniques in squamous cell carcinomas, and the virus has been identified using electron microscopy in cultured cervical carcinoma cells.[28] In *in vitro* model systems, HSV-2 can transform cultured cell lines and HSV-2 can cause the formation of tumors when injected into newborn hamsters and nude mice.[29,30] However, more recent studies have failed to consistently detect HSV-2 genomic sequences in invasive cervical carcinomas.

Numerous case control studies have compared the prevalence of HSV-2 antibodies in women with invasive cervical cancer and in various control groups. Early HSV-2 serologic studies were flawed because the serological tests that were used were unable to discriminate between common antibodies to HSV-1 and HSV-2. However, more recent studies have utilized better serological assays for HSV-2 and have found that patients with invasive cancers of the cervix have a higher prevalence of neutralizing antibodies against HSV-2 than do controls matched for age, race, and socioeconomic studies.[31,32] One of the largest of these case-control studies was of women with invasive cervical cancer from Latin America.[32] This study identified a possible interaction between HPV 16, HPV 18, and HSV-2 in the development of cervical cancer. Seropositivity for HSV-2, by itself, was associated with a relative risk of 1.6 for the development of invasive cervical cancer, which was considerably less than the risk associated with the detection of HPV 16 or 18 DNA. However, patients who had antibodies against HSV-2 and who were also HPV 16 or 18 DNA-positive had a twofold greater risk for developing invasive cervical cancer than did those patients who were only HPV 16 or 18 positive.[32] In another case-controlled study of cervical cancer in women enrolled from Spain and Colombia, Munoz et al. found that HSV-2 seropositivity was associated with an increased risk for invasive cancer in the women enrolled from Spain, but not from Colombia.[33]

3.2.2 Human Papillomavirus (HPV)

There is now extensive laboratory and epidemiologic evidence indicating a role for human papillomavirus (HPV) in the pathogenesis of invasive cervical cancer and its precursors.[34,35]

3.2.2.1 Classification of Papillomaviruses

Papillomaviruses are classified as members of the family Papovaviridae.[36] This family of viruses include simian virus (SV-40) and polyoma virus as well as the papillomaviruses. All members of this family are classified as DNA tumor viruses and, although they lacked shared antigens, they have transforming proteins with similar biologic effects. However, the papillomaviruses have a relatively low level of nucleotide or amino acid sequence homology with other members of the family Papovaviridae and it has been recommended that they be placed in a separate family.[37]

The papillomaviruses are widely distributed throughout nature and papillomaviruses which infect humans, cows, dogs, rabbits, birds and deer have all been identified.[38] The viruses which infect one species do not generally infect other species. Therefore, papillomaviruses are classified on the basis of the species which they infect. The viruses are further subclassified on the basis of their extent of DNA relatedness. A novel HPV genome is classified as a new type if the nucleotide sequence of the E6, E7 and L1 open reading frames differ by more than 10% with the sequence of any previously described HPV type.[39]

3.2.2.2 Genomic Organization of Papillomaviruses

Papillomaviruses have a double-stranded DNA genome of approximately 8000 base pairs in length. The viral genome can be divided into three sections, Figure 1. One is referred to as the *upstream regulatory region* (URR). The URR is a noncoding portion of the viral genome that is approximately 400 base-pairs in length. This region is adjacent to the origin of viral replication and contains a complex array of overlapping binding sites for both virally-encoded and host cell-derived transcriptional activators and transcriptional repressors.[40] The URR plays an important role in the viral life cycle by regulating transcription from the early and late regions of the viral genome and controls the production of viral proteins and virions.

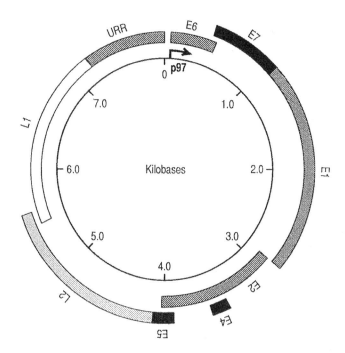

Figure 1 Genomic organization of HPV. (Modified with permission from Reference 46.)

The *early region* is downstream of the URR and contains six open reading frames (ORFs). These are designated E1, E2, E4, E5, E6, and E7. These open reading frames play important roles in viral replication. The E6 and the E7 ORFs encode for oncoproteins which are critical for viral replication and transformation of the host cell by HPV (see Chapter 16). The E1 and E2-encoded proteins are also important for viral replication. E1-encoded proteins of some papillomaviruses have ATPase and helicase activities[41] and E2-encoded proteins can bind to the HPV URR and act as transcriptional activators and repressors.[42] E4-encoded proteins associate with the cellular cytoskeleton and induce a collapse of the cytoplasmic cytokeratin network.[43] This collapse may produce the characteristic cytopathic effects observed in HPV infections and be important for virion release from infected cells.

The *late region* includes 2 ORFs designated L1 and L2 which are transcribed late during the viral lifecycle. These ORFs encode for the major and minor capsid proteins, respectively.

3.2.2.3 Association of Specific Types of Human Papillomavirus with Specific Lesions

HPV are epitheliotropic viruses and preferentially infect skin and the epithelium of mucosal surfaces, including the cervix, conjunctiva, oral cavity, larynx, esophagus, bladder, and anus. Papillomaviruses characteristically produce benign epithelial proliferations or papillomas at the site of infection. To date more than 70 types of human papillomaviruses have been identified, Table 2. These 70 HPV types can be divided into three general groups. One group is a mucocutaneous group which contains those viruses which infect the skin and the oral epithelium. These include viruses such as 1, 2, 3, 4, 7, 10, 13, 28, 29, 32, and 38 which cause lesions such as juvenile and plantar warts. The second group includes HPVs isolated from patients with epidermodysplasia verruciformis, which is a rare genetic disorder of cellular

Table 2 Classification of Human Papillomaviruses

Type	Lesion	Type	Lesion	Type	Lesion
		Associated with Mucocutaneous Lesions			
1	Verruca plantaris	7	Butcher's warts	34	Squamous cell carcinoma
2	Verruca vulgaris	10	Verruca plana	37	Keratoacanthoma
	Verruca plantaris	13	Focal epithelial hyperplasia	38	Verruca vulgaris
3	Verruca plana	28	Verruca plana	41	Squamous cell carcinoma
4	Verruca vulgaris	29	Verruca vulgaris	48	Squamous cell carcinoma
	Verruca plantaris	32	Focal epithelial hyperplasia	49	Verruca plana
		Associated with Genital Lesions			
6	Condyloma acuminatum	40	SIL	61	VAIN
	Low-grade SIL (CIN 1)	42	Low-grade SIL (CIN 1)	62	VAIN
11	Condyloma acuminatum	43	Low-grade SIL (CIN 1)	64	VAIN
	Low-grade SIL (CIN 1)	44	Low-grade SIL (CIN 1)	66	Squamous cell carcinoma
	Squamous cell carcinoma	45	All grades of SIL	67	VAIN
16	All grades of SIL		Squamous cell carcinoma	69	SIL
	Squamous cell carcinoma	51	All grades of SIL	70	Vulvar papilloma
18	All grades of SIL		Squamous cell carcinoma		
	Adeno- & squamous carcinoma	52	All grades of SIL		
30	All grades of SIL		Squamous cell carcinoma		
31	All grades of SIL	53	Normal cervical epithelium		
	Squamous cell carcinoma	54	Condyloma acuminatum		
33	All grades of SIL	55	Bowenoid papulosis		
	Squamous cell carcinoma	56	All grades of SIL		
35	All grades of SIL		Squamous cell carcinoma		
	Squamous cell carcinoma	58	All grades of SIL		
39	All grades of SIL	59	VIN		
	Squamous cell carcinoma				
		Associated with Epidermodysplasia Verruciformis			
Types 5, 8, 9, 12, 15, 17, 19, 20, 21, 23, 24, 25, 36, 46, 47					

immunity associated with cutaneous HPV lesions which can progress to invasive squamous cell carcinomas after sun exposure. This group of viruses include types 5, 8, 9, 15, 17 and others. The third group of viruses is classified as genital tract viruses. This group includes those viruses which cause genital warts (e.g., condyloma acuminatum), squamous intraepithelial lesions (SIL) of the cervix, intraepithelial neoplasias of the vulva, penis, anus, and perirectal region, as well as invasive carcinomas of the cervix, anus, and vulva.

Based on their associations with specific types of genital tract lesions, the most common anogenital human papillomaviruses can be subdivided into three "oncogenic risk" groups.[44] One group is referred to as a low "oncogenic risk" group and includes HPV 6 and 11. These two HPV types are considered to be of low oncogenic risk because they are frequently associated with benign condyloma acuminata and occasionally with low-grade SIL of the cervix, but are rarely found in association with high-grade SIL of the cervix or with invasive squamous cell carcinomas. HPV types 42, 43, and 44 have classically been included in the low "oncogenic risk" group because they are associated with the same type of low-grade lesions as are HPV 6 and 11. The intermediate "oncogenic risk" group consists of HPVs which are found frequently in association with all grades of cervical SIL, but are only infrequently found in association with invasive anogenital cancers. HPV types 31, 33, 35, 51, and 52 are the most common members of the intermediate "oncogenic risk" group. The high oncogenic risk HPV group includes types 16, 18, 45, and 56. These viruses are characterized as being of high "oncogenic risk" because they are the viruses most frequently found in association with invasive cancers of the lower genital tract, including cervical cancer.[44,45]

3.2.2.4 Epidemiologic Evidence Linking HPV to the Development of Cervical Cancer and Its Precursor Lesions

It is now well established that HPV is a causative agent for the development of both invasive cervical cancer and its precursor lesions.[34,45] Recent epidemiological studies that have used highly sensitive and reproducible HPV DNA detection methods consistently report that: (1) the vast majority of SIL and invasive cancers are associated with specific types of human papillomaviruses, (2) HPV DNA positivity is associated with a significantly increased risk for the concomitant presence of SIL and invasive cancer, and (3) many of the traditional epidemiologic risk factors for SIL and invasive cancer, including number of lifetime sexual partners, lower socioeconomic status, and exposure to "high risk" males appear to be risk factors because they are surrogates of HPV infection.

Numerous studies have clearly demonstrated a strong association between specific types of HPV and invasive squamous cell and adenocarcinomas of the cervix.[34] The largest study of the association of specific types of HPV and invasive cervical cancers was recently published by Bosch and co-workers who analyzed almost 900 invasive cervical cancers from a variety of geographic locations, including Africa, South and North America, Southeast Asia and Europe.[45] In this study, a sensitive polymerase chain reaction (PCR) based method for detecting HPV was used, and it was demonstrated that over 90% of invasive cervical cancers are associated with HPV DNA. This study also found that the specific types of HPV associated with cancers were similar, irrespective of geographic location. HPV types 16 and 18 were the most commonly detected viruses and together were detected in approximately two-thirds of all invasive cervical cancers. Strong associations between specific types of HPV and SIL have also been clearly documented.[46-51] For example, in a recent study that used PCR and cervicovaginal lavages to detect HPV DNA in women with SIL, Wright et al. detected HPV DNA in 75% of women with low-grade SIL and 100% of women with high-grade SIL.[51]

3.3 Other Risk Factors

In addition to HPV, several other risk factors, including hormonal and nutritional risk factors, are recognized for invasive cervical cancer and its precursor lesions. Oral contraceptive use is associated with a significant increase in the risk for SIL and invasive cancer.[52-54] Similarly, cigarette smoking confers a significantly increased risk for SIL that is independent of sexual activity and appears to be duration- and dose-dependent.[55,56] Deficiencies of vitamins A, C and folate have been found to be significant risk factors for development of SIL in several studies.[57,58] The mechanisms by which oral contraceptives, cigarette smoking, and micronutrient deficiencies increase the risk for cervical disease is unknown, but all of these have been hypothesized to act via interactions with HPV. Smoking, for example, has been shown to reduce the number of Langerhan cells in the cervical mucosa which may result in local defects in cell-mediated immunity.[59] Oral contraceptives containing progesterone could potentially act by increasing transcription of HPV-encoded oncogenes.

4. CERVICAL CANCER PRECURSORS

Invasive cervical cancers develop from well-defined precursor lesions which have been referred to by a variety of names, including dysplasia/carcinoma *in situ*, cervical intraepithelial neoplasia, and squamous intraepithelial lesions. Irrespective of the terminology used, these lesions demonstrate both a temporal and a spatial association with invasive cervical cancer.

4.1 HISTOPATHOLOGIC CLASSIFICATION OF CERVICAL CANCER PRECURSORS

Several different histopathological classification systems are currently used to refer to cervical cancer precursors.[46,60] The WHO classification uses the term "dysplasia" and "carcinoma *in situ*" to refer to putative precursor lesions.[61] Dysplasia is defined as a squamous epithelial lesion that is characterized by disordered maturation and nuclear abnormalities. The abnormalities include nuclear pleomorphism, a loss of nuclear polarity, coarsening of nuclear chromatin, irregularities of nuclear membrane, and mitotic figures at various levels in the epithelium. Carcinoma *in situ* (CIS) is used to refer to a subset of precursor lesions in which nuclear abnormalities involve the full thickness of the epithelium. Dysplastic lesions are usually divided into mild, moderate, and severe forms based on whether these nuclear abnormalities are confined to the lower one-third, two-thirds, or extend into the upper third of the epithelium.

In the late 1960s, a number of studies demonstrated that the cellular changes of dysplastic and carcinoma *in situ* were qualitatively similar and remained throughout the histological spectrum. Therefore, Richart introduced the concept that all histologic grades of precursor represented a single disease process, which he termed cervical intraepithelial neoplasia (CIN).[62] Richart's CIN terminology divided cervical cancer precursors into three groups. CIN 1 corresponded to those lesions previously diagnosed as mild dysplasia. CIN 2 corresponded to lesions previously diagnosed as moderate dysplasia, and CIN

3 to severe dysplasia as well as carcinoma in situ. The CIN terminology was based on the concept that all cervical cancer precursors represent a single disease process with a shared common etiology, biology, and natural history.[62] It is now recognized, however, that cervical cancer precursors are heterogenous with respect to their behavior, associated HPV types, and clonal status.[46,60,63,64] Therefore, the basic premise underlying the CIN terminology is incorrect, and it has been suggested that the terminology for cervical cancer precursors be changed to better reflect the biological processes that underlie the histological patterns.[60,65,66] Richart has recommended that the CIN classification system be modified from a three-tiered to a two-tiered classification. In Richart's modified CIN classification system, the term *low-grade CIN* is used to refer to lesions which were previously classified as CIN 1 and flat condylomas. The term *high-grade CIN* is used to refer to lesions which were previously classified as CIN 2 and CIN-3.[65] The Bethesda System of cytologic diagnoses also uses two terms: *low-grade squamous intraepithelial lesion* (LoSIL) and *high-grade squamous intraepithelial lesion* (HiSIL).[67] LoSIL is synomous with low-grade CIN and is used for those lesions previously classified as mild dysplasia (CIN 1); the term HiSIL is synonomous with high-grade CIN and is used for lesions previously classified as moderate, severe dysplasia and carcinoma in-situ (CIN 2 and CIN 3). Currently in the United States there is no consensus as to which histopathologic terminology should be used and some laboratories are using the WHO terminology, others the original CIN terminology, and some a two-tiered classification system. However, since the Bethesda classification is now almost universally used to report cervical cytologic diagnoses, we feel that it should also be adopted for reporting cervical histopathologic diagnoses.[46,60]

4.2 Low-Grade Squamous Intraepithelial Lesions (LoSIL)

4.2.1 Histopathology of LoSIL

LoSIL lesions are histologically low-grade. The lesions are characterized by a loss of cellular polarity, crowding and overlapping of the cells, and disorganization of the basal cell and parabasal cell layers of the epithelium (Figure 2). LoSIL is the morphological manifestation of a productive HPV infection and many viral particles can be identified by either electron microscopy, immunohistochemical staining, or in-situ hybridization. It should be stressed that because LoSIL represents a productive viral infection, there are marked HPV cytopathic effects. These include perinuclear cytoplasmic cavitation that is accompanied by nuclear atypia and anisocytosis, Figure 3. The nuclear atypia is characterized by nuclear enlargement and irregularity, as well as hyperchromasia. Nuclear atypia appears to be secondary to heteroploidy. The combination of cytoplasmic cavitation and nuclear atypia is referred to as *koilocytosis*. Bi- and multinucleated cells are more frequently identified in low-grade SIL.

4.2.2 LoSIL — Associated HPV Type, DNA Content, and Clonality

LoSIL are heterogenous with regard to their associated HPV types. They can be associated with any of the known anogenital HPVs, as well as with novel or unknown HPV types.[64,68,69] One study, by Lungu et al. used a PCR method to detect and type HPV DNA in LoSIL and found that 34% of the lesions were associated with low-oncogenic risk types of HPV, such as 6 and 11, or novel HPV types. HPV 16 was associated with 16% of LoSIL and HPV 18 with 3% of the lesions.[64,68,69] More than one HPV type was identified in 22% of LoSIL. Similar results have been reported by others.[69,70] Therefore, it appears that LoSIL is quite heterogenous with respect to associated HPV types, is associated with high-oncogenic risk HPV types such as 16 and 18 in only about 25% of cases, and is frequently associated with more than one type of HPV.

A number of studies have analyzed nuclear DNA content in LoSIL. Using computerized imaging cytometry, Fu and co-workers have determined that most LoSIL have either a diploid or polyploid DNA pattern (e.g., are heteroploid).[71,72] Similar results were obtained by Jacobson et al. who used flow cytometry to measure DNA content of LoSIL.[73]

Recently, Park et al. analyzed the clonality of LoSIL using a PCR method based on methylation differences adjacent to a polymorphic region of the androgen receptor gene, which is located on the X chromosome.[74] The androgen receptor gene is highly polymorphic due to the presence of a hypervariable trinucleotide repeat which is referred to as a $(CAG)_n$ polymorphism.[75] Almost all individuals have two androgen receptor alleles which can be distinguished on the basis of the size of the amplified DNA product obtained by amplification with PCR primers that flank the hypervariable trinucleotide repeat, Figure 4. The hypervariable trinucleotide repeat of the androgen receptor gene is adjacent to Hpa II restriction enzyme digestion sites which become methylated on the inactive X-chromosome, but are not methylated on the active X-chromosome. Therefore after digestion with Hpa II, the active (unmethylated) allele becomes cleaved and cannot be amplified whereas the inactive (methylated) androgen receptor

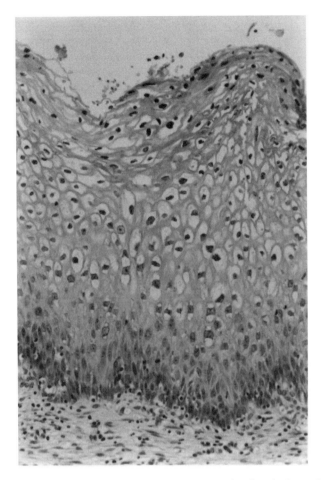

Figure 2 Histopathology of LoSIL. Immature basaloid cells are restricted to the lower third of the epithelium. The superficial layers of the epithelium contain mature keratinocytes that demonstrate HPV's cytopathic effect.

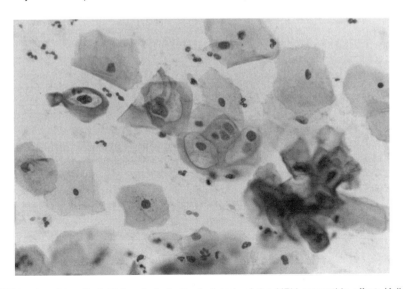

Figure 3 HPV cytopathic effect. Koilocytosis is the hallmark of the HPV cytopathic effect. Koilocytosisis is defined as perinuclear halos, multinucleation, nuclear enlargement, and hyperchromacity.

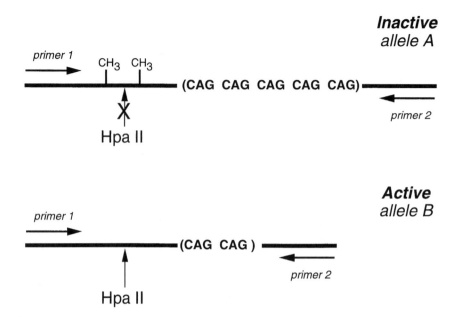

Figure 4 PCR method for determining clonality. The PCR method for assessing clonality amplifies a segment of the X-chromosome located androgen receptor gene. The amplified segment of the androgen receptor gene is highly polymorphic due to the presence of a hypervariable trinucleotide $(CAG)_n$ repeat and therefore varies in length. The hypervariable trinucleotide $(CAG)_n$ repeat is located adjacent to a Hpa II restriction endonuclease cleavage site. This restriction endonuclease cleavage site becomes methylated on the inactive X-chromosome but is not methylated on the active X-chromosome. Therefore after predigestion with Hpa II, only the inactive (methylated) androgen receptor allele can be amplified since the active (unmethylated) allele is cleaved by Hpa II. (From Park, T.J., Richart, R.M., Sun, X.W., and Wright, T.C., *J. Natl. Cancer Inst.*, in press. With permission.)

allele is not digested and therefore can be amplified. This preferential digestion and amplification allows clonality to be assessed by comparing the ratio of PCR products in Hpa II digested and undigested lesional tissue from the same patient. Polyclonal tissues are composed of cells with a random X-chromosome inactivation pattern and therefore two amplified DNA products of different sizes are identified after PCR of the Hpa II digested tissue. In contrast, monoclonal (i.e., neoplastic) tissues are composed of a single clone of cells which have an identical X-chromosome inactivation pattern, and only a single amplified DNA product is identified. Using this novel PCR method, LoSIL were found to be heterogenous with respect to their clonality, Figure 5.[74] LoSIL which were associated with intermediate and high "oncogenic risk" HPV types were invariably monoclonal lesions, whereas LoSIL that were associated with low or novel oncogenic risk HPV types were usually polyclonal lesions.

4.3 High-Grade Squamous Intraepithelial Lesions (HiSIL)
4.3.1 Histopathology of HiSIL
HiSIL lesions are histologically high-grade. HiSIL is characterized by immature basal-type cells which extend above the lower third of the epithelium. These cells have lost their normal cellular polarity and demonstrate nuclear crowding and pleiomorphism, Figure 6. The nuclei of these immature basal-type cells are frequently enlarged and the nuclear chromatin is coarsely granular. Mitotic figures are common and include abnormal mitotic forms, Figure 7. The basal-type cells have scant cytoplasm resulting in an increase in the nuclear: cytoplasmic ratio. Variabilty in nuclear size (anisonucleosis) is often present, although the subset of HiSIL referred to as carcinoma *in situ* typically has relatively uniform nuclei. The cytopathic effects of HPV (e.g., koilocytosis and multinucleation) are usually less prominent in HiSIL than in LoSIL.

4.3.2 HiSIL-Associated HPV Type, DNA Content, and Clonality
Compared to LoSIL, HiSIL is relatively homogeneous with respect to its associated HPV types.[64,70] Most HiSIL are associated with HPV types classified as being of intermediate- or high-oncogenic risk. HPV 16 has been detected in over 50% of HiSIL in most studies. Multiple HPV types are infrequently detected

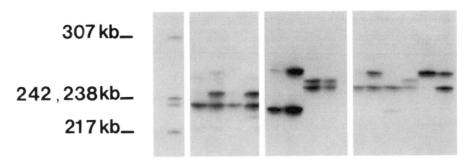

1   2   3 4   5   6   7

M L C L C L C L C   L C L C L C

307 kb—

242 , 238 kb—

217 kb—

Figure 5 Clonality of SIL lesions. Clonal analysis of cervical squamous intraepithelial lesions (SIL). Autorad-iogram of amplified DNA from 7 patients. Lanes L: Microdissected epithelium from cervical SILs. Lanes C: Normal control tissues prepared from stroma underlying the epithelial lesion. In lesion 1,2,3,5,6 and 7, one of the two androgen receptor alleles was selectively eliminated by Hpa II digestion prior to PCR, indicating monoclonality. In lesion 4, both androgen receptor alleles are present indicating polyclonality. Lesion 1, 5: High-grade SILs containing "cancer-associated" HPV type. Lesion 2,3,6,7: Low-grade SILs containing "cancer- asso-ciated" HPV type. Lesion 4: Low-grade SIL containing low "oncogenic" risk HPV type. (From Park, T.J., Richart, R.M., Sun, X.W., and Wright, T.C., *J. Natl. Cancer Inst.*, in press. With permission.)

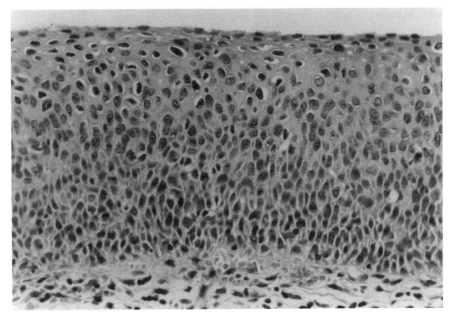

Figure 6 Histopathology of HiSIL. The lesion is characterized by immature basal-type cells which extend into the upper two-thirds of the epithelium. The cytopathic effects of HPV are usually less prominent than observed in LoSIL.

in HiSIL.[64,69] Chromosomal karyotyping studies as well as studies directly measuring the DNA content of lesional tissue using either Fulgen microspectroscopy or flow cytometry have demonstrated that most HiSIL are aneuploid.[72] A histological correlate of aneuploidy is the presence of abnormal mitotic figures which can often be found in HiSIL, Figure 7.[76] Clonal studies indicate that HiSIL are invariably monoclonal lesions.[74]

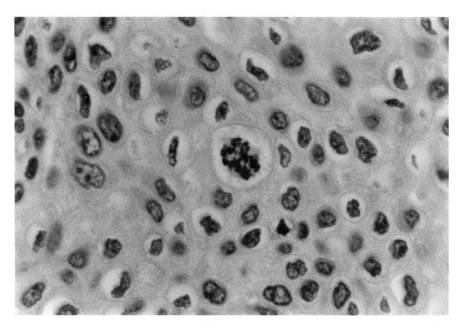

Figure 7 Abnormal mitotic figures in HiSIL. The lesion contains an abnormal mitotic figure (AMF).

4.4 Natural History of Squamous Intraepithelial Lesions

Numerous studies have analyzed the natural history of SIL. Although the specific results of these studies have varied, depending on the population being studied, the length of follow-up and whether or not biopsies of the lesions have been performed prior to following them, the conclusions of the studies are relatively consistent.[46,77]

LoSIL is characterized by high rates of spontaneous regression in the absence of therapy and low rates of progression to higher-grade lesions. In the absence of therapy, approximately half of LoSIL will spontaneously regress, approximately one third will persist as LoSIL and approximately 10 to 20% will progress to HiSIL. In contrast, HiSIL is much less likely to regress in the absence of therapy and is much more likely to persist or progress to higher grade lesions. In one of the lengthiest long-term follow-up studies of women with moderate dysplasia (CIN 2), Nasiell et al. followed 894 women with moderate dysplasia without biopsy for an average of 51 months.[78] Spontaneous regression was documented in 28% and progression to severe dysplasia or carcinoma in-situ occurred in 50%. Even higher rates of progression have been documented in studies of carcinoma in-situ. Kottmeir followed 30 patients with carcinoma in-situ of the cervix for more than 12 years and found that 72% of these patients developed invasive carcinoma.[79] McIndoe et al. reported a series of patients with carcinoma in situ who were followed either without treatment or with inadequate treatment for 1 to 20 years.[80] Carcinoma *in situ* progressed to cancer in 29% of the patients and the rate of progression increased directly with the length of follow-up, peaking to 35% in patients followed for 14 years. These prospective follow-up data agree with the results of population-based Pap smear screening data which indicate that the prevalence of LoSIL decreases proportionally with age and that there is a 1,000 to 2,000 × higher annual incidence of HiSIL in women with previously documented LoSIL than in women with normal cytologic findings.[81]

Whether or not HiSIL develops directly from LoSIL and, if so, whether this is the only pathway for the development of HiSIL is controversial. In animal models, cervical lesions have been shown to develop in progressive stages from LoSIL, to HiSIL, to frankly invasive cancer.[82] In colpomicroscopic studies, Richart and co-workers demonstrated that HiSIL usually begins as a small focus within an LoSIL and gradually expands and replaces the LoSIL.[83] This suggests that the transition from an LoSIL to an HiSIL lesion represents a monoclonal event within an HPV-infected epithelium. However, others have argued that HiSIL does not develop by direct transformation of cells within a LoSIL lesion. Mapping studies of the distribution of SIL lesions in the cervix by Burghardt and co-workers as well as by Koss suggest that HiSIL develops *de novo* from non-lesional epithelium adjacent from LoSIL.[84-86] Recently, the entire concept that LoSIL acts as a precursor to HiSIL and invasive cancer has been questioned by Kiviat and colleagues. In a long-term follow-up study of women attending an STD clinic, they found that most cases of HiSIL arose *de novo* in the absence of a previous LoSIL lesion.[66,87]

4.5 Treatment of Squamous Intraepithelial Lesions

Over the last several decades the clinical management of women with cervical cancer precursors has changed in parallel with the histopathologic terminology used to refer to these precursors. Prior to the introduction of the continuum concept by Richart in the late 1960s, many clinicians treated dysplasia and carcinoma *in situ* differently.[88,89] Patients with carcinoma in-situ were usually managed surgically with either a cone biopsy, or in some cases, a hysterectomy. Women diagnosed as having dysplasia were frequently followed up with repeat Pap smears or simply ignored. The CIN terminology introduced by Richart changed this approach since it stressed that all histologic grades of precursor lesions had the potential to progress to invasive cancer.[90] Therefore many clinicians began treating all grades of CIN similarly. Treatment decisions were based on the size and distribution of CIN rather than the grade of the lesion.[91] Lesions that could be fully visualized using a colposcope were treated with local ablative therapies such as cryosurgery or laser ablation. Lesions that extended into the endocervical canal, or lesions for which invasive cancer could not be ruled-out after colposcopy, were usually excised using a cone biopsy.

With the reintroduction of a two-tiered classification system for cervical cancer precursors and the introduction of rapid outpatient loop electrosurgical excision procedures in the 1980s, our approach to managing women with cervical cancer precursors has again changed.[92,93] Many clinicians now follow women with low-grade SIL once colposcopy has been performed and cervical cancer has been ruled out. This approach is felt to be preferable to treating all women with low-grade SIL since many of these lesions are associated with low or no "oncogenic-risk" HPV types and many will spontaneously regress in the absence of therapy.[94] High-grade SIL in which cancer has been ruled out after colposcopy are either ablated using cryosurgery, cold coagulatory, or laser ablation or excised using loop electrosurgical methods. Both ablative and excisional methods have overall success rates approaching 95%.[93,95] SIL lesions for which invasive cancer has not been ruled out after colposcopy continue to be treated using conization, although in many instances conization is performed in an outpatient setting using an electrosurgical approach.[96]

5. INVASIVE CERVICAL CANCER

5.1 Histopathologic Classification of Cervical Cancer

Three general categories of epithelial tumors of the cervix are recognized by the World Health Organization (WHO), Table 3.[61] These are squamous cell carcinomas, adenocarcinomas and "other" epithelial tumors (including less common types such as adenosquamous carcinoma, glassy cell carcinoma, adenoid basal cell carcinoma, as well as carcinoid-like and small cell carcinoma). The most common histologic type of cervical carcinoma is squamous cell carcinoma which accounts for 60 to 80% of all cervical cancers. Adenocarcinomas account for approximately 20% of invasive cervical carcinomas and display a variety of histological types and subtypes.

5.2 FIGO Staging of Invasive Cervical Cancer

The most widely accepted staging system for carcinomas of the cervix is that of the International Federation of Gynecologists and Obstetricians (FIGO), Table 4.[97,98] The FIGO staging system divides invasive cervical cancer into four stages. Stage I includes tumors which are confined to the cervix and is subdivided into microinvasive tumors and more deeply invasive carcinomas. Stage II tumors are those which extend beyond the cervix, but have not extended to the pelvic wall. If these tumors involve the vagina, they have not extended as far as the lower third. Stage III tumors have extended to the pelvic wall or involve the lower third of the vagina, or are associated with hydronephrosis or nonfunctioning kidneys. Stage IV includes carcinomas which have extended beyond the true pelvis or involve the mucosa of the rectum or bladder.

5.3 Microinvasive Squamous Cell Carcinoma

The new 1995 FIGO staging for cervical cancer defines stage IA1 as invasive carcinomas identified only microscopically with measured invasion of stroma no greater than 3.0 mm in depth and no wider than 7 mm.[97] Stage IA1 tumors are generally referred to as *microinvasive carcinomas.*

Microinvasive squamous cell carcinomas of the cervix are generally found in women in their early 40s and are associated with large HiSIL lesions that involve either the surface of the cervix or the endocervical crypts.[99,100] The prevalence of microinvasive squamous cell carcinoma of the cervix in the British Columbia cervical cancer registry is 4.8: 100,000 women screened.[101] Most women with

Table 3 World Health Organization Histologic
Classification of Invasive Carcinomas of the Cervix

Squamous Cell Carcinoma

Keratinizing carcinoma
Nonkeratinizing carcinoma
Verrucous carcinoma
Warty (condylomatous) carcinoma
Papillary squamous cell (transitional) carcinoma
Lymphoepithelioma-like carcinoma

Adenocarcinoma

Mucinous adenocarcinoma
Endocervical type
    Adenoma malignum (minimal deviation adenocarcinoma)
    Villoglandular papillary adenocarcinoma
Intestinal type
Endometrial adenocarcinoma
Endometrial adenocarcinoma with squamous metaplasia
Clear cell adenocarcinoma
Serous adenocarcinoma
Mesonephric carcinoma

Other Epithelial Tumors

Adenosquamous carcinoma
Glassy cell carcinoma
Adenoid cystic carcinoma
Adenoid basal carcinoma
Carcinoid tumor
Small cell carcinoma
Undifferentiated carcinoma

microinvasive squamous cell carcinoma have no symptoms and the lesion is detected during routine Pap smear screening. Microinvasion can easily be missed during colposcopy for the evaluation of an abnormal Pap smear.[99,101] In one histopathologic analysis of shallow laser excisional conization specimens, micro-invasive squamous cell carcinoma of the cervix was detected in 1% of patients thought to have only SIL on colposcopy.[102] Other studies of specimens obtained by loop electrosurgical excision have detected microinvasive squamous cell carcinoma of the cervix in 0.4 to 3.0% of patients being treated with loop excision for SIL.[103-107]

Microinvasive squamous cell carcinoma is characterized by the presence of one or more tongues of malignant cells which penetrate through the basement of the squamous epithelium into the stromal tissue. A SIL lesion is invariably present. Cells in the microinvasive foci tend to be better differentiated with abundant eosinophilic cytoplasm and prominent nucleoli as compared with the associated SIL, Figure 8. There is frequently a conspicuous lymphoplasmocytic infiltrate surrounding the tips of the invasive epithelial prongs and there is often an adjacent desmoplastic response in the stroma.

The reason for recognizing microinvasive squamous cell carcinoma of the cervix as a distinct clinical entity is that this lesion is associated with a low rate of recurrence and nodal metastases.[108-110] The data on risk factors for nodal metastases, recurrence and death, although incomplete, suggest that lesions with 3 mm or less stromal penetration and without lymphvascular space involvement have virtually no potential for metastases or recurrence.[111,112] Therefore, these patients can be managed with less radical procedures than required by more deeply invasive lesions.

5.4 Invasive Squamous Cell Carcinoma
5.4.1 Presenting Symptoms

The presenting systems of invasive squamous cell carcinoma depend on the size and stage of the lesion.[113] Patients with large, bulky, late-stage lesions frequently complain of abnormal vaginal bleeding or discharge. Patients with small, stage I lesions are frequently detected on the basis of an abnormal Pap smear. Weight loss, weakness, edema of the lower extremities and hematuria are symptoms of locally advanced or metastatic disease.

Table 4 1995 Modification of FIGO Staging of Carcinoma of the Cervix Uteri

Stage 1	The Carcinoma Is Strictly Confined to the Cervix
IA	Invasive cancer identified only microscopically. All gross lesions even with superficial invasion are stage IB cancers. Invasion is limited to measured stromal invasion with maximum depth of 5.0 mm and no wider than 7.0 mm[1]
IA1	Measured invasion of stroma no greater than 3.0 mm in depth and no wider than 7.0 mm
IA2	Measured invasion of stroma greater than 3.0 mm and no greater than 5.0 mm and no wider than 7.0 mm
IB	Clinical lesions confined to the cervix or preclinical lesions greater than stage IA
IB1	Clinical lesions no greater than 4.0 cm in size
IB2	Clinical lesions greater than 4.0 cm in size
Stage II	**The Carcinoma Extends Beyond the Cervix but Has Not Extended to the Pelvic Wall. The Carcinoma Involves the Vagina but not as far as the Lower Third**
IIA	No obvious parametrial involvement
IIB	Obvious parametrial involvement
Stage III	**The Carcinoma Has Extended to the Pelvic Wall. On Rectal Examination, There Is No Cancer-Free Space between the Tumor and Pelvic Wall. The Tumor Involves the Lower Third of the Vagina. All Cases with Hydronephrosis or Nonfunctioning Kidney Are Included unless They Are Known to Be Due to Other Causes**
IIIA	No extension to the pelvic wall
IIIB	Extension to the pelvic wall and/or hydronephrosis or nonfunctioning kidney
Stage IV	**The Carcinoma Has Extended beyond the True Pelvis or Has Clinically Involved the Mucosa of the Bladder or Rectum. A Bullous Edema as such Does Not Permit a Case to Be Allotted to Stage IV**
IVA	Spread of the growth to adjacent organs
IVB	Spread to distant organs

[1] The depth of invasion should not be more than 5 mm taken from the base of the epithelium, either surface or glandular, from which it originates. Vascular space involvement, either venous or lymphatic, should not alter the staging.

Modified from International Federation of Gynecology and Obstetrics. Annual Report on the Results of Treatment in Gynecological Cancer. Vol. 20. Stockholm: FIGO, 1988 and FIGO News, *Int. J. Gynecol. Obstet.*, 50, 215-216, 1995.

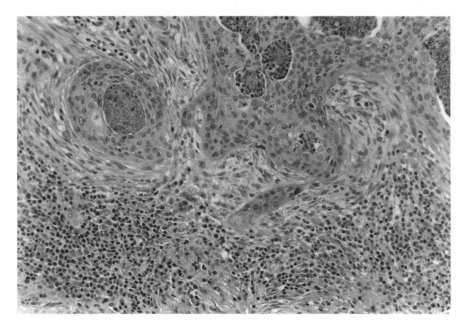

Figure 8 Microinvasive squamous cell carcinoma of the cervix. The lesion is characterized by an irregular tongue of squamous epithelium extending into the cervical stroma.

5.4.2 Gross and Microscopic Pathology

Invasive squamous cell carcinoma of the cervix can be exophytic or endophytic. Endophytic lesions tend to develop within the endocervical canal and frequently are either ulcerated or nodular. In some cases they invade deeply into the endocervical stroma and produce an enlarged, hard, barrel-shaped cervix. In other cases, the cervix appears grossly normal. Exophytic squamous cell carcinomas have a polypoid or papillary appearance and frequently bleed on contact.

The characteristic microscopic feature of invasive squamous cell carcinoma of the cervix is anastomosing tongues or cords of tumor cells that infiltrate through the cervical stroma, Figure 9. The current WHO classification of female genital tract tumors recognizes six different histological variants of squamous cell carcinomas.[61] The two most common forms are keratinizing and non-keratinizing. Keratinizing squamous cell carcinomas are characterized by the presence of well differentiated squamous cells which are arranged in nests or cords and demonstrate the presence of keratin "pearls" within the epithelium. Keratin pearls are clusters of squamous epithelial cells which have undergone keratinization and are arranged in a concentric nest, Figure 9. Non-keratiniziung squamous cell carcinomas of the cervix contain individually keratinized cells but lack keratin pearls. This is the most common subtype of squamous cell carcinoma of the cervix.

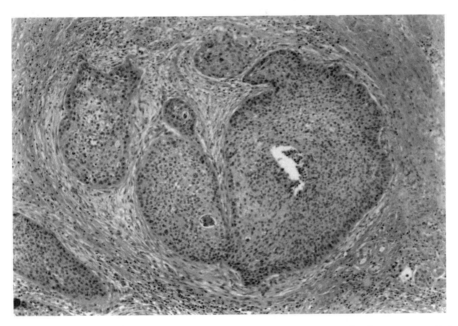

Figure 9 Squamous cell carcinoma of the cervix. Anastomosing tongues or cords of squamous cells infiltrate the stroma. A keratin pearl is identified indicating that this is a well differentiated tumor.

5.4.3 Prognostic Indicators

The most important prognostic factor in squamous cell carcinoma is FIGO stage.[114,115] Within a particular stage other prognostic factors of importance include tumor size, the presence of vascular space involvement, depth of invasion and the size of bulk of lymph node metastases.[116-120] Histological grade is of little prognostic significance. The prognostic importance of HPV DNA status is presently controversial. Earlier studies reported correlations between the presence of HPV 18 and tumor grade and lymph node involvement.[121,122] Those studies also reported worse outcomes in tumors associated with HPV 18. However, more recent studies have failed to confirm an association between specific types of HPV and clinical outcome or the presence of adverse prognostic indicators.[123-125] Tumors that appear to be HPV DNA negative have had a worse prognosis than those which are HPV DNA positive in several studies.[126,127]

5.5 Invasive Adenocarcinoma of the Cervix
5.5.1 Risk Factors and Associated HPV Types

Adenocarcinomas of the cervix are similar to squamous cell carcinomas of the cervix with respect to the age distribution of the two types of cancers and the epidemiologic risk factors.[128-130] The mean age of patients with invasive adenocarcinoma of the cervix is 50 years. Both adenocarcinomas and squamous cell carcinomas of the cervix are associated with SIL, multiple sexual partners, a young age of first intercourse and with HPV DNA. Although HPV 16 and 18 are the most commonly detected types of HPV in adenocarcinomas and squamous cell carcinomas of the cervix, HPV 18 is more frequently detected in adenocarcinomas than in squamous cell carcinomas.[122,131]

5.5.2 *Microscopic Pathology*

The WHO histological classification of female genital tract tumors recognizes seven different histological variants of cervical adenocarcinoma.[61] These tumors are subdivided on the basis of the type of glandular differentiation which is present. Adenocarcinomas with mucinous differentiation (mucinous adenocarcinoma) are the most common subtypes of invasive cervical adenocarcinomas. These mucinous adenocarcinomas are further subdivided into tumors with endocervical-type differentiation and tumors with intestinal epithelial-type differentiation. Endocervical-type adenocarcinomas are composed of cells that have basal nuclei and pale granular cytoplasm that stains with a mucicarmine stain (Figure 10). Most of these tumors are either well- or moderately-differentiated, and the glandular elements are arranged in complex patterns. The intestinal-type mucinous adenocarcinomas histologically resemble adenocarcinomas of the large intestine. The second most common form of endocervical-type adenocarcinomas are tumors which show endometrioid differentiation (endometrioid adenocarcinomas). Endometrioid adenocarcinomas account for approximately 30% of endocervical adenocarcinomas and have a histological appearance similar to that of endometrioid adenocarcinomas in either the endometrium or the ovary.

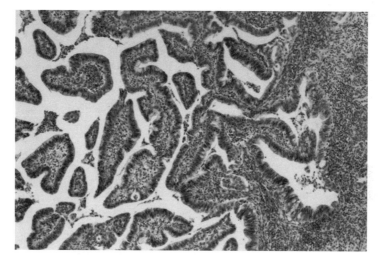

Figure 10 Mucinous adenocarcinoma of the cervix, endocervical-type.

5.5.3 *Prognostic Indicators*

The major prognostic indicators for invasive cervical adenocarcinoma include FIGO stage, depth of invasion, tumor diameter, involvement of lymphvascular spaces and presence or absence of lymph node metastases.[61,132-134] As with invasive squamous cell carcinomas of the cervix, stage is the most important prognostic indicator. In addition to stage, poor prognostic features include a tumor diameter greater than 3 cm, uterine enlargement and high histological grade. The clinical behavior of cervical adenocarcinoma is similar to that of squamous cell carcinoma and both are treated similarly.

5.6 Management of Invasive Cervical Cancer
5.6.1 *Diagnosis of Invasive Cervical Cancer*

Although Papanicolaou smears are widely used to screen women for cervical cancer and a diagnosis of invasive cervical cancer is usually first suggested by an abnormal Papanicolaou smear, they are not considered to be a diagnostic test, Cervical cancers are frequently associated with ulceration, bleeding, infection and necrosis which causes the false negative rate of Papanicolaou smears to be unacceptably high for a diagnostic modality. Therefore whenever a cervix is abnormal appearing or a patient has an abnormal Papanicolaou smear, a site-directed cervical biopsy which is usually obtained during a colposcopic examination is indicated.

5.6.2 Clinical Staging

The FIGO staging classification of cervical cancer is described in Section 5.2. Clinical staging may include examination (usually performed under anesthesia), chest and skeletal X-rays, intravenous pyelogram, cystoscopy and proctoscopy. The FIGO system does not mandate evaluation of the aortic lymph nodes. However, since approximately 20% of women with stage IIB/IIIB disease have metastatic disease in this area, many centers perform some type of diagnostic procedure to determine whether the radiation fields should include the aortic lymph nodes.[136] Techniques for evaluating this area include lymphangiogram, imaging studies such as CT scan followed by fine needle aspiration of suspicious lymph nodes and retroperitoneal staging laparotomy or laparoscopy with selective lymphadenectomy.

5.6.3 Treatment Modalities

The basic tenet for successful management of women with invasive cervical cancer is that not only the primary lesion, but also the regional lymph node groups to which the cancer may metastasize must be treated. Radical surgical therapy and radiation therapy either alone or in combination are the primary therapeutic modalities used. Chemotherapy currently plays a minor therapeutic role and is usually reserved for women whose cancer is too advanced to be treated with radical surgery or radiation therapy.

Radical surgery has potential advantages over radiotherapy including retention of normal ovarian function in younger women, better preservation of sexual function and elimination of radiation effects on the intestine and bladder. Radical surgery is most successful when restricted to patients whose tumors have not spread beyond the cervix to the parametria or regional lymph nodes. Thererfore radical hysterectomy with pelvic lymphadenectomy is generally reserved for healthy patients with FIGO stage Ia (ii), Ib and IIa tumors. Tumor size is important when determining whether or not a patient is suitable for radical hysterectomy. Tumors larger than 3 cm are more difficult to excise and are associated with a greater risk for lymph node metastases. Therefore patients with large tumors are frequently treated with radiation therapy.

In appropriately selected patients, 5 year survival rates after radical hysterectomy with pelvic lymphadenectomy are between 87 to 92%.[136] The major complications of radical surgery are related to the urinary and intestinal tract.[135] Although the incidence of post-operative fistula formation has dropped significantly over the last several decades, approximately 2 to 3% of patients still develop a vesicovaginal or ureterovaginal fistula after radical hysterectomy. Approximately 1% develop intestinal obstruction and bladder dysfunction develops in about 4%. Pulmonary embolism, which is frequently life-threatening, occurs postoperatively in about 1% of patients.

Radiation therapy is an effective method for treating women with all stages of invasive cervical cancer and is the preferred method for women with large tumors or stage IIb or greater cancers. Since effective outcome depends on delivering tumoricidal dosages to the cancer while not exceeding the tissue tolerance of the adjacent structures, the best outcomes are obtained in centers where there is close cooperation between the radiation oncologist, gynecologic oncologist, and pathologist. Radiation is delivered to the pelvis using a combination of external beam and intracavitary therapy. External beam irradiation is usually produced by a linear accelerator and is given first since it is intended to both reduce the tumor volume and to eliminate distortion of the pelvic anatomy by infiltrating tumor. This allows optimization of the tumoricidal dose that can be obtained using intracavitary radiation therapy which is given after the external beam irradiation. Five-year success rates after radiation therapy depend on FIGO stage and range from 40 to 60% in stage II/III disease. Complications of radiation therapy include acute effects such as perforation of the uterus at the time of insertion of intracavitary radiation sources, radiation-induced proctosigmoiditis which occurs in approximately 8% of patients and acute hemorrhagic cystitis which occurs in approximately 3% of patients.[135] Common chronic complications include vaginal stenosis (in up to 70% of patients), vesicovaginal or rectovaginal fistulas (each about 1% of patients) and obstruction of the small bowel (about 2% of patients).

Since a significant fraction of women with advanced stage disease are not cured with radiation therapy there has been interest in using concomitant chemotherapy.[135] Support for this approach is provided by *in vitro* studies that have shown that several drugs including hydroxyurea, cisplatin, and 5-fluorouracil increase the sensitivity of cells to the cytotoxic effect of radiation. Although several uncontrolled studies have suggested higher response rates with chemo-radiation, prospective randomized studies to date have failed to demonstrate a clinically significant survival advantage for multimodal therapy.

REFERENCES

1. **Wright, T.C.,** Invasive cervical cancer and cervical intraepithelial neoplasia in women infected with the human immunodeficiency virus, *Clin. Consul. Obstet. Gynecol.*, 6, 11, 1994.

2. **Parkin, D.M., Pisani, P., and Ferlay, J.,** Estimates of the worldwide incidence of eighteen major cancers in 1985, *Int. J. Cancer*, 54, 594, 1993.

3. **Parkin, D.M.,** Estimates of the worldwide frequency of sixteen major cancers in 1980, *Int. J. Cancer*, 41, 184, 1988.

4. **Tomatis, L.,** *Cancer: Causes, Occurrence and Control*, Vol. 100, World Health Organization International Agency for Research on Cancer, Lyon, 1990.

5. **Beral, V. and Booth, M.,** Predictions of cervical cancer incidence and mortality in England and Wales, *Lancet*, 1, 495, 1986.

6. **Beral, V., Hermon, C., Munoz, N., and Devesa, S.S.,** Cervical cancer, *Cancer Surv.*, 19/20, 265, 1994.

7. **Larsen, N.,** Invasive cervical cancer arising in young white females, *J. Natl. Cancer Inst.*, 86, 6, 1994.

8. **Boring, C., Squires, T., and Tong, T.,** Cancer statistics, 1992, *CA Cancer J. Clinicians*, 43, 7, 1993.

9. **Devesa, S.S.,** Descriptive epidemiology of cancer of the uterine cervix, *Obstet. Gynecol.*, 63, 605, 1984.

10. **Devesa, S.S., Silverman, D.T., Young, J.L., Pollack, E.S., Brown, C.C., Horm, J.W., Percy, C.L., Myers, M.H., McKay, F.W., and Fraumeni, J.F., Jr.,** Cancer incidence and mortality trends among whites in the United States, 1947–1984, *J. Natl. Cancer Inst.*, 79, 701, 1987.

11. **Baquet, C.R., Horm, J.W., Gibbs, T., and Greenwald, P.,** Socioeconomic factors and cancer incidence among Blacks and Whites, *J. Natl. Cancer Inst.*, 83, 551, 1991.

12. **Devesa, S.S. and Diamond, E.L.,** Association of breast cancer and cervical cancer incidences with income and education among Whites and Blacks, *J. Natl. Cancer Inst.*, 65, 515, 1980.

13. **Apple, R.J., Erlich, H.A., Klitz, W., Manos, M.M., Becker, T.M., and Wheeler, C.M.,** HLA DR-DQ associations with cervical carcinoma show papillomavirus-type specificity, *Nat. Genet.*, 6, 157, 1994.

14. **Glew, S.S., Stern, P.L., Davidson, J.A., and Dyer, P.A.,** HLA antigens and cervical carcinoma, *Nature*, 356, 22, 1992.

15. **Wank, R. and Thomssen, C.,** High risk of aquamous cell carcinoma of the cervix for women with HLA-DQw3, *Nature*, 352, 723, 1991.

16. **Parkin, D.M.,** Screening for cervix cancer in developing countries, *Cancer Screening*, Miller, A.B., Chamberlain, J., Day, N.E., Hakama, M. and Prorok, P.C., Eds., Cambridge University Press, Cambridge, England, 184, 1994.

17. **Brinton, L.A. and Fraumeni, J.F.,** Epidemiology of uterine cervical cancer, *J. Chron. Dis.*, 39, 1051, 1986.

18. **Brinton, L.A., Tashima, K., Lehman, H.F., Levine, R.S., Mallin, K., Savitz, D.A., Stolley, P.D., and Fraumeni, J.F., Jr.,** Epidemiology of cervical cancer by cell type, *Cancer Res.*, 47, 1706, 1987.

19. **Brinton, L.A.,** Epidemiology of cervical cancer — an overview, *The Epidemiology of Cervical Cancer and Human Papillomavirus*, Vol. 119, Munoz, N., Bosch, F.X., Shah, K. and Meheus, A., Eds., IARC Scientific Publications, Lyon, 3, 1992.

20. **Slattery, M.L., Overall, J.C., Abbott, T.M., French, T.K., Robison, L.M., and Gardner, J.,** Sexual activity, contraception, genital infections and cervical cancer: support for a sexually transmitted disease hypothesis, *Am. J. Epidemiol.*, 130, 248, 1989.

21. **Martinez, I.,** Relationship of squamous cell carcinoma of the cervix uteri to squamous cell carcinoma of the penis, *Cancer*, 24, 777, 1969.

22. **Kurihara, M. and Asano, K.,** The correlation between cancer of uterus and cancer of penis, *Tohoku J. Exp. Med.*, 64, 104, 1956.

23. **Kessler, I.I.,** Venereal factors in human cervical cancer: evidence from marital clusters, *Cancer*, 39, 1912, 1977.

24. **Brinton, L.A., Reeves, W.C., Brenes, M.M., Herrero, R., Gaitan, E., Tenorio, F., de Britton, R.C., Garcia, M., and Rawls, W.E.,** The male factor in the etiology of cervical cancer among sexually monogamous women, *Int. J. Cancer*, 44, 199, 1989.

25. **Kjaer, S.K., de Villiers, E.M., Dahl, C., Engholm, G., Bock, J.E., Vestergaard, B.F., Lynge, E., and Jensen, O.M.,** Case-control study of risk factors for cervical neoplasia in Denmark. I: role of the "male factor" in women with one lifetime sexual partner, *Int. J. Cancer*, 48, 39, 1991.

26. **Reeves, W.C., Brinton, L.A., Brenes, M.M., Quiroz, E., Rawls, W.E., and de Britton, R.C.,** Case-control study of cervical cancer in Herrera Province, Republic of Panama, *Int. J. Cancer*, 36, 55, 1985.

27. **Fenoglio, C.N. and Ferenczy, A.,** Etiologic factors in cervical neoplasia, *Sem. Oncology*, 9, 349, 1982.

28. **Royston, I. and Aurelian, L.,** Immunofluorescent detection of herpes virus antigens in exfoliated cells from human cervical carcinoma, *Proc. Natl. Acad. Sci. U.S.A.*, 67, 204, 1970.

29. **Jariwalla, R.J., Aurelian, L., and Ts'O, P.O.,** Neoplastic transformation of cultured Syrian hamster embryo cells by DNA of herpes simplex virus type 2, *J. Virol.*, 30, 404, 1979.

30. **Galloway, D.A. and McDougall, J.K.,** The oncogenic potential of herpes simplex viruses: evidence for a "hit and run" mechanism, *Nature*, 302, 21, 1983.

31. **Dale, G.E., Coleman, R.M., Best, J.M., Benetato, B.B., Drew, N.C., Chinn, S., Papacosta, A.O., and Nahmias, A.J.,** Class-specific herpes simplex virus antibodies in sera and cervical secretions from patients with cervical neoplasia: a multi-group comparison, *Epidem. Inf.*, 100, 455, 1988.

240

32. **Hildescheim, A., Mann, V., Brinton, L.A., Szklo, M., Reeves, W.C., and Rawls, W.E.,** Herpes simplex virus type 2: a possible interaction with human papillomavirus types 16/18 in the development of invasive cervical cancer, *Int. J. Cancer*, 49, 335, 1991.

33. **Munoz, N., Kato, I., Bosch, F.X., De Sanjose, S., Sundquist, V.A., Izarzugaza, I., Gonzalez, L.C., Tafur, L., Gili, M., and Viladiu, P.,** Cervical cancer and herpes simplex virus type 2: case-control studies in Spain and Colombia with special reference to immunoglobulin-G sub-classes, *Int. J. Cancer*, 60, 438, 1995.

34. **IARC,** *Human Papillomaviruses*, 64, IARC, Lyon, 407, 1995.

35. **Park, T.W., Fujiwara, H., and Wright, T.C.,** Molecular biology of cervical cancer and its precursors, *Cancer*, 76, 1902, 1995.

36. **Pfister, H. and Fuchs, P.G.,** Anatomy, taxonomy and evolution of papillomaviruses, *Intervirology*, 37, 143, 1994.

37. **Bernard, H.-U., Chan, S.-Y., and Delius, H.,** Evolution of papillomaviruses, *Curr. Topics in Microbiol. Immunol.*, 186, 33, 1994.

38. **Pfister, H.,** *Papillomaviruses: General Description, Taxonomy and Classification*, 2, Plenum Press, New York, 1, 1987.

39. **de Villiers, E.M.,** Human pathogenic papillomavirus types: an update, *Curr. Topics Microbiol. Immunol.*, 186, 1, 1994.

40. **Turek, L.P.,** The structure, function and regulation of papillomaviral genes in infection and cervical cancer, *Adv. Viral Res.*, 44, 305, 1994.

41. **Bream, G.L., Ohmstede, C.A., and Phelps, W.C.,** Characterization of human papillomavirus type 11 E1 and E2 proteins expressed in insect cells, *J. Virol.*, 67, 2655, 1993.

42. **Ward, P., Coleman, D.V., and Malcolm, D.B.,** Regulatory mechanisms of the papillomaviruses, *Trends Genet.*, 5, 97, 1989.

43. **Doorbar, J.,** An emerging function for E4, *Papillomavirus Reports*, 2, 145, 1991.

44. **Lorincz, A.T., Reid, R., Jenson, A.B., Greenberg, M.D., Lancaster, W., and Kurman, R.J.,** Human papillomavirus infection of the cervix: relative risk associations of 15 common anogenital types, *Obstet. Gynecol.*, 79, 328, 1992.

45. **Bosch, F.X., Manos, M.M., Munoz, N., Sherman, M., Jansen, A.M., Peto, J., Schiffman, M.H., Moreno, V., Kurman, R., and Shah, K.V.,** Prevalence of human papillomavirus in cervical cancer: a worldwide perspective international biological study on cervical cancer (IBSCC) study group, *J. Natl. Cancer Inst.*, 87, 779, 1995.

46. **Wright, T.C., Ferenczy, A.F., and Kurman, R.J.,** Precancerous lesions of the cervix, in *Blaustein's Pathology of the Female Genital Tract*, Kurman, R.J., Ed., Springer-Verlag, New York, 229, 1994.

47. **Schiffman, M.H.,** New epidemiology of human papillomavirus infection and cervical neoplasia, *J. Natl. Cancer Inst.*, 87, 1345, 1995.

48. **Schiffman, M.H.,** Recent progress in defining the epidemiology of human papillomavirus infection and cervical neoplasia, *J. Natl. Cancer Inst.*, 84, 394, 1992.

49. **Schiffman, M.H., Bauer, H.M., Hoover, R.N., Glass, A.G., Cadell, D.M., Rush, B.B., Scott, D.R., Sherman, M.E., Kurman, R.J., and Wacholder, S.,** Epidemiologic evidence that human papillomavirus infection causes most cervical intraepithelial neoplasia, *J. Natl. Cancer Inst.*, 85, 958, 1993.

50. **Cuzick, J., Terry, G., Ho, L., Hollingworth, T., and Anderson, M.,** Type-specific HPV DNA as a predictor of high grade cervical intraepithelial neoplasia, *Br. J. Cancer*, 69, 167, 1994.

51. **Wright, T.C., Ellerbrock, T.V., Chiasson, M.E., Sun, X., and Vandervanter, N.,** New York Cervical Disease, Cervical intraepithelial neoplasia in women infected with human immunodeficiency virus: prevalence, risk factors and validity of Papanicolaou smears, *Obstet. Gynecol.*, 84, 591, 1994.

52. **Brinton, L.A.,** Oral contraceptives and cervical neoplasia, *Contraception*, 43, 581, 1991.

53. **Schlesselman, J.J.,** Net effect of oral contraceptive use on the risk of cancer in women in the United States, *Obstet. Gynecol.*, 85, 793, 1995.

54. **Ursin, G., Peters, R.K., Henderson, B.E., d'Ablaing, G., Monroe, K.R., and Pike, M.C.,** Oral contraceptive use and adenocarcinoma of cervix, *Lancet*, 344, 1390, 1994.

55. **Gram, I.T., Austin, H., and Stalsberg, H.,** Cigarette smoking and the incidence of cervical intraepithelial neoplasia, grade III, and cancer of the cervix, *Am. J. Epidemiol.*, 135, 341, 1992.

56. **Brinton, L.A., Schairer, C., Haenszel, W., Stolley, P., Lehman, H.F., Levine, R., and Savitz, D.A.,** Cigarette smoking and invasive cervical cancer, *JAMA*, 255, 3265, 1986.

57. **Butterworth, C.E.J., Hatch, K.D., Macaluso, M., Cole, P., Sauberlich, H.E., Soong, S.J., Borst, M., and Baker, V.V.,** Folate deficiency and cervical dysplasia, *JAMA*, 267, 528, 1992.

58. **Block, G.,** Epidemiologic evidence regarding vitamin C and cancer, *Am. J. Clin. Nutr.*, 54, 1310, 1991.

59. **Barton, S.E., Jenkins, D., Cuzick, J., Maddox, P.H., Edwards, R., and Singer, A.,** Effect of cigarette smoking on cervical epithelial immunity: a mechanism for neoplastic change?, *Lancet*, 2, 652, 1988.

60. **Wright, T.C. and Kurman, R.J.,** A critical review of the morphologic classification systems of preinvasive lesions of the cervix: the scientific basis of the paradigm, *Papillomavirus Rep.*, 5, 175, 1994.

61. **Scully, R.E., Bonfiglio, T.A., Kurman, R.J., Silverberg, S.G., and Wilkinson, E.J.,** *Histological Typing of Female Genital Tract Tumors*, Springer-Verlag, Berlin, 189, 1994.

62. **Richart, R.M.,** A theory of cervical carcinogenesis, *Obstet. Gynecol. Survey*, 24, 874, 19xx.

63. **Wright, T.C. and Nuovo, G.,** Syllabus, Cervical cancer and its precursors; neoplasms of the uterus and vulva; tumors of the ovary,

64. **Lungu, O., Sun, X.W., Felix, J., Richart, R.M., Silverstein, S., and Wright, T.C.,** Relationship of human papillomavirus type to grade of cervical intraepithelial neoplasia, *JAMA*, 267, 2493, 1992.

65. **Richart, R.M.,** A modified terminology for cervical intraepithelial neoplasia, *Obstet. Gynecol.*, 75, 131, 1990.

66. **Kiviat, N.B., Critchlow, C.W., and Kurman, R.J.,** Reassessment of the morphological continuum of cervical intraepithelial lesions; does it reflect different stages in the progression to cervical carcinoma? *The Epidemiology of Cervical Cancer and Human Papillomavirus*, Vol. 119, Munoz, N., Bosch, K., Shah, K., and Mehens, A., Eds., I.A.R.C., Lyons, 59, 1992.

67. **Luff, R.D.,** The Bethesda System for reporting cervical/vaginal cytologic diagnoses: report of the 1991 Bethesda Workshop, *Hum. Pathol.*, 23, 719, 1992.

68. **Genest, D.R., Stein, L., Cibas, E., Sheets, E., Zitz, J.C., and Crum, C.C.,** A binary (Bethesda) system for classifying cervical cancer precursors: criteria, reproducibility, and viral correlates, *Hum. Pathol.*, 24, 730, 1993.

69. **Bergeron, C., Barrasso, R., Beaudenon, S., Flamant, P., Croissant, O., and Orth, G.,** Human papillomaviruses associated with cervical intraepithelial neoplasia. Great diversity and distinct distribution in low- and high-grade lesions, *Am. J. Surg. Pathol.*, 16, 641, 1992.

70. **Wright, T.C., Sun, X.W., and Koulos, J.,** Comparison of management algorithms for the evaluation of women with low-grade cytologic abnormalities, *Obstet. Gynecol.*, 85, 202, 1995.

71. **Fu, Y.S., Reagan, J.W., and Richart, R.M.,** Precursors of cervical cancer, *Cancer Surv.*, 2, 359, 1983.

72. **Fu, Y.S., Huang, I., Beaudenon, S., Ionesco, M., Barrasso, R., de Brux, J., and Orth, G.,** Correlative study of human papillomavirus, DNA, histopathology and morphometry in cervical condyloma and intraepithelial neoplasia, *Int. J. Gynecol. Pathol.*, 7, 297, 1988.

73. **Jakobsen, A., Kristensen, P.B., and Poulsen, H.K.,** Flow cytometric classification of biopsy specimens from cervical intraepithelial neoplasia, *Cytometry*, 4, 166, 1983.

74. **Park, T.J., Richart, R.M., Sun, X.W., and Wright, T.C.,** Associations between HPV type and clonal status of cervical squamous intraepithelial lesions (SIL), *J. Natl. Cancer Inst.*, (in press), 1996.

75. **Mutter, G.L., Chaponot, M.L., and Fletcher, J.A.,** A polymerase chain reaction assay for non-random X chromosome inactivation identifies monoclonal endometrial cancers and precancers, *Am J. Pathol.*, 146, 501, 1995.

76. **Winkler, B., Crum, C.P., and Fujii, T.,** Koilocytotic lesions of the cervix: the relationship of mitotic abnormalities to the presence of papillomavirus antigens and nuclear DNA content, *Cancer*, 53, 1081, 1984.

77. **Ostor, A.G.,** Natural history of cervical intraepithelial neoplasia: a critical review, *Int. J. Gynecol. Pathol.*, 12, 186, 1993.

78. **Nasiell, K., Nasiell, M., and Vaclavinkova, V.,** Behavior of moderate cervical dysplasia during long-term follow-up, *Obstet. Gynecol.*, 61, 609, 1983.

79. **Kottmeier, H.L.,** Evolution et traitement des epitheliomas, *Rev. Fran. Gynecol. d'Obstet.*, 56, 821, 1961.

80. **McIndoe, W.A., McLean, M.R., Jones, R.W., and Mullins, P.R.,** The invasive potential of carcinoma in situ of the cervix, *Obstet. Gynecol.*, 64, 451, 1984.

81. **Miller, A.B., Knight, J., and Narod, S.,** The natural history of cancer of the cervix and the implications for screening policy, *Cancer Screening*, Miller, A.B., Chamberlain, J., Day, N.E., Hakama, M. and Prorok, P.C., Eds., Cambridge University Press, Cambridge, 141, 1991.

82. **Rubio, C.A. and Lagerlof, B.,** Studies on the histogenesis of experimentally induced cervical carcinoma, *Acta Pathol. Microbiol. Scand.*, 82, 153, 1974.

83. **Richart, R.M.,** Colpomicroscopic studies of cervical intraepithelial neoplasia, *Cancer*, 19, 395, 1966.

84. **Koss, L. and Durfee, G.R.,** Unusual patterns of squamous epithelium of uterine cervix: cytologic and pathologic study of koilocytotic atypia, *Ann. N.Y. Acad. Sci.*, 63, 1245, 1956.

85. **Koss, L.G.,** *Diagnostic Cytology and Its Histopathologic Basis*, Vol. 1, J.B. Lippincott Company, New York, 1992.

86. **Burghardt, E. and Ostor, A.G.,** Site and origin of squamous cervical cancer: a histomorphologic study, *Obstet. Gynecol.*, 62, 117, 1983.

87. **Koutsky, L.A., Holmes, K.K., Critchlow, C.W., Stevens, C.E., Paavonen, J., Beckmann, A.M., De Rouen, T.A., Galloway, D.A., Vernon, D., and Kiviat, N.B.,** A cohort study of the risk of cervical intraepithelial neoplasia grade 2 or 3 in relation to papillomavirus infection, *New Engl. J. Med.*, 327, 1272, 1992.

88. **Wright, T.C., Richart, R.M., and Ferenczy, A.F.,** *Electrosurgery for HPV-Related Lesions of the Anogenital Tract*, Arthurvision, New City, N.Y., 272, 1992.

89. **Wright, T.C. and Richart, R.M.,** Etiology, diagnosis and management of cervical intraepithelial neoplasia, *Curr. Prob. Obstet. Gynecol. Fert.*, 163, 1992.

90. **Richart, R.M.,** Cervical intraepithelial neoplasia: a review, *Pathology Annual*, 8, Sommers, S.C., Ed., Appleton-Century-Croft, East Norwalk, CT, 301, 1973.

91. **Townsend, D.E. and Richart, R.M.,** Cryotherapy and carbon dioxide laser management of cervical intraepithelial neoplasia: a controlled comparison, *Obstet. Gynecol.*, 61, 75, 1983.

92. **Richart, R.M. and Wright, T.C.,** The American View, Large loop excision of the transformation zone, in *A Practical Guide to LLETZ*, Prendeville, W., Ed., Chapman and Hall Medical, London, 88, 1993.

93. **Richart, R.M. and Wright, T.C.,** Controversies in the management of low grade cervical intraepithelial neoplasia, *Cancer*, 71, 1413, 1993.

94. **Kurman, R.J., Henson, D.E., Herbst, A.L., Noller, K.L., and Schiffman, M.H.,** Interim guidelines for management of abnormal cervical cytology, The 1992 National Cancer Institute workshop, *JAMA*, 271, 1866, 1994.

95. **Richart, R.M.,** Causes and management of cervical intraepithelial neoplasia, *Cancer*, 60, 1951, 1987.

96. **Wright, T.C. and Richart, R.M.,** Loop excision of the uterine cervix, *Curr. Opinion Obstet. Gynecol.*, 7, 30, 1995.

97. FIGO, Modifications in the staging for stage I vulvar and stage I cervical cancer, *Int. J. Gynecol. Obst.*, 50, 215, 1995.

98. **Petterson, F.,** *Annual Report on Results of Treatment in Gynecologic Cancer*, 20, FIGO, Stockholm, 1988.

99. **Anderson, M.C.,** Are we vapourising microinvasive lesions? *Colpo Gynecol. Laser Surg.*, 3, 33, 1987.

100. **Tidbury, P., Singer, A., and Jenkins, D.,** CIN 3: the role of lesion size in invasion, *Br. J. Obstet. Gynecol.*, 99, 583, 1992.

101. **Benedet, J.L., Anderson, G.H., and Boyes, D.A.,** Colposcopic accuracy in the diagnosis of microinvasive and occult invasive carcinoma of the cervix, *Obstet. Gynecol.*, 65, 5557, 1985.

102. **McIndoe, G.A., Robinson, M.S., Tidy, J.A., Mason, W.P., and Anderson, M.C.,** Laser excision rather than vaporization: the treatment of choice for cervical intraepithelial neoplasia, *Obstet. Gynecol.*, 74, 165, 1989.

103. **Bigrigg, M.A., Haffenden, D.K., Sheehan, A.L., Codling, B.W., and Read, M.D.,** Efficacy and safety of large-loop excision of the transformation zone, *Lancet*, 343, 32, 1994.

104. **Bigrigg, M.A., Codling, B.W., Pearson, P., Read, M.D., and Swingler, G.R.,** Colposcopic diagnosis and treatment of cervical dysplasia at a single clinic visit: experience of low-voltage diathermy loop in 1000 patients, *Lancet*, 336, 229, 1990.

105. **Hallam, N.F., West, J., and Harper, C.,** Large loop excision of the transformation zone (LLETZ) as an alternative to both ablative and cone biopsy treatment: a series of 1000 patients, *J. Gynecol. Surg.*, 9, 77, 1993.

106. **Murdoch, J.B., Grimshaw, R.N., and Monaghan, J.M.,** Loop diathermy excision of the abnormal cervical transformation zone, *Int. J. Gynecol. Cancer*, 1, 105, 1991.

107. **Murdoch, J.B., Grimshaw, R.N., Morgan, P.R., and Monaghan, J.M.,** The impact of loop diathermy on management of early invasive cervical cancer, *Int. J. Gynecol. Cancer*, 2, 129, 1992.

108. **Smiley, L.M., Burke, T.W., Silva, E.G., Morris, M., Gershenson, D.M., and Wharton, J.T.,** Prognostic factors in stage IB squamous cervical cancer patients with low risk for recurrence, *Obstet. Gynecol.*, 77, 271, 1991.

109. **Hopkins, M.P. and Morley, G.W.,** Radical hysterectomy versus radiation therapy for stage IB squamous cell cancer of the cervix, *Cancer*, 68, 272, 1991.

110. **Copeland, L.J., Silva, E.G., Gershenson, D.M., Morris, M., Young, D.C., and Wharton, J.T.,** Superficially invasive squamous cell carcinoma of the cervix, *Gynecol. Oncol.*, 45, 307, 1992.

111. **Sevin, B.U., Nadji, M., Averette, H.E., Hilsenbeck, S., Smith, D., and Lampe, B.,** Microinvasive carcinoma of the cervix, *Cancer*, 70, 2121, 1992.

112. **Wright, T.C., Ferenczy, A.F., and Kurman, R.J.,** Carcinoma and other tumors of the cervix, in *Blaustein's Pathology of the Female Genital Tract*, Kurman, R.J., Ed., Springer-Verlag, New York, 279, 1994.

113. **Pretorius, R., Semrad, N., Watring, W., and Fotheringham, N.,** Presentation of cervical cancer, *Gynecol. Oncol.*, 42, 48, 1991.

114. **Lanciano, R.M., Won, M., and Hanks, G.E.,** A reappraisal of the International Federation of Gynecology and Obstetrics staging system for cervical cancer, A study of patterns of care, *Cancer*, 69, 482, 1992.

115. **Rutledge, F.N., Mitchell, M.F., Munsell, M., Bass, S., McGuffee, V., and Atkinson, E.N.,** Youth as a prognostic factor in carcinoma of the cervix: a matched analysis, *Gynecol. Oncol.*, 44, 123, 1992.

116. **Inoue, R.,** Prognostic significance of the depth of invasion relating to nodal metastases, parametrial extension and cell types, *Cancer*, 54, 1714, 1984.

117. **Inoue, T. and Okumura, J.,** Prognostic significance of parametrial extension in patients with cervical carcinoma stages Ib, IIa and IIb, *Cancer*, 54, 3035, 1984.

118. **Inoue, T. and Morita, K.,** The prognostic significance of number of positive nodes in cervical stages IB, IIA and IIB, *Cancer*, 65, 1923, 1990.

119. **Inoue, T., Chihara, R., and Morita, K.,** The prognostic significance of the size of the largest nodes in metastatic carcinoma from the uterine cervix, *Gynecol. Oncol.*, 19, 187, 1984.

120. **Zaino, R.J., Ward, S., Delgado, G., Bundy, B., Gore, H., Fetter, G., Ganjei, P., and Frauenhoffer, E.,** Histopathologic predictors of the behavior of surgically treated stage IB squamous cell carcinoma of the cervix, *Cancer*, 69, 1750, 1992.

121. **Barnes, W., Delgado, G., Kurman, R.J., Petrilli, E.S., Smith, D.M., Ahmed, S., Lorincz, A.T., Temple, G.F., Jenson, A.B., and Lancaster, W.D.,** Possible prognostic significance of human papillomavirus type in cervical cancer, *Gynecol. Oncol.*, 29, 267, 1988.

122. **Walker, J., Bloss, J.D., Liao, S.Y., Berman, M., Bergen, S., and Wilczynski, S.P.,** Human papillomavirus genotype as a prognostic indicator in carcinoma of the uterine cervix, *Obstet. Gynecol.*, 74, 781, 1989.

123. **Jarrell, M.A., Heintz, N., Howard, P., Collins, C., Badger, G., Belinson, J., and Nason, F.,** Squamous cell carcinoma of the cervix: HPV 16 and DNA ploidy as predictors of survival, *Gynecol. Oncol.*, 46, 361, 1992.

124. **Kenter, G.G., Cornelisse, C.J., Jiwa, N.M., Aartsen, E.J., Hermans, J., Mooi, W., Heintz, A.P., and Fleuren, G.J.,** Human papillomavirus type 16 in tumor tissue of low-stage squamous carcinoma of the uterine cervix in relation to ploidy grade and prognosis, *Cancer*, 71, 397, 1993.

125. **Sebbelov, A.M., Kjorstad, K.E., Abeler, V.M., and Norrild, B.,** The prevalence of human papillomavirus type 16 and 18 DNA in cervical cancer in different age groups: a study on the incidental cases of cervical cancer in Norway in 1983, *Gynecol. Oncol.*, 41, 141, 1991.

126. **Riou, G., Favre, M., Jeannel, D., Bourhis, J., Doussal, V.L., and Orth, G.,** Association between poor prognosis in early stage invasive cervical carcinomas and non-detection of HPV DNA, *Lancet*, 335, 1171, 1990.

127. **Higgins, G.D., Davy, M., Roder, D., Uzelin, D.M., Phillips, G.E., and Burrell, C.J.,** Increased age and mortality associated with cervical carcinomas negative for human papillomavirus RNA, *Lancet*, 338, 910, 1991.

128. **Anton-Culver, H., Bloss, J.D., Bringman, D., Lee-Feldstein, A., DiSaia, P., and Manetta, A.,** Comparison of adenocarcinoma and squamous cell carcinoma of the uterine cervix: a population based epidemiologic study, *Am. J. Obstet. Gynecol.*, 186, 1507, 1992.

129. **Vesterinen, E., Forss, M., and Nieminen, U.,** Increase of cervical adenocarcinoma: a report of 520 cases of cervical carcinoma including 112 tumors with glandular elements, *Gynecol. Oncol.*, 33, 49, 1989.

130. **Hopkins, M.P. and Morley, G.W.,** A comparison of adenocarcinoma and squamous cell carcinoma of the cervix, *Obstet. Gynecol.*, 77, 912, 1991.

131. **Wilczynski, S.P., Walker, J., Liao, S.Y., Bergen, S., and Berman, M.,** Adenocarcinoma of the cervix associated with human papillomavirus, *Cancer*, 62, 1331, 1988.

132. **Saigo, P.E., Cain, J.M., Kim, W.S., Gaynor, J.J., Johnson, K., and Lewis, J.L.,** Prognostic factors in adenocarcinoma of the uterine cervix, *Cancer*, 57, 1584, 1986.

133. **Eifel, P.J., Morris, M., Oswald, J., Wharton, J.T., and Delclos, L.,** Adenocarcinoma of the uterine cervix, *Cancer*, 65, 2507, 1990.

134. **Eifel, P.J., Burke, T.W., Delclos, L., Wharton, J.T., and Oswald, M.J.,** Early stage I adenocarcinoma of the uterine cervix: treatment results in patients with tumors < 4 cm in diameter, *Gynecol. Oncol.*, 41, 199, 1991.

135. **van Nagell, J.R., Higgins, R.V., and Powell, D.E.,** Invasive cervical cancer, in *Gynecologic Oncology*, Knapp, R.C. and Berkowitz, R.S., Eds., McGraw-Hill Inc., New York, 192, 1993.

136. **Nelson, J.,** Clinical invasive carcinoma of the cervix: place of radical hysterectomy, *Gynecologic Oncology*, Coppleson, M., Ed., Churchill Livingstone, Edinburgh, 697, 1992.

Chapter 15

Cervical Cancer: Hormones, Growth Factors, and Cytokines

Robin Leake

CONTENTS

1. INTRODUCTION

Cervical cancer is associated with early age at first intercourse and multiple sexual partners.[1,2] Further, long-term use of oral contraceptives[3] may increase the risk of, especially, adenocarcinomas; parity has also been found to have a strong association.[4] The human papilloma viruses (HPV) 16 and 18 are major risk factors. Taken together, these observations imply that, although they may act in conjunction with HPV, hormones (and associated local growth factors and cytokines) are involved in tumor promotion. It has been argued elsewhere[5] that hormones and growth factors which influence levels of functional p53 and retinoblastoma protein (Rb), will alter the ability of HPV 16 and 18 to promote cancer formation since the ability of the viral proteins to interact with and neutralise these two growth regulators is thought to be their principal mechanism of action in the cervix. This section will review the known abilities of extracellular factors to modify growth regulation within cervical epithelial cells.

2. HORMONAL REGULATION OF CERVICAL CANCER

2.1 Normal Tissue and Steroid Receptors

The proliferation of normal squamous epithelial cells of the cervix is under the control of steroid hormones during the menstrual cycle. For reproductive tissues to be sensitive to steroid hormones, it is assumed that there should be a sufficient concentration of the relevant, functional steroid receptor present in the target tissue. An immunohistochemical study[6] showed that estrogen receptor (ER) content of the squamous epithelium varies as the relative levels of estradiol and progesterone change. In the early, proliferative phase, cells of all layers are negative for ER staining. As plasma estradiol levels rise, basal and parabasal layers become positive and, in the secretory phase, nuclei in the superficial layers also become stained. This study found no staining for progesterone receptor (PR) in any squamous cells. However, further work[7] is required because another study suggested that PR was detectable in normal cells but decreased as tissue became pre-invasive.

2.2 Tumor Tissue and Steroid Receptors

Because ER and, probably, PR are present in normal tissue, it is essential that measurement of ER and PR levels in cervical cancer should be done in such a way that the large majority of tissue studied is, in fact, tumor tissue. Perhaps, because of this problem, the literature contains conflicting studies on steroid receptors in cervical cancer. In a study with careful histological controls, 71% of cervical cancers were found by Soutter to be ER-positive.[8] Peyrat[9] supported this high figure when he reported 75% of cervical cancers were ER-positive, but Martin[10] found only 33% to be ER-positive. Part of the explanation may lie in the observations of Darne[11] and co-workers, who found that, after defining receptors as only being functional if both soluble and tightly DNA-bound receptor could be detected in an exchange assay, then only 35% were ER-positive. Interestingly, in this latter study, 19% of biopsies contained only soluble receptor, 23% contained only DNA-bound receptor and 23% contained no detectable receptor. These results may well explain why the reported figures for ER in cervical cancer are so variable (overall, value range from 12% in Toppila's study[12] to 100% in Syrjala's study[13]) but the real question is how to measure steroid receptors so that they are of value in predicting either prognosis or potential response to therapy. It is essential to establish standard conditions for tissue handling and assay methodology, as well as determining an appropriate clinical cut-off value for cervical cancer. This can only be done through international cooperation and restriction to studies with large numbers of patients and good follow-up.

As already implied, one important facet in any study of receptors (whether for steroid hormones or for peptide hormones and growth factors) in a clinical context, is some demonstration of receptor functionality, since mutant receptors with binding ability but no signaling function have been found in many tumors. In addition to demonstrating ER functionality through presence of both soluble ("empty") and DNA-bound (activated) receptor, there have been studies which have measured (1) presence of PR as a product of an estrogen-activated gene, (2) pS2 and epithelial cell cathepsin D as other examples of estrogen-induced proteins, and (3) estrogen-induced components of DNA synthesis and cell division (thymidine labeling index, thymidine kinase activity, changes in S-phase fraction, etc). Of these possibilities, only presence of both soluble and tightly DNA-bound ER, or of PR, has been applied to a reasonable number of cervical cancer studies. The published results are summarized in Table 1. On this basis, the proportion of cervical cancer patients who might be expected to get objective benefit from endocrine therapy is only about 30% at most.

Table 1 Presence of "Functional" Estrogen Receptor in Squamous Cervical Cancer

Ref.	Definition of "Functionality"	Proportion "Functional"		Total Patients
		n	%	
Ford et al. (1983)	ER/PR	5	21	24
Peyrat et al. (1983)	ER/PR	12	37	32
Soutter et al. (1983)	ER_c/ER_n	11	19	58
Toppila et al. (1983)	ER/PR	2	12	17
Hunter et al. (1987)	ER/PR	16	29	56
Harding et al. (1990)	ER/PR	29	32	89
Darne et al. (1990)	ER_c/ER_n	45	34	131

Note: ER/PR = presence of both ER and PR (cut-off varies between studies); ER_c/ER_n = presence of both soluble and tightly DNA-bound ER.

There have been suggestions that adenocarcinoma of the cervix might be more likely to contain functional estrogen receptor and so be a better target for endocrine therapy. Although the published data (Table 2) confirm that functional receptor is more likely to be detected, the patient numbers are too small to make any worthwhile conclusions. Any study of endocrine therapy in adenocarcinoma of the cervix would need to be multicenter in order to amass sufficient patients. Receptor assays would, therefore, have to be carefully controlled to ensure that the quantitative determinations were strictly comparable in each center.

Table 2 Presence of "Functional" Estrogen Receptor in Adenocarcinoma of the Cervix

Ref.	Definition of "Functionality"	Proportion "Functional"		Total Patients
		n	%	
Ford et al. (1983)	ER/PR	3	50	6
Peyrat et al. (1983)	ER/PR	6	67	9
Toppila et al. (1983)	ER/PR	4	67	6
Harding et al. (1990)	ER/PR	8	62	13

Note: ER/PR = presence of both ER and PR (cut-off varies between studies).

2.3 Prognostic Value

As with studies of steroid receptors in other tissues, an attempt has been made to determine whether presence of steroid receptors can be used as a sign of better prognosis (stage for stage and grade for grade). Martin et al.[14] followed up 246 cases of primary carcinoma of the cervix. The survival curves showed no significant difference between the ER-positive and ER-negative groups, nor between the PR-positive and PR-negative groups. Hunter et al.[15] found a weak correlation between presence of PR and increased survival but no correlation between ER and survival — no attempt was made to relate amount of receptor to survival because of the small (n = 56) total number of patients. Twiggs et al.[16] suggested that ER had a worthwhile prognostic role specifically in Stage IB cervical cancer. Work from Fujimoto et al.[17] showed no value for either PR or androgen receptor in prognosis but claimed that an ER content > 5 fmol/µg was associated with longer survival. In our experience, this is a fairly ER-rich value and would limit the observation to only a very small group, though a good prognostic index can be of value for a small group, as long as it is reproducible. Harding et al.[18] reported a study of steroid receptors in relation to follow-up of 102 patients with cervical carcinoma. Larger tumors and poorer grade of differentiation correlated with ER-negativity. Despite this, there was no significant difference in progression-free survival in the ER-positive and ER-negative groups. Darne et al.[11] had come to a similar conclusion, though they noted that total absence of either soluble or DNA-bound ER was associated with poorer prognosis. It is, therefore, reasonable to conclude that, on its own, biochemical determination of steroid receptor status has little to contribute to cervical cancer management. Larger studies may eventually show that quantitative values can give better discrimination.

Before making a final conclusion on the value, or otherwise, of looking at hormones and their receptors in cervical cancer, it is important to consider what the relatively simple immuno-histochemistry assays can offer. An initial study by Mosny et al.[19] concluded that neither ER nor PR can be detected in cervical cancer cells. This is obviously in contrast to the biochemical studies. Use of alternative anti-ER antibodies (Dako) or amplification methods have led to positive findings but a proper study of comparison of biochemical and immunohistochemical values on comparable pieces of tumor is still required.

Studies of ER in breast cancer have shown that cancer cells can often contain mutated receptor which may lack essential functional elements. This could be a possible explanation of the difference in detection rate of ER by the biochemical and immunohistochemical methods. On the one hand, binding of ligand does not necessarily imply that a full DNA-binding domain is present, nor that the transcriptional activation regions are intact. On the other hand, loss of the epitope that is detected by a specific antibody does not always reflect loss of functional receptor. The solution to this dilemna is to demonstrate response to endocrine therapy which is an ideal biological confirmation of the presence of functional receptors.

Because presence of both soluble and nuclear ER (or of ER and PR) in the same cells is only demonstrated in a small proportion of cervical cancers, response rates to endocrine therapy would be expected to be low. Early studies on endocrine therapy for cervical cancer were limited to patients with advanced disease.[20-22] None of these studies reported anything other than small effects of endocrine therapy.

Combination of radiotherapy with estrogen has been reported[23] to improve 5-year survival and other studies have shown that estrogen-induced increased sensitivity to irradiation is only demonstrable in ER-positive cells, which would suggest that some receptor mediated responses may be expected in cervical cancer. One study[24] reported that there was no response to tamoxifen but, again, the patients had already failed on chemotherapy and so firm conclusions cannot be made. Realistically, a sound conclusion on the proportion of cervical cancers which would benefit from endocrine therapy will only be determined by a prospective study of first-line endocrine therapy. Perhaps this might be more realistic now that pathology laboratories are more experienced in measuring ER levels with a semi-quantitative immunohistochemical scoring system.

Another approach to the estimation of steroid sensitivity of cervical cells is to look at secretion of estrogen-induced proteins. Cathepsin D is one such protein, being released in an estrogen-controlled manner by both ER-positive breast and ovarian cancer cells.[25,26] However, cathepsin D can also be secreted by ER-negative breast cancer cells[27] and is secreted by invading macrophages. A new study[28] has shown no correlation between cathepsin D and ER or PR expression, implying that cathepsin D is not estrogen-regulated in cervical cancer. However, there was no separate assessment of macrophage secretion of cathepsin D and such an analysis might have shed further light on the question. Complete or partial response to chemotherapy was associated with higher levels of cathepsin D, yet 3-year survival was 52% for patients whose tumors had high cathepsin D levels and 78% for those with low levels of cathepsin D. There are obviously some clear questions that must still be answered about the regulation of cathepsin D secretion by cervical cancer cells.

3. GROWTH FACTOR AND CYTOKINE REGULATION

Loss of sensitivity to transforming growth factor-β (TGF-β) has been associated with increased aggressiveness in colon and breast cancer. A recent study[29] of the effects of immortalizing keratinocytes with HPV16 has suggested that the loss of sensitivity to down regulation by retinoids, subsequent to introduction of HPV16, is due to loss of TGF-β production by the cells. This has been used as evidence to suggest that chemoprevention of cervical cancer with retinoids needs to be administered prior to the establishment of HPV16.

A further study[30] has concluded that interferon-γ acts on cervical cancer cells to reduce expression of E6 and E7 (the early proteins of HPV16 and 18 which are thought to promote tumor growth by inactivating retinoblastoma protein and p53). This study also provides evidence that retinoic acid, in addition to reducing cervical cell growth by stimulating TGF-β production, also causes down regulation of the cell signaling pathways of both the epidermal growth factor (EGF) and insulin-like growth factor-I (IGF-I). This biological role for interferon action on reduction of E6 and E7 protein production has led to renewed interest in interferon use in cytotoxic cocktails for cervical cancer therapy.[31]

Although standard plasma tumor markers have contributed little to the management of cervical cancer,[32] one study[33] has suggested that sufficient soluble basic fibroblast growth factor (bFGF) is released by cervical cancers that detection of bFGF in serum may be an effective way of screening for primary cancers, as well as for monitoring response to therapy or growth of a recurrent disease.

Another aspect of potential promotion of cervical cancer by EGF (EGF receptors are present in high numbers in many squamous cervical cancers) is that, in addition to standard growth stimulation, EGF[34] has been shown to have a direct stimulatory effect on expression of the E6 and E7 genes (through activation of the AP-1 transcription factor). Further, both EGF and TGF-β have been shown to have direct effects on the integrins which normally anchor cervical cells into the immediate basement membrane. Both alpha-4 and alpha-5 have been noted[35] to be absent from epithelial cells within cancer (although retained in the stromal cells).

Given the epidemiological data that smoking alters the risk of cervical cancer, it is interesting to note that nicotine stabilizes both TGF-β and tumor necrosis factor-α within cervical cancer cells. This stabilization is reported to reverse the effects of TGF-β on growth.

Overall, there is less literature on growth factors and cytokines in cervical cancer than might be expected. However, the picture emerges that EGF, IGF-I and bFGF might be seen as the growth driving forces, whereas TGF-β is the normal growth inhibitor. Chemoprevention with retinoids seems a sound possibility since they reduce the effectiveness of the EGF and IGF-I signaling pathways, thereby reducing the effectiveness of HPV 16 or 18 (since production of the critical early proteins, E6 and E7, will no longer be promoted by EGF). Such reduction in the role of HPV 16 may be further enhanced by interferons. Promotion of secretion of TGF-β is not only directly effective in down-regulating growth but seems also to stabilize those integrins that anchor cervical epithelium to its basement membrane, thereby reducing invasive potential. These concepts are summarized in Figure 1.

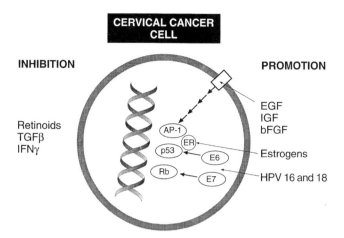

Figure 1 Factors promoting and inhibiting the development of cervical cancer. Positive growth factors act initially through their specific membrane-bound receptors but then use common intracellular pathways to influence the early proteins such as fos and jun (AP-1). Estrogens act through their nuclear receptor which can also interact with AP-1 at DNA level. The early proteins of HPV 16 and 18 (E6 and E7) promote growth inactivating p53 and retinoblastoma protein and their effect is co-promoted by EGF. TGF-ß, retinoids, and interferons can all reduce/prevent cervical cancer promotion by directly or indirectly blocking the effects of growth factors and virus early proteins. (See text for the relevant references.)

REFERENCES

1. **La Vecchia, C., Franceschi, S., Decarli, A., Fasoli, M., Gentile, A., Parazinni, F., and Regallo, M.,** Sexual factors, venereal diseases, and the risk of intraepithelial and invasive cervical neoplasia, *Cancer*, 58, 935, 1986.

2. **Kjaer, S.K., Poll, P., Jensen, H., Engholm, G., Haugaard, B.J., Teisen, C., and Christensen, R.B.,** Abnormal Papanicolau smear: a population-based study of risk factors in Greenlandic and Danish women, *Acta Obstet. Gynecol. Scand.*, 69, 79, 1990.

3. **Brinton, L.A.,** Editorial commentary and cervical cancer — current status, *Am. J. Epidemiol.*, 131, 958, 1990.

4. **Brinton, L.A., Reeves, W.C., Brenes, M.M., Herrero, R., Gaitan, E., Tenorio, F., de Britton, R.C., Garcia, M., and Rawls, W.E.,** The male factor in the etiology of cervical cancer among sexually monogamous women, *Int. J. Cancer*, 44, 199, 1989.

5. **Leake, R.E.,** Cell Cycle, in *Scientific Essentials of Reproductive Medicine*, Hillier, S.G., Kitchener, H. and Neilson, J.P., Eds., W.B. Saunders, London, 1995.

6. **Mosny, D.S., Herolz, J., and Bender, H.G.,** Immunohistochemical investigations of steroid receptors in normal and neoplastic squamous epithelium of the uterine cervix, *Gynecol. Oncol.*, 35, 373, 1989.

7. **Sadan, O., Frohlich, R.P., Driscoll, J.A., Apostoleris, A., Savage, N., and Zakut, H.,** Is it safe to prescribe hormonal contraception and replacement therapy to patients with premalignant and malignant uterine cervices?, *Gynecol. Oncol.*, 34, 159, 1989.

8. **Soutter, W.P., Pegoraro, R.J., Green-Thompson, R.W., Naidoo, D.V., Joubert, S.M., and Philpott, R.H.,** Nuclear and cytoplasmic estrogen receptors in squamous carcinoma of the cervix, in *Recent Clinical Developments in Gynecologic Oncology*, Eds., Morrow, C.P., Bonnar, J., O'Brian, T.J. and Gibbons, W.E., New York, p23, 1983.

9. **Peyrat, J.P., Vandewalle, B., and Gougeon, E.,** Second International Congress on Hormones and Cancer, *J. Steroid Biochem.*, 19, 74s, 1983.

10. **Martin, J.D., Hahnel, R., McCartney, A.J., and Woodings, T,** Prognostic value of estrogen receptors in cancer of the uterine cervix, *New Engl. J. Med.*, 306, 485, 1982.

11. **Darne, J., Soutter, W.P., Ginsberg, R., and Sharp, F.,** Nuclear and cytoplasmic estrogen and progesterone receptors in squamous cell carcinoma of the cervix, *Gynecol. Oncol.*, 38, 216, 1990.

12. **Toppila, M., Willcocks, D., and Tyler, J.P.P.,** Sex steroid receptors in gynecological malignancy, *J. Obstet. Gynecol.*, 4, 53, 1983.

13. **Syrjala, P., Kontula, K., Janne, O., Kuppila, A., and Vihko, R.,** Steroid receptors in normal and neoplastic human uterine tissue, in *Endometrial Cancer*, Brush, M.G., King, R.J.B., Taylor, R. , Eds., Bailliere Tindall, London, 1978, 242.

14. **Martin, J.D., Hanel, R., McCartney, A.J., and De Klerk,** The influence of estrogen and progesterone receptors on survival in patients with carcinoma of the uterine cervix, *Gynecol. Oncol.*, 23, 329, 1986.

15. **Hunter, R.E., Longcope, C., and Keough, P.,** Steroid hormone receptors in carcinoma of the cervix, *Cancer*, 60, 392, 1987.

16. **Twiggs, L.B., Potish, R.A., Leung, B.S., Carson, L.F., Adcock, L.L., Savage, J.E., and Prem, J.E.,** Cytosolic estrogen and progesterone receptors as prognostic parameters in stage IB cervical carcinoma, *Gynecol. Oncol.*, 28, 156, 1987.

17. **Fujimoto, J., Fujita, H., Horoda, S., Okada, H., and Tamaya, T.,** Prognosis of cervical cancers with reference to steroid receptors, *Nippon Gan Chiriyo Cakhui Shi*, 24, 21, 1989.

18. **Harding, M., McIntosh, J., Paul, J., Symonds, P., Reed, N., Habeshaw, T., Stewart, M., and Leake, R.E.,** Oestrogen and progesterone receptors in carcinoma of the cervix, *Clin. Oncol.*, 2, 213, 1990.

19. **Mosny, D.S., Herholtz, J., Degan, W., and Bender, H.G.,** Immunohistochemical investigations of steroid receptors in normal and neoplastic squamous epithelium of the uterine cervix, *Gynecol. Oncol.*, 35, 373, 1989.

20. **Malkasia, G.D., Decker, D.G., Jorgenson, E.O., and Webb, M.J.,** Evaluation of 6,17-α-dimethyl-6-dehydroprogesterone for treatment of recurrent and metastatic gynecologic malignancy, *Am. J. Obstet. Gynecol.*, 118, 461, 1974.

21. **Van Exter, C. and Pauls, F.,** Traitement des cancer avances du col par la medroxyprogesterone, *J. Gynecol. Obstet. Biol. Reprod.*, (suppl 2), 1, 383, 1972.

22. **Varga, A. and Henriksen, E.,** Effects of 17α-hydroxyprogesterone-17n-caproate on various pelvic malignancies, *Obstet. Gynecol.*, 23, 51, 1964.

23. **Sugimori, H., Taki, I., and Koga, K.,** Adjuvant hormone therapy to radiation treatment for cervical cancer, *Acta Obstet. Gynecol. Jpn.*, 23, 77, 1976.

24. **Capony, F., Morisset, M., Barrett, A.J., Capony, J.P., Broquet, P., Vignon, F., Chambon, M., Louisot, P., and Rochefort, H.,** Phosphorylation, glycosylation and proteolytic activity of the 52-kD estrogen-induced protein secreted by MCF-7 cells, *J. Cell. Biol.*, 104, 253, 1987.

25. **Galtier-Dereure, F., Capony, F., Muadelonde, T., and Rochefort, H.,** Estradiol stimulates cell growth and secretion of procathepsin D and a 120-kilodalton protein in the human ovarian cancer cell line BG-1, *J. Clin. Endocrinol. Metab.*, 75, 1497, 1992.

26. **Garcia, M., Lacombe, M.J., Duplay, H., Cavailles, V., Deracq, D., Delarue, J.C., Krebs, B., Contesso, G., Sancho-Garnier, H., and Richer, G.,** Immunohistochemical distribution of the 52-kDa protein in mammary tumours: a marker associated with cell proliferation rather than hormone responsiveness, *J. Steroid Biochem.*, 27, 439, 1987.

27. **Scambia, G., Beneditti Panici, P., and Ferrandini, M.,** Significance of Cathepsin D expression in uterine tumours, *Eur. J. Cancer*, 31A, 1449, 1995.

28. **Creek, K.E., Geslani, G., Batova, A., and Pirisi, L.,** Progressive loss of sensitivity to growth control by retinoic acid and transforming growth factor-β at late stages of human papillomavirus type 16-initiated transformation of human keratinocytes, *Adv. Exp. Med. Biol.*, 365, 117, 1995.

29. **Eckert, R.L., Agarwal, C., Hembree, J.R., Choo, C.K., Sizemore, N., Andreatta-van-Leyen, S., and Rorke, E.A.,** Human cervical cancer: retinoids, interferon and papillomavirus, *Adv. Exp. Med. Biol.*, 375, 31, 1995.

30. **Gonzalesde, L.C., Lippman, S.M., Kudelka, A.P., Edwards, C.L., and Kavanagh, J.J.,** Phase II study of cisplatin, 5-fluorouracil and interferon-alpha in recurrent carcinoma of the cervix, *Invest. New Drugs*, 13, 73, 1995.

31. **Bichler, A. and Giegerl, E.,** Tumour markers in cervical carcinoma, *Clin. Consult. Obstet. Gynecol.*, 7, 94, 1995.

32. **Sliutz, G., Tempfer, C., Obermair, A., Reinthaller, A., Gitsch, G., and Kainz, Ch.,** Serum evaluation of basic fibroblast growth factor in cervical cancer patients, *Cancer Lett.*, 94, 227, 1995.

33. **Peto, M., TolleErsu, I., Kreysch, H.G., and Klock, G.,** Epidermal growth factor induction of human papillomavirus 16 E6/7 mRNA in tumour cells involves two AP-1 binding sites in the viral enhancer, *J. Gen. Virol.*, 76, 1945, 1995.

34. **Valea, F.A., Haskill, S., Moore, D.H., and Fowler, W.C., Jr.,** Immunohistochemical analysis of alpha-integrins in cervical cancer, *Am. J. Obstet. Gynecol.*, 173, 808, 1995.

35. **Rakowicz-Szulczynska, E.M., McIntosh, D.G., and Smith, M.,** Growth factor-mediated mechanisms of nicotine-dependent carcinogenesis, *Carcinogenesis*, 15, 1839, 1994.

Chapter 16

Oncogenes and Tumor-Suppressor Genes in Cervical Cancer

Camille Busby-Earle

CONTENTS

1. INTRODUCTION

Cervical cancer is globally the most common gynecological malignancy and annually accounts for over 400,000 new cases of invasive cancer. Its etiology remains unknown, but there is reason to suspect that, like other neoplasms, both intrinsic (genetic) and extrinsic (environmental) factors play a role. There is a wealth of evidence for an association between cervical carcinoma and various behavioral risk factors, particularly indicators of sexual activity; and the roles of some nonsexually related variables have also been examined.[1] Dysplastic lesions of the cervix known as cervical intraepithelial neoplasia (CIN) appear to follow a continuum, with the risk of progression to invasive carcinoma being related to the severity of the dysplasia across a spectrum from CIN1 to CIN3. However, until fairly recently, there have been relatively few data on the role of genetic events, susceptibility, or predisposition. While unequivocal evidence for a causal relationship between cervical cancer (or any of its preinvasive states) and any of the risk factors identified is still lacking, recent studies have highlighted some of the molecular events — involving viral oncogenes, cellular oncogenes, and tumor-suppressor genes — which may be central to cervical carcinogenesis.

The recent development of elegant and sophisticated molecular biological techniques for DNA analysis has enhanced our understanding of genetic alterations in carcinogenesis. Some general conclusions have emerged. Carcinogenesis is a multifactorial and multistep process involving several alterations in specific genes, in particular the activation of oncogenes, and the inactivation of tumour suppressor genes.[2-5] These changes are manifest as loss of control of growth, replication and differentiation, with the acquisition of invasive and metastatic potential — features which are characteristic of transformed and malignant cells. Tumorigenesis is believed to result from such genetic changes occurring in a single cell, which then has the potential for unlimited clonal expansion and development of a malignant lesion.

Evidence for the role of specific oncogenes and tumor-suppressor genes in cervical cancer is still rather scanty, but some interesting data have been generated. Studies have taken two main forms — those using cell lines and those in which human tissues were analyzed. Unfortunately, results from the former cannot always safely be extrapolated to the *in vivo* situation. It is difficult to entirely separate the role of oncogenes from that of tumor-suppressor genes, as there appear to be many areas of interaction. This chapter will review the commonly emerging themes which exist with respect to the transforming genes of human papillomaviruses, particularly with regard to their interaction with the tumor-suppressor gene products p53 and pRb; examine the general search for tumor-suppressor genes involved in cervical carcinogenesis with specific reference to chromosomes 3, 11 and 17; and finally review work which has been conducted on cellular oncogenes, of which most has involved the *myc* and *ras* oncogenes.

2.   HUMAN PAPILLOMAVIRUSES

There is substantial evidence implicating the oncogenicity of papillomaviruses, and an equally large body of evidence suggesting a major etiological role for human papillomaviruses (HPVs) in the development of cervical neoplasia. Support for this association, and for their oncogenic potential has essentially come from two sources — first, from clinico-epidemiological observations and studies; and second, from laboratory based experimental studies.

Over 70 types of HPV have been identified, but site specificity has dictated that only a few types, namely, 6, 11, 16, 18, 31, 33, 35, 39, 42-44, 51, and 52 are associated with the human genital tract. Types 6, 11, 16, 18, 31, and 33 are most frequently isolated, and of these only the latter four types have consistently been associated with malignancy in the cervix. However, most of the data on HPV in cervical cancer focus on types 16 and 18, which together account for the HPV status of 80 to 90% of cervical carcinomas and high grade CIN.

The papillomavirus genome comprises a single closed circular double-stranded DNA core. The viral genome is approximately 8,000 base pairs in length with a molecular weight of about 5.2×10^6 D. The linear organization of all the papillomaviruses characterized to date has been remarkably similar, consisting of at least 10 potential protein coding sequences or open reading frames (ORFs) all on the same strand; the other DNA strand contains only small unconserved ORFs and is assumed to be non-coding.[6-8] Functionally, the genome can be divided into 2 coding regions — early (E) regions of which there are between 6 and 8 and late (L) regions of which there are two. The early genes E1-8 are important in viral replication and cellular transformation while the late genes L1 and L2 code for structural proteins of the virion. The non-coding region, located between L1 and the start of the E6 ORF, termed the long control region (LCR) or upstream regulatory region (URR) contains transcriptional control sequences involved in the control of viral gene expression.[6-8]

Transforming activity for different HPV types, including types 16 and 18, has been demonstrated *in vitro*. Cloned HPV viral DNA has been introduced successfully into rodent cell lines to effect transformation defined by altered morphology, loss of serum- and anchorage-dependence, and tumorigenicity in nude mice. While the transforming efficiency of HPVs is lower than that of their bovine counterparts, transforming activity of HPV has confidently been assigned to the E6 and E7 ORFs. It is still uncertain, however, whether differences exist between the *in vitro* transforming capacity of HPVs differing in *in vivo* oncogenic potential.[6,7] Immortalization has been demonstrated in human foreskin keratinocytes that have been transfected with HPV 16, 18, 31 or 33 but not with types 6 or 11; and when allowed to stratify into multilayer epithelia, those transfected with HPV16 show characteristics of CIN3. Immortalized cell lines contain integrated, transcriptionally active viral genomes but are not tumorigenic *in vivo*. The viral DNA remains episomal in those transfected with HPV types 6 or 11, and their lifespan is the same as that of non-transfected cells.[9] Transfection of NIH 3T3 (cultured fibroblast) cells with HPV 16 DNA resulted in stimulation of their growth properties, but prolonged incubation resulted in recognizable morphological changes in approximately 1 cell per 1000, with no apparent alteration in the state of expression of the viral DNA — a change thought to result from a second, probably mutational, cellular event.[10]

Several studies have examined viral expression and interaction with host cellular genes in primary cervical tumors and cervical carcinoma cell lines. In all HPV 16 or 18 positive cell lines (e.g., CaSki, SiHa, HeLa) and in most cervical cancers and high grade CIN lesions, the viral genome has been detected as integrated into the cellular DNA.[11,12] No specific integration sites have yet been identified in the host genome[13,14] but integration at fragile chromosome sites and near protooncogenes has been recognized. However, the opening of the viral circular DNA shows remarkable specificity during integration. The opening usually occurs within the 3′ end of E1 or the 5′ end of E2, thereby disrupting the genome in such a way that the E2 transactivator gene is no longer functional.[6] This leads to loss of viral self-regulation, with the viral transforming genes coming under the control of host cellular genes.

The E6 and E7 genes encode the major transforming proteins comprising 158 and 98 amino acids respectively. These gene products are the viral proteins most consistently detected in HPV-positive cervical carcinoma cell lines and in cells transformed *in vitro*. Their transforming properties may be required for the initiation and maintenance of the transformed state. Tissue-specific constitutive enhancers have been identified in the regulatory regions of HPV 16 and 18. It has also been shown that their promoting influence on expression of E6 and E7 is not affected by the disruption of the E2 region, which accompanies viral integration, and may imply an important E6 and E7 enhancing role in cervical carcinogenesis.[15] The combination of E6 and E7 genes, normally retained and expressed in cervical carcinomas, is sufficient for immortalization; but, the E7 gene alone induces immortality less efficiently. Immortalized cells are resistant to terminal differentiation; and host regulation of viral gene expression by endogenous cytokines and by cellular transcriptional activators and suppressors is perhaps also relevant to the involvement of human papillomaviruses in cervical carcinogenesis.[16]

The site of cellular integration of HPV appears to be variable. However, a single HPV integration site of HPV 16 in the SW 756 cell line has been identified as a heritable fragile site on chromosome 12 in the 12q13 region. Studies on the sites of integration of HPV 18 in the host cell genome of cervical carcinoma derived HeLa and C4-1 cell lines have demonstrated integration at chromosome 8q24 in the region of the c-*myc* oncogene, and the detection of elevated *myc* mRNA in both cell lines has suggested that activation of this cellular oncogene may occur by integration of viral promoter/enhancer sequences.[17] Resected HPV-positive specimens of CIN and invasive carcinoma, as well as cell lines, have demonstrated alterations in host cellular oncogene expression. Experiments on primary rat cells have identified a co-operative effect between HPV 16 and the activated *ras* oncogene;[18] treatment of HPV-immortalized cells with *ras* or a subfragment of herpes simplex virus type 2 resulted in locally invasive carcinomas when the cells were implanted subcutaneously in nude mice;[16] and *in vitro* experiments have demonstrated the binding of HPV 16 and 18 E6 and E7 gene products to the protein products of the p53 and retinoblastoma (Rb) tumor suppressor genes respectively, resulting in their inactivation (see below).

Despite all of these observations of associations between HPV and cervical cancer, and demonstrations of their oncogenic potential both *in vitro* and *in vivo*, there are data to suggest that HPV is unlikely to be the only factor. Firstly, HPV-negative cervical carcinomas have been reported repeatedly, and the high prevalence of latent HPV infection in the general population indicates that HPV alone may not be an absolute requirement for neoplastic transformation. This is supported by the finding that cells immortalized by HPV (even "high risk" types) *in vitro* are rarely oncogenic *in vivo* (so far reported only for HPV18), implying that a second event may be required.[16,19]

3. TUMOR-SUPPRESSOR GENES

Loss or inactivation of specific genes has been identified as a characteristic of many malignancies, and has led to the identification of a group of genes now known as tumor-suppressor genes, which appear to contribute to the suppression of tumorigenesis.[20-23] Based on Knudson's "two-hit" hypothesis of carcinogenesis,[24] it has been deduced that tumor-suppressor gene behavior is recessive (both copies needing to be inactivated) with respect to tumorigenesis, but dominant (only one copy being required) in terms of tumor suppression.[25-27] Most of the recent data on tumor-suppressor genes have come from studies on solid tumors, and molecular genetic analyses have identified regions of certain chromosomes that are consistently inactivated (mutated or deleted) in specific cancers.

There is now evidence, from restriction fragment length polymorphism analyses, for allele loss in the vicinity of tumor-suppressor genes in many human malignancies, and the association between allele losses on specific chromosomes in particular cancers has proved highly significant. Loss of heterozygosity (LOH), indicating allele loss, has been used to detect chromosome regions that harbor these genes, and it has now been demonstrated in diverse familial and sporadic cancers. While allele loss at some putative tumor suppressor gene loci has been identified in several different types of tumor; others have been relatively tumor type specific, and LOH at multiple tumor-suppressor gene loci has been found in some cancers. There have been a limited number of attempts to find tumor-suppressor genes important in the genesis of cervical carcinoma, but results remain inconclusive. Although cervical carcinoma, with its identifiable premalignant phases, is in some respects an attractive malignancy in which to analyze genetic events such as allele loss (events can be analyzed and their importance evaluated at various steps in the progression of the disease); other factors have made it less than ideal. For example, cervical carcinoma is a sporadic cancer for which no familial form has been identified, and which so far, has not been associated with any inherited syndrome. Few clues therefore exist to the location of candidate tumor-suppressor genes. In the developed world, where most of the laboratory based research is conducted, the diagnosis is often made in the early stages of the disease. Tumor size is frequently small, and this severely limits tumor tissue (and hence DNA) availability in many individual cases. The use of techniques such as polymerase chain reaction (PCR) has helped to overcome this difficulty to some extent. Further, a dense inflammatory infiltrate is common in these lesions, and this together with the presence of admixed normal stroma throughout the malignant tissue, can pose problems in interpretation of LOH data. Difficulty may arise in distinguishing an informative case in which there is no LOH, from one in which contaminating normal inflammatory or stromal cellular DNA obscures the LOH. The results of most available reports on allele loss in cervical cancer have therefore been derived from small series, with analyses restricted to specific and perhaps randomly selected regions of the genome, and although evidence for involvement of tumor-suppressor genes remains rather scanty, some interesting data have been generated.

Loss of heterozygosity in primary cervical carcinomas was initially reported in all nine cases on chromosome 3 in the region 3p14-21 in a small series of nine heterozygous cases,[28] but subsequent studies have not shown such a high incidence of loss. A common region of deletion was identified at chromosome 3p13-14.3 with 6 of 8 (75%) cervical carcinomas showing LOH at this region.[29] Using a probe which mapped to the 3p21 region, Busby-Earle et al. could demonstrate no significant loss in a series in which only four cases were informative.[30] Chung et al.[31] detected LOH in 10 of 12 and 10 of 11 informative cases at 3p25 and 3p14 respectively, and further study showed LOH at both loci in 4 of 5 cases of CIN. As no LOH was demonstrated at these loci in normal cervical tissues, the authors inferred that deletional events occurring on the short arm of chromosome 3 at 3p25 and 3p14 were early events in the development of carcinoma of the cervix.[31] Kohno et al.[32] reported LOH at one or more loci on chromosome 3p in 21 of 47 cases (45%) of cervical cancer. Four of the tumors showed partial or interstitial deletions, and the common region of LOH in these tumors was 3p13-21.1 suggesting the presence of a previously unidentified tumor-suppressor gene in this region.[32] Results remain inconclusive, but may suggest the presence of a tumor-suppressor gene involved in cervical carcinogenesis on chromosome 3p.

Chromosome 11 appears to hold the promise of another tumor-suppressor gene important in cervical carcinogenesis. Loss of heterozygosity has been described in 36% of cases at the c-Ha-*ras* locus on chromosome 11p;[33] in 30% of cases on chromosome 11 using markers on both the short and the long arms;[34] and in 26% of cases using three markers on chromosome 11p.[30] Research on cervical carcinoma cell lines has also suggested a role for genes on chromosome 11,[35] and microcell transfer of a single copy of fibroblast chromosome 11 into tumorigenic HeLa cells converted them into a non-tumorigenic state.[36] Further, a putative tumor suppressor gene identified in HeLa cells has been mapped to the chromosome 11q13 region.[34] More recently, Hampton and colleagues identified a common region (11q22-24) of LOH in 14 of 32 (44%) cervical carcinomas using material derived from the UK and USA.[37]

Other tumor-suppressor gene loci which have been examined in cervical carcinoma have included sites on chromosome 5q in the vicinity of the MCC and APC genes; chromosome 13q within the Rb locus; chromosome 17p in the region of the p53 gene; and chromosome 18q in the vicinity of the DCC gene. None has shown a significant incidence of LOH when compared to other solid tumors, in which these genes are commonly inactivated. This led to the conclusion that tumor suppressor genes commonly implicated in other solid tumors appear not to be involved in cervical carcinogenesis.[30]

The consistent finding of chromosome 17-derived markers in karyotypic analyses of cervical carcinomas led to the suggestion that the importance of chromosome 17 to the development of this cancer might lie in the loss of recessive genes on chromosome 17p.[38] However, using a series of probes on chromosome 17p in the vicinity of the p53 gene, Busby-Earle et al. detected LOH in only 3 of 20 (15%) informative cases[30] — a similar proportion to that identified (14%) in a similar series.[39] No allele loss was detected within the Rb gene in a series of 14 informative cases.[30] Chromosomes 17 and 13 are of particular interest because of the location of the tumor suppressor genes p53 and Rb, respectively. While no significant allele losses have been detected on either of these chromosomes, the interaction of the products of these two tumor, suppressor genes with the products of the transforming genes E6 and E7 of oncogenic HPV types has excited interest in the role of these tumor suppressors, particularly that of the p53 gene, in cervical carcinogenesis.

p53 is a tumor-suppressor gene located on the short arm of chromosome 17 at 17p13.1. Highly conserved across species and initially classified as an oncogene, mutation in this gene is the most common genetic abnormality identified to date in human malignancy. Its function is believed to lie in its accumulation in cells with DNA damage, resulting in arrest of the cell cycle at G1 phase, in order to permit either DNA repair prior to the resumption of replication or commitment to apoptosis.[40] It thereby prevents the replication of cells with damaged DNA and hence the genesis of a population of DNA-damaged cells. The binding of the E6 gene product of oncogenic HPVs such as HPV 16 and 18 to the p53 gene product results in its inactivation and rapid degradation by a ubiquitin-dependent pathway.[41,42] This process is significantly more efficient with the oncogenic virus types than with non-oncogenic types such as HPV 6 and HPV 11.[43] Mutation in the "hot-spot regions" of the p53 gene (regions in which mutations occur most frequently in diverse human cancers), a common mechanism for its inactivation in other common solid tumors,[44] appears to be a much rarer event in cervical carcinoma[45,46] and in some series, mutations in the p53 gene have been restricted to HPV-negative tumors and cell lines.[47,48] This finding in cervical carcinoma cell lines and tumors in some series, initially led researchers to believe that cervical carcinomas segregated into two categories — those that were HPV and wild-type p53 positive, and those that were HPV negative with mutated p53. A central role for p53 in cervical carcinogenesis was postulated when Crook and colleagues suggested its inactivation by alternative pathways — complexing with HPV E6 and rapid degradation, in HPV positive carcinomas; p53 gene mutation, in HPV negative tumors.[47,48] However, the picture emerging seems much more complex, as cases of HPV negative cervical cancer lacking p53 mutation and vice versa have been reported.[45,46,49-51] This has been supported by immunohistochemical studies in which positive staining of accumulated (presumed mutated p53) in HPV negative cervical carcinomas has been reported as infrequent, sparse and faint.[46] While this does not entirely refute the hypothesis, it suggests that either alternative mechanisms of p53 inactivation such as stabilization by the *MDM-2* oncogene product may operate, or that there may be other routes to cervical carcinogenesis that are independent of p53 inactivation.

Inactivation of the retinoblastoma gene product, pRb, provides a possible alternative route. Rb1 encodes a ubiquitously expressed 105 kDa DNA binding nuclear phosphoprotein that appears to be involved in transcriptional regulation and cell cycle control. Like p53, pRb is capable of forming complexes with the oncoproteins of some oncogenic DNA viruses including the E7 gene of HPV.[42,52,53] pRb has received less attention than p53 in studies on cervical carcinogenesis, but in complexing with E7, the latter binds preferentially to the underphosphorylated "active" form of pRb, leading to its functional inactivation. One possibility is that this binding, and hence crippling of pRb function, may account for the oncogenicity of these viruses, and for the role of this tumor-suppressor gene in cervical carcinogenesis.

4. ONCOGENES

The roles of viral oncogenes, particularly the transforming genes, E6 and E7, of the oncogenic human papillomaviruses, in cervical carcinoma have already been discussed. This section will briefly review the available data on cellular oncogenes and their relevance to cervical carcinogenesis. Studies on oncogenes in cervical carcinoma have focused mainly on the *ras* and *myc* oncogenes; others, including *MDM-2*, *erbB-2*, *mos*, *jun*, *fm*s and LA-1 have received very limited attention.

It is now clear that *ras* gene mutations (particularly activating mutations in codons 12, 13, and 61 of Ha-*ras*, Ki-*ras* and N-*ras*) have been found in a variety of tumor types.[54] In experimental models, HPV 16 DNA has been shown to cooperate with the mutated c-Ha-*ras* gene in transforming primary cells.[18] The transfection of activated Ha-*ras* to HPV 16-immortalized human cervical cells resulted in the development of squamous carcinomas when these cells were injected into nude mice.[19] In an immunohistochemical study *ras* p21 protein overexpression was related to prognosis in cervical cancers, but the correlation was found to be dependent on the histological type.[55] Restriction fragment length polymorphism analyses of cervical carcinomas (90% of which were HPV positive) revealed LOH in tumor versus constitutional DNA at the c-Ha-*ras* locus in 36% of heterozygous cases, but there was no correlation with tumor aggressiveness.[33] In this study, 24% of advanced tumors were found to have somatic mutations at codon 12, affecting either one or both c-Ha-*ras* alleles. The presence of *ras* codon 12 mutation was found to be significantly associated with poor prognosis, and was associated with loss of the remaining c-Ha-*ras* allele in 40% of tumors.[33] In another study, analysis of Ki-*ras* mutation at codon 12 in tumors of the female genital tract failed to demonstrate mutation in the four cervical carcinomas examined.[56] Riou et al. further reported that 100% of tumors containing mutated *ras*, and 70% of those with a deleted *ras* allele, also contained an amplified or overexpressed *myc* gene.[33,57] The possibility of cooperation between c-*myc* and c-Ha-*ras* in the progression of cervical carcinoma has been suggested.[57] However, the results of molecular genetic studies on c-*myc* in cervical carcinoma remain somewhat controversial and inconclusive.

A study of the chromosomal location of cellular sequences flanking integrated HPV DNA sequences in cervical carcinoma cell lines and tumor tissue has revealed that in some cases, sites of integration were within proximity of, or linked to the c-*myc* oncogene. In HPV 18-positive HeLa and C4-1 cell lines, the HPV DNA was found integrated within 40kb of the c-*myc* gene at chromosome 8q24; and elevated steady state levels of c-*myc* RNA were found in these relative to other cervical carcinoma cell lines.[17] c-*myc* gene amplification and/ or overexpression have been reported.[58,59] Riou et al. detected a 4-20-fold increase in c-*myc* expression in approximately one third of tumors examined, and this, together with c-*myc* amplification were significantly associated with more aggressive cancers — patients having an eightfold increased incidence of early relapse. In another study, Ocadiz et al. reported c-*myc* amplification (some up to 60-fold) in 48% of tumor samples, with rearrangement of the gene in 90%.[58] This was contrary to the findings of Riou et al. who failed to detect any c-*myc* rearrangement, and Choo et al. who detected neither rearrangement nor amplification of c-*myc* in a study in a Chinese population.[60]

The latter study also described altered levels of c-*myc* expression, thought to be related to HPV presence, in some cervical carcinoma cell lines.

Most of the more recent studies on oncogenes have been clinically oriented and have attempted to correlate their expression with prognosis. One such study examined the immunohistochemical expression of both c-*myc*, *ras* and c-*jun* (encodes a DNA-binding protein) in tumor tissue derived from 55 patients (responders and nonresponders to neo-adjuvant chemotherapy prior to radical radiotherapy) with advanced (Stage III or IV) cervical carcinoma. The frequency of oncogene expression in these tumors ranged from 39.2% for c-*jun* to 80.4% for *ras*, but no correlation was found between *ras*, c-*myc* and c-*jun* proto-oncogene expression and prognosis; and there was no statistically significant association between oncogene expression and time to local recurrence or development of metastases.[61]

More recently, Mitra et al. evaluated a panel of 22 proto-oncogenes (including c-*myc*, *bcl*-1, c-Ha-*ras* and *erb*B-2) for amplification in 50 primary, untreated, Stage II and III, moderately well differentiated, squamous cell carcinomas of cervix. Amplification (represented by 5 or more copies) of the gene was observed in very few cases; the majority (seven cases — 14%) being observed for *erb*B-2, with amplification ranging from 5–68 copies. In addition, two of the tumors with *erb*B-2 amplification showed possible mutation or rearrangement of the gene, suggesting a possible role for this gene encoding the p185 protein in cervical tumorigenesis.[62]

c-*mos* is a proto-oncogene with a tyrosine kinase product about which little is known. Western blot analyses have shown c-*mos* proteins to be expressed in cervical carcinoma derived cell lines.[63] However, the ubiquitous expression of low levels of this oncoprotein in a variety of human tissues examined[63] would suggest a fundamental role for this proto-oncogene rather than one that was specific to cervical carcinogenesis. The c-*fms* oncogene product, a transmembrane protein, was similarly detected in a variety of normal as well as malignant tissues, including cervical carcinomas, and no significant implications were suggested by its ubiquitous expression.[64]

The product of the amplified *MDM*-2 oncogene located at chromosome 12q13-14 has been found to bind and inactivate the p53 protein. As *MDM*-2 amplification and p53 mutation seem rarely to occur together in the same tumor, this may provide an alternative mechanism for p53 inactivation, particularly in HPV-negative cervical carcinomas with WT-p53. However, in a series of 35 cervical carcinomas (four of which were HPV-negative with WT-p53), *MDM*-2 amplification could not be identified, and led to the conclusion that, like p53 mutation, *MDM*-2 amplification is uncommon in primary cervical carcinomas.[45]

Studies using an antibody to a synthetic peptide designated LA-1, have identified a 60 kDa protein present in CIN and invasive cervical carcinoma tissue. Terzano et al.[65] used the LA-1 antibody with an immunoperoxidase method to compare expression of this oncogene product in 223 cases of CIN and invasive carcinoma with 39 specimens of normal cervix and 52 specimens of squamous carcinoma from non-cervical sites. A higher frequency of positive staining was noted with increasing grade of CIN, and the highest incidence of positivity (64.7%) was observed in the invasive cervical carcinomas. In a follow-up study, positive staining was observed in 9 of 12 patients whose lesions had progressed, in comparison to 9 of 11 patients with static lesions which were LA-1 negative.[65] However, 10.3% of the normal cervical tissue samples examined also stained positively, suggesting that while this may be a useful biological marker in cervical neoplasia, its expression is not necessarily predictive of underlying disease or prognosis.

The roles of all of these oncogenes in cervical carcinogenesis have not been completely resolved and it remains uncertain whether oncogenes play any part in the development of this tumor.

5. CONCLUSIONS

Carcinogenesis in the cervix is a multistep process involving viral oncogenes of the human papilloma-viruses, tumor suppressor genes and cellular oncogenes. However, the molecular mechanisms involved are complex and it is likely that there are multiple genetic pathways to the malignant phenotype in the cervix. The analysis of these mechanisms, and the identification of critical genetic events may, in addition to enhancing our understanding of tumor biology, have the potential for clinical impact in the form of more selective, less radical treatment in the preinvasive phases of this disease.

REFERENCES

1. **Davey Smith, G., and Phillips, A.N.,** Confounding in epidemiological studies: why "independent" effects may not be all they seem, *Br. Med. J.*, 305, 757, 1992.

2. **Bodmer, W.F.,** Somatic cell genetics and cancer, *Cancer Surv.*, 7, 239, 1988.

3. **Solomon, E., Borrow, J., and Goddard, A.D.,** Chromosome aberrations and cancer, *Science*, 254, 1153, 1991.

4. **Steel, C.M.,** Oncogenes and anti-oncogenes in human cancer, *Proc. R. College Physicians Edinburgh*, 19, 413, 1989.

5. **Wynford-Thomas, D.,** Oncogenes and anti-oncogenes; the molecular basis of tumour behaviour, *J. Pathol.*, 165, 187, 1991.

6. **Arends, M.J., Wyllie, A.H., and Bird, C.C.,** Papillomaviruses and human cancer, *Hum. Pathol.*, 21, 686, 1990.

7. **Chang, F.,** Role of papillomaviruses, *J. Clin. Pathol.*, 43, 269, 1990.

8. **Sousa, R., Dostatni, N., and Yaniv, M.,** Control of papillomavirus gene expression, *Biochim. Biophys. Acta*, 1032, 19, 1990.

9. **Woodworth, C.D. Doniger, J., and Di Paolo, J.A.,** Immortalisation of human foreskin keratinocytes by various human papillomavirus DNAs corresponds to their association with cervical carcinoma, *J. Virol.*, 63, 159, 1989.

10. **Noda, T., Yajima, H., and Ito, Y.,** Progression of the phenotype of transformed cells after growth stimulation of cells by a human papillomavirus type 16 gene function, *J. Virol.*, 62, 313, 1988.

11. **Durst, M., Kleinheinz, A., Hotz, M., and Gissmann, L.,** The physical state of human papillomavirus type 16 DNA in benign and malignant tumors, *J. Gen. Virol.*, 66, 1515, 1985.

12. **Pfister, H.,** Human papillomaviruses and genital cancer, *Adv. Cancer Res.*, 48, 113, 1987.

13. **Dalgleish, A.G.,** Viruses and cancer, *Br. Med. Bull.*, 47, 21, 1991.

14. **zur Hausen, H.,** Viruses in human cancers, *Science*, 254, 1167, 1991.

15. **Shah, K.V. and Gissmann, L.,** Experimental evidence on oncogenicity of papillomaviruses, in *Human Papillomaviruses and Cervical Cancer*, 94, Munoz, N., Bosch, F.X., Jensen, O.M., Eds., IARC Scientific Publications, Lyon, 1989, Chap. 5, 105.

16. **Di Paolo, J. A., Popescu, N.C., Alvarez, L., and Woodworth, C.D.,** Cellular and molecular alterations in human epithelial cells transformed by recombinant human papillomavirus DNA, *Crit. Rev. Oncogen.*, 4, 337, 1993.

17. **Durst, M., Croce, C.M., Gissmann, L., Schwarz, E., and Huebner, K.,** Papillomavirus sequences integrate near cellular oncogenes in some cervical carcinomas, *Proc. Natl. Acad. Science U.S.A.*, 84, 1070, 1987.

18. **Matlashewski, G., Osborn, K., Banks, L., Stanley, M., and Crawford, L.,** Transformation of primary human fibroblast cells with human papillomavirus type 16 DNA and EJ-*ras*, *Int. J. Cancer*, 42, 232, 1988.

19. **Di Paolo, J.A., Woodworth, C.D., Popescu, N.C., Notario, V., and Doniger, J.,** Induction of human cervical squamous cell carcinoma by sequential transfection with human papillomavirus 16 DNA and viral Harvey *ras*, *Oncogene*, 4, 395, 1989.

20. **Green, A.R. and Wyke, J.A.,** Anti-oncogenes, a subset of regulatory genes involved in carcinogenesis?, *Lancet*, 2, 475, 1985.

21. **Friend, S.H., Dryja, T.P., and Weinberg, R.A.,** Oncogenes and tumor-suppressing genes, *N. Engl. J. Med.*, 318, 618, 1988.

22. **Goudie, R.B.,** (Editorial) What are antioncogenes?, *J. Pathol.*, 154, 297, 1988.

23. **Green, A.R.,** Recessive mechanisms of malignancy, *Br. J. Cancer*, 58, 115, 1992.

24. **Knudson, A.G.,** Mutation and cancer: Statistical study of retinoblastoma, *Proc. Natl. Acad. Science U.S.A.*, 68, 820, 1971.

25. **Ponder, B.,** Gene losses in human tumours, *Nature*, 335, 400, 1988.

26. **Mitchell, C.D.,** Recessive oncogenes, antioncogenes and tumour suppression, *Br. Med. Bull.*, 47, 136, 1991.

27. **Weinberg, R.A.,** tumor suppressor genes, *Science*, 254, 1138, 1991.

28. **Yokota, J., Tsukada, Y., Nakajima, T., Gotoh, M., Shimosato, Y., Mori, N., Tsunokawa, Y., Sugimura, T., and Terada, M.,** Loss of heterozygosity on the short arm of chromosome 3 in carcinoma of the uterine cervix, *Cancer Res.*, 49, 3598, 1989.

29. **Jones, M.H. and Nakamura, Y.,** Deletion mapping of chromosome 3p in female genital tract malignancies using microsatellite polymorphisms, *Oncogene*, 7, 1631, 1992.

30. **Busby-Earle, R.M.C., Steel, C.M., and Bird, C.C.,** Cervical carcinoma: low frequency of allele loss at loci implicated in other common malignancies, *Br. J. Cancer*, 67, 71, 1993.

31. **Chung, G.T., Huang, D.P., Lo, K.W., Chan, M.K., and Wong, F.W.,** Genetic lesion in the carcinogenesis of cervical cancer, *Anticancer Res.*, 12, 1485, 1992.

32. **Kohno, T., Takayama, H., Hamaguchi, M., Takano, H., Yamaguchi, N., Tsuda, H., Hirohashi, S., Vissing, H., Shimizu, M., Oshimura, M., and Yokota, J.,** Deletion mapping of chromosome 3p in human uterine cervical cancer, *Oncogene*, 8, 1825, 1993.

33. **Riou, G., Barrois, M., Sheng, Z., Duvillard, P., and Lhomme, C.,** Somatic deletions and mutations of c-Ha-*ras* in human cervical cancers, *Oncogene*, 3, 329, 1988.

34. **Srivatsan, E.S., Misra, B.C., Venugopalan, M., and Wilczynski, S.P.,** Loss of heterozygosity for alleles on chromosome 11 in cervical carcinoma, *Am. J. Hum. Genet.*, 49, 868, 1991.

35. **Kaelbling, M., Roginski, R.S., and Klinger, H.P.,** DNA polymorphisms indicate loss of heterozygosity for chromosome 11 of D98AH2 cells, *Cytogenet. Cell Genet.*, 41, 240, 1986.

36. **Saxon, P.J., Srivatsan, E.S., and Stanbridge, E.J.,** Introduction of human chromosome 11 via microcell transfer controls tumorigenic expression of HeLa cells, *EMBO J.*, 5, 3461, 1986.

37. **Hampton, G.M., Penny, L.A., Baergen, R.N., Larson, A., Brewer, C., Liao, S., Busby-Earle, R.M.C., Williams, A.R.W., Steel, C.M., Bird, C.C., Stanbridge, E.J., and Evans, G.A.,** Loss of heterozygosity in cervical carcinoma: Subchromosomal localization of a putative tumor-suppressor gene to chromosome 11q22-q24, *Proc. Natl. Acad. Science U.S.A.*, 91, 6953, 1994.

38. **Atkin, N.B. and Baker, M.C.,** Chromosome 17p loss in carcinoma of the cervix uteri, *Cancer Genet. Cytogenet.*, 37, 229, 1989.

39. **Kaelbling, M., Burk, R.D., Atkin, N.B., Johnson, A.B., and Klinger, H.P.,** Loss of heterozygosity on chromosome 17p and mutant p53 in HPV-negative cervical carcinomas, *Lancet*, 340, 140, 1992.

40. **Lane, D.P.,** p53, guardian of the genome, *Nature*, 358, 15, 1992.

41. **Scheffner, M., Werness, B.A., Hulbregtse, J.M., Levine, A.J., and Howley, P.M.,** The E6 oncoprotein encoded by human papillomavirus types 16 and 18 promotes the degradation of p53, *Cell*, 63, 1129, 1990.

42. **Scheffner, M., Munger, K., Byrne, J.C., and Howley, P.M.,** The state of p53 and retinoblastoma genes in human cervical carcinoma cell lines, *Proc. Natl. Acad. Science U.S.A.*, 88, 5523, 1991.

43. **Crook, T., Tidy, J.A., and Vousden, K.H.,** Degradation of p53 can be targeted by HPV E6 sequences distinct from those required for p53 binding and transactivation, *Cell*, 67, 547, 1991.

44. **Nigro, J.M., Baker, S.J., Preisinger, A.C., Jessup, J.M., Hostetter, R., Cleary, K., Bigner, S.H., Davidson, N., Baylin, S., Devilee, P., Glover, T., Collins, F.S., Weston, A., Modali, R., Harris, C.C., and Vogelstein, B.,** Mutations in the p53 gene occur in diverse human tumour types, *Nature*, 342, 705, 1989.

45. **Kessis, T.D., Slebos, R.J., Han, S.M., Shah, K., Bosch, X.F., Munoz, N., Hedrick, L., and Cho, K.R.,** p53 gene mutations and MDM2 amplification are uncommon in primary carcinomas of the uterine cervix, *Am. J. Pathol.*, 143, 1398, 1993.

46. **Busby-Earle, R.M.C., Steel, C.M., Williams, A.R.W., Cohen, B., and Bird, C.C.,** p53 mutations in cervical carcinogenesis — low frequency and lack of correlation with human papillomavirus status, *Br. J. Cancer*, 69, 732, 1994.

47. **Crook, T., Wrede, D., and Vousden, K.H.,** p53 point mutation in HPV negative human cervical carcinoma cell lines, *Oncogene*, 6, 873, 1991.

48. **Crook, T., Wrede, D., Tidy, J.A., Mason, W.P., Evans, D.J., and Vousden, K.H.,** Clonal p53 mutation in primary cervical cancer: association with human-papillomavirus-negative tumours, *Lancet*, 339, 1070, 1992.

49. **Busby-Earle, R.M.C., Steel, C.M., Williams, A.R.W., Cohen, B., and Bird, C.C.,** Papillomaviruses, p53, and cervical cancer, *Lancet*, 339, 1350, 1992.

50. **Borresen, A.L., Helland, A., Nesland, J., Holm, R., Trope, C., and Kaern, J.,** Papillomaviruses, p53, and cervical cancer, *Lancet*, 339, 1350, 1992.

51. **Fujita, M., Inoue, M., Tanizawa, O., Iwamoto, S., and Enomoto, T.,** Alterations of the p53 gene in human primary cervical carcinoma with and without human papillomavirus infection, *Cancer Res.*, 52, 5323, 1992.

52. **Dyson, N., Howley, P.M., Munger, K., and Harlow, E.,** The human papillomavirus-16 E7 oncoprotein is able to bind to the retinoblastoma gene product, *Science*, 243, 934, 1989.

53. **Banks, L., Edmonds, C., and Vousden, K.H.,** Ability of HPV 16 E7 protein to bind Rb and induce DNA synthesis is not sufficient for efficient transforming activity in NIH 3T3 cells, *Oncogene*, 5, 1383, 1990.

54. **Bos, J.L.,** *ras* oncogenes in human cancer: a review, *Cancer Res.*, 49, 4682, 1989.

55. **Sagae, S., Kuzumaki, N., Hisada, T., Mugikura, Y., Kudo, R., and Hashimoto, M.,** *ras* oncogene expression and prognosis of invasive squamous cell carcinomas of the uterine cervix, *Cancer*, 63, 1577, 1989.

56. **Enomoto, T., Inoue, M., Perantoni, A.O., Terakawa, N., Tanisawa, O., and Rice, J.M.,** K-*ras* activation in neoplasms of the human female reproductive tract, *Cancer Res.*, 50, 6139, 1990.

57. **Riou, G.F.,** Proto-oncogenes and prognosis in early carcinoma of the uterine cervix, *Cancer Surv.*, 7, 441, 1988.

58. **Ocadiz, R., Sauceda, R., Cruz, M., Graef, A.M., and Gariglio, P.,** High correlation between molecular alterations of the c-*myc* oncogene and carcinoma of the uterine cervix, *Cancer Res.*, 47, 4173, 1987.

59. **Riou, G., Barrois, M., Le, M.G., George, M., Le Doussal, V., and Haie, C.,** C-*myc* proto-oncogene expression and prognosis in early carcinoma of the uterine cervix, *Lancet*, 1, 761, 1987.

60. **Choo, K., Chong, K., Chou, H., Liew, L., and Liou, C.,** Analysis of the structure and expression of the c-*myc* oncogene in cervical tumors and in cervical tumor-derived cell lines, *Biochem. Biophys. Res. Commun.*, 158, 334, 1989.

61. **Symonds, R.P., Habeshaw, T., Paul, J., Kerr, D.J., Darling, A., Burnett, R.A., Sotsiou, F., Linardopoulos, S., and Spandidos, D.A.,** No correlation between *ras*, c-*myc* and c-*jun* proto-oncogene expression and prognosis in advanced carcinoma of cervix, *Eur. J. Cancer*, 28A, 1615, 1992.

62. **Mitra, A.B., Murty, V.V., Pratap, M., Sodhani, P., and Chaganti, R.S.,** Erb-B2 (HER2/*neu*) oncogene is frequently amplified in squamous cell carcinoma of the uterine cervix, *Cancer Res.*, 54, 637, 1994.

63. **Li, C.C., Chen, E., O'Connell, C.D., and Longo, D.L.,** Detection of c-*mos* proto-oncogene expression in human cells, *Oncogene*, 8, 1685, 1993.

64. **Storga, D., Pecina-Slaus, N., Pavelic, J., Pavelic, Z.P., and Pavelic, K.,** c-fms is present in primary tumours as well as in their metastases in bone marrow, *Int. J. Exp. Pathol.*, 73, 527, 1992.

65. **Terzano, P., Martini, F., Costa, S., and Martinelli, G.N.,** Immunohistochemistry with antibody to the LA-1 oncogene as a prognostic marker in cervical intraepithelial neoplasia, *Gynecol. Oncol.*, 48, 317, 1993.

Part VI.
Other Gynecological Cancers

Chapter 17

Vulval Neoplasia

George E. Smart, Mark J. Arends, and Awatif Al-Nafussi

CONTENTS

1. INTRODUCTION

Cancer of the vulva is the 4th most common malignancy of the female genital tract, accounting for approximately 4% of all gynecological malignancies. Ninety percent of these cancers are squamous with an annual incidence of between 1 and 2 per 100,000 in the Western world.[1] The incidence is at least 10 times higher among women over the age of 75 and it has always been regarded as a malignancy predominantly of older age groups. It is of interest, therefore, that over the past 20 years there has been a marked increase reported in the occurrence of vulval intra-epithelial neoplasia (VIN) in younger women, as there has with cervical intra-epithelial (CIN).[2] Whether this will eventually be followed by a corresponding increase in incidence of invasive cancer remains to be seen, although a recent report does suggest a trend towards younger women presenting with the disease.[3] There is also evidence that the progression rate of VIN 3 to invasive cancer is probably in the region of 10% per year,[4] which is considerably higher than rates of between 2 and 4% previously reported.[5,6]

There is a marked association (up to about 50% according to Jones, 1995)[2] between VIN and neoplasia in other parts of the genital tract, including cervix, vagina, perineum, and anal canal,[7] such "multicentricity" of involvement being seen particularly in immunosuppressed patients (e.g., post-transplant patients or those with AIDS). This highlights the importance of regarding the lower genital tract, vulva, and anal canal as being part of the same "at risk" region within the so-called "anogenital unit." It is both convenient and logical to regard the stratified squamous epithelium of the labia majora, minora, perineum, and lower part of the anorectal canal as constituting a single "anogenital unit." Not only does it share a common embryological origin from the ectoderm of the genital folds and proctodeum, but it appears to be subjected to similar oncogenic influences thereafter and to have a propensity for undergoing similar neoplastic changes.[8,9] Whether the vagina and cervix should also be considered as part of this unit is more contentious. Although there is a well known association between virally related neoplasms of all these structures, and indeed the metaplastic activity of the epithelium of the cervical transformation zone shows many of the characteristics of that of the so-called "anal transitional zone,"[10] nevertheless their embryological ancestries are different and for this reason the term anogenital unit should probably exclude these parts of the internal genitalia (being derived from the endoderm of the urogenital sinus and caudal part of the paramesonephric duct).

2. ETIOLOGICAL FACTORS IN VULVAL NEOPLASMS

The cause of vulval carcinoma is not known. In contrast to what has been stated for many years, recent case-control studies have found no evidence for diabetes mellitus, obesity, vascular diseases, nulliparity, an early menopause, or syphilis being risk factors in the development of vulval carcinoma. The factors that probably do play important roles in the etiology of vulval carcinoma are: vulval intraepithelial neoplasia (VIN), lichen sclerosus (LS), a history of genital human papillomavirus (HPV) infection including warts, number of sexual partners, a history of cervical neoplasia, increasing age, smoking and compromised immunity.[8,11-15] In addition to warts, other sexually transmitted diseases such as gonorrhoea, HSV2[16] and lymphogranuloma venereum have been associated, albeit rarely, with vulval squamous carcinoma.

2.1 Human Papillomavirus and Vulval Intraepithelial Neoplasia

Most studies investigating the association between HPV infection and vulval intraepithelial neoplasia (VIN) have focused on high grade or VIN III disease, in which up to 92% positivity for HPV DNA has been documented.[17] This is mostly accounted for by HPV type 16, which has been found in over 80% of VIN III lesions.[18] HPV 33 has also been found in VIN III, but in less than 10% of patients.

2.2 Human Papillomavirus and Vulval Carcinoma

It has been shown that vulval cancer in young women is often HPV related in contrast to a very much reduced association in older women.[19] On the basis of several studies of HPV DNA prevalence in vulval carcinoma, it has been proposed that there are at least two different etiologies for vulval squamous carcinoma.[20,21] The first etiological group is related to human papillomaviruses of high risk type (usually types 16 or 33) which are found in association with 95% of the "warty" and "basaloid" types of vulval carcinomas, and generally occur in younger women and correspond to their respective warty and basaloid VIN 3 precursors. The second and most common etiological group relates to lichen sclerosus and/or squamous hyperplasia, which are very often found adjacent to the more common "keratinizing" squamous carcinoma seen predominantly in older women, and which is less frequently associated with human papillomavirus (about 40% of cases).[22,23] It is of interest that one study suggests that HPV negative vulval carcinomas are associated with an increased risk of recurrence and death.[24]

3. PREMALIGNANT CONDITIONS

Although the modern simplified classification (Table 1, Reference 25) of disorders of the vulval epithelium suggests the existence of only three potentially premalignant conditions, namely VIN, Paget's disease, and melanoma-*in situ*, there is evidence that other intraepithelial changes can, albeit rarely, predispose to the development of invasive disease. Thus one of the so-called non-neoplastic conditions, lichen sclerosus (LS), can, occasionally become aggressive, especially if inadequately treated, and apparently lead to invasive squamous carcinoma in about 3 to 6% of cases.[17]

Table 1   ISSVD Classification of Epithelial Disorders of Vulval Skin

Non-Neoplastic
Lichen sclerosus
Squamous cell hyperplasia
Other dermatoses
Squamous Intraepithelial Neoplasia
VIN 1
VIN 2
VIN 3
Nonsquamous Intraepithelial Neoplasia
Paget's disease
Melanoma *in situ*

Modified from Wilkinson, E., Kneale, B., and Lynch, P.J., *J. Reprod. Med.*, 31, 973, 1986.

3.1 Vulval Intraepithelial Neoplasia

Vulval Intraepithelial Neoplasia (VIN) is probably becoming increasingly common especially in young women;[2] it is usually asymptomatic, but may cause pruritus and may also involve the peri-anal skin and squamous mucosa of the anal canal.[26] As already outlined in the introduction, VIN has until recently usually been considered to be a relatively benign condition with a progression rate from VIN 3 to invasive cancer in the region of 2 to 4%.[5,6,27] However, a recent New Zealand study suggests that the progression rate may be in excess of 10% per year.[4] Certainly, older women and those with immunosuppression are at an increased risk of developing invasive cancer. One quarter of women with VIN 3 have persistent or recurrent disease after local excision, and one fifth to one third of patients with invasive vulval squamous carcinoma have adjacent foci of VIN 3 in their vulval skin.[28] VIN 3 is often multifocal, appearing as multiple papules around the vulva, and multicentric with high-grade intraepithelial neoplasia found at other anogenital sites such as cervix, vagina, perineum, and anal canal, indicating part of a field change associated with human papillomavirus infection.[18] Those lesions not related to HPV infection are more typically unifocal in distribution. VIN 3 may be subdivided into different types which may mirror and be associated with progression to respective subtypes of vulval carcinoma.[20] One study found that 46% of cases of VIN 3 were of the bowenoid or warty subtype, 17% of the basaloid subtype, 35% mixed and only 2% of the differentiated subtype.[18]

3.2 Paget's Disease of the Vulva

Paget's disease of the vulva is an uncommon neoplasm, and is an intraepidermal adenocarcinoma in the majority of cases, occurring predominantly in postmenopausal white women.[29] The clinical behavior of this neoplasm is extremely variable, reflecting the various histological patterns that have been reported with this lesion. It has been shown that Paget's cells with aneuploid DNA stem lines may be associated with the potential for aggressive biological behavior.[30]

3.3 Malignant Melanoma Precursors

Vulval melanomas account for 2 to 9% of all vulval malignancies,[31] and may arise either in normal skin or in the junctional component of a preexisting melanocytic naevus. A proportion of vulval naevi in premenopausal women may exhibit atypical features.[32] The premalignant potential of these naevi is unclear. Atypical melanocytic proliferation is found in some cutaneous melanocytic lesions elsewhere in the body and these may eventually lead to malignant melanoma in situ, and then be followed by invasive malignant melanoma.[33] Currently however, no data exist regarding atypical melanocytic proliferation in the vulva.

3.4 Lichen Sclerosus

Although classified as a non-neoplastic condition (Table 1), lichen sclerosus nevertheless can, as previously stated, progress in about 3 to 6% of cases to invasive cancer. It can affect any area of the body in both sexes but is most commonly seen in the vulva of postmenopausal women. Factors involved with its origin and pathogenesis are uncertain, but a link with autoimmune disease has been established.[34] Squamous cell carcinoma has also been reported in association with lichen sclerosus in extragenital locations.[35]

Although several studies have failed to demonstrate HPV in LS lesions by Southern blotting, HPV 16 has been detected in 4 out of 18 patients by PCR.[36] Whereas the authors consider the possibility of this HPV infection being superimposed upon the LS and not necessarily etiologically linked, nevertheless, three out of the four women were postmenopausal and it is suggested that such patients may be subsequently at risk from developing vulval cancer. There is evidence of significantly increased p53 immunoreactivity in LS lesions, some of which also show increased staining for PCNA.[37,38]

4. SQUAMOUS CELL CARCINOMA

Vulval carcinoma has been subdivided into four major types: keratinizing, bowenoid, basaloid, and verrucous carcinomas. The classical keratinizing squamous cell carcinoma is found in older patients (the majority being more than 65 years of age and usually white) of whom 83% have adjacent vulval squamous hyperplasia, with or without associated lichen sclerosus, and only 39% show positivity for HPV DNA.[23] The bowenoid or warty carcinoma occurs in younger patients (the majority being less than 60 years of age) and is strongly associated with human papillomavirus.[39,40] It commonly displays an exophytic condylomatous appearance.[41] The basaloid carcinoma also tends to occur in younger patients and again is strongly associated with human papillomavirus.[21] The verrucous carcinoma is a locally destructive, and non-metastasizing form of well-differentiated squamous cell carcinoma which often has an association with HPV infection.[42] On rare occasions tumor recurrences show changes to a high grade squamous carcinoma.[43]

5. GENETIC ALTERATIONS IN VULVAL CARCINOMAS

In a study of p53 mutations and HPV in vulval carcinomas, 57% of 21 vulval carcinomas were found to be positive for HPV DNA out of which only 1 of 12 (8%) showed a p53 mutation, in contrast to 44% which were HPV negative and showed p53 mutations.[44,45] This is broadly in line with much work on the relationship between p53 mutation and human papillomavirus presence in cervical neoplasia, which has suggested in general an inverse correlation between the two, explicable by the fact that high-risk HPV types 16 and 18 express a viral oncogene, E6, capable of binding to p53 and directing its rapid degradation.[46] Hence, p53 function may be altered either by presence of HPV E6 directed p53 degradation or, in the absence of HPV, by mutation of one p53 gene often with allele loss of the other wild type p53 gene. Abnormal p53, determined by immunohistochemical studies of p53 protein accumulation, has been found in a small proportion of patients with lichen sclerosus and this finding is often associated with increased evidence of proliferation such as an increased mitotic index or increased staining with PCNA (proliferation cell nuclear antigen, a marker only expressed during the cell cycle).[37,38]

Although very few studies have been performed on other genetic changes in vulval carcinomas, one study has shown increased expression of the oncogene c-myc in some cases. However, no evidence of amplification of the c-myc gene was found.[47] Another study has shown increased expression of PRAD-1 (cyclin D1, a cell cycle regulator) in 3 vulval carcinoma cell lines, and this has been confirmed by amplification of the gene.[48] PRAD-1 is known to complex to the retinoblastoma protein Rb and may represent one possible mechanism of deregulation of cell cycle control. Other studies have shown no evidence for significant numbers of ras mutations in vulval carcinomas.[49]

Although presumably other genetic changes occur during neoplastic progression from precursor lesions to vulval cancer, only limited studies of these changes have so far been reported.

6. IMMUNOSUPPRESSION AND ANOGENITAL NEOPLASIA

The interesting association between immunosuppression and epithelial cancer and precancer of the genital tract and anogenital unit, in both males and females, is well documented,[9,50-54] and attests to the effect of deficiencies in cell-mediated immunity on the development of neoplasia known to be causally associated with oncogenic viruses. Thus in transplant recipients, the incidence of cancers arising *de novo* is reported as being as high as 18%, although this is thought to be an underestimate.[50] Cancers of the anogenital region (including vulva, perineum, scrotum, penis, perianal skin, and anus) occur approximately 100 times as frequently in immunosuppresssed patients as in controls, with a female to male ratio of 2.5:1 compared with a ratio of 1:2 in other post-transplant cancers.[50] It seems that in transplant patients, the depression of cell-mediated immunity by immunosuppressive drugs (mainly azothioprine, cyclosporine, and steroids) can lead to defects in immunosurveillance, that includes the normal processing of oncogenic viral antigens by the Class II MHC expressing epithelial dendritic (Langerhans) cells, antigen presentation to cytotoxic T lymphocyte (CTL) cells and subsequent anti-viral CTL-mediated target cell killing.

Malignancies that are predominantly increased in frequency in immunocompromised individuals are those known to be related to oncogenic viruses, namely non-Hodgkin's lymphoma (EBV related), Kaposi's sarcoma (KSV related) and epithelial cancers particularly in the genital and anogenital regions (HPV related). In HIV positive individuals, epithelial neoplasms of the lower genital tract, perineum, and anus are also markedly increased.[40,50,52] There is evidence that in HIV positive women anal HPV infection and associated epithelial abnormalities (e.g., warts, VIN, AIN) are at least as common as cervical HPV infection.[55] There seems to be a not unexpected association between the duration as well as degree of immunosuppression and the propensity for epithelial abnormalities. The higher prevalence of anogenital warts, for example, in HIV positive homosexual men appeared to be related to their significantly lower CD4 lymphocyte counts and their increased duration of immunosuppression, when compared with their HIV positive heterosexual counterparts.[56] Moreover, in renal transplant recipients, the frequency of skin cancers and warts (not only of the vulval area) increased from approximately 20% in the first 5 years after transplantation to 70% after 5 years.[57]

There remain a number of unresolved questions relating to the reasons for the development of cancers and precancers of the genital tract in transplant recipients and HIV+ individuals, and why the natural history of these neoplasms is so foreshortened with their propensity for persistence and recurrence after treatment together with relatively rapid progression from the premalignant to the malignant states.[9,52] Certainly, there is impaired immune surveillance as mentioned above, in which presumably neoplastic cells or the viral antigens which are partly responsible for their transformation, are not efficiently recognized, but to what extent a reduction in local, as against general immune competence contributes towards malignant transformation is not understood. It is not uncommonly experienced that withdrawal of immunosuppression after graft rejection or removal can result in marked amelioration or even disappearance of the precancer or HPV associated lesions.[58] Whether in HIV positive women there is an interaction between HIV and HPV or other cofactors to cause pre-invasive disease is uncertain, but the poor survival rate of these patients presumably explains why the incidence of invasive genital tract cancers is not as proportionately high as their preinvasive counterparts.

It may be significant that seminal fluid exhibits some of the most powerful immunosuppressive properties known to exist in any biological system.[59,60] This is by virtue of its high concentration of prostaglandins of the E series (PGE and 19-OH PGE) and of its particulate prostasomal component. Among numerous other immunosuppressive actions, PGEs are powerful suppressors of lymphocyte proliferation and activation, and also of NK cell activity, together with causing a marked decrease in cytokine production and phagocytic properties of macrophages.[61,62] Moreover, prostasomes are a source of powerful complement inhibitors.[62] It has also been demonstrated that seminal plasma has a powerful suppressor effect on antitumor cytotoxicity,[63] and that PGEs can stimulate herpes replication[64,65] and HIV replication in permissive cells.[66]

It has been demonstrated in animal models that semen deposition per rectum results in prostaglandin uptake into the peripheral circulation which in turn can give rise to quite marked depression of peripheral immune response.[67-69] Moreover, anal intercourse in man has been shown to be followed by an increase in circulating CD8 lymphocytes, probably attributable to PGE absorption. The multifactorial effects therefore of HPV infection and smoking against a background of relative immunosuppression (general and/or local) may well be the main factors related to the etiology of some cases of anogenital neoplasia, and may help to explain part of the increase in pre-invasive and possibly invasive disease in younger women.

REFERENCES

1. **Beilby, J.O.W. and Ridley, C.M.,** Pathology of the vulva, in *Obstetrical and Gynaecological Pathology*, Fox, H., Ed, Churchill Livingstone, 1987, pp 64.
2. **Jones, R.W.,** The natural history of vulvar intraepithelial neoplasia, *Br. J. Obstet. Gynaecol.*, 102, 764, 1995.
3. **Messing, M.J. and Gallup, D.G.,** Carcinoma of the vulva in young women, *Obstet. Gynec.*, 86, 51, 1995.
4. **Jones, R.W. and Rowan, D.M.,** Vulvar intraepithelial neoplasia III: a clinical study of the outcome in 113 cases with relation to the late development of invasive vulvar carcinoma, *Obstet. Gynecol.*, 84, 741, 1994.
5. **Woodruff, J.D.,** Carcinoma in situ of the vulva, *Clin. Obstet. Gynecol.*, 34, 669, 1991.
6. **Singer, A. and Monaghan, J.M.,** Vulvar intraepithelial neoplasia in lower genital tract precancer, Singer, A. and Monaghan, J.M., Eds., Blackwell Science, Boston, 177, 1994.
7. **Scholefield, J.H., Hickson, W.G.E., Smith, J.H.F., Rogers, K., and Sharp, F.,** Anal intraepithelial neoplasia: Part of a multifocal disease process, *Lancet*, 340, 1271, 1992.
8. **Penn, I.,** Cancers of the anogenital region in renal transplant recipients. *Cancer*, 58, 611, 1986.
9. **Sillman, F.H. and Sedlis, A.,** Anogenital papillomavirus infection and neoplasia in immunodeficient women: an update, *Dermatol. Clin.*, 9, 353, 1991.
10. **Fenger, C.,** The anal transitional zone, *Acta Pathol. Immunol. Scand. Sect. A*, 95 Suppl. 289, 10, 1987.
11. **Ansink, A.C. and Heint, P.M.,** Epidemiology and etiology of squamous cell carcinoma of the vulva, *Eur. J. Obstet. Gynecol. Reprod. Biol.*, 48, 111, 1993.
12. **Caterson, R.J., Furber, J., Murray, J., McCarthy, W., Mahony, J.F., and Sheil, A.G.R.,** Carcinoma of the vulva in two young renal allograft recipients, *Transplant Proc.*, 16, 559, 1984.
13. **Daling, J.R., Sherman, K.J., Hislop, T.G., Maden, O., Mandelson, M.T., Beckmann, A.J., and Weiss, N.S.,** Cigarette smoking and the risk of anogenital cancer, *Am. J. Epidemiol.*, 135, 180, 1992.
14. **Brinton, L.A., Nascap, C., Mallin, K., Baptiste, M.S., Wilbanks, G.D., and Richart, R.M.,** Case-control study of cancer of the vulva, *Obstet. Gynecol.*, 75, 859, 1990.
15. **Mabuchi, K., Bross, D.S., and Kessler, I.,** Epidemiology of cancer of the vulva. A case-control study, *Cancer*, 55,1843, 1985.

16. **Sherman, K.J., Daling, J.R., Chu, J., Weiss, N.S., Ashley, R.L., and Corey, L.,** Genital warts, other sexually transmitted diseases and vulvar cancer, *Epidemiol.*, 2, 257, 1991.

17. **Hofmann, U.,** Squamous cell carcinoma and lichen sclerosus et atrophicus of the vulva. *H+G Zeitsch. Hautkrankheiten*, 68, 409, 1993.

18. **Van Beurden, M., Ten Kate, F.J.W., Smits, H.L., Berkhout, R.J.M., De Craen, A.J.M., Van der Vange, N., Lammes, F.B. and Ter Schegget, J.,** Multifocal vulvar intraepithelial neoplasia grade III and multicentric lower genital tract neoplasia is associated with transcriptionally active human papillomavirus, *Cancer*, 75, 2879, 1995.

19. **Toki, T., Kurman, R.J., Park, J.S., Kessis, T., Daniel, R.W., and Shah, K.V.,** Probable nonpapillomavirus etiology of squamous cell carcinoma of the vulva in older women: a clinicopathologic study using in situ hybridization and polymerase chain reaction, *Int. J. Gynecol. Pathol.*, 10, 107, 1991.

20. **Kurman, R.J., Toki, T., and Schiffman, M.H.,** Basaloid and warty carcinomas of the vulva. Distinctive types of squamous cell carcinoma frequently associated with human papillomaviruses, *Am. J. Surg. Pathol.*, 17, 133, 1993.

21. **Hording, U., Junge, J., Daugaard, S., Lundvall, F., Poulsen, H., and Bock, J.E.,** Vulvar squamous cell carcinoma and papillomaviruses: Indications for two different etiologies, *Gynecol. Oncol.*, 52, 241, 1994.

22. **Bloss, J.D., Liao, S.Y., Wilczynski, S.P., Macri, C., Walker, J., Peake, M., and Berman, M.L.,** Clinical and histologic features of vulvar carcinomas analyzed for human papillomavirus status: Evidence that squamous cell carcinoma of the vulva has more than one etiology, *Hum. Pathol.*, 22, 711, 1991.

23. **Andersen, W.A., Franquemont, D.W., Williams, J., Taylor, P.T., and Crum, C.P.,** Vulvar squamous cell carcinoma and papillomaviruses: Two separate entities?, *Am. J. Obstet. Gynecol.*, 165, 329, 1991.

24. **Monk, B.J., Burger, R.A., Lin, F., Parham, G., Vasilev, S.A., and Wilczynski, S.P.,** Prognostic significance of human papillomavirus DNA in vulvar carcinoma, *Obstet. Gynecol.*, 85, 709, 1995.

25. **Wilkinson, E., Kneale, B., and Lynch, P.J.,** International Society for the study of vulvar disease: Report of the ISSVD Terminology Committee, *J. Reprod. Med.*, 31, 973, 1986.

26. **Andreasson, B. and Bock, J.E.,** Intraepithelial neoplasia in the vulva region, *Gynecol. Oncol.*, 21, 300, 1985.

27. **Narayan, H., Cullimore, J., Brown, J., and Byrne, P.,** Vulvar intraepithelial neoplasia, *Contemp. Dev. Obstet. Gynaecol.*, 5, 43, 1993.

28. **Haefner, H.K., Tate, J.E., McLachlin, C.M., and Crum, C.P.,** Vulvar intraepithelial neoplasia: Age, morphological phenotype, papillomavirus DNA, and coexisting invasive carcinoma, *Hum. Pathol.*, 26, 147, 1995.

29. **Geisler, J.P., Stowell, M.J., Melton, M.E., Maloney, C.D., and Geisler, H.E.,** Extramammary Paget's disease of the vulva recurring in a skin graft, *Gynecol. Oncol.*, 56, 446, 1995.

30. **Michael, H., Roth, L.M., and Sutton, G.P.,** Flow cytometric DNA analysis of extramammary Paget's disease of the vulva, *Int. J. Gynecol. Pathol.*, 14, 324, 1995.

31. **Bradgate, M.G., Rollason, T.P., McConkey, C.C., and Powell, J.,** Malignant melanoma of the vulva: a clinicopathological study of 50 women, *Br. J. Obstet. Gynaecol.*, 97, 124, 1990.

32. **Friedman, R.J. and Ackerman, A.B.,** Difficulties in the histological diagnosis of melanocytic naevi on the vulva of premenopausal women, in *Pathology of Malignant Melanoma*, Ackerman, A.B., Ed., Masson, New York, p 119, 1981.

33. **Clark, W.H., Elder, D.R., and Guerry, D.,** A study of tumor progression: the precursor lesions of superficial spreading and nodular melanoma, *Hum. Pathol.*, 15, 1147, 1984.

34. **Ridley, C.M.,** Lichen sclerosus, *Dermatol. Clinics*, 10, 309, 1992.

35. **Dores, J.A., De Almeida Goncalves, J.C., and Carmo Carvalho, M.,** Squamous-cell carcinoma in extra-genital lesion of lichen sclerosus et atrophicus in a man. A case report, *Skin Cancer*, 9, 33, 1994.

36. **Kiene, P., Milde-Langosch, K., and Loning, T.,** *Arch. Dermatol. Res.*, 283, 445, 1991.

37. **Soini, Y., Paakko, P., Vahakangas, K., Vuopala, S., and Lehto, V.P.,** Expression of p53 and proliferating cell nuclear antigen in lichen sclerosus et atrophicus with different histological features, *Int. J. Gynecol. Pathol.*, 13, 199, 1994.

38. **Tan, S.H., Derrick, E., Mckee, P.H., Hobbs, C., Ridley, M., and Neill, S.,** Altered p53 expression and epidermal cell proliferation is seen in vulval lichen sclerosus, *J. Cutan. Pathol.*, 21, 316, 1994.

39. **Hording, U., Junge, J., Poulsen, H., and Lundvall, F.,** Vulvar intraepithelial neoplasia III: A viral disease of undetermined progressive potential, *Gynecol. Oncol.*, 56, 276, 1995.

40. **Petry, Ku., Kochel, H., Bode, U., Schedel. I., Niesert, S., Glaubitz, M., Maschek, H., and Kuhnle, H.,** Human papillomavirus is associated with the frequent detection of warty and basaloid high-grade neoplasia of the vulva and cervical neoplasia among immunocompromised women, *Gynecol. Oncol.*, 60, 30, 1996.

41. **Zemtsov, A., Koss, W., Dixon, L., Tyring, S., and Rady, P.,** Anal verrucous carcinoma associated with human papillomavirus type 11: magnetic resonance imaging and flow cytometry evaluation, *Arch. Dermatol.*, 128, 564, 1992.

42. **NobleTopham, S.E., Fliss, D.M., Hartwick, R.W.J., McLachlin, C.M., Freeman, J.L., Noyek, A.M., and Andrulis, I.L.,** Detection and typing of human papillomavirus in verrucous carcinoma of the oral cavity using the polymerase chain reaction, *Arch. Otolaryngol. Head Neck Surg.*, 119, 1299, 1993.

43. **Levitan, Z., Kaplan, A.L., and Kaufman, R.H.,** Advanced squamous cell carcinoma of the vulva after treatment for verrucous carcinoma: A case report, *J. Reprod. Med. Obstet. Gynecol.*, 37, 889, 1992.

44. **Pilotti, S., Donghi, R.D., Amato, L., Giarola, M., Longoni, A., Della Torre, G., De Palo, G., Pierotti, M.A., and Rilke, F.,** Papillomavirus, p53 alteration and primary carcinoma of the vulva, *Eur. J. Cancer*, 29, 924, 1993.

45. **Lee, Y.Y., Wilczynski, S.P., Chumakov, A., Chih, D., and Koeffler, H.P.,** Carcinoma of the vulva: HPV and p53 mutations, *Oncogene*, 9, 1655, 1994.
46. **Scheffner, M., Werness, B.A., Huibregtse, J.M., Levine, J.M., and Howley, P.M.,** The E6 oncoprotein encoded by human papillomavirus types 16 and 18 promotes the degradation of p53, *Cell*, 63, 1129, 1990.
47. **MildeLangosch, K., Becker, G., and Loning, T.,** Human papillomavirus and c-myc/c-erbB2 in uterine and vulvar lesions, *Virchows Archiv — A Pathological Anatomy and Histopathology*, 419, 479, 1991.
48. **Kurzrock, R., Ku, S., and Talpaz, M.,** Abnormalities in the PRAD1 (CYCLIN D1/BCL-1) oncogene are frequent in cervical and vulvar squamous cell carcinoma cell lines, *Cancer*, 75, 584, 1995.
49. **Tate, J.E., Mutter, G.L., Prasad, C.J., Berkowitz, R., Goodman, H., and Crum, C.P.,** Analysis of HPV-positive and -negative vulvar carcinomas for alterations in c-myc, Ha-, Ki-, and N-ras genes, *Gynecol. Oncol.*, 53, 78, 1994.
50. **Penn, I.,** Depressed immunity and the development of cancer, *Cancer Detection Prevention*, 18, 241, 1994.
51. **Alloub, M.I., Barr, B.B.B., McLaren, K.M., Smith, I.W., Bunney, M.H., and Smart, G.E.,** Human papillomavirus infection and cervical intraepithelial neoplasia in women with renal allografts, *Br. Med. J.*, 298, 153, 1989.
52. **Johnstone, F.D., McGoogan, E., Smart, G.E., Brettle, R.P., and Prescott, R.J.,** A population-based controlled study of the relation between HIV infection and cervical neoplasia, *Br. J. Obst. Gyn.*, 101, 986, 1994.
53. **Ogunbiyi, O.A., Scholefield, J.H., Raftery, A.T., Smith, J.H.F., Duffy, S., Sharp, F., and Rogers, K.,** Prevalence in anal human papillomavirus infection and intraepithelial neoplasia in renal allograft recipients, *Br. J. Surg.*, 81, X, 1994.
54. **Laga, M., Icenogle, J.P., Marsella, R., Manoka, A.T., Nzila, N., Ryder, R.W., Vermund, S.H., Heyward, W.L., Nelson, A., and Reeves, W.C.,** Genital papillomavirus infection and cervical dysplasia: opportunistic complications of HIV infection, *Int. J. Cancer*, 50, 45, 1992.
55. **Williams, A.B., Darragh, T.M., Vranizan, K., Ochia, C., Moss, A.R., and Palefshy, J.M.,** Anal and cervical human papillomavirus infection and risk of anal and cervical epithelial abnormalities in human immunodeficiency virus-infected women, *Obstet. Gynecol.*, 83, 205, 1994.
56. **McMillan, A. and Bishop, P.E.,** Clinical course of anogenital warts in men infected with human immunodeficiency virus, *Genitourin. Med.*, 65, 225, 1989.
57. **Barr, B.B.B., Benton, E.C., McLaren, K., Binney, M.H., Smith, I.W., Blessing, K., and Hunter, J.A.A.,** Human papillomavirus infection and skin cancer in renal allograft recipients, *Lancet*, 1, 124, 1989.
58. **Benton, C., Shahidullah, H., and Hunter, J.A.A.,** Human papillomavirus in the immunosuppressed, *Papilloma Virus Report*, 3, 23, 1992.
59. **Stites, D.P. and Erickson, R.P.,** Suppressive effect of seminal plasma on lymphocyte activation, *Nature*, 253, 727, 1975.
60. **Shivazi, S., Sheit, K.H., and Bhargava, D.M.,** *Proteins of Seminal Plasma*, John Wiley, New York, 1990.
61. **Ohmari, Y., Strassman, G., and Hamilton, T.A.,** cAMP differentially regulates expression of mRNA encoding 1L - 1a and lL - b in murine peritoneal macrophages, *J. Immunol.*, 145, 3333, 1990.
62. **Kelly, R.W.,** Immunosuppression mechanisms in semen: implications for contraception, *Hum. Reprod.*, 10, 1686, 1995.
63. **Rees, R.C., Vallely, P.J., Clegg, A., and Potter, C.W.,** Suppression of natural and activators antitumor cytotoxicity by human seminal plasma, *Clin. Exp. Immunol.*, 63, 687, 1986.
64. **Hill, T.J. and Blyth, W.A.,** An alternative theory of herpes simplex recurrence and possible role of prostaglandins, *Lancet*, 1, 397, 1976.
65. **Baker, D.A., Thomas, J., Epstein, J., Possilico, D., and Stone, M.L.,** The effect of prostaglandins in the multiplication and cell-to-cell spread of herpes simplex virus type 2 in vitro, *Am. J. Obstet. Gynecol.*, 144, 346, 1982.
66. **Kuno, S., Veno, R., Hayaishi, O., Nakashima, M., Harada, S., and Yamamuto, N.,** Prostaglandin E2, a seminal constituent facilitates the replication of acquired immune deficiency syndrome virus in vitro, *Prod. Natl. Acad. Sci. U.S.A.*, 83, 3487, 1986.
67. **Alexander, N.J., Tarter, T.H., Fulghamn, D.L., Ducsay, C.A., and Novy, M.J.,** Rectal infusion of semen results in transient elevation of blood prostaglandins, *Am. J. Reprod. Immunol.*, 15, 47, 1987.
68. **Richards, J.M., Bedford, J.M., and Witkin, S.S.,** Rectal insemination modifies immune response in rabbits, *Science*, 224, 390, 1984.
69. **Veselsky, L., Dostal, J., and Drahorad, J.,** Effect of intra-rectal administration of boar seminal immunosuppressive fraction on mouse lymphocytes, *J. Reprod. Fertil.*, 101, 519, 1994.

Chapter 18

Gestational Trophoblastic Disease

George L. Mutter

CONTENTS

1. INTRODUCTION

Gestational trophoblastic diseases are a heterogenous pool of clinically and histopathologically defined entities with two clinically relevant features: reproductive failure and a high neoplastic potential. Here are reviewed recent advances in understanding of the biology and natural history of the most common forms of trophoblastic disease with an emphasis on clinical implications. Of particular interest is the behavior of complete hydatidiform moles, a potentially aggressive lesion which has been associated with an elevated risk for development of malignant choriocarcinoma. There are no reliable genetic markers for predicting which subset of complete moles will behave aggressively, nor are there molecular diagnostic methods at present that offer advances over traditional histopathology. The predominant genetic finding in complete and partial hydatidiform moles is an imbalance of parental chromosomes, completely androgenetic (paternal) in the former and an extra paternal haploid set in the latter. This lack or imbalance of a maternal genomic contribution probably changes the gene expression since there is no evidence of gene mutation in these lesions.

1.1 Historical and Clinical Background

The neoplastic potential of trophoblastic diseases has been recognized for millennia. The Roman naturalist Pliny the Elder in the first century A.D. described moles as lacking a fetus, having swollen placental villi and symptoms referable to high human chorionic gonadotropin (HCG) levels, such as loss of menses. With the advent of diagnostic light microscopy in the late 19th century, villous expansion was confirmed to be caused by an accumulation of fluid in large mesenchymal cisternae. While several early pathogenic theories of molar disease focused on changes in the vascular system that might explain these accumulations, it became apparent that there are quantitative and qualitative abnormalities of the trophoblast

0-8493-9443-0/97/$0.00+$.50
© 1997 by CRC Press LLC

itself, the very component that defines choriocarcinoma. Abnormal trophoblast is the pathognomonic feature shared by all trophoblastic diseases, and the biology of this tissue defines their neoplastic potential. Invasive qualities of normal placenta are now recognized to be conferred by elaboration of proteolytic enzymes (gelatinase B and urokinase-type plasminogen activator), and switch in cell surface adhesion molecules (integrins), permitting migration into the maternal compartment.[1] While these insights provide exciting possibilities for manipulation of the invasive qualities of trophoblast, their relevance to pathological trophoblast is unclear. In hydatidiform moles there are infrequent single gene mutations of tumor suppressor genes or proto-oncogenes such as is often seen in spontaneous non-gestational tumors, but rather gross errors of fertilization and/or zygotic division which result in an excess of paternal chromosomes. Thus insights into the basis of parent-specific differences in DNA, genomic imprinting, are of relevance to this story. HCG, a compound elaborated by trophoblast, is a convenient serologic marker for surveillance and management of individual patients. There are limitations, however, to this seemingly straightforward practice. First, HCG is a useful marker only as long as the patient refrains from getting pregnant again, necessitating birth control during the surveillance interval — often a full year. This may be a substantial hardship for the woman in her late thirties or early forties who has limited time to achieve a desired pregnancy. Second, the schedule of surveillance may affect the apparent progression/persistence rate, most notably in the case of PHM.[2] Third, not all trophoblastic lesions elaborate equivalent amounts of HCG, and PSTT in particular may persist with low or undetectable levels of HCG.[3] For these reasons, accuracy and thoroughness in initial diagnosis are essential prior to committing a patient to HCG follow up which generally demands reflexive intervention if abnormal.

2. COMPLETE HYDATIDIFORM MOLES (CHM)

2.1 General

Complete hydatidiform moles consist solely of extra-embryonic tissues characterized by atypical hyperplastic trophoblast and cavitated hydropic villi.[4,5] Histopathologic criteria for diagnosis of CHM are straightforward (Table 1), with the majority of diagnostic dilemmas posed by incomplete sampling or poor tissue preservation.

2.2 Origin of Complete Moles

Analysis of polymorphic chromosomal markers, such as chromosomal banding heteromorphisms[4] and hypervariable DNA sequences,[6] has shown that CHMs are composed solely of paternal DNA (Figure 1), either by fertilization of an egg devoid of a maternal pronucleus, or by loss of the maternal chromosomes in the first cleavage division.[4,5] That fertilization has occurred in molar gestations is confirmed by the presence of maternally derived cytoplasmic components such as mitochondrial DNA.[7,8] Approximately 84% are products of monospermy,[6,9,10] having a 46,XX genotype (46,YY moles have not been described, probably because the X chromosome is required for cell survival). The remaining cases are dispermic, with either a 46,XX or 46,XY genotype. Features of the molar phenotype may be present when only a part of the genome is paternally uniparental. Paternal uniparental disomy in the region of chromosome 11, which contains the imprinted genes H19 and IFG2,[11,12] produces placentomegaly with hydropic villi, including cistern formation, mimicking hydatidiform mole.[13] These changes are only part of a larger constellation of features in these patients and are known as the Beckwith-Weidemann syndrome.[14] There are reports of live-born, diploid, singleton, molar gestations delivered at term, which probably are related to this syndrome.[15] Other genes or genetic regions may prove to be important in trophoblast/placental development that, if mutated or inherited in an isodisomic manner, may mimic molar disease.

2.3 Repetitive Molar Disease

The chances of having another molar gestation following an initial molar pregnancy is about 1%, whereas this risk skyrockets to 25% after two episodes of trophoblastic disease.[16] There are data to suggest that some patients maintain an inherently high risk of recurrence despite complete ablation of earlier lesions. Comparison of hypervariable genetic markers between occurrences has shown repeat molar pregnancies were from different fertilization events[17] (one patient with repeat CHM, and two with repeat PHM). One can only speculate whether this reflects an underlying defect of gametogenesis or fertilization.

Table 1 Features of Trophoblastic Diseases

	CHM	PHM	Choriocarcinoma	PSTT
Histopathology				
Embryo	None	Sometimes present	None	None
Placental villi	Extensive cisternae	Patchy edema	None	None
Tropohoblast type	All present	All present	Mononuclear and syncytial	Intermediate
Trophoblast amount	Globally increased	Mild focal hyperplasia	Abundant	Abundant
Trophoblast atypia	Present, moderate	Mild, if present	Severe	Moderate-severe
Clinical				
Incidence, per 1000 pregnancies[a]	1.3[40]	2.9[40]	0.02–0.35[31]	Unknown (rare)
Post-curettage persistent HCG[b]	17% (13/78)[18]	0[2]–5.5[41]%	—	HCG insensitive as progression marker
Extrauterine progression	8.9% (7/78) metastatic[18]	Very rare case reports[42-45]	Frequent	23%(12/52)[52] lethal
Genetics				
Ploidy	Diploid	Triploid	Diploid or tetraploid[46,47]	Diploid[50]
Genetic parentage	Paternal[4,5]	Biparental[34,40]	Biparental or paternal[48]	Biparental or paternal[51]

[a] Number per 1,000 naturally occurring pregnancies (live births and spontaneous abortions).

[b] Percentage of patients presenting with histologically diagnosed molar disease who, after evacuation, have persistent or progressive HCG levels. Includes cases with and without evidence of extrauterine progression. The two figures for PHM used different criteria for HCG surveillance.

Modified and expanded after Roberts, D. and Mutter, G. L., *J. Reprod. Med.*, 39, 201, 1994.

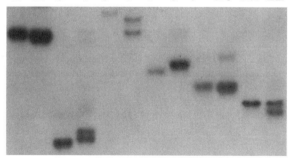

Figure 1 Paternal origin of complete hydatidiform mole. Comparison of complete hydatiform mole lesional (odd numbered lanes) and maternal (even numbered lanes) polymorphic microsatellite genetic markers. PCR primers flanking six unlinked hypervariable loci (IGF1, lanes 1–2; TRCD, lanes 3-4; ADA, lanes 5–6; SST, lanes 7-8; APOC2, lanes 9–10; and APOA2, lanes 11–12)[26,67] were used to amplify variable length alleles, and the radiolabeled products resolved by polyacrylamide gel electrophoresis. The maternal tissues were demonstrably heterozygous, as expected with these markers that have polymorphism information contents of 0.46–0.79. In contrast, only a single allele was seen in the molar tissue with all six markers, indicating a homozygous genotype (probability of heterozygous genotype with all markers identical is 0.0023), or derivation from a haploid chromosomal complement. The haploid-derived molar genome had several unique alleles not present in the host maternal tissues, confirming its paternal derivation. Flow cytometric analysis showed a 2n (diploid) DNA content in the molar cells (data not shown). These findings allow characterization of this particular complete hydatidiform mole as the product of monospermy, with exclusion or loss of the maternal genome, and subsequent endoreduplication. 18 hours exposure autoradiogram. (From Mutter, G. L., Stewart, C. L., Chaponot, M. L., and Pomponio, R. J., *Am. J. Hum. Genet.*, 53, 1096, 1993. With permission.)

2.4 Progression Risk in CHM

CHM have a high risk for progression to more aggressive forms of trophoblastic disease, with approximately 9% of patients developing extrauterine spread of choriocarcinoma after uterine evacuation.[18] This represents a several thousand-fold increase in risk of choriocarcinoma following CHM compared to a normal pregnancy. Management of patients with CHM includes serologic monitoring for persistently elevated HCG levels which, if present, initiate a metastatic workup to be followed by effective chemotherapy. An outstanding question is whether there is any feature of the mole itself which will predict which fraction of cases will persist or progress to metastasis. Histologic grading has been claimed to have such predictive value,[19] but this has not been confirmed in several studies.[20,21]

2.5 Zygosity of CHM

In some series, complete moles with a heterozygous genotype have a higher risk of progression than homozygous moles. One study[22] reported a persistence rate of 50% following heterozygous complete moles, contrasting with 4% for homozygous moles. This suggests a 12- to 14-fold increased risk associated with heterozygosity. What follows is a summary of recent studies which have failed to confirm an increased neoplastic potential for heterozygous CHM, whether 46,XY or 46,XX.

Using the polymerase chain reaction to objectively identify XY CHM heterozygotes (Figure 2),[18] there is no evidence that heterozygous Y chromosome-containing moles have a higher chance of metastasizing than do XX moles (Table 2). This conclusion had been subject to some uncertainty in older studies which used insensitive and nonspecific indicators of sex chromosome genotype such as quinacrine fluorescence of the Y body.[23,24] Studies using Southern blot ascertained restriction fragment length polymorphisms have not shown heterozygous 46,XX moles to have a higher risk for progression than 46,XX homozygous CHM,[6] although it must be noted that there are some technical and interpretive barriers in distinction between heterozygous and homozygous 46,XX CHM. Required autosomal marker systems[6,25] are available in PCR format,[26] theoretically capable of tapping into a large pool of archival material needed to achieve statistical significance in a series of cases. Unfortunately, application to paraffin embedded material is difficult because: (1) accurate interpretation of maternal contamination requires marker studies on normal maternal and paternal tissues for comparison[27]; (2) the DNA quantity and quality from paraffin sections requires high cycle numbers to detect a strong signal — a process that with repetitive sequences may induce artifactual bands subject to misinterpretation[28]; and (3) hypervariable repeat markers are frequently subject to rearrangement during gametogenic recombination events,[29,30] generating novel alleles absent in both parents.[27] For these reasons, care must be exercised in interpretation of PCR-based microsatellite marker data.

3. PARTIAL HYDATIDIFORM MOLES (PHM)

Partial hydatidiform moles are abnormal gestations which, unfortunately, remain shrouded in controversy. Their diagnosis, neoplastic potential, and treatment are all without consensus.

3.1 Diagnosis

Histopathologic criteria for diagnosis of PHM include a patchy distribution of mildly hyperplastic trophoblast with scalloping, and villous stromal edema.[31] Diagnostic molar villi may coexist with abundant or even predominant populations of normally configured villi, increasing the possibility of underdiagnosis due to incomplete sampling, or overdiagnosis due to intensive scrutiny of a minor population of villi.

DNA ploidy has been suggested as a parameter useful in differentiating PHMs from other gestations. With rare exceptions, cytogenetic analysis of molar tissue has shown that CHMs are diploid (46 chromosomes) and PHMs are triploid (69 chromosomes).[32-34] While differential diagnosis between CHM and PHM is frequently discussed as a significant interpretive problem, this is rarely the case. A more frequent dilemma faced by practicing pathologists is distinction between PHM and a (normal or miscarried) conceptus with superimposed degenerative changes. Ploidy information may be of some use where the histology is strongly suggestive of a PHM, as a triploid result may provide additional evidence for the diagnosis, although it should be remembered that not all PHM have this genotype.[15,35,36] In the absence of diagnostic histopathologic features of a PHM, the benefit of ploidy analysis diminishes, as triploid genotypes are frequently encountered in non-molar gestations.[37-39]

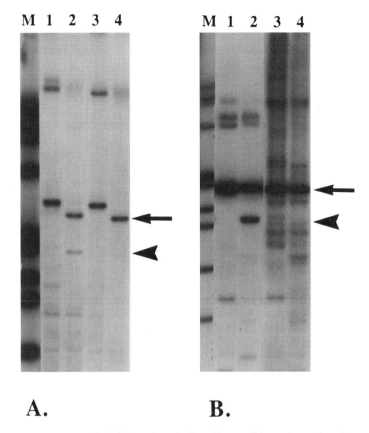

Figure 2 Sex chromosome specific PCR products in Y-positive and Y-negative moles. Autoradiogram of radiolabeled PCR products from Y-chromosome positive (lanes 1 and 2) and Y-chromosome negative complete hydatidiform moles (Lanes 3 and 4). Primer pairs zf-a/c (Panel A) and st-a/b (Panel B) co-amplify homologous X and Y chromosome alleles for the ZFX/ZFY and STS genes, respectively, in a region of high nucleotide homology, which upon digestion with the restriction endonuclease HaeIII, generates X (arrow) and Y (arrowhead) chromosome specific restriction fragments. Uncut PCR products are in lanes 1, 3, HaeIII digested PCR products in lanes 2,4. DNA isolation from paraffin blocks and PCR amplification as described in Reference 18. M: molecular weight marker, phi-X DNA digested with HaeIII. Exposure time 16 hours for both panels. (From Mutter, G. L., Pomponio, R., Berkowitz, R., and Genest, D., *Am. J. Obstet. Gynecol.*, 168, 1547, 1993. With permission.)

Table 2 Sex Chromosome Composition of Complete Hydatidiform Moles in 22 Cases without Metastasis, and 13 Cases with Metastasis

| | | Metastasis | | |
		No	Yes	Totals
	Negative	20	12	32
Y-Chromosome #	Positive	2	1	3
	% Positive	9.1%	7.7%	8.6%

Note: # Sex chromosome composition determined by PCR using primers zf-a/c and st-a/b as illustrated in Figure 2.

From Mutter, G. L., Pomponio, R., Berkowitz, R., and Genest, D., *Am. J. Obstet. Gynecol.*, 168, 1547, 1993. With permission.

3.2 Origin

Genetic studies have provided valuable insights into the mechanisms of origin of PHM. Most triploid conceptuses have a biparental genome, and the parental origin of the extra haploid set cosegregates well with phenotypes: a PHM if paternal,[34,40] and a more developed fetus without a molar placenta if maternal.[37,38] The extra paternal genomic dose in a PHM might be introduced either by an abnormal fertilization event involving two sperm (dispermy), or normal fertilization by an abnormally diploid sperm.

3.3 Natural History

Disease persistence or progression following uterine evacuation of a PHM may be defined either by serologic monitoring of HCG, or anatomic demonstration of physical spread of lesional tissue, lending themselves to very different conclusions. HCG defined persistence rates vary as a function of the monitoring protocol used, with a high of 5.5% (17/310)[41] based on weekly determinations beginning immediately after evacuation that show either elevation of HCG between any two determinations, or plateau for 3 successive weeks, to a low of 0% (0/51)[2] based on every other week HCG determinations that show increasing HCG levels in the post-evacuation period or persistent HCG 4 to 6 months after evacuation. This suggests that many of the persistent or increased HCG levels seen in the recent postcurettage interval are transient, with ability to recognize them a function of the sampling frequency.[2] There is no compelling histopathologic or anatomic evidence that the risk for developing choriocarcinoma following PHM differs from that of a normal pregnancy, as supported by the extreme rarity of reports (only five cases in the literature[42-45]) of choriocarcinoma following PHM. In all three cases in which lesion ploidy was determined,[44,45] the PHM was triploid and the choriocarcinoma was either diploid (two cases) or diploid/triploid (one case). This discrepancy between genotypes is consistent with origin of the choriocarcinoma from an intervening subclinical pregnancy rather than the PHM itself.

Divergent management strategies for treatment of PHM reflect the differing notions of its potential for aggressiveness, although all rely on serological HCG monitoring as a trigger for additional diagnostic or therapeutic interventions. Persistent disease detected by HCG screening is then staged by additional diagnostic tests before administration of curative chemotherapy.

4. CHORIOCARCINOMA

4.1 General

The pathogenesis of choriocarcinoma is complex since it may arise in a nongestational setting as a primary germ cell tumor of the ovary, or as postgestational sequelae of a molar or normal pregnancy. Approximately 50%[31] of gestational choriocarcinomas are associated with a prior complete hydatidiform mole. While some choriocarcinomas may have a diploid DNA content, most are in the tetraploid range.[46,47]

4.2 Choriocarcinoma Precursors

Much of the clinical and investigational focus in trophoblastic disease assumes that one type of lesion evolves into another in a progressive fashion, such as transformation of a pre-existing complete hydatidiform mole into choriocarcinoma. Ablation of existing lesions remains the priority in management. Discrimination between progression of a single lesion, and elevated risk of developing a new lesion requires the ability to track the fate of individual tissues. Fortunately, random chromosomal sorting and crossover during gametogenesis produce unique haplotypes contributed by each parent to a particular conceptus (Figure 1). The ability to "fingerprint" the genomic makeup of a conceptus is limited only by availability of informative polymorphic markers, and the technology to score them. The past decade has seen dramatic advances on both fronts, and yielded some surprises in our understanding of trophoblastic disease.

Clinical history is itself inadequate to establish continuity between two trophoblastic proliferations. There are well-documented cases in which a CHM precedes the development of choriocarcinoma by many years, with intervening normal pregnancies. By simply analyzing the sex chromosome composition of moles, and comparison with the gender of antecedent normal pregnancies, it has been shown that at least some clinically postgestational choriocarcinomas are not derived from the antecedent normal pregnancy.[17] Similarly, restriction fragment length polymorphisms have been used to show that a choriocarcinoma was not derived from the antecedent normal gestation, but rather from an earlier complete mole.[48]

5. PLACENTAL SITE TROPHOBLASTIC TUMOR (PSTT)

Placental site trophoblastic tumors[31,49] are known only through scattered case reports or small series. They are characterized by masses of human placental lactogen-containing intermediate trophoblast in the myometrium, and are usually diploid.[50] There is usually a history of a prior pregnancy in the preceding 1 to 2 years. Genetic marker studies have shown PSTT to share the genotype with either a preceding diploid normal gestation or complete hydatidiform mole.[51] They represent a particular clinical challenge due to poor response to existing chemotherapeutic regimens and a high (about one quarter) lethality rate.[52]

6. IMPRINTING AND TROPHOBLASTIC DISEASE

6.1 Introduction

PHM and CHM share an imbalance of parental genetic contributions, favoring the paternal genome, potentially altering the effective dose of genes which are differentially expressed from the two parental genomes. Since androgenetic CHMs contain two paternal alleles, for example, these tissues might express a "double dose" of genes, such as IFG2, which are usually expressed specifically from the paternal copy. Genomic imprinting is the mechanism whereby DNA derived from the sperm and egg are inequivalent, and an understanding of the imprinting process may explain why diploid androgenetic gestations (CHM) are so different from diploid biparental (normal) gestations.

6.2 Inequivalence of the Parental Genomes

Completion of normal development requires a genetic contribution from both parents. Perhaps the most compelling evidence of this fact is that asexual reproduction (parthenogenesis and androgenesis) has never produced a viable mammalian offspring. More elaborate evidence of requisite biparental contributions comes from experiments by Surani,[53] in which androgenetic or gynogenetic mouse embryos were created by microtransfer of the zygotic pronuclei of fertilized eggs. Gynogenetic embryos transplanted to a foster mother developed only a small placenta with a secondarily stunted embryo, whereas comparable androgenetic embryos had a bulky, hypertrophic placenta, and a rudimentary embryo,[54] mimicking a human complete molar gestation. Such a functional divergence of maternally and paternally derived uniparental cells was supported further by the observation that murine aggregation chimeras created by fusion of androgenetic and parthenogenetic blastomeres were followed by physical segregation of the two cell lineages — androgenetic cells in the extraembryonic tissues and parthenogenetic cells in the embryo.[53] The general impression is that both parental genomes are needed for complete development, with the paternal DNA required for extraembryonic differentiation and maternal DNA for embryonic components.

The failure of normal development of uniparental mammalian conceptuses may be due to the existence of an, as yet, undefined number of genes that are only expressed from either the maternal or paternal locus. The major part of these observations were established by using mouse embryos, and have resulted in the identification of an expanding list of imprinted genes, including: $Igf2$,[55] H19,[56] and a small nuclear ribonuclear protein ($Snrpn$).[57]

It is feasible that the imprinting process or status is altered during differentiation into specific cell types such as trophoblast. Non-random paternal X chromosome inactivation in the placenta,[58] a process determined by a parental imprint acting directly or indirectly on the control mechanism of X inactivation, is not seen in the embryo itself. One proposed mechanism of imprinting, DNA methylation, varies globally[59] and with specific endogenous genes such as FMR-1,[60] and transgenes[61] between embryonic and extraembryonic compartments.

6.3 Pathology of Imprinting

Loss of the normal imprint of two genes, H19 and IGF2, in uniparental gestations hints at the possibility of a global failure of the imprinting process in CHMs. The embryos of normal gestations only express H19 from the maternal allele and IGF2 from the paternal allele, whereas neither is expressed from the maternal alleles of parthenogenetic teratomas (ovarian teratomas), and both paternal alleles are expressed in androgenetic CHMs (Figure 3).[62,63] This argues that the imprinting process itself may require a biparental genome, becoming stabilized in the zygote by interaction between maternal and paternal factors.

Modification of parental imprints has been associated with tumor formation, leading to the hypothesis that relaxation or loss of an imprint might contribute to the neoplastic phenotype. While the imprinted

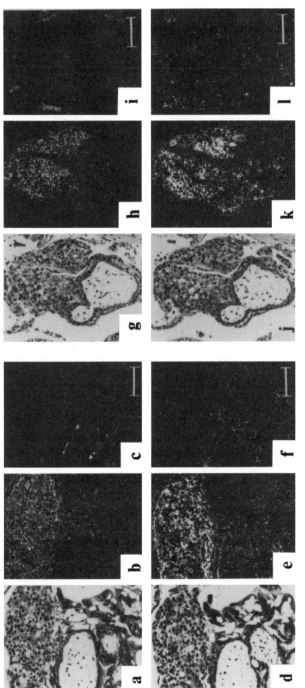

Figure 3 Coexpression of oppositely imprinted genes H19 and IGF2 in mononuclear trophoblast of androgenetic and biparental gestations. Complete hydatiform mole (Panels a–f) and normal 4.5 week gestational age human placenta (Panels g–l) tissue sections were hybridized in situ with H19 (Panels b, h antisense; c, i sense) or IGF2 (Panels e, k antisense; f, l sense) oligonucleotide probes. Mononuclear cytotrophoblast and intermediate trophoblast (top of each panels) coexpress H19 and IGF2 in biparental and androgenetic conceptions. Expression of IGF2 is widespread in normal placenta, including mesenchymal cells in the villous core (Panel k), and areas of maternal decidua (not shown). Multinuclear syncytiotrophoblast (lower right, Panels a, d; surrounding villous in lower half of g, j) do not express either gene. Exposure interval 20 and 15 days for Panels a–f and g–l. Photos taken using T-Max 100 film and transmitted brightfield (Panels a, d, g, j) or darkfield (Panels b–c, e–f, h–i, j–l) optics. Brightfield panels are the same field as the panel to their right. Bar is 100 μm. (From Mutter, G. L., Stewart, C. L., Chaponot, M. L., and Pomponio, R. J., *Am. J. Hum. Genet.*, 53, 1096, 1993. With permission.)

genes H19 and IFG2 are expressed in normal fetal tissues from different parental chromosomes,[55,64] biallelic expression has been found in Wilms' tumor[65,66] and choriocarcinoma.[67] In other embryonic tumors the effects of chromosomal deletion may in part be dictated by the parental origin of the lost DNA, such as selective deletion of maternally derived chromosome 11p15.5.[68] Thus, loss of normal transcriptional asymmetry between imprinted parental alleles, either through erasure of the imprint or mutation/deletion of one of the alleles, may contribute to a change in neoplastic potential.

REFERENCES

1. **Cross, J., Werb, Z., and Fisher, S.,** Implantation and the placenta: key pieces of the developmental puzzle, *Science*, 266, 1508, 1994.
2. **Bagshawe, K., Dent, J., and Webb, J.,** Hydatidiform mole in England and Wales 1973-1983, *Lancet*, 2, 673, 1986.
3. **Finkler, N., Berkowitz, R., Driscoll, S., Goldstein, D., and Bernstein, M.,** Clinical experience with placental site trophoblastic tumors at the New England Trophoblastic Disease Center, *Obstet. Gynecol.*, 71, 854, 1988.
4. **Kajii, T. and Ohama, K.,** Androgenetic origin of hydatidiform mole, *Nature*, 268, 633, 1977.
5. **Lawler, S.D., Povey, S., Fisher, R.A., and Pickthall, V.J.,** Genetic studies on hydatidiform moles. II. The origin of complete moles, *Ann. Hum. Genet.*, 46, 209, 1982.
6. **Lawler, S., Fisher, R., and Dent, J.,** A prospective genetic study of complete and partial hydatidiform moles, *Am. J. Obstet. Gynecol.*, 164, 1270, 1991.
7. **Azuma, C., Saji, F., Tokugawa, Y., Kimura, T., Nobunaga, T., Takemura, M., Kameda, T., and Tanizawa, O.,** Application of gene amplification by polymerase chain reaction to genetic analysis of molar mitochondrial DNA: the detection of anuclear empty ovum as the cause of complete mole, *Gynecol. Oncol.*, 40, 29, 1991.
8. **Wallace, D.C., Surti, U., Adams, C.W., and Szulman, A.E.,** Complete moles have paternal chromosomes but maternal mitochondrial DNA, *Hum. Genet.*, 61, 145, 1982.
9. **Ohama, K., Kajii, T., Okamoto, E., Fukuda, Y., Imaizumi, K., Tsukahara, M., Kobayashi, K., and Hagiwara, K.,** Dispermic origin of XY hydatidiform moles, *Nature*, 292, 551, 1981.
10. **Fisher, R.A., Povey, S., Jeffreys, A.J., Martin, C.A., Patel, I., and Lawler, S.D.,** Frequency of heterozygous complete hydatidiform moles, estimated by locus-specific minisatellite and Y chromosome-specific probes, *Hum. Genet.*, 82, 259, 1989.
11. **Henry, I., Bonaiti-Pellie, C., Chehensse, V., Beldjord, C., Schwartz, C., Utermann, G., and Junien, C.,** Uniparental paternal disomy in a genetic cancer-predisposing syndrome, *Nature*, 351, 665, 1991.
12. **Little, M., Van Heyningen, V., and Hastie, N.,** Dads and disomy and disease, *Nature*, 351, 609, 1991.
13. **Lage, J.M.,** Placentomegaly with massive hydrops of placental stem villi, diploid DNA content and fetal omphaloceles: possible association with Beckwith-Wiedemann syndrome, *Hum. Pathol.*, 22, 591, 1991.
14. **Beckwith, J.B.,** Macroglossia, omphalocele, adrenal cyomegaly, gigantism, and hyperplastic visceromegaly, *Birth Defects*, 5, 188, 1969.
15. **Feinberg, R.F., Lockwood, C.J., Salafia, C., and Hobbins, J.C.,** Sonographic diagnosis of a pregnancy with a diffuse hydatidiform mole and coexistent 46,XX fetus: A case report, *Obstet. Gynecol.*, 72, 485, 1988.
16. **Sand, P., Lurain, J., and Brewer, J.,** Repeat gestational trophoblastic disease, *Obstet. Gynecol.*, 142, 140, 1984.
17. **Roberts, D. and Mutter, G.L.,** Molecular biology of trophoblastic disease, *J. Reprod. Med.*, 39, 201, 1994.
18. **Mutter, G.L., Pomponio, R., Berkowitz, R., and Genest, D.,** Sex chromosome composition of complete hydatidiform moles: relationship to metastasis, *Am. J. Obstet. Gynecol.*, 168, 1547, 1993.
19. **Hertig, A.T. and Sheldon, W.H.,** Hydatidiform mole: a pathologico-clinical correlation of 200 cases, *Am. J. Obstet. Gynecol.*, 53, 1, 1947.
20. **Elston, C. and Bagshawe, K.,** The value of histological grading in the management of hydatidiform mole, *J. Obstet. Gynaecol. Br. Commw.*, 79, 717, 1972.
21. **Genest, D., Laborde, O., Berkowitz, R., Goldstein, D., Bernstein, M., and Lage, J.,** A clinicopathologic study of 153 cases of complete hydatidiform mole (1980-1990): histologic grade lacks prognostic significance, *Obstet. Gynecol.*, 78, 402, 1991.
22. **Wake, N., Fujino, T., Hoshi, S., Shinkai, N., Sakai, K., Kato, H., Hashimoto, M., Yasuda, T., Yamada, H., and Ichinoe, K.,** The propensity to malignancy of dispermic heterozygous moles, *Placenta*, 8, 319, 1987.
23. **Davis, J.R., Surwit, E.A., Garay, J.P., and Fortier, K.J.,** Sex assignment in gestational trophoblastic neoplasia, *Am. J. Obstet. Gynecol.*, 148, 722, 1984.
24. **Thomsen, J.L. and Niebuhr, E.,** The frequency of false-positive and false negative results in the detection of Y-chromosomes in interphase nuclei, *Hum. Genet.*, 73, 27, 1986.
25. **Wake, N., Seki, T., Fujita, H., Okubo, H., Sakaai, K., Okuyama, K., Hayashi, H., Shiina, Y., Sato, H., Kuroda, M., and Ichinoe, K.,** Malignant potential of homozygous and heterozygous complete moles, *Cancer Res.*, 44, 1226, 1984.
26. **Weber, J.L. and May, P.E.,** Abundant class of human DNA polymorphisms which can be typed using the polymerase chain reaction, *Am. J. Hum. Genet.*, 44, 388, 1989.

27. **Kovacs, B., Shahbahrami, B., Tast, D., and Curtin, J.,** Molecular genetic analysis of complete hydatidiform mole, *Cancer Genet. Cytogenet.*, 54, 143, 1991.

28. **Bell, D.A. and DeMarini, D.M.,** Excessive cycling converts PCR products to random-length higher molecular weight fragments, *Nucleic Acids Res.*, 19, 5079, 1991.

29. **Armour, J., Patel, I., Thein, S., Fey, M., and Jeffreys, A.,** Analysis of somatic mutations at human minisatellite loci in tumors and cell lines, *Genomics*, 4, 318, 1989.

30. **Wahls, W.P., Wallace, L.J., and Moore, P.D.,** Hypervariable minisatellite DNA is a hotspot for homologous recombination in human cells, *Cell*, 60, 95, 1990.

31. **Silverberg, S. and Kurman, R.,** Classification and pathology of gestational trophoblastic disease, in Tumors of the Uterine Corpus and Gestational Trophoblastic Disease, Armed Forces Inst. of Pathology, Washington, D.C., 1991, p. 233.

32. **Szulman, A.E. and Surti, U.,** The syndromes of hydatidiform mole. I. Cytogenetic and morphologic correlations, *Am. J. Obstet. Gynecol.*, 131, 665, 1978.

33. **Vassilakos, P., Riotton, G., and Kajii, T.,** Hydatidiform mole: two entities, *Am. J. Obstet. Gynecol.*, 127, 167, 1977.

34. **Lawler, S.D., Fisher, R.A., Pickthall, V.J., Povey, S., and Evans, M.W.,** Genetic studies on hydatidiform moles. I. The origin of partial moles, *Cancer Genet. Cytogenet.*, 5, 309, 1982.

35. **Vejerslev, L.O., Fisher, R.A., Surti, U., and Walke, N.,** Hydatidiform mole: cytogenetically unusual cases and their implications for the present classification, *Am. J. Obstet. Gynecol.*, 157, 180, 1987.

36. **Vejerslev, L.O., Sunde, L., Hansen, B.F., Larsen, J.K., Christensen, I.J., and Larsen, G.,** Hydatidiform mole and fetus with normal karyotype: support of a separate entity, *Obstet. Gynecol.*, 77, 868, 1991.

37. **Fryns, J.P., van de Kerckhove, A., Goddeeris, P., and Van Den Berghe, H.,** Unusually long survival in a case of full triploidy of maternal origin, *Hum. Genet.*, 38, 147, 1977.

38. **McFadden, D.E. and Kalousek, D.K.,** Two different phenotypes of fetuses with chromosomal triploidy. Correlation with parental origin of the extra haploid set, *Am. J. Med. Genet.*, 38, 535, 1991.

39. **McFadden, D.E., Kwong, L.C., Yam, I.Y.L., and Langlois, S.,** Parental origin of triploidy in human fetuses: evidence for genomic imprinting, *Hum. Genet.*, 92, 465, 1993.

40. **Jacobs, P.A., Hunt, P.A., Matsuura, J.S., Wilson, C.C., and Szulman, A.E.,** Complete and partial hydatidiform mole in Hawaii: cytogenetics, morphology, and epidemiology, *Br. J. Obstet. Gynaecol.*, 89, 258, 1982.

41. **Lage, J., Berkowitz, R., Rice, L., Goldstein, D., Bernstein, M., and Weinberg, D.,** Flow cytometric analysis of DNA content in partial hydatidiform moles with persistent gestational trophoblastic tumor, *Obstet. Gynecol.*, 77, 111, 1991.

42. **Vejerslev, L., Larsen, G., and Jacobsen, M.,** Partial hydatidiform mole with subsequent trophoblastic tumor; a case report, *Eur. J. Obstet. Gynecol.*, 40, 73, 1991.

43. **Looi, L.,** Malignant evolution with fatal outcome in a patient with partial hydatidiform mole, *Austr. N.Z. J. Obstet. Gynaecol.*, 21, 51, 1981.

44. **Gardner, H.A. and Lage, J.M.,** Choriocarcinoma following a partial hydatidiform mole: a case report, *Hum. Pathol.*, 23, 468, 1992.

45. **Bagshawe, K.D., Lawler, S.D., Paradinas, F.J., Dent, J., Brown, P., and Boxer, G.M.,** Gestational trophoblastic tumours following initial diagnosis of partial hydatidiform mole, *Lancet*, 335, 1074, 1990.

46. **Sekiya, S., Shirotake, S., Kaiho, T., Iwasawa, H., Kawata, M., Higaki, K., Ishige, H., Takamizawa, H., and Minamihisamatsu, M.,** A newly established human gestational choriocarcinoma cell line and its characterization, *Gynecol. Oncol.*, 15, 413, 1983.

47. **Sasaki, S., Katayama, P.K., Roesler, M., Pattillo, R.A., Mattingly, R.F., and Ohkawa, K.,** Cytogenetic analysis of choriocarcinoma cell lines, *Acta Obstet. Gynaecol. Jpn.*, 34, 2253, 1982.

48. **Fisher, R., Newlands, E., Jeffreys, A., Boxer, G., Begent, R., Rustein, G., and Bagshawe, K.,** Gestational and nongestational trophoblastic tumors distinguished by DNA analysis, *Cancer*, 69, 839, 1992.

49. **Kurman, R., Scully, R., and Norris, H.,** Trophoblastic pseudotumor of the uterus: an exaggerated form of syncitial endometritis simulating a malignant tumor, *Cancer*, 38, 1214, 1976.

50. **Fukunaga, M. and Ushigome, S.,** Malignant trophoblastic tumors: immunohistochemical and flow cytometric comparison of choriocarcinoma and placental site trophoblastic tumors, *Hum. Pathol.*, 24, 1098, 1993.

51. **Fisher, R.A., Paradinas, F.J., Newlands, E.S., and Boxer, G.M.,** Genetic evidence that placental site trophoblastic tumours can originate from a hydatidiform mole or a normal conceptus, *Br. J. Cancer*, 65, 355, 1992.

52. **McClellan, R., Buscema, J., Currie, J., and Woodruff, J.,** Placental site trophoblastic tumor in a postmenopausal woman, *Am. J. Clin. Pathol.*, 95, 670, 1991.

53. **Surani, M.A., Barton, S.C., and Norris, M.L.,** Influence of parental chromosomes on spatial specificity in andro-genetic-parthenogenetic chiameras in the mouse, *Nature*, 326, 395, 1987.

54. **Barton, S., Surani, M., and Norris, M.,** Role of paternal and maternal genomes in mouse development, *Nature*, 311, 374, 1984.

55. **DeChiara, T., Robertson, E., and Efstratiadis, A.,** Parental imprinting of the mouse insulin-like growth factor II gene, *Cell*, 64, 849, 1991.

56. **Bartolomei, M., Zemel, S., and Tilghman, S.,** Parental imprinting of the mouse H19 gene, *Nature*, 351, 153, 1991.

57. **Leff, S., Brannan, C., Reed, M., Ozcelik, T., Francke, U., Copeland, N., and Jenkins, N.,** Maternal imprinting of the mouse Snrpn gene and conserved linkage homology with the human Prader-Willi syndrome region, *Nature Genet.*, 2, 259, 1992.

58. **Harrison, K.,** X-Chromosome inactivation in the human trophoblast, *Cytogenet. Cell Genet.*, 52, 37, 1989.

59. **Monk, M., Boubelik, M., and Lehnert, S.,** Temporal and regional changes in DNA methylation in the embryonic, extraembryonic and germ cell lineages during mouse embryo development, *Development*, 99, 371, 1987.

60. **Sutcliffe, J., Nelson, D., Zhang, F., Pieretti, M., Caskey, C., Saxe, D., and Warren, S.,** DNA methylation represses FMR-1 transcription in fragile X syndrome, *Hum. Mol. Genet.*, 1, 397, 1993.

61. **Gundersen, G., Kolsto, A.-B., Larsen, F., and Prydz, H.,** Tissue-specific methylation of a CpG island in transgenic mice, *Gene*, 113, 207, 1992.

62. **Mutter, G.L., Stewart, C.L., Chaponot, M.L., and Pomponio, R.J.,** Oppositely imprinted genes H19 and insulin-like growth factor 2 are coexpressed in human androgenetic trophoblast, *Am. J. Hum. Genet.*, 53, 1096, 1993.

63. **Walsh, C., Miller, S.J., Flam, F., Fisher, R.A., and Ohlsson, R.,** Paternally derived H19 is differentially expressed in malignant and nonmalignant trophoblast, *Cancer Res.*, 55, 1111, 1995.

64. **Brunkow, M. and Tilghman, S.,** Ectopic expression of the H19 gene in mice causes prenatal lethality, *Genes Dev.*, 5, 1092, 1991.

65. **Ogawa, O., Eccles, M.R., Szeto, J., McNoe, L.A., Yun, K., Maw, M.A., Smith, P.J., and Reeve, A.E.,** Relaxation of insulin-like growth factor II gene imprinting implicated in Wilms' tumour, *Nature*, 362, 749, 1993.

66. **Rainier, S., Johnson, L., Dobry, C., Ping, A., Grundy, P., and Feinberg, A.,** Relaxation of imprinted genes in human cancer, *Nature*, 362, 747, 1993.

67. **Hashimoto, K., Azuma, C., Koyoma, M., Ohashi, K., Kamiura, S., Nobunaga, T., Kimura, T., Tokugawa, Y., Kanai, T., and Saji, F.,** Loss of imprinting in choriocarcinoma, *Nat. Genet.*, 9, 109, 1995.

68. **Hao, Y., Crenshaw, T., Moulton, T., Newcomb, E., and Tycko, B.,** Tumour-suppressor activity of H19 RNA, *Nature*, 365, 764, 1993.

69. **Economou, E., Bergen, A., Warren, A., and Antonarakis, S.,** The polydeoxadenylate tract of Alu repetitive elements is polymorphic in the human genome, *Proc. Natl. Acad. Sci. U.S.A.*, 87, 2951, 1990.

Part VII.
Hereditary Aspects of
Gynecological Malignancies

Chapter 19

Hereditary Aspects of Female Cancers

Anne P. Shapter, Andrew Berchuck, and Beth Y. Karlan

CONTENTS

1. INTRODUCTION

It has long been suspected that heredity contributes to the development of some female cancers. Initially, the existence of autosomal dominant cancer-causing genes was postulated to explain the observation of striking familial clusters of specific types of cancers.[1-3] Familial patterns of cancer inheritance such as breast/ovarian cancer syndrome and the Lynch syndrome I and II colon cancer syndromes were defined on the basis of pedigree analysis. Population-based epidemiologic studies which showed that a family history of certain cancers increased one's risk of developing cancer further strengthened the hypothesis that heredity contributes to the development of a fraction of cancers. For example, analysis of the Cancer and Steroid Hormone (CASH) data base revealed that the risk of ovarian cancer in first and second degree relatives of women with ovarian cancer was increased 3.6 and 2.9-fold respectively compared to women with no family history of ovarian cancer.[4] In addition, a family history of either ovarian or breast cancer increased the risk of both cancers in first degree relatives.[5-7] Currently, it is estimated that as many as 5 to 10% of breast and ovarian cancers have a hereditary basis.

In the 1990s, many of the genes responsible for autosomal dominant familial cancer syndromes have been identified (Table 1). With the exception of the *ret* oncogene, which causes MEN2 syndrome, all the other hereditary cancer syndromes appear to be due to inheritance of a mutant copy of a tumor suppressor gene. Because inactivation of both copies of a tumor suppressor gene is required to elicit tumorigenesis, the transformation process is dependent on acquired damage to the remaining normal copy of the gene in a single cell. Thus, although inheritance of familial cancer syndromes follows an autosomal "dominant" Mendelian pattern, tumor suppressor genes often are referred to as "recessive" cancer causing genes because of the requirement for inactivation of both gene copies.

In this chapter, the discussion will focus primarily on those genes in which inherited mutations are characterized by the development of gynecologic or breast cancers. In this regard, the BRCA1 gene on chromosome 17q has been shown to be responsible for both breast/ovarian cancer syndrome and site-specific ovarian cancer.[8] Other families in which hereditary breast cancer occurs usually in the absence of ovarian cancer have been shown to have inherited mutations in the BRCA2 gene on chromosome 13q.[9] Finally, endometrial cancer is a prominent feature of Lynch syndrome II or hereditary nonpolyposis colorectal cancer syndrome, which has been shown to be caused by mutations in DNA mismatch repair genes.[10,11]

Table 1 Hereditary Cancer Syndromes

Syndrome	Gene	Chromosome	Predominant Cancers
Familial breast/ovarian cancer	BRCA1	17q21	Breast, ovary
Familial breast cancer	BRCA2	13q12	Breast
Hereditary non-polyposis colon cancer	MSH2	2p	Colon, endometrium, ovary and others
	MLH1	3p21	
	PMS1	2q31	
	PMS2	7p22	
Ataxia-Telangectasia	AT	11q22	Breast
Familial polyposis coli	APC	5q21	Colonic polyps and cancers
Li-Fraumeni syndrome	p53	17p13	Sarcomas, leukemias, breast, brain and others
von Hippel-Lindau	VHL	3p25	Kidney, and others
Neurofibromatosis	NF1	17q11	Neurofibromas
	NF2	22q12	
Retinoblastoma	Rb	13q14	Retinoblastoma, sarcoma
Wilm's tumor	WT1	11p13	Kidney
Familial melanoma	MLM	9p21	Melanoma
Multiple-endocrine neoplasia-type 2	ret	10q11	Thyroid, adrenal, parathyroid

Although there is definitive evidence for the existence of autosomal dominant cancer syndromes involving the breast, ovary, and endometrium, the high penetrance single gene mechanism of disease appears to have little or no role in the development of cervical, vaginal, or vulvar cancers or gestational trophoblastic disease. There is now evidence to suggest that more subtle or weakly penetrant genetic effects also may contribute to cancer susceptibility. Polymorphisms that change the amino acid sequence of proteins have been described that may alter function sufficiently to either increase or decrease susceptibility to cancer. For example, polymorphisms associated with epoxide hydrolase[12] and the progesterone receptor[13,14] have been linked with an increased incidence of ovarian cancer in some studies. Because these polymorphisms are not strongly associated with cancer families, the relationship between rare alleles and cancer risk is difficult to prove conclusively and requires large population-based studies. In addition, their contribution to cancer susceptibility may be influenced by diet and other environmental factors.

2. THE HEREDITARY FEMALE CANCER SYNDROMES

The hereditary ovarian cancer syndromes, which include familial breast/ovarian cancer syndrome, site-specific ovarian cancer, and the Lynch syndrome II, comprise only 5 to 10% of the approximately 24,000 new cases of ovarian cancer per year in the United States.[15] The remainder of cases are felt to be sporadic. Similarly, only 5 to 10% of all breast cancers are attributed to the inheritance of a susceptibility gene. As many as 14% of all endometrial cancers may occur in association with Lynch syndrome II. Of note, several authors have recently suggested that a small percentage of endometrial carcinomas may be inherited as site-specific tumors, not in conjunction with other cancers.[16,17] Although sporadic malignancies are much more common, hereditary cancer syndromes provide an opportunity to identify genes that may be causative or contributory to both inherited and sporadic tumors.

In addition to the well established hereditary cancer syndromes noted above that include breast, ovarian and endometrial cancers, sex cord stromal tumors, especially granulosa theca cell tumors, have been noted to occur in association with the Peutz-Jeghers syndrome.[18] Familial clustering of germ cell tumors rarely occurs, but there have been several reports in the literature including one family in which dysgerminomas were observed in three relatives.[19] Shulman and colleagues recently reviewed 78 pedigrees of patients with ovarian germ cell malignancies and found no evidence of heritability of these tumors.[20] Germ cell tumors may be a part of the Li-Fraumeni syndrome, however.[21]

2.1 Familial Breast and Ovarian Cancer

Lynch and colleagues first noted an association between breast and ovarian cancer in 1978 in their review of 12 family pedigrees and therefore suggested the existence of a genetic factor.[2] Since that time, several other studies have demonstrated such an association. In a large case-control study, Schildkraut and colleagues estimated the heritability for ovarian, breast, and endometrial cancers to be 40%, 56%, and 52% respectively in patients with a positive family history.[5] In addition, they noted a significant genetic association between breast and ovarian cancers, although neither of these was significantly associated with endometrial cancer.

Breast cancers in patients with the breast-ovary cancer syndrome have a greater tendency towards bilaterality and are associated with better survival rates than occurs with sporadic breast cancers. The syndrome is also characterized by an earlier age at cancer diagnosis than that which is seen for sporadic breast cancers. In twelve families with hereditary breast-ovarian cancer, Lynch and colleagues reported an average age at diagnosis of 52.4 years for ovarian cancer in contrast to an average age of 59 for ovarian cancer in the general population.[22] Like the ovarian cancer which occurs in conjunction with the hereditary breast-ovary cancer syndrome, site-specific ovarian cancer is also characterized by an earlier age at diagnosis than occurs with sporadic ovarian cancers. In their analysis of four families with this syndrome with 18 total cases, Lynch and colleagues reported that the average age at diagnosis was 48.9 years in contrast to 59 years in the general population.[22]

In 1990, Hall et al. reported linkage of familial breast/ovarian cancer syndrome to the long arm of chromosome 17 using marker D17S574.[23] Narod and colleagues later used the same marker and demonstrated linkage in five families with breast-ovary cancer syndrome.[24] This breast-ovarian cancer susceptibility gene became designated as "BRCA1." Further research eventually localized the BRCA1 gene more specifically to chromosome 17q12-21. In 1994, BRCA1 was cloned and the presence of germline mutations demonstrated in affected members of breast/ovarian cancer families.[8,25] Women in these families have approximately a 90 to 100% lifetime risk of developing breast and/or ovarian cancer.[26] The incidence of prostate and colon cancers has been found to be increased by approximately three- and fourfold, respectively,[27] but these are the only other malignancies in which a higher than expected incidence has been noted in families with BRCA1 mutations.

Although the function of BRCA1 in the breast and ovary remains essentially unknown, the recent report of a phenotypically normal woman who inherited two mutant copies of this gene indicates that BRCA1 is not requisite for growth and development.[28] As is the case for other genes that cause hereditary cancers, BRCA1 is expressed in tissues other than those in which it causes cancer.[8] It is thought that BRCA1 is a member of the tumor suppressor gene family, since the normal copy of BRCA1 is invariably deleted in breast and ovarian cancers that arise in women who inherit a mutant BRCA1 gene.[29] Once the mechanism by which mutant BRCA1 predisposes to breast and ovarian cancer has been determined, perhaps this will facilitate the development of preventive strategies aimed at replacing its function.

Although the discovery of BRCA1 has been accompanied by tremendous excitement, there are several significant obstacles that must be overcome before BRCA1 testing becomes widely available. Because mutations occur throughout this relatively large gene, the most reliable method of detecting BRCA1 mutations is to sequence the entire coding region. DNA sequencing of the entire BRCA1 gene is a highly labor intensive process, however. In view of this fact, several short cuts have been developed for screening the gene including single stranded conformation polymorphism analysis[25,30-32] and protein truncation tests.[33] The protein truncation test is particularly appealing as approximately 85% of BRCA1 mutations are either nonsense mutations in which a single nucleotide substitution produces a stop codon, or frameshift mutations in which one or more nucleotides are deleted producing a stop codon somewhere down stream.[25] In either case, the result is production of a smaller than usual BRCA1 protein that can be detected on an electrophoretic gel. Reliance on methods other than DNA sequencing as an initial screen results in lowered sensitivity for detecting mutations.

Because a straightforward BRCA1 test with high sensitivity and specificity is not available and the estimated carrier frequency in the general population is estimated to be 1 in 800, population screening for BRCA1 mutations is impractical. A possible exception is suggested by a recent study of 850 Ashkanazi Jews, which revealed that approximately 1 in 100 carried an identical mutation in codon 185 (deletion AG) of the gene.[34] If the results of confirmatory studies now underway are similar, it may be reasonable to offer screening for this specific mutation to Jewish individuals.

Although it is believed that over 90% of hereditary ovarian cancer is due to mutations in BRCA1,[35,36] a small fraction of familial ovarian cancer is probably due to other genes. Occasionally, ovarian cancer occurs in Lynch syndrome II cancer families, but colon and endometrial cancer are the hallmarks of this syndrome. The Lynch syndrome II is due to inherited mutations in a family of DNA repair genes described below.[10,11] Analysis of these genes is appropriate in pedigrees in which colon cancer predominates. A significant fraction of breast cancer families do not have BRCA1 mutations, but do show linkage to chromosome 13q.[37] Recently, this second breast cancer susceptibility gene (BRCA2) on chromosome

13q12 was identified.[9] Notable attributes of BRCA2 families include early age of onset and the occurrence of male breast cancer. Ovarian cancer does not appear to be a prominent feature of these families, but there may be some exceptions. It is estimated that BRCA1 is responsible for approximately half of hereditary breast cancer, whereas mutations in BRCA2 cause an additional 40%. The remaining 10% of hereditary breast cancer is associated with germline mutations in p53, the Ataxia-Telangectasia gene, the HNPCC genes and rarely others.[38]

2.2 Hereditary Nonpolyposis Colorectal Cancer

Families with Hereditary Nonpolyposis Colorectal Cancer (HNPCC), also known as Lynch syndrome II, are primarily affected by colorectal cancer at a young age, but endometrial cancers also frequently occur in women in these kindreds. In addition, there is also an excess of ovarian, gastric, breast and other cancers,[11,39] (Figure 1) but endometrial cancer is the second most frequent cancer in this syndrome. In contrast, Lynch syndrome I consists of familial colorectal cancer in the absence of other primary malignancies that are seen in the Lynch syndrome II. Colonic and other cancers occur at a young age in HNPCC families, but tend to have different clinical characteristics than sporadic cancers of the same type. The colonic cancers usually are proximal to the splenic flexure, have a diploid DNA content and more favorable prognosis.[40] Similarly, we found that endometrial cancers in HNPCC families were diploid and had a favorable prognosis.[41]

The Patient's Pedigree

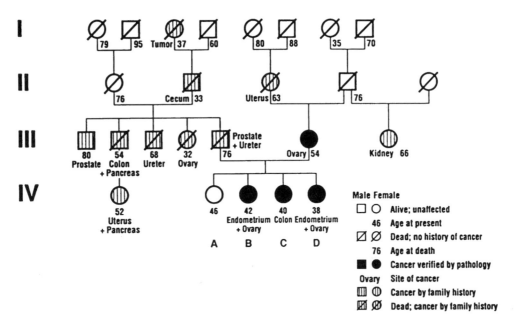

Figure 1 A Lynch syndrome II pedigree. Prior to obtaining the paternal history, this pedigree was suspected to represent a breast/ovarian cancer syndrome family. In retrospect, the ovarian cancer that occurred in the mother likely represents a sporadic case. (Reprinted with permission from the American College of Obstetricians and Gynecologists. *Obstet. Gynecol.*, 1991, 78, 1023).

Like the other hereditary cancer syndromes, pedigree analysis of HNPCC families suggests an autosomal dominant mode of transmission. It has now been shown that HNPCC is due to inherited mutations in one of several genes involved in DNA mismatch repair (Table 1).[10] The first gene to be associated with HNPCC, known as MSH2, was isolated in 1993 and may account for as many as 60% of cases.[40,42] More recently, a second gene that causes HNPCC was demonstrated on chromosome 3p21-23.[10] This gene, known as MLH1, accounts for approximately 30% of HNPCC cases. Finally, two other genes, PMS1 and 2 are thought to account for most of the remaining cases.[11] It has been shown in bacteria and yeast that these genes act in concert to repair DNA mismatch mutations. Mutational inactivation of these DNA repair genes leads to a mutator phenotype with accumulation of unrepaired mutations. The inability to repair mutations is particularly problematic in repetitive microsatellite sequences, which appear to be more susceptible to mutations. In fact, the instability of these repetitive microsatellite sequences in HNPCC associated cancers was the first clue that alterations in DNA repair genes might represent the underlying cause.[43] Since most human cancers occur due to accumulation of multiple mutations throughout the genome, it is thought that inactivation of the DNA repair process serves to accelerate the rate of malignant transformation.

2.3 Site-Specific Endometrial Cancer

Several authors have recently suggested the existence of a site-specific heritable endometrial carcinoma that occurs in the absence of other primary malignancies. Schildkraut and colleagues in a large case-control study involving mothers and sisters with ovarian, breast, and endometrial cancers reported a heritability rate of 52% for endometrial cancer.[5] Although endometrial cancer in this study was genetically unrelated to either breast or ovarian cancer, there was strong evidence of heritability suggesting that a hereditary site-specific endometrial cancer may indeed exist.

Boltenberg and colleagues reviewed pedigrees of 51 consecutive cases of endometrial cancer and concluded that there may be two forms of hereditary cancer of the endometrium: the cancer family syndrome or Lynch II syndrome and hereditary endometrial adenocarcinoma in the absence of a strong association with other primary malignancies.[16] Sandles and colleagues also reported data consistent with the existence of a site- specific endometrial cancer.[17] In their study involving 64 probands, in index cases of endometrial adenocarcinoma there were four families in which endometrial cancer was diagnosed in at least one first-degree relative without evidence of colon or ovarian cancers in other relatives. Future research will determine whether or not site-specific endometrial cancer is a separate entity. As hereditary site-specific endometrial cancer has only recently been described, there are no specific genetic loci which have been linked to susceptibility to this disease.

2.4 Hereditary Aspects of Cervical Carcinoma

Most research in the field of cervical cancer has been directed towards the close association with sexually transmitted human papillomavirus infection. Because of the lack of a striking familial pattern, relatively little effort has been directed towards hereditary or genetic aspects of carcinoma of the cervix. Brinton and colleagues conducted a large multicenter case-control study involving 481 patients with invasive cervical cancer and 801 controls.[44] In addition to the traditional risk factors, they noted a familial tendency for all cell types of cervical cancer. However, it is difficult to determine the role of hereditary factors in this study as detailed pedigrees were not obtained. Further work is needed to explore the possibility that hereditary factors interact with the well-established environmental factors in the pathogenesis of carcinoma of the cervix.

Several case reports and case-control studies have identified familial clustering of carcinoma of the cervix and preinvasive cervical disease.[16,45-47] However, it is difficult to know how much environmental and cultural factors contribute to these clusters. Case-control studies have been inadequate and infrequent with regard to the question of a possible genetic factor. In one case-control study in Malmo, Sweden, the families of male partners of patients with cervical cancer served as controls.[16] Cervical carcinoma was noted more frequently in the mothers and sisters of patients (7.9% and 7.5%, respectively) than in the mothers and sisters of consorts (1.0% and 1.1%, respectively).

2.5 Hereditary Aspects of Gestational Trophoblastic Neoplasia

Gestational trophoblastic tumors include complete and partial hydatidiform moles, invasive mole, choriocarcinoma, and placental site trophoblastic tumor. The precise contribution of genetics to this spectrum of tumors is unclear but Parazzini and colleagues in a case report in 1984 suggest that there may be a familial tendency.[48] The tendency for these tumors to occur in Asian women also suggests a possible genetic component in addition to environmental influences.

There have been numerous reports of familial clusters of patients with trophoblastic disease. In 1980, Ambani and colleagues noted clusters of trophoblastic neoplasia in sisters in three families.[49] Parazzini and colleagues reported two sisters each of whom had two complete molar pregnancies.[48] In a case-control study involving 100 women with trophoblastic tumors, LaVecchia and colleagues noted a familial tendency in addition to other risk factors including relatively older maternal and paternal age and smoking.[50] Further research is needed to clarify the role of hereditary and genetic factors in the development of gestational trophoblastic tumors.

3. CLINICAL IMPLICATIONS OF HEREDITARY CANCER GENES

The discovery of specific genes that are responsible for most hereditary cancers provides the opportunity for interventions aimed at decreasing cancer incidence and mortality in these families. At many university affiliated medical centers, clinics are opening to meet the need for familial cancer genetic testing. Multidisciplinary teams in these clinics will need to address challenges in the areas of cancer diagnosis, treatment and prevention, genetics, psychology, as well as social implications. The involvement of formally trained genetic counselors is critical as they are specifically trained in the process of educating patients regarding the complex principles of medical genetics and patterns of inheritance.[51]

The protocol for managing individuals with suspected genetic susceptibility to cancer should be similar to the approach long used by genetic counselors for other inherited disorders. This includes: (1) performing a pedigree analysis to determine if a genetic disorder likely exists in a family, (2) nondirective pre-counseling assessment and education to arrive at a decision whether the individual wishes to undergo genetic testing, (3) genetic testing, (4) disclosure of results by a multidisciplinary team, (5) determination of cancer risk for both carriers and noncarriers, (6) discussion of options for management, (7) psychological counseling with regard to carrier status, and (8) follow-up counseling. The importance of follow-up counseling by a genetic counselor or other team member cannot be overemphasized as profound psychological consequences are not unusual following genetic testing. Severe depression and fatalism often are experienced by those found to be carriers. In addition, rather than feeling a sense of relief, those who are found not to be carriers often are overwhelmed by guilt because they have been spared while other family members are affected.

3.1 Pedigree Analysis

Careful pedigree analysis is the first step in determining whether a given woman has an increased risk of developing breast or gynecologic cancer due to an inherited mutation. Although construction of a pedigree can be challenging, a detailed and thorough family history is essential if one is to accurately determine the probability of a heritable cancer syndrome. Houlston and colleagues emphasized the importance of pedigree analysis in their study of 391 patients with positive "family histories" of ovarian cancers.[52] Detailed investigation revealed that only 82 pedigrees actually were suggestive of a familial cancer syndrome. Karlan et al. also demonstrated the importance of pedigree analysis in the identification of individuals at risk for one of the hereditary cancer syndromes.[53] In their study of 597 women with "family histories" of ovarian cancer more than half had a first- or second-degree relative with the disease, but only 76 (12.8%) had pedigrees consistent with a hereditary cancer syndrome.

There may be inconsistencies and uncertainties in the cancer histories obtained from patients and other family members. Review of actual pathology slides or reports usually will clarify the diagnosis, however. In addition, age of onset of disease should be specified whenever possible as hereditary cancers tend to occur at an earlier age than do sporadic cancers at the same site. The total number of affected individuals is also an important feature in determining the likelihood of a hereditary cancer syndrome. On the other hand, relying on the total number of cases can lead to ascertainment bias in which small families are less likely to be suspected of having a hereditary cancer syndrome. Finally, because the incidence of breast cancer is high in our society, it is likely that some familial clusters occur sporadically in the absence of any genetic susceptibility.

When constructing a pedigree it is important to remember that maternal and paternal transmission of these genes occurs with equal probability. Therefore, the paternal side of the patient's family must be studied closely for a history of breast, ovarian, and endometrial cancers in female relatives. In the case of the Lynch syndrome II, male relatives as well as female relatives may exhibit colorectal cancer as part of this syndrome. In one report, initial maternal pedigree analysis of four sisters whose mother had ovarian cancer was very suggestive of a site-specific ovarian cancer syndrome (Figure 1).[54] Review of the paternal pedigree, however revealed an excess of primary cancers of the colon, pancreas, and bladder. Three of the four sisters had combinations of endometrial, ovarian, and colon cancers. The paternal pedigree was consistent with the Lynch syndrome II and the mother's case of ovarian cancer was felt to be sporadic.

3.2 Genetic Testing

After a comprehensive family history is obtained, in some cases it is determined that the likelihood of that individual carrying a mutation in one of the hereditary cancer genes is too low to justify genetic testing. For example, based on experience to date, it is estimated that the probability of finding a BRCA1

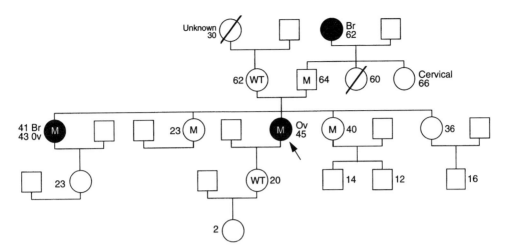

Figure 2 Familial ovarian cancer pedigree with BRCA1 mutation. The age of each family member and type of cancer is noted. Slashes denote individuals who have died of cancer. The arrow denotes the proband who was treated for ovarian cancer. Her 40-year-old sister elected to have prophylactic oophorectomy and mastectomy. Individuals denoted M have a germline BRCA1 mutation that leads to a truncated 75 amino acid protein product, whereas individuals denoted WT have normal wild-type BRCA1 genes.

mutation in a woman over age 50 who is the only individual in her family with ovarian or breast cancer is significantly less than 3%.[25] In contrast, in families with two cases of breast cancer and two cases of ovarian cancer, the probability of finding a BRCA1 mutation is approximately 90%.[25] Although the decision as to who should undergo mutational screening is straightforward at these extremes, most patient's family histories place them in an intermediate risk group. One significant opportunity in the familial cancer clinic is to reassure women with an insignificant family history or a negative BRCA1 test that they do not appear to be at high risk of developing ovarian or breast cancer.

Many of the women who seek genetic testing have not yet developed cancer. It is not optimal to begin analysis of a family by testing these women, however. Failure to find a mutation in an unaffected individual might mean that the familial aggregation of cancer occurred by chance or due to other genetic alterations. Alternatively, if these cancers were due to a mutation in a cancer causing gene, this would escape detection if the individual in whom the test was performed did not inherit a mutant copy of the gene. Although less desirable, testing unaffected individuals may be unavoidable in some circumstances if affected individuals in a family are unavailable or unwilling to be tested. It is not unusual for some family members to decline genetic testing. Sometimes refusal to be tested is due to irrational fears or ignorance, but individual decisions to decline testing after having received appropriate educational material and counseling must be respected. Health care providers should not lose sight of the fact that until it has been clearly proven that genetic testing translates into decreased cancer mortality, the value of such testing remains speculative.

3.3 Cancer Prevention

The ability to identify mutations in the BRCA1 gene represents a major advance in the management of familial ovarian cancer. The breast cancer linkage consortium has shown that over 90% of cancer families that contained cases of ovarian cancer in addition to breast cancer were linked to chromosome 17q.[35,36,55,56] In most cases it should now be possible to identify mutations in the BRCA1 gene in these families. Prevention of ovarian cancer can be focused on those women who carry the mutant gene while those who carry two normal copies of BRCA1 can be reassured that they are not at increased risk. Only a minority of cases of familial ovarian cancer should have to be managed as we have in the past, by simply recommending prophylactic oophorectomy on the basis of a strong family history.[15,57,58]

There is some debate in the literature as to whether the age of onset of ovarian cancer is actually younger in women who inherit a mutant BRCA1 gene.[26] It is clear, however, that familial ovarian cancer often strikes during the fourth decade of life and the estimated lifetime risk of ovarian cancer is 60 to 70% in BRCA1 carriers.[26] It has been suggested that the likelihood of developing ovarian cancer may differ between various BRCA1 mutations, with ovarian cancer being less prevalent in families that carry mutations affecting the 3' end of the gene,[25,59] but too few women have been studied thus far to allow firm conclusions. In view of the failure of screening tests to detect ovarian cancer while it is still localized and curable, prophylactic oophorectomy is a reasonable approach to decreasing ovarian cancer mortality in BRCA1 carriers.

Because oophorectomy can now be performed laparascopically in an outpatient setting, surgical prophylaxis of ovarian cancer seems a reasonable strategy. In addition, most women do not view removal of the ovaries as cosmetically mutilating and oophorectomy causes only modest changes in body image and self esteem. Finally, estrogen replacement can be administered either orally or transdermally, thereby avoiding the deleterious side effects of premature menopause. Although there is some concern that estrogen replacement might increase the risk of breast cancer in these women, this risk already approaches 85% in women with BRCA1 mutations.[26]

One concern regarding prophylactic oophorectomy is the observation that a small fraction of women subsequently have developed intraperitoneal carcinomatosis that was indistinguishable macroscopically or microscopically from advanced ovarian cancer.[60,61] Although oophorectomy does not provide absolute protection, in the Gilda Radner Registry carcinomatosis developed in only 2% of 324 women who had undergone prophylactic oophorectomy.[61] Another strategy that has been suggested to decrease risk of ovarian cancer in women with germline BRCA1 mutations is use of oral contraceptives.[57] It has been shown that long term oral contraceptive pill use decreases the risk of ovarian cancer in the general population by as much as 60%.[62] Oral contraceptives might be a particularly attractive alternative for young women who have not yet completed childbearing, but as with estrogen replacement there is some concern that oral contraceptive use might increase the risk of breast cancer.

Prevention of breast cancer mortality in BRCA1 and BRCA2 carriers presents different issues because these cancers are much more readily detected at an early stage than ovarian cancers. As a result, breast cancer 5-year survival in the United States is approximately 70% compared to only 30% for ovarian cancer. Furthermore, unlike oophorectomy, mastectomy with reconstruction causes marked alterations in self esteem and body image. Many prophylactic mastectomies performed in the past have been subcutaneous mastectomies in which most of the breast tissue is removed but the nipple is preserved. Advocates of prophylactic mastectomy today generally recommend total mastectomy as malignancy can potentially form in the nipple if it is not removed. Even if a total mastectomy is performed, there is no guarantee that all of the breast tissue will be successfully excised. Unfortunately, the risk of breast cancer after prophylactic mastectomy is unknown. Although some women will continue to choose mastectomy, close surveillance with mammography and breast self exam may prove equally effective in reducing mortality in view of the good prognosis for women with early breast cancer.[63] Chemoprophylaxis of breast cancer using antiestrogens such as tamoxifen is another strategy being considered.

Lynch has recommended that members of families with HNPCC should undergo colonoscopy every 2 years from age 20 to 35 and then yearly after age 35.[11] Colonoscopy rather than sigmoidoscopy is appropriate as most of the colon cancers in this syndrome are right-sided.[64] In addition, subtotal colectomy is an option that some have advocated in patients documented to have HNPCC on the basis of inherited mutations in one of the DNA repair genes. In addition, yearly endometrial sampling is suggested beginning at age 30. Although pelvic ultrasound studies and serum CA 125 levels have been advocated for ovarian cancer screening in HNPCC and familial ovarian cancer,[11] the utility of these techniques for diagnosing ovarian cancer while it is still confined to the ovaries is unproven.

3.4 Social Implications

Prevention of cancer in women who are found to carry mutations in BRCA1 and other genes is an important endeavor, but we also must participate in discussions regarding associated non-medical implications. Misuse of genetic information potentially could have devastating consequences including difficulty in securing employment and life or health insurance.[65] Since society has not resolved these issues, the right to confidentiality of genetic test results must be respected. Health care professionals should actively advocate that genetic testing for cancer susceptibility mutations be used constructively to modify risk, rather than to stigmatize individuals or deprive them of appropriate care. For example, some insurers have denied payment for prophylactic oophorectomy in the setting of familial ovarian cancer because it is a preventative measure rather than treatment of an illness. In at least one case, a strongly proactive group of physicians helped a woman to contest such a judgment and the Nebraska Supreme Court eventually ruled in her favor.[66]

As the number of individuals who are identified as carriers of cancer susceptibility mutations rises, we undoubtedly will receive requests for prenatal testing.[67] With knowledge of the precise mutation in a parent, prenatal testing using fetal cells is relatively straightforward and a result could be obtained in a few days allowing for pregnancy termination if desired. Whether a fetus with a cancer susceptibility mutation should be aborted is a matter that the parents should decide after receiving educational material and nondirective counseling.

REFERENCES

1. **Fraumeni, J.F., Grundy, G.W., Creagan, E.T., and Everson, R.B.,** Six families prone to ovarian cancer, *Cancer*, 36, 364, 1975.
2. **Lynch, H.T., Harris, R.E., Guirgis, H.A., Maloney, K., Carmody, L.L., and Lynch, J.F.,** Familial association of breast/ovarian carcinoma, *Cancer*, 41, 1543, 1978.
3. **Lynch, H.T., Albano, W., Black, L., Lynch, J.F., Recabaren, J., and Pierson, R.,** Familial excess of cancer of the ovary and other anatomic sites, *JAMA*, 245, 261, 1981.
4. **Schildkraut, J.M. and Thompson, W.D.,** Familial ovarian cancer: A population-based case-control study, *Am. J. Epidemiol.*, 128, 456, 1988.
5. **Schildkraut, J.M., Risch, N., and Thompson, W.D.,** Evaluating genetic association among ovarian, breast, and endometrial cancer: evidence for a breast/ovarian relationship, *Am. J. Hum. Genet.*, 45, 521, 1989.
6. **Claus, E.B., Risch, N., and Thompson, W.D.,** Genetic analysis of breast cancer in the cancer and steroid hormone study, *Am. J. Hum. Genet.*, 48, 232, 1991.
7. **Claus, E.B., Risch, N., and Thompson, W.D.,** The calculation of breast cancer risk for women with a first degree family history of ovarian cancer, *Breast Cancer Res. Treat.*, 28, 115, 1993.
8. **Miki, Y., Swensen, J., Shattuck-Eidens, D., Futreal, A., Harshman, K., Tavtigian, S., Liu, Q., Cochran, C., Bennett, L.M., Ding, W., Bell, R., Rosenthal, J., Hussey, C., Tran, T., McClure, M., Frye, C., Hattier, T., Phelps, R., Haugen-Strano, A., Katcher, H., Yakumo, K., Gholami, Z., Shaffer, D., Stone, S., Bayer, S., Wray, C., Bogden, R., Dayananth, P., Ward, J., Tonin, P., Narod, S., Bristow, P.K., Norris, F.H., Helvering, L., Morrison, P., Rosteck, P., Lai, M., Barrett, J.C., Lewis, C., Neuhausen, S., Cannon-Albright, L., Goldgar, D., Wiseman, R., Kamb, A., and Skolnick, M.H.,** A strong candidate for the breast ovarian cancer susceptibility gene BRCA1, *Science*, 266, 66, 1994.
9. **Wooster, R., Bignell, G., Lancaster, J., Swift, S., Seal, S., Manglon, J., Collins, N., and Micklem, G.,** Identification of the breast cancer susceptibility gene BRCA2, *Nature*, 378, 789, 1995.
10. **Nystrom-Lahti, M., Parsons, R., Sistonen, P., Pylkkanen, L., Aaltonen, L.A., Leach, F.S., Hamilton, S.R., Watson, P., Bronson, E., and Fusaro, R.,** Mismatch repair genes on chromosomes 2p and 3p account for a major share of hereditary nonpolyposis colorectal cancer families evaluable by linkage, *Am. J. Hum. Genet.*, 55, 659, 1994.
11. **Lynch, H.T. and Lynch, J.,** Genetic counseling for hereditary cancer, *Oncology*, 10, 27, 1996.
12. **Schisselbauer, J.C., Hogan, W.M., Buetow, K.H., and Tew, K.D.,** Heterogeneity of glutathione S-transferase enzyme and gene expression in ovarian carcinoma, *Pharmacogenetics*, 2, 63, 1992.
13. **McKenna, N.J., Kieback, D.G., Carney, D.N., Fanning, M., McLinden, J., and Headon, D.R.,** A germline TaqI restriction fragment length polymorphism in the progesterone receptor gene in ovarian carcinoma, *Br. J. Cancer*, 71, 451, 1995.
14. **Rowe, S.M., Coughlan, S.J., McKenna, N.J., Garrett, E., Kieback, D.G., Carney, D.N., and Headon, D.R.,** Ovarian carcinoma-associated TaqI restriction fragment length polymorphism in intron G of the progesterone receptor gene is due to an Alu sequence insertion, *Cancer Res.*, 55, 2743, 1995.
15. **Lynch, H.T., Albano, W.A., Lynch, J.F., Lynch, P.M., and Campbell, A.,** Surveillance and management of patients at high genetic risk for ovarian carcinoma, *Obstet. Gynecol.*, 59, 589, 1982.
16. **Boltenberg, A., Furgyik, S., and Kullander, S.,** Familial cancer aggregation in cases of adenocarcinoma corporis uteri, *Acta Obstet. Gynecol. Scand.*, 69, 249, 1990.
17. **Sandles, L.G., Shulman, L.P., Elias, S., Photopulos, G.J., Smiley, L.M., Posten, W.M., and Simpson, J.L.,** Endometrial adenocarcinoma: genetic analysis suggesting heritable site-specific uterine cancer, *Gynecol. Oncol.*, 47, 167, 1992.
18. **Giardiello, F.M., Welsh, S.B., Hamilton, S.R., Offerhaus, G.J., Gittelsohn, A.M., Booker, S.V., Krush, A.J., Yardley, J.H., and Luk, G.D.,** Increased risk of cancer in the Peutz-Jeghers syndrome, *N. Engl. J. Med.*, 316, 1511, 1987.
19. **Jackson, S.M.,** Ovarian dysgerminoma in three generations?, *J. Med. Genet.*, 4, 112, 1967.
20. **Shulman, L.P., Muram, D., Marina, N., Jones, C., Portera, J.C., Wachtel, S.S., Simpson, J.L., and Elias, S.,** Lack of heritability in ovarian germ cell malignancies, *Am. J. Obstet. Gynecol.*, 170, 1803, 1994.
21. **Hartley, A.L., Birch, J.M., Kelsey, A.M., Marsden, H.B., Harris, M., and Teare, M.D.,** Are germ cell tumors part of the Li-Fraumeni cancer family syndrome?, *Cancer Genet. Cytogenet.*, 42, 221, 1989.
22. **Lynch, H.T., Watson, P., Bewtra, C., Conway, T.A., Hippee, C.R., Kaur, P., Lynch, J.F., and Ponder, B.A.,** Hereditary ovarian cancer. Heterogeneity in age at diagnosis, *Cancer*, 67, 1460, 1991.
23. **Hall, J., Lee, M., Newman, B., Morrow, J., Huey, B., and King, M.C.,** Linkage of early onset familial breast cancer to chromosome 17q21, *Science*, 250, 1684, 1990.

24. **Narod, S.A., Feunteun, J., Lynch, H.T., Watson, P., Conway, T., Lynch, J., and Lenoir, G.M.,** Familial breast-ovarian cancer locus on chromosome 17q12-q23, *Lancet*, 338, 82, 1991.

25. **Shattuck-Eidens, D., McClure, M., Simard, J., Labrie, F., Narod, S., Couch, F., Hoskins, K., Weber, B., Castilla, L., and Erdos, M.,** A collaborative survey of 80 mutations in the BRCA1 breast and ovarian cancer susceptibility gene. Implications for presymptomatic testing and screening, *JAMA*, 273, 535, 1995.

26. **Easton, D.F., Ford, D., and Bishop, D.T.,** Breast and ovarian cancer incidence in BRCA1-mutation carriers. Breast Cancer Linkage Consortium, *Am. J. Hum. Genet.*, 56, 265, 1995.

27. **Ford, D., Easton, D.F., Bishop, D.T., Narod, S.A., and Goldgar, D.E.,** Risks of cancer in BRCA1-mutation carriers. Breast Cancer Linkage Consortium, *Lancet*, 343, 692, 1994.

28. **Boyd, M., Harris, F., McFarlane, R., Davidson, H.R., and Black, D.M.,** A human BRCA1 gene knockout, *Nature*, 375, 541, 1995.

29. **Smith, S.A., Easton, D.F., Evans, D.G., and Ponder, B.A.,** Allele losses in the region 17q12-21 in familial breast and ovarian cancer involve the wild-type chromosome, *Nat. Genet.*, 2, 128, 1992.

30. **Simard, J., Tonin, P., Durocher, F., Morgan, K., Rommens, J., Gingras, S., Samson, C., Leblanc, J.F., Belanger, C., and Dion, F.,** Common origins of BRCA1 mutations in Canadian breast and ovarian cancer families, *Nat. Genet.*, 8, 392, 1994.

31. **Friedman, L.S., Ostermeyer, E.A., Szabo, C.I., Dowd, P., Lynch, E.D., Rowell, S.E., and King, M.C.,** Confirmation of BRCA1 by analysis of germline mutations linked to breast and ovarian cancer in ten families, *Nat. Genet.*, 8, 399, 1994.

32. **Castilla, L.H., Couch, F.J., Erdos, M.R., Hoskins, K.F., Calzone, K., Garber, J.E., Boyd, J., Lubin, M.B., Deshano, M.L., and Brody, L.C.,** Mutations in the BRCA1 gene in families with early-onset breast and ovarian cancer, *Nat. Genet.*, 8, 387, 1994.

33. **Hogervorst, F.B., Cornelis, R.S., Bout, M., van Vliet, M., Oosterwijk, J.C., Olmer, R., Bakker, B., Klijn, J.G., Vasen, H.F., and Meijers-Heijboer, H.,** Rapid detection of BRCA1 mutations by the protein truncation test, *Nat. Genet.*, 10, 208, 1995.

34. **Struewing, J.P., Abeliovich, D., Peretz, T., Avishar, N., Kaback, M.M., Collins, F.S., and Brody, L.C.,** The carrier frequency of the BRCA1 185delAG mutation is approximately 1 percent in Ashkenazi Jewish individuals, *Nat. Genet.*, 11, 198, 1995.

35. **Steichen-Gersdorf, E., Gallion, H.H., Ford, D., Girodet, C., Easton, D.F., Evans, G., Ponder, M.A., Pye, C., Mazoyer, S., et al.,** Familial site-specific ovarian cancer is linked to BRCA1 on 17q12-21, *Am. J. Hum. Genet.*, 55, 870, 1994.

36. **Narod, S., Ford, D., Devilee, P., Barkardottir, R.B., Eyfjord, J., Lenoir, G., Serova, O., Easton, D., and Goldgar, D.,** Genetic heterogeneity of breast-ovarian cancer revisited, *Am. J. Hum. Genet.*, 57, 957, 1995.

37. **Wooster, R., Neuhausen, S.L., Mangion, J., Quirk, Y., Ford, D., Collins, N., Nguyen, K., Seal, S., Tran, T., and Averill, D.,** Localization of a breast cancer susceptibility gene, BRCA2, to chromosome 13q12-13, *Science*, 265, 2088, 1994.

38. **Ford, D. and Easton, D.F.,** The genetics of breast and ovarian cancer, *Br. J. Cancer*, 72, 805, 1995.

39. **Lynch, H.T., Bronson, E.K., Strayhorn, P.C., Smyrk, T.C., Lynch, J.F., and Ploetner, E.J.,** Genetic diagnosis of Lynch syndrome II in an extended colorectal cancer-prone family, *Cancer*, 66, 2233, 1990.

40. **Peltomaki, P., Aaltonen, L.A., Sistonen, P., Pylkkanen, L., Mecklin, J., Jarvinen, H., Green, J.S., Jass, J.R., Weber, J.L., Leach, F.S., Petersen, G.M., Hamilton, S.R., de la Chapelle, A., and Vogelstein, B.,** Genetic mapping of a locus predisposing to human colorectal cancer, *Science*, 260, 810, 1993.

41. **Risinger, J.I., Berchuck, A., Kohler, M.F., Watson, P., Lynch, H.T., and Boyd, J.,** Genetic instability of microsatellites in endometrial carcinoma, *Cancer Res.*, 53, 5100, 1993.

42. **Fishel, R., Lescoe, M.K., Rao, M.R., Copeland, N.G., Jenkins, N.A., Garber, J., Kane, M., and Kolodner, R.,** The human mutator gene homolog MSH2 and its association with hereditary nonpolyposis colon cancer, *Cell*, 75, 1027, 1993.

43. **Parsons, R., Li, G.M., Longley, M.J., Fang, W.H., Papadopoulos, N., Jen, J., de la Chapelle, A., Kinzler, K.W., Vogelstein, B., and Modrich, P.,** Hypermutability and mismatch repair deficiency in RER+ tumor cells, *Cell*, 75, 1227, 1993.

44. **Brinton, L.A., Tashima, K.T., Lehman, H.F., Levine, R.S., Mallin, K., Savitz, D.A., Stolley, P.D., and Fraumeni, J.F.,** Epidemiology of cervical cancer by cell type, *Cancer Res.*, 47, 1706, 1987.

45. **Bender, S.,** Carcinoma in-situ of cervix in sisters, *Br. Med. J.*, 1, 502, 1976.

46. **Way, S.,** Letter: Carcinoma-in-situ of cervix in sisters, *Br. Med. J.*, 1, 834, 1976.

47. **Bruinse, H.W., Velde, E.R., and de Gast, B.C.,** Human leukocyte antigen patterns in a family with cervical cancer, *Gynecol. Oncol.*, 12, 249, 1981.

48. **Parazzini, F., La Vecchia, C., Franceschi, S., and Mangili, G.,** Familial trophoblastic disease: case report, *Am. J. Obstet. Gynecol.*, 149, 382, 1984.

49. **Ambani, L.M., Vaidya, R.A., Rao, C.S., Daftary, S.D., and Motashaw, N.D.,** Familial occurrence of trophoblastic disease: report of recurrent molar pregnancies in sisters in three families, *Clin. Genet*, 18, 27, 1980.

50. **La Vecchia, C., Franceschi, S., Parazzini, F., Fasoli, M., Decarli, A., Gallus, G., and Tognoni, G.,** Risk factors for gestational trophoblastic disease in Italy, *Am. J. Epidemiol.*, 121, 457, 1985.

51. **Biesecker, B.B., Boehnke, M., Calzone, K., Markel, D.S., Garber, J.E., Collins, F.S., and Weber, B.L.,** Genetic counseling for families with inherited susceptibility to breast and ovarian cancer, *JAMA*, 269, 1970, 1993.

52. **Houlston, R., Bourne, T.H., Davies, A., Whitehead, M.I., Campbell, S., Collins, W.P., and Slack, J.,** Use of family history in a screening clinic for familial ovarian cancer, *Gynecol. Oncol.*, 47, 247, 1992.

53. **Karlan, B.Y., Raffel, L.J., Crvenkovic, G., Smrt, C., Chen, M.D., Lopez, E., Walla, C.A., Garber, C., Cane, P., Sarti, D.A., Rotter, J.I., and Platt, L.D.,** A multidisciplinary approach to the early detection of ovarian carcinoma: rationale, protocol design, and early results, *Am. J. Obstet. Gynecol.*, 169, 494, 1993.

54. **Trimble, E.L., Karlan, B.Y., Lagasse, L.D., and Hoskins, W.J.,** Diagnosing the correct ovarian cancer syndrome, *Obstet. Gynecol.*, 78, 1023, 1991.

55. **Narod, S.A., Ford, D., Devilee, P., Barkardottir, R.B., Lynch, H.T., Smith, S.A., Ponder, B.A., Weber, B.L., Garber, J.E., and Birch, J.M.,** An evaluation of genetic heterogeneity in 145 breast-ovarian cancer families. Breast Cancer Linkage Consortium, *Am. J. Hum. Genet.*, 56, 254, 1995.

56. **Easton, D.F., Bishop, D.T., Ford, D., and Crockford, G.P.,** Genetic linkage analysis in familial breast and ovarian cancer: results from 214 families. The Breast Cancer Linkage Consortium, *Am. J. Hum. Genet.*, 52, 678, 1993.

57. **Piver, M.S., Baker, T.R., Jishi, M.F., Sandecki, A.M., Tsukada, Y., Natarajan, N., and Blake, C.A.,** Familial ovarian cancer. A report of 658 families from the Gilda Radner Familial Ovarian Cancer Registry 1981-1991, *Cancer*, 71, 582, 1993.

58. **Kerlikowske, K., Brown, J.S., and Grady, D.G.,** Should women with familial ovarian cancer undergo prophylactic oophorectomy?, *Obstet. Gynecol.*, 80, 700, 1992.

59. **Gayther, S.A., Warren, W., Mazoyer, S., Russell, P.A., Harrington, P.A., Chiano, M., Seal, S., Hamoudi, R., van Rensburgh, E.J., Dunning, A.M., Love, R., Evans, G., Easton, D., Clayton, D., Stratton, M.R., and Ponder, B.A.J.,** Germline mutations of the BRCA1 gene in breast and ovarian cancer families provide evidence for genotype-phenotype correlation, *Nat. Genet.*, 11, 428, 1995.

60. **Tobacman, J.K., Greene, M.H., Tucker, M.A., Costa, J., Kase, R., and Fraumeni, J.F.,** Intra-abdominal carcinomatosis after prophylactic oophorectomy in ovarian-cancer-prone families, *Lancet*, 2, 795, 1982.

61. **Piver, M.S., Jishi, M.F., Tsukada, Y., and Nava, G.,** Primary peritoneal carcinoma after prophylactic oophorectomy in women with a family history of ovarian cancer. A report of the Gilda Radner Familial Ovarian Cancer Registry, *Cancer*, 71, 2751, 1993.

62. **Whittemore, A.S., Harris, R., and Itnyre, J.,** Characteristics relating to ovarian cancer risk. Collaborative analysis of twelve US case-control studies: IV. The pathogenesis of epithelial ovarian cancer, *Am. J. Epidemiol.*, 136, 1212, 1992.

63. **Hoskins, K.F., Stopfer, J.E., Calzone, K.A., Merajver, S.D., Rebbeck, T.R., Garber, J.E., and Weber, B.L.,** Assessment and counseling for women with a family history of breast cancer. A guide for clinicians, *JAMA*, 273, 577, 1995.

64. **Vasen, H.F., Taal, B.G., Nagengast, F.M., Griffioen, G., Menko, F.H., Kleibeuker, J.H., Offerhaus, G.J., and Meera Khan, P.,** Hereditary nonpolyposis colorectal cancer: results of long-term surveillance in 50 families, *Eur. J. Cancer*, 31A, 1145, 1995.

65. The ad hoc committee on genetic testing/insurance issues, Genetic testing and insurance, *Am. J. Hum. Genet.*, 56, 327, 1995.

66. **Lynch, H.T., Severin, M.J., Mooney, M.J., and Lynch, J.,** Insurance adjudication favoring prophylactic surgery in heritable breast-ovarian cancer syndrome, *Gynecol. Oncol.*, 57, 23, 1995.

67. **Lancaster, J.M., Wiseman, R.W., and Berchuck, A.,** An inevitable dilemma: Prenatal testing for mutations in the BRCA1 breast/ovarian cancer susceptibility gene, *Obstet. Gynecol.*, 87, 306 1996.

Index

INDEX